AF598255

Integrated Environmental Modelling to Solve Real World Problems: Methods, Vision and Challenges

Geological Society books refereeing procedures

The Society makes every effort to ensure that the scientific and production quality of its books matches that of its journals. Since 1997, all book proposals have been refereed by specialist reviewers as well as by the Society's Books Editorial Committee. If the referees identify weaknesses in the proposal, these must be addressed before the proposal is accepted.

Once the book is accepted, the Society Book Editors ensure that the volume editors follow strict guidelines on refereeing and quality control. We insist that individual papers can only be accepted after satisfactory review by two independent referees. The questions on the review forms are similar to those for *Journal of the Geological Society*. The referees' forms and comments must be available to the Society's Book Editors on request.

Although many of the books result from meetings, the editors are expected to commission papers that were not presented at the meeting to ensure that the book provides a balanced coverage of the subject. Being accepted for presentation at the meeting does not guarantee inclusion in the book.

More information about submitting a proposal and producing a book for the Society can be found on its website: www.geolsoc.org.uk.

It is recommended that reference to all or part of this book should be made in one of the following ways:

RIDDICK, A. T., KESSLER, H. & GILES, J. R. A. (eds) 2017. *Integrated Environmental Modelling to Solve Real World Problems: Methods, Vision and Challenges*. Geological Society, London, Special Publications, **408**.

KINGDON, A., GILES, J. R. A. & LOWNDES, J. P. 2017. Future of technology in NERC data models and informatics: outputs from InformaTEC. *In:* RIDDICK, A. T., KESSLER, H. & GILES, J. R. A. (eds) *Integrated Environmental Modelling to Solve Real World Problems: Methods, Vision and Challenges*. Geological Society, London, Special Publications, **408**, 245–253. First published online September 17, 2014, https://doi.org/10.1144/SP408.5

GEOLOGICAL SOCIETY SPECIAL PUBLICATION NO. 408

Integrated Environmental Modelling to Solve Real World Problems: Methods, Vision and Challenges

EDITED BY

A.T. RIDDICK, H. KESSLER and J.R.A. GILES
British Geological Survey, UK

2017
Published by
The Geological Society
London

THE GEOLOGICAL SOCIETY

The Geological Society of London (GSL) was founded in 1807. It is the oldest national geological society in the world and the largest in Europe. It was incorporated under Royal Charter in 1825 and is Registered Charity 210161.

The Society is the UK national learned and professional society for geology with a worldwide Fellowship (FGS) of over 10 000. The Society has the power to confer Chartered status on suitably qualified Fellows, and about 2000 of the Fellowship carry the title (CGeol). Chartered Geologists may also obtain the equivalent European title, European Geologist (EurGeol). One fifth of the Society's fellowship resides outside the UK. To find out more about the Society, log on to www.geolsoc.org.uk.

The Geological Society Publishing House (Bath, UK) produces the Society's international journals and books, and acts as European distributor for selected publications of the American Association of Petroleum Geologists (AAPG), the Indonesian Petroleum Association (IPA), the Geological Society of America (GSA), the Society for Sedimentary Geology (SEPM) and the Geologists' Association (GA). Joint marketing agreements ensure that GSL Fellows may purchase these societies' publications at a discount. The Society's online bookshop (accessible from www.geolsoc.org.uk) offers secure book purchasing with your credit or debit card.

To find out about joining the Society and benefiting from substantial discounts on publications of GSL and other societies worldwide, consult www.geolsoc.org.uk, or contact the Fellowship Department at: The Geological Society, Burlington House, Piccadilly, London W1J 0BG: Tel. +44 (0)20 7434 9944; Fax +44 (0)20 7439 8975; E-mail: enquiries@geolsoc.org.uk.

For information about the Society's meetings, consult *Events* on www.geolsoc.org.uk. To find out more about the Society's Corporate Affiliates Scheme, write to enquiries@geolsoc.org.uk.

Published by The Geological Society from:
The Geological Society Publishing House, Unit 7, Brassmill Enterprise Centre, Brassmill Lane, Bath BA1 3JN, UK

The Lyell Collection: www.lyellcollection.org
Online bookshop: www.geolsoc.org.uk/bookshop
Orders: Tel. +44 (0)1225 445046, Fax +44 (0)1225 442836

British Library Cataloguing in Publication Data

A catalogue record for this book is available from the British Library.
ISBN 978-1-86239-687-6
ISSN 0305-8719

Distributors
For details of international agents and distributors see:
www.geolsoc.org.uk/agentsdistributors

Typeset by Nova Techset Private Limited, Bengaluru & Chennai, India
Printed and bound by CPI Group (UK) Ltd, Croydon CR0 4YY

Contents

Preface

This publication has its roots in a number of developments in Integrated Environmental Modelling (IEM) over the past few years. There has been an increasing groundswell of opinion, predominantly it has to be said from North America, that the complex environmental problems that humans are faced with in the 21st Century require appropriate modelling solutions. This has been reflected in the number of meetings that have been convened both in the United States and the UK to further develop the concepts of integrated modelling and particularly the methodologies available for linking together models to solve specific problems – 'model fusion'. These include meetings convened by the British Geological Survey (BGS) as a result of their Data and Research for Environmental Applications and Models (DREAM) programme which formed part of the BGS strategy from 2010–2015.

There is a growing community of IEM practitioners who are striving to solve the problems presented by this step change in how modelling takes place. The challenges are many and various, but predominantly involve 'professionalising' the modelling approach, making models discoverable, linkable, reproducible and the results easily visualized all in a framework of assessing the uncertainty. The models themselves are applied to a wide range of environmental problems. It is the hope of the editors that this volume represents the current landscape of IEM by presenting papers on the vision and challenges, case studies of how IEM techniques are being applied, social aspects of IEM, and software solutions and methodology.

A volume of this size cannot be put together without the efforts of the reviewers and the following are thanked for their efforts: James Breck, Nina Burkhardt, Paul Conrads, Dan Cornford, Rachel Dearden, Paul Duller, Jon Ford, Pierre Glynn, Patricia Gober, James Griffiths, Rachel Heaven, Andrew Hughes, Gerry Laniak, Murray Lark, John Laxton, Jonathan Mackay, Christos Makropoulos, Majdi Mansour, John Rees, Carolyn Rosten, Jonty Rougier, Jan Stafleu, Jana Stewart, James Sutherland and Kyle Young.

Introduction to integrated environmental modelling to solve real world problems: methods, vision and challenges

ANDREW RIDDICK*, ANDREW HUGHES, HOLGER KESSLER & JEREMY GILES

British Geological Survey, Nicker Hill, Keyworth, Nottingham NG12 5GG, UK

**Correspondence: atr@bgs.ac.uk*

Across the world, stakeholders are asking questions of their governments and decision makers to quantify the risks of environmental threats to their well-being. These questions manifest themselves as 'deceptively simple questions', which are easy to articulate but difficult to solve. An example of which is: 'how much will the eruption of an Icelandic volcano cost the UK economy'. Answering these questions requires predictions of the interaction of multiple environmental processes, this requires the development and maintenance of systems that allow these processes to be simulated, and that is the nascent science of integrated environmental modelling (IEM). Such processes may be long-term (e.g. those that are impacted by climate change) or short-term threats, such as the impact of drought on UK agriculture or the impact of space weather on energy supply systems.

These questions require holistic solutions drawn from different disciplines (Laniak *et al.* 2013). Importantly, the technology has progressed significantly such that models from different disciplines can be linked and integrated assessments of problems can now be made (Kelly *et al.* 2013). Now that it is possible to link models, consideration has to be given to the opportunities, and also the science issues, that are generated by studying interacting processes. These include:

- Understanding interactions within and across disciplines: for example, how best to simulate the interaction between agricultural policy and farming practice (e.g. the EU-funded SEAMLESS project: Van Ittersum *et al.* 2008).
- Reconciling the differing functions used by different disciplines when evaluating similar interacting processes (e.g. sewers and river flooding (Van Assel *et al.* 2010); and sediment transport in rivers (Shrestha *et al.* 2013)).
- Reconciling the different languages that different scientific disciplines use (e.g. soil scientists, ecologists and geologists all have subtly different definitions for what constitutes 'soil'), thereby reducing the opportunity for error in finding and linking models (e.g. Knapen *et al.* 2013).
- Defining model metadata so that developers can describe, and users can find, evaluate and validly link, the models needed to resolve their problem. Whilst no internationally recognized standard exists, workers such as Harpham & Danovaro (2015) have suggested possible approaches based on existing standards such as ISO 19115.
- Encouraging the development of standardization both in input–output data and the modelling process itself: an example being the use of ontologies and 'controlled vocabularies' to identify similarities in the concepts being modelled. Work has been undertaken on metadata standards to assist in the coupling of models (e.g. Peckham *et al.* 2013).
- Understanding the propagation of uncertainty in model chains; projects such as UncertWeb have proposed possible solutions (Bastin *et al.* 2013).
- Providing suitable computational resources to run linked model compositions. Here, high-performance computing (HPC) holds the key, as espoused by the CSDMS project (Peckham *et al.* 2013).
- Quality assurance, and issues of being auditable.

But progress is being made and a number of different initiatives have been undertaken in different science areas: for example, climate, hydrology, environmental regulation, insurance and human health. Climate models can be linked using a number of different frameworks, such as BFG (Barkwith *et al.* 2014), and with models from other disciplines (e.g. Goodall *et al.* 2013). Hydrology has well-developed model linking approaches such as OpenMI and FluidEarth, which provides a toolkit to implement this standard (Harpham & Danovaro 2015). This OpenMI implementation (.NET/in-memory) is complimented by the HPC-based CSDMS (Peckham *et al.* 2013). The United States Environmental Protection Agency (US EPA) has developed their system (FRAMES_3MRA) to enable rapid assessment of the environmental

From: Riddick, A. T., Kessler, H. & Giles, J. R. A. (eds) 2017. *Integrated Environmental Modelling to Solve Real World Problems: Methods, Vision and Challenges*. Geological Society, London, Special Publications, **408**, 1–5.
First published online September 30, 2016, https://doi.org/10.1144/SP408.17

impact of potentially polluting activities (Whelan *et al.* 2014). The insurance industry, which relies on catastrophe models (Royse *et al.* 2014) to assess risk, has spawned its own approach to providing a framework to link models, that of OASIS-LMF (Barkwith *et al.* 2014). Even a discipline area seemingly far removed, such as modelling the human body to safely test new drugs, has its integrated modelling approaches (e.g. Virtual Physiological Human: Hunter *et al.* 2010).

Despite these diverse approaches, several trends exist that provide some confidence that interoperability should be easier in the future. There are a lot of resources and initiatives available, not just dedicated to a single discipline, to ease access: there have been significant recent developments in interoperability – both in data and models; access to HPC is becoming increasingly available (Peckham *et al.* 2013); web services to serve data and model results over the Internet are easy to access (Goodall *et al.* 2013); smart phones have a huge amount of uptake and development behind them (Moore & Hughes 2016); the science of user uptake for both model results and presentation of future risk is maturing (Beven & Lamb 2014); and the number of pilot projects to enable users to run relatively simple models (e.g. EVOp http://www.evo-uk.org) is rapidly increasing. This Special Publication summarizes the progress made in linking data and models, and how it can be used to solve some of the difficult environmental problems discussed above.

This volume first introduces and expands on a number of the key themes described above that underpin the future vision and strategy for IEM, focusing on the experiences of a number of organizations that are promoting these technologies. For example, a common theme in the experience of the British Geological Survey (**Peach *et al.* 2016**) and the Environment Agency in the UK (**Farrell *et al.* 2016**) is the requirement to be able to utilize and link existing models that have often required significant investment in resources to create, and so to establish methods of linking together and adapting existing modelling systems (**Sutherland *et al.* 2014**). A key area of research highlighted by several authors is the building upon a well-developed data and information framework in which appropriate standards of data quality and semantic interoperability are maintained (**Sutherland *et al.* 2014**; **Laxton 2016**). Whilst the resulting model is often more widely reported, the underlying data and information structure sometimes receives less attention in terms of organizational investment but forms the basis upon which successful integrated modelling is built. The need for linking mechanisms that cross discipline, and also national, boundaries has been particularly highlighted by **Moore & Hughes (2016)** and **Peach *et al.* (2016)**.

Sutherland *et al.* (2014) review various approaches to hydraulic modelling and discuss the historical development of linked modelling, exploring the implications of fusion and emphasizing that the fusion of models involves not only linking them together, but also providing easier access to information about the models and software tools to facilitate the linking process. These authors also raise another very pertinent issue – the potential blurring between what is considered raw data and what constitutes a model. For example, a monitoring instrument may use an established relationship (in effect, a conceptual model) to derive the value of a parameter (e.g. water level calculated from pressure measurements), and so the water-level data is also modelled to some extent. These authors argue that this blurring of the boundaries is a strong driver for the further development of standards for interoperability between data and models.

Many of the technical requirements (i.e. those relating to data, software and IT infrastructure) to facilitate integrated modelling are already either in place or at an advanced stage of development. These, for example, include protocols for linking models and software, and data standards to promote linked modelling. However, the development of research and user communities to take IEM forwards is also an important requirement, as highlighted by **Sutherland *et al.* (2014)**, who describe the development of the Fluid Earth network to facilitate the development of the Open MI protocol. Both **Gober *et al.* (2014)** and **Glynn (2015)** further emphasized the importance of the community and human dimension in the development of IEM. Through enabling a better understanding of environmental processes and systems, IEM has the potential to assist communities in adapting to increasingly complex environmental stresses: although **Glynn (2015)** indicates that to be understandable and usable by a broader community, IEM will need to involve simplifications (e.g. of the processes involved) and may be subjected to inherent human biases. As a counterpoint to this, **Gober *et al.* (2014)** present examples of how modelling has been used to facilitate the fair allocation of water resources in drought-prone regions of the United States.

Further, Moore & Hughes (2016) point out there is an overall tendency for much of IEM technology to be developed within research organizations, and there is a requirement for an increasing adoption of such technologies for commercial applications, and this process will also require the increasing take-up of IEM by user communities

Sutherland *et al.* (2014) and Moore & Hughes (2016) also describe the development of the Open Model Interface (OpenMI) protocol to link existing models at run-time. Sutherland *et al.* (2014)

further describe the Fluid Earth network, which provides tools for running integrated modelling using OpenMI and also promotes the development of a user community.

The present volume also contains a number of papers that describe the application of IEM to specific environmental problems, including large-scale groundwater modelling for regulatory purposes, smaller-scale groundwater modelling to resolve more local problems and also an increasing opportunity to utilize integrated modelling techniques in catastrophe modelling.

Farrell *et al.* (2016) describe the development of the existing National Groundwater Modelling System (NGMS) at the Environment Agency in the UK from a groundwater modelling system to a system that can also perform recharge modelling. This has clearly provided the opportunity for increased efficiency and time-saving in a regulatory environment, and suggests that other types of models (e.g. river flow) may also be integrated with the system in future. This work also further highlights the need for tools and model-linking methods that are easy to use for those who are not software or programming experts. A frequent requirement in groundwater modelling is to be able to constrain the detailed geological structure to better understand groundwater movement. Such a case study is described by **Pasanen & Okkonen (2016)**, where geological modelling using the GSI3D software is used to constrain groundwater modelling using GMS and FEFlow, and in this case the model fusion is facilitated by the integration of data between these different tools.

Being able to apply integrated modelling methodologies to addressing the simulation of natural hazards in catastrophe modelling involves not only linking different environmental models together, but also linking them to financial models representing the extent of financial loss that may occur. **Royse *et al.* (2014)** cite a useful example of such a catastrophe model, which highlights the importance of modelling the whole environmental system and, therefore, the continued development of 'plug and play' integrated modelling in which different modelling components can be rapidly interchanged. The example focused on calculating the financial losses resulting from groundwater flooding (Hughes *et al.* 2011).

One of the developments becoming increasingly important in IEM is the consideration and interaction of social aspects in relation to the environment within models. Human interaction with the modelling process is addressed by **Glynn (2015)**. **Gober *et al.* (2014)** emphasize the importance of the integration of human behaviour into environmental decision-making in drought-affected areas of the United States and Canada, and suggests that the value of large-scale climate models can be limited in uncertainties in downscaling these to specific local areas. The use of exploratory modelling, scenario planning and risk assessment is advocated, allowing policy makers to investigate the likely result of policy decisions before committing to them.

The theme of integrating social and behavioural considerations is further developed by **Makropoulos (2014)**, who describes the development of a toolkit to assist in the integration of social concepts into the technical understanding of water resource management. A case study focusing on the city of Athens is described, in which several water resources models are linked together currently by directly integrating the model code via tools such as MATLAB, which provides the basis for enabling social factors to be included.

The need to be able to integrate environmental models into decision support systems to facilitate access to modelled outputs to a range of stakeholders is also discussed by **Rowe *et al.* (2014)** and **Conrads & Roehl (2015)**. **Rowe *et al.* (2014)** suggest that the process of model fusion can have the effect of accelerating model growth as further models are added, even though a simpler model may possibly be more appropriate and provide results that are more usable by decision makers. **Conrads & Roehl (2015)** describes the development of neural-network-based models linked with data mining of related datasets in order to develop the PRISM-2 decision-support system and its use to better understand planning for salinity incursions on the SE coast of the United States.

The papers by **Beven & Lamb (2014)** and **Kingdon *et al.* (2014)** focus on underlying technical developments that are supporting model fusion and which are, to some extent, driven by the need to develop more integrated models of natural systems. **Beven & Lamb (2014)** outline the various sources of uncertainly in IEM, including uncertainties in the input data, the model structure, uncertainties in various parameters used to constrain the model and uncertainties in observations upon which the model is based. **Kingdon *et al.* (2014)** describe informatics techniques that support and utilize IEM, these include the development of environmental sensor networks to monitor changes in environmental systems in real time, the use of semantic interoperability to assist interdisciplinary modelling and the increased availability of cloud computing. Integrated modelling and the development of fused and linked models also have significant implications on understanding the constraints regarding uncertainty. These authors discuss methods for making realistic estimates of uncertainty within linked modelling ensembles.

The issues in estimation of uncertainty are further discussed by **Wildhaber *et al.* (2015*a*)**, who outline a process of downscaling from regional climate models to the river scale, and the impact of the uncertainties involved in linking hydrological and temperature models with a bioenergetics model (**Wildhaber *et al.* 2015*b***) for the pallid sturgeon species of fish. This work provides an excellent case study of the application of uncertainty considerations. A future extension of this work would be the development of a framework for understanding the impact of climate models on large river ecosystems.

The paper by **Laxton (2016)** describes the process of fusion of different geological maps within the OneGeology-Europe project and draws some parallels with the process of model fusion: in particular, the semantic relationships between concepts are important, as are an understanding of the differences in scale.

IEM is too large a problem to be solved by one organization, and requires an increased level of collaboration between organizations and not just within their own country or geographical region. The collection of papers presented in this volume represents a representative cross-section of initiatives in both Europe and the United States. By bringing together these good works, the future for IEM is very promising. Its ability to solve complex environmental problems so that decision makers at all levels can be better informed of the consequences of their actions is becoming a reality, and the deceptively simple questions that are, in reality, very complex can be properly addressed.

References

Barkwith, A.K.A.P., Pachocka, M., Watson, C. & Hughes, A.G. 2014. *Couplers for Linking Environmental Models: Scoping Study and Potential Next Steps*. British Geological Survey Internal Report, OR/14/022 British Geological Survey, Nottingham.

Bastin, L., Cornford, D. *et al.* 2013. Managing uncertainty in integrated environmental modelling: the UncertWeb framework. *Environmental Modelling & Software*, **39**, 116–134.

Beven, K. & Lamb, R. 2014. The uncertainty cascade in model fusion. *In*: Riddick, A.T., Kessler, H. & Giles, J.R.A. (eds) *Integrated Environmental Modelling to Solve Real World Problems: Methods, Vision and Challenges*. Geological Society, London, Special Publications, **408**. First published online June 30, 2014, https://doi.org/10.1144/SP408.3

Conrads, P.A. & Roehl, E.A. 2015. The use of data-mining techniques for developing effective decision support systems: a case study of simulating the effects of climate change on coastal salinity intrusion. *In*: Riddick, A.T., Kessler, H. & Giles, J.R.A. (eds) *Integrated Environmental Modelling to Solve Real World Problems: Methods, Vision and Challenges*. Geological Society, London, Special Publications, **408**. First published online February 16, 2015, https://doi.org/10.1144/SP408.8

Farrell, R., Ververs, M., Davison, P., Howlett, P. & Whiteman, M. 2016. Splicing recharge and groundwater flow models in the Environment Agency National Groundwater Modelling System. *In*: Riddick, A.T., Kessler, H. & Giles, J.R.A. (eds) *Integrated Environmental Modelling to Solve Real World Problems: Methods, Vision and Challenges*. Geological Society, London, Special Publications, **408**. First published online September 30, 2016, https://doi.org/10.1144/SP408.14

Glynn, P.D. 2015. Integrated Environmental Modelling: human decisions, human challenges. *In*: Riddick, A.T., Kessler, H. & Giles, J.R.A. (eds) *Integrated Environmental Modelling to Solve Real World Problems: Methods, Vision and Challenges*. Geological Society, London, Special Publications, **408**. First published online May 21, 2015, https://doi.org/10.1144/SP408.9

Gober, P., White, D.D., Quay, R., Sampson, D.A. & Kirkwood, C.W. 2014. Socio-hydrology modelling for an uncertain future, with examples from the USA and Canada. *In*: Riddick, A.T., Kessler, H. & Giles, J.R.A. (eds) *Integrated Environmental Modelling to Solve Real World Problems: Methods, Vision and Challenges*. Geological Society, London, Special Publications, **408**. First published online June 23, 2014, https://doi.org/10.1144/SP408.2

Goodall, J.L., Saint, K.D. *et al.* 2013. Coupling climate and hydrological models: interoperability through Web Services. *Environmental Modelling & Software*, **46**, 250–259.

Harpham, Q. & Danovaro, E. 2015. Towards standard metadata to support models and interfaces in a hydro-meteorological model chain. *Journal of Hydroinformatics*, **17**, 260–274.

Hughes, A.G., Vounaki, T. *et al.* 2011. Flood risk from groundwater: examples from a Chalk catchment in southern England. *Journal of Flood Risk Management*, **4**, 143–155, https://doi.org/10.1111/j.1753-318X.2011.01095.x

Hunter, P., Coveney, P.V. *et al.* 2010. A vision and strategy for the virtual physiological human in 2010 and beyond. *Philosophical Transactions of the Royal Society of London A: Mathematical, Physical and Engineering Sciences*, **368**, 2595–2614.

Kelly, R.A., Jakeman, A.J. *et al.* 2013. Selecting among five common modelling approaches for integrated environmental assessment and management. *Environmental Modelling & Software*, **47**, 159–181.

Kingdon, A., Giles, J.R.A. & Lowndes, J.P. 2014. Future of technology in NERC data models and informatics: outputs from InformaTEC. *In*: Riddick, A.T., Kessler, H. & Giles, J.R.A. (eds) *Integrated Environmental Modelling to Solve Real World Problems: Methods, Vision and Challenges*. Geological Society, London, Special Publications, **408**. First published online September 17, 2014, https://doi.org/10.1144/SP408.5

Knapen, R., Janssen, S., Roosenschoon, O., Verweij, P., De Winter, W., Uiterwijk, M. & Wien, J.E. 2013. Evaluating OpenMI as a model integration platform across disciplines. *Environmental Modelling & Software*, **39**, 274–282.

LANIAK, G.F., OLCHIN, G. ET AL. 2013. Integrated environmental modeling: a vision and roadmap for the future. *Environmental Modelling & Software*, **39**, 3–23.

LAXTON, J.L. 2016. Geological map fusion: OneGeology-Europe and INSPIRE. *In*: RIDDICK, A.T., KESSLER, H. & GILES, J.R.A. (eds) *Integrated Environmental Modelling to Solve Real World Problems: Methods, Vision and Challenges*. Geological Society, London, Special Publications, **408**. First published online September 28, 2016, https://doi.org/10.1144/SP408.16

MAKROPOULOS, C. 2014. Thinking platforms for smarter urban water systems: fusing technical and socio-economic models and tools. *In*: RIDDICK, A.T., KESSLER, H. & GILES, J.R.A. (eds) *Integrated Environmental Modelling to Solve Real World Problems: Methods, Vision and Challenges*. Geological Society, London, Special Publications, **408**. First published online July 17, 2014, https://doi.org/10.1144/SP408.4

MOORE, R.V. & HUGHES, A.G. 2016. Integrated environmental modelling: achieving the vision. *In*: RIDDICK, A.T., KESSLER, H. & GILES, J.R.A. (eds) *Integrated Environmental Modelling to Solve Real World Problems: Methods, Vision and Challenges*. Geological Society, London, Special Publications, **408**. First published online May 23, 2016, https://doi.org/10.1144/SP408.12

PASANEN, A.H. & OKKONEN, J.S. 2016. 3D geological models to groundwater flow models: data integration between GSI3D and groundwater flow modelling software GMS and FeFlow®. *In*: RIDDICK, A.T., KESSLER, H. & GILES, J.R.A. (eds) *Integrated Environmental Modelling to Solve Real World Problems: Methods, Vision and Challenges*. Geological Society, London, Special Publications, **408**. First published online September 28, 2016, https://doi.org/10.1144/SP408.15

PEACH, D.W., RIDDICK, A.T., HUGHES, A.G., KESSLER, H., MATHERS, S., JACKSON, C.R. & GILES, J.R.A. 2016. Model fusion at the British Geological Survey: experiences and future trends. *In*: RIDDICK, A.T., KESSLER, H. & GILES, J.R.A. (eds) *Integrated Environmental Modelling to Solve Real World Problems: Methods, Vision and Challenges*. Geological Society, London, Special Publications, **408**. First published online September 15, 2016, https://doi.org/10.1144/SP408.13

PECKHAM, S.D., HUTTON, E.W. & NORRIS, B. 2013. A component-based approach to integrated modeling in the geosciences: the design of CSDMS. *Computers & Geosciences*, **53**, 3–12.

ROWE, E.C., WRIGHT, D.G., BERTRAND, N. & REIS, S. 2014. Fusing and disaggregating models, data and analysis tools for a dynamic science–society interface. *In*: RIDDICK, A.T., KESSLER, H. & GILES, J.R.A. (eds) *Integrated Environmental Modelling to Solve Real World Problems: Methods, Vision and Challenges*. Geological Society, London, Special Publications, **408**. First published online June 23, 2014, https://doi.org/10.1144/SP408.1

ROYSE, K.R., HILLIER, J.K., HUGHES, A., KINGDON, A., SINGH, A. & WANG, L. 2014. The potential for the use of model fusion techniques in building and developing catastrophe models. *In*: RIDDICK, A.T., KESSLER, H. & GILES, J.R.A. (eds) *Integrated Environmental Modelling to Solve Real World Problems: Methods, Vision and Challenges*. Geological Society, London, Special Publications, **408**. First published online December 3, 2014, https://doi.org/10.1144/SP408.7

SUTHERLAND, J., TOWNEND, I.H., HARPHAM, Q.K. & PEARCE, G.R. 2014. From integration to fusion: the challenges ahead. *In*: RIDDICK, A.T., KESSLER, H. & GILES, J.R.A. (eds) *Integrated Environmental Modelling to Solve Real World Problems: Methods, Vision and Challenges*. Geological Society, London, Special Publications, **408**. First published online December 1, 2014, https://doi.org/10.1144/SP408.6

SHRESTHA, N.K., LETA, O.T., DE FRAINE, B., VAN GRIENSVEN, A. & BAUWENS, W. 2013. OpenMI-based integrated sediment transport modelling of the river Zenne, Belgium. *Environmental Modelling & Software*, **47**, 193–206.

VAN ASSEL, J., WATERSCHOOT, G., DEVROEDE, N., RONSE, Y., ANDERSON, S. & MILLINGTON, R. 2010. Modelling bidirectional interactions between sewer and river systems using OpenMI – a case study in the Scheldt River Basin (Belgium). *In*: *NOVATECH 2010. Proceedings of the 7th International Conference on sustainable techniques and strategies for urban water management*. Groupe de Recherche Rhône Alpes sur les Infrastructures et l'Eau (GRAIE), Lyon, France, https://documents.irevues.inist.fr/handle/2042/35709

VAN ITTERSUM, M.K., EWERT, F. ET AL. 2008. Integrated assessment of agricultural systems–a component-based framework for the European Union (SEAMLESS). *Agricultural Systems*, **96**, 150–165.

WHELAN, G., KIM, K. ET AL. 2014. Design of a component-based integrated environmental modeling framework. *Environmental Modelling & Software*, **55**, 1–24.

WILDHABER, M.L., DEY, R., WIKLE, C.K., MORAN, E.H., ANDERSON, C.J. & FRANZ, K.J. 2015*a*. A stochastic bioenergetics model-based approach to translating large river flow and temperature into fish population responses: the pallid sturgeon example. *In*: RIDDICK, A.T., KESSLER, H. & GILES, J.R.A. (eds) *Integrated Environmental Modelling to Solve Real World Problems: Methods, Vision and Challenges*. Geological Society, London, Special Publications, **408**. First published online May 28, 2015, updated March 10, 2016, https://doi.org/10.1144/SP408.10

WILDHABER, M.L., WIKLE, C.K., MORAN, E.H., ANDERSON, C.J., FRANZ, K.J. & DEY, R. 2015*b*. Hierarchical stochastic modelling of large river ecosystems and fish growth across spatio-temporal scales and climate models: the Missouri River endangered pallid sturgeon example. *In*: RIDDICK, A.T., KESSLER, H. & GILES, J.R.A. (eds) *Integrated Environmental Modelling to Solve Real World Problems: Methods, Vision and Challenges*. Geological Society, London, Special Publications, **408**. First published online October 12, 2015, https://doi.org/10.1144/SP408.11

Model fusion at the British Geological Survey: experiences and future trends

DENIS PEACH, ANDREW RIDDICK, ANDREW HUGHES*, HOLGER KESSLER, STEVE MATHERS, CHRISTOPHER JACKSON & JEREMY GILES

British Geological Survey, Environmental Science Centre, Nicker Hill, Keyworth, Nottingham NG12 5GG, UK

**Correspondence: aghug@bgs.ac.uk*

Abstract: The British Geological Survey (BGS) is developing integrated environmental models to address the grand challenges that face society. Here we describe the BGS vision for an Environmental Modelling Platform (BGS 2009) that will allow integrated models to be built, and describe case studies of emerging models in the United Kingdom.

This Environmental Modelling Platform will be founded on the data and information that the BGS holds. This will have to be made as accessible and interoperable as possible to both the academic and stakeholder decision-making community. The geological models that have been built in an *ad hoc* way over the last 5–10 years will be encompassed in a National Geological Model that will be multi-scaled, beginning with onshore UK and eventually including the offshore continental shelf. The future will be characterized by the routine delivery of 3D model products from a multi-scaled and scalable 3D geological model of the UK that can be dynamically updated. The deployment of this model will generate further significant requirements across the Information and Knowledge Exchange spectrum, from applications development (database, GIS, web and mobile device), data management, information product development, to delivery to a growing number of publics and stakeholders.

There is now a growing realization in the environmental and social sciences that to address the grand challenges that face the world, a whole system approach is required. These challenges, including climate change, natural resource and energy security, and environment vulnerability, raise multi- and transdisciplinary issues that require integrated understanding and analysis. Not only must we model the whole physical Earth system, bringing together climate, ecological, hydrological, hydrogeological and geological models to name but a few, we must link them to socio-economic models. Model fusion may well be the only adequate way to provide the necessary coupled process framework through which predictions and planning or management decisions can most appropriately be made.

A scoping study (Giles *et al.* 2010) assessed the current situation and made some preliminary recommendations in order to create a more integrated and semantically harmonized future in environmental modelling. The only viable option is a 'linked models' approach, which enables models to pass parameters between each other at runtime. This solution can bring together the best and most appropriate scientific models, and allows the various scientific disciplines to continue the development of their current models. This linking approach is also relevant for integrating models that have been largely built and optimized individually, with appropriate configuration. The European Union has funded multi-national, multi-disciplinary research into 'linked modelling', using the Open Model Interchange (OpenMI) standard. This software used, in conjunction with critical underpinning activities such as data management, semantics and ontologies, understanding of uncertainty, and visualization, offers a rapidly maturing solution, the creation of an Environmental Modelling Platform, with the potential to fulfil this vision.

The British Geological Survey has a long history of developing 3D geological framework models, as well as groundwater flow models and other process models such as those to assess impacts of carbon storage in the subsurface. Increasingly, the survey is now producing 3D geological models alongside or as an alternative to 2D geological maps. It is clear that there has been a proliferation of modelling activities across many other environmental sciences disciplines. Thus, a significant number of environmental models are now available, including models to predict environmental hazards and habitat quality, and those focused on environmental resource sustainability. Whilst such models represent significant advances in technology, and frequently considerable investments in time and intellectual effort, they are often discipline-specific and developed to address a specific issue or problem. The ability to link such models together allows a holistic

From: Riddick, A. T., Kessler, H. & Giles, J. R. A. (eds) 2017. *Integrated Environmental Modelling to Solve Real World Problems: Methods, Vision and Challenges*. Geological Society, London, Special Publications, **408**, 7–16.
First published online September 16, 2016, https://doi.org/10.1144/SP408.13

understanding of Earth system processes. These linkages and the systems-level understanding that they provide are increasingly important in helping us to solve overarching environmental problems facing the world today, such as: (1) predicting and responding to environmental change; (2) ensuring the security of natural resources for the future; and (3) understanding and predicting natural hazards. It is only through harnessing some of the existing large and very complex models available that further progress can be made in addressing these issues.

Specific examples of the application of linked and integrated modelling could include linking groundwater models with models predicting climate and future land-use trends in order to understand better the impact of climate change on agricultural policies and planning regulations. In such a scenario, the BGS clearly has an important role in contributing to our knowledge of the 3D distribution of rock units, groundwater movement and the impact of geology on land use. The integration of large-scale climate models with possibly more localized models of groundwater, rainfall and land use brings us to the importance of being able to deal with heterogeneity and scale between different models. Understanding the interaction between groundwater levels and changes in sea level due to climate change is also a topical issue in a number of parts of the UK, and requires fusing together with groundwater, rainfall and other climate models. There are also numerous instances in the subsurface where being able understand the interaction between different flow regimes is critical: for example, in planning carbon dioxide sequestration.

In addition to drivers from the wider research community, the BGS strategy requires the interpretation and analysis of a range of environmental and resource observations in an integrated way. In order to accomplish this, we are aware of an increasing need to be able to integrate different models and datasets to answer science questions, and to provide the outputs from these modelling activities to users in a form that they can use directly. This objective involves creating both the IT infrastructure components to support this work, and developing the organizational culture and the ways of working needed. The importance of addressing the requirements of a number of stakeholders (not only the modelling specialist creating the model) in the modelling process, particularly in appreciating the inputs to a model and the outputs from it, has also been pointed out by Glynn (2015) and Voinov *et al.* (2016). The proposed solution is the creation of an open environmental modelling platform that will provide the methodologies, software tools and standards upon which to undertake integrated multi-scaled environmental modelling, providing a sound and reproducible basis for decision-making.

Building the Environmental Modelling Platform

In order to design the Environmental Modelling Platform, the BGS undertook a scoping study to understand the future requirements for integrated modelling both within the organization and more widely. One of the key recommendations of this study was that since many environmental organizations have already invested significant time, cost and effort in developing their models, then the focus for onward development should be to harness this investment, rather than to try to propose an entirely new software system. Thus, the BGS vision for the Environmental Modelling Platform consists of a portfolio of methodologies, software tools and standards to allow us to model the environment in an integrated way, and to link with relevant models outside the geoscience domain.

The key components within the Environmental Modelling Platform include BGS's data and information resources, as well as existing geological models, conceptual models and process models (e.g. for groundwater modelling). It can be thought of as the entire research and knowledge base of the BGS. Figure 1 shows the conceptual relationship between these components. Our data and information resources form the foundation for the platform represented by the base of the triangle in this figure. Three-dimensional geological framework models are then built using this input data, and forms one of a number of means of constraining conceptual models and process models. Progressing from 3D geological framework modelling through the development of a conceptual model and then the process modelling itself allows predictive modelling and scenario planning.

There are a number of challenges to be overcome when constructing the Environmental Modelling Platform. These include issues concerning underpinning technologies, challenges in developing appropriate modelling methods (e.g. for model parameterization) and uncertainty quantification. There are also challenges in developing the required workflows and cultural practices, both within the organization and externally, and in the delivery of products.

The main challenges and underpinning areas of the work are discussed in the following.

Standards

There are a wide variety of software data and modelling standards that are applicable to environmental modelling. The BGS's approach is to support and adopt existing standards wherever possible, and to create new standards only where this is absolutely necessary. The OpenMI methodology for linking

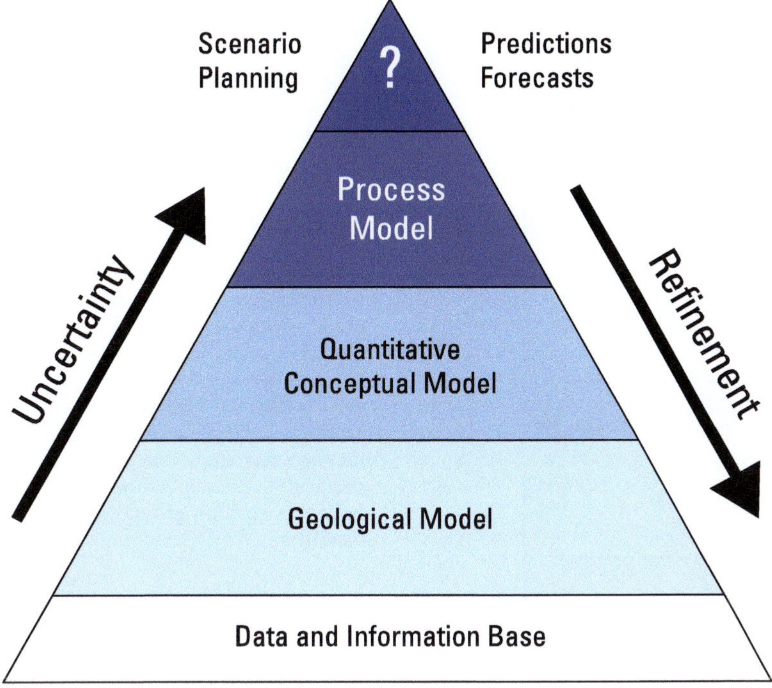

Fig. 1. Relationships between modelling concepts.

models at runtime has been described above, and this has now been adopted as an Open Geospatial Consortium (OGC) standard (OGC 2013). The drivers for adoption of the Open MI as an OGC standard included an interest by the OGC in addressing the time dimension, and also the value of the OpenMI in the increasing efforts to link sensors, datasets and models within the OGC's community of users. The OpenMI Standard can be implemented using a software development toolkit downloadable from the OpenMI Association website or, alternatively, via open-source tools also downloadable from the web, or users can write their own code using information provided by the OpenMI Association. The use of these tools and implementation methods is optional and does not form part of the standard.

Other standards employed by the BGS in developing integrated modelling include GeoSciML for representing data vocabularies (Sen & Duffy 2005), and the ISO metadata standards (ISO 19115-1 2014) for data quality and discovery metadata.

Software

A number of methodologies to link environmental models, particularly at runtime, have been proposed. One approach is offered by the Open Modelling Interface (OpenMI) Association (Gregersen *et al.* 2007), which has produced an open standard for exchanging information between OpenMI compliant models at runtime. A demonstration project, financed by the European Commission – Life Programme, was centred on the transnational Scheldt River Basin. Water management in the basin is distributed among many different authorities and operators in three countries: Belgium, France and The Netherlands. The introduction of the European Water Framework Directive requires water management to be integrated. Existing models have been developed independently, so that integration is far from straightforward. The OpenMI Standard has provided a mechanism to enable the existing models to work together. Important features of Open MI are that it allows parameters to be passed between models at runtime, it is also implemented as a software 'wrapper' around modelling code developed, for example, in languages such as C++ and, therefore, permits ready re-use of code.

Gregersen *et al.* (2005, 2007) described how OpenMI provides a standardized interface to define, describe and transfer data on a time basis between software components that run simultaneously. This supports systems where feedback between the modelled processes is necessary in order to achieve physically sound results. The way the OpenMI works is by modifying models into three distinct

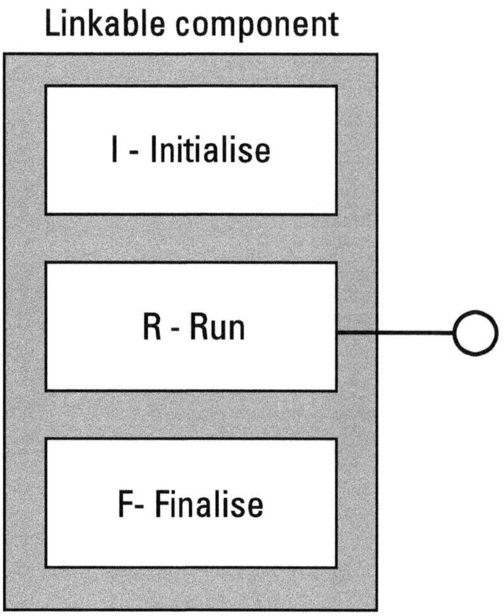

Fig. 2. The structure of a linkable model component within the OpenMI framework.

parts: initialize, run and finalize. The initialize part sets up the model, reads in data and so on, the run part allows the control of one time step at a time and the finalize part then 'tidies' up the model at the end of the run. This could mean writing out the results and releasing memory, amongst other things (see Fig. 2). The model with the smallest time step controls the simulation. It marches forward one time step and then requests data from the model components linked to it (Fig. 3). This model ('component B') then interpolates the value and returns it to model component A. This is repeated until the end of model component B's time step, at which a 'real' value is returned. This whole process is repeated until the simulation is finished.

In this way, the OpenMI allows the linking of models with different spatial and temporal representations: for example, linking river models and groundwater models, where the river model typically uses a 1D grid and a short time step, and the groundwater model uses a 2D or 3D grid and a longer time step. The OpenMI method has now been applied extensively within the BGS, particularly for work on groundwater and climate impact modelling.

Semantic concepts and the use of ontologies

The capability of software to integrate models depends, to a great extent, upon how relationships between important underlying concepts are communicated between models by the software. There is a requirement for such concepts to be communicated in both a human-understandable form to facilitate model design and also in machine-readable form.

The Semantic Web (Berners-Lee *et al.* 2001) uses standard formats to integrate and combine

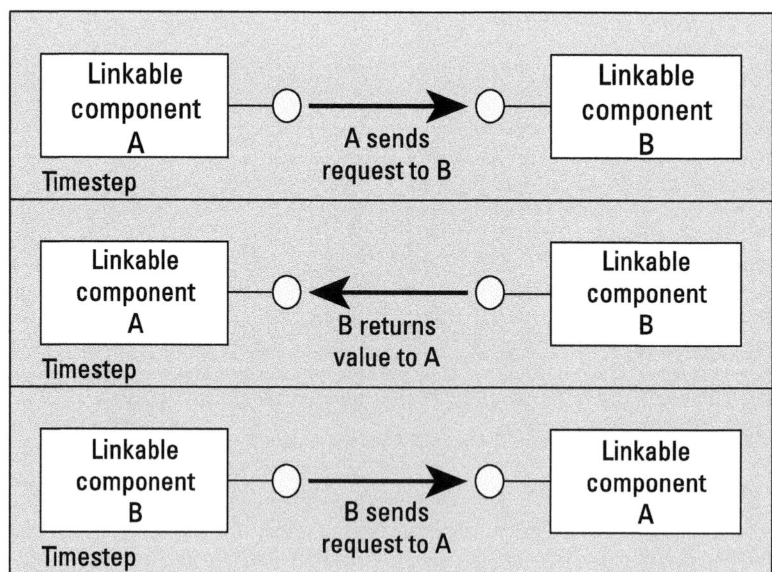

Fig. 3. Mechanism of passing data at runtime between linkable model components in OpenMI.

resources from diverse sources. This allows a person, or a machine, to follow a trail of links through multiple data sources. One of the results of this is to enable natural language querying through establishing unique resource identifiers, a standard way to describe resources (resource description format – RDF), and the use of ontologies. An ontology is a structured way of representing real-world concepts. Semantics enable us to exchange information and knowledge about an object or concept that exists in an ontology. The BGS has been developing a set of standardized database dictionaries that represent vocabularies of key geological terms and concepts as a basis for such ontology development. A key onwards development will be to establish relationships and equivalences between geological vocabularies and those used in other environmental science disciplines. This is actually less of a technical task and more of a community building challenge in bringing relevant researchers together to discuss the concepts involved.

Data management and quality

Well-managed datasets are those that are easily accessible, contain timely data and are stored in a secure environment. In general, scientists spend considerable amounts of time searching for and formatting datasets so that they are usable. Well-managed datasets are said to be accessible when the dataset is easy to locate and retrieve from a data store, they are available in the format in which they are normally used, and the intellectual property rights are clearly understood and articulated.

The BGS has directed significant resources over recent years to create a well-documented data architecture that supports environmental modelling at a range of scales. Our borehole and geological property datasets form a key element of this, as does the adoption of internationally recognized metadata standards. We have also created a range of applications and tools to permit browsing and searching of our datasets in a manner that fits with the workflow of geological and environmental modellers.

Specifically, we are developing a database storage system to store our 2D and 3D data used in modelling. This is currently in progress and is discussed in more detail below. We are also involved in a management process to improve data quality following the NERC Science Information Strategy (2009).

Modelling-scale heterogeneity and uncertainty

In the development of multi-scaled models of various types, attributed with subsurface properties, there are significant challenges to be overcome. In order, for example, to develop a 3D geological model of the UK, based on the integration of a large number of regional models, it is necessary to integrate models at different scales. In addition to constructing the framework model, providing decision-making functionality requires that the models can be attributed with parameters to represent the physical properties of the subsurface (e.g. geotechnical properties, porosity and permeability). There is also a considerable difficulty in delivering models to users (often outside the BGS) in a form in which they can be easily used. Clearly, many users do not have access to the specialist modelling software used within the BGS, and so delivery mechanisms need to take this into account. Delivery of models must also be built with an understanding of the specific requirements of the user community. Perhaps, the most critical issue to be addressed in model delivery is how to communicate the inevitable uncertainties embodied within a multi-dimensional model.

The BGS is addressing the issues surrounding integrating models at a range of scales through the development of a National 3D Geological Model of the UK, built upon various regional models. Parameterization of models, and the uncertainty and delivery problems are being solved through a number of test-bed implementations of the Environmental Modelling Platform, which are discussed in more detail in the following section.

Building a research community

The technical underpinning work to enable many of the issues described above is well established or, at least, underway (e.g. software to support model coupling, and in ensuring sound management and organization of the input data for modelling). One of the critical factors in enabling integrated environmental modelling to progress further is more cultural in nature. There is a great need to facilitate the creation of a networked and linked research community tackling these issues across various environmental science disciplines, so that the barriers between disciplines can be crossed. This research community should include end users and other stakeholders, not only researchers (Glynn 2015; Voinov *et al.* 2016).

Progress and implementations

In order to progress the Environmental Modelling Platform, we have sought to base development on several key larger projects that need to include integrated modelling approaches. Two projects, in particular, are described here: an integrated model of the groundwater systems in the Thames Basin in the SE of the UK; and a programme of integrated modelling in the Glasgow area of Scotland, also investigating the interactions between hydrology and geology. Both of these research projects involve the linking together of environmental process models within the framework of the BGS National Geological Model.

The Thames integrated model

BGS' Thames integrated model (TIM) links the detailed geology of the Thames catchment with groundwater and surface water hydrology, including rainfall, runoff, and recharge, and to a limited extent the source and resource management applied in the Thames catchment. There has been particular emphasis on the geology of the greater London area. The project brings together a unique combination of geological, hydrogeological, environmental and socio-economic challenges that are intrinsically linked and impacted by climate change. To address these challenges requires fully attributed 3D models that incorporate information and processes from all of these disciplines so that accurate representations, simulations, forecasts and predictions can be made. These forecasts and 'predictions' are required to enable informed decision making and planning for sustainability. Within this project integrated modelling is being applied to understand the interaction between rainfall evaporation and runoff, together with river flows for the River Thames and its tributaries, and groundwater flow.

The importance of interaction between the surface water system and abstraction has been demonstrated by using the TIM linked model composition developed by Mansour *et al.* (2013). A model composition which allows an appropriate representation of the hydrological system has been developed and is shown diagrammatically (Fig. 4). Two groundwater models have been developed: one of the Chalk using ZOOMQ3D (Jackson & Spink 2004) and another of the Jurassic limestones using BGSGW (Mansour *et al.* 2013). The former is a distributed groundwater flow model and the latter is a semi-distributed lumped parameter model that can be used to simulate groundwater behaviour. The boundaries of both models are shown on Figure 4 and it should be noted that for the Chalk model these extend outside of the Thames Basin. This is to ensure that sensible groundwater flow boundaries are defined so as not to erroneously calculate baseflow to rivers whose surface catchments are different to groundwater catchments and unduly affect the impacts of groundwater abstractions.

The two groundwater models are linked via a river model (MCRouter) developed using the Muskingham-Cunge approach (Chadwick & Morfett 1986). A simple hydraulic river model was chosen to ensure the composition can be tested before including a more complex river model. These three models are driven by run-off and recharge generated by the recharge model ZOODRM (Mansour & Hughes 2004; Hughes *et al.* 2008). The run-off is routed to the river model whilst the recharge is passed to the two groundwater flow models (Fig. 5). As described above, the Jurassic limestones and Chalk aquifer are not linked via the sub-surface, only via interaction with the River Thames and its tributaries. The two groundwater models and the

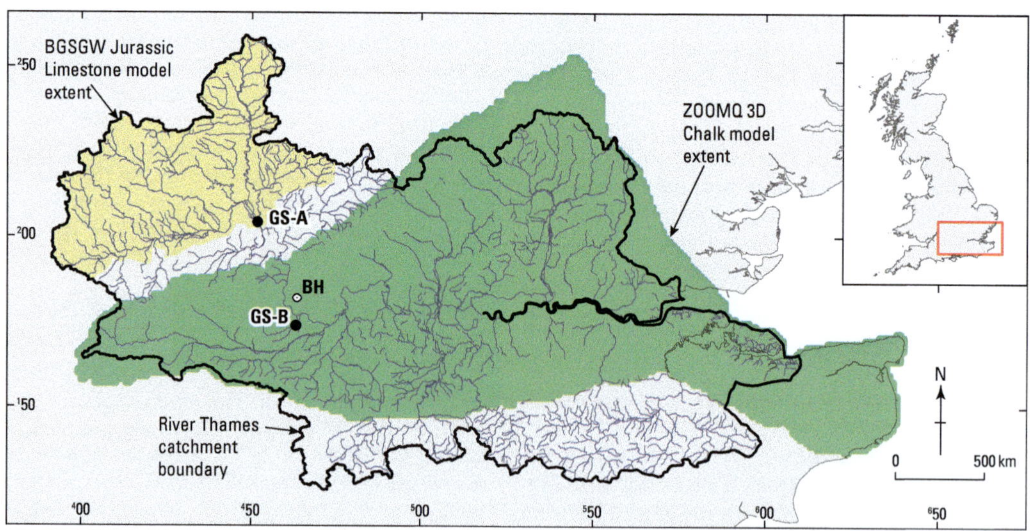

Fig. 4. The River Thames catchment boundary, and the location and extent of models within the TIM composition.

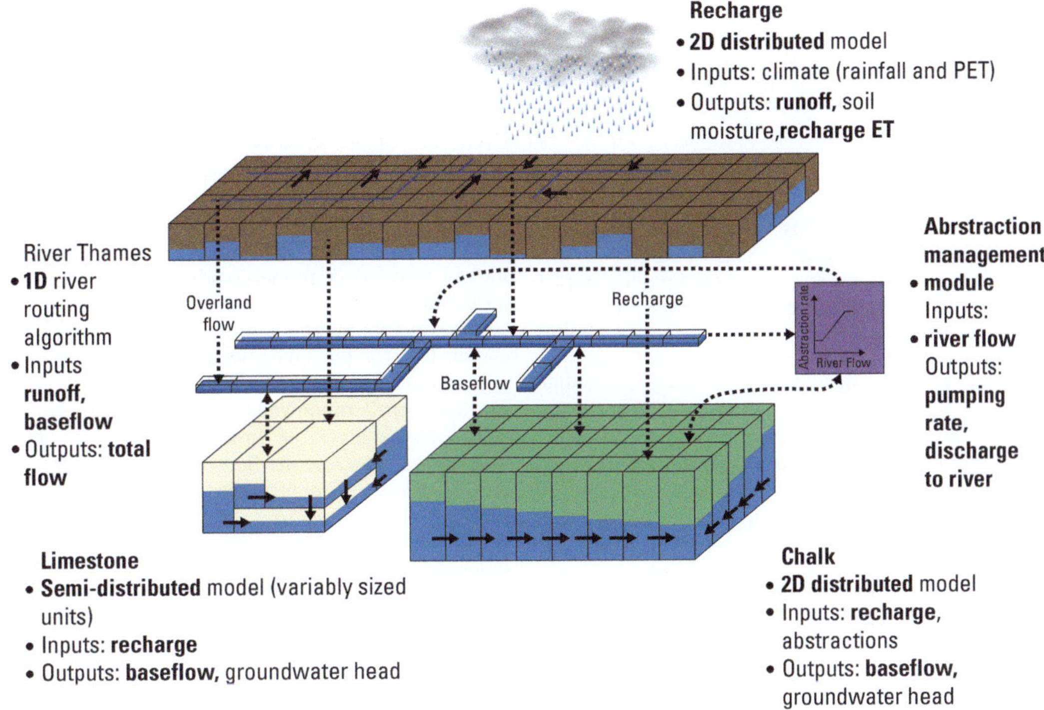

Fig. 5. Model linkages and input/output parameters within the TIM composition. PET, potential evapotranspiration; WM, water management.

river model are, therefore, dynamically linked by the model linkage standard OpenMI (Gregersen *et al.* 2007). This linkage has been facilitated by using the Fluid Earth software development toolkit (SDK) and the associated editor, Pipistrelle to construct compositions (Harpham *et al.* 2014).

The model composition has been applied to a hypothetical situation to test the applicability of the models and what benefit is accrued from linking them. A hypothetical groundwater abstraction of 150 Mlday^{-1} has been used to investigate the impacts of groundwater abstraction on the flows in the River Thames. The groundwater abstraction was simulated from a point on the banks of the River Thames (labelled BH on Fig. 4). This abstraction was linked to riverflow as measured at a downstream gauging station (labelled GS-B on Fig. 4). To investigate the effect of different management regimes two scenarios were simulated: One with a fixed groundwater abstraction of 150 Mlday^{-1} and the other with an abstraction that varies from 50 to 150 Mlday^{-1} depending on river flow at GS-B (Fig. 4). The relationship between river flow and magnitude of groundwater withdrawal was achieved by the inclusion of an abstraction management component in the composition which modified groundwater abstraction during runtime. This component related groundwater abstraction by a simple rule and was included in the composition and linked using OpenMI. The abstracted water is returned to the river some 35 km upstream (labelled GS-A on Fig. 4) of the groundwater abstraction as this represents typical water use, from abstraction, supply to the city of Oxford and after use, discharge to the sewerage system and hence finally sewage effluent returned to the river.

The model composition was simulated for a 30 year period and groundwater heads and the river hydrograph downstream of the groundwater abstractions plotted (see figs 3 & 4 in Mansour *et al.* 2013). During conditions of low flow, i.e. during the 1975/6 drought the model simulation resulted in a river flow that was lower for the scenario with fixed abstraction than for the scenario where groundwater abstraction decreased. However, as groundwater abstraction was reduced, groundwater levels surrounding the abstraction were higher. Whilst the variable abstraction reduced the impact on groundwater heads the overall impact on river flows increased due to the reduction in return flows to the River Thames. Whilst this is a hypothetical example, the results are contrary to expectation and the utility of the modelling composition in the development of management policies for droughts, and its potential for other scenarios was demonstrated.

Glasgow model

The availability of detailed 3D geological framework models for the Glasgow area (Merritt *et al.* 2007) provides an accurate representation of the subsurface within which to evaluate linked groundwater models of the complex Quaternary deposits of the Glasgow area. This study evaluated the use of 3D modelling to constrain groundwater flow predictions, building on earlier work in Glasgow that used interpretations of the geology based on 2D maps. An understanding of the 3D geometry of the lithostratigraphical units has allowed a very detailed conceptualization of the likely groundwater regime. Recharge (Mansour *et al.* 2008) and groundwater flow models (Turner *et al.* 2015) have been developed to test this conceptual understanding.

The results from numerical modelling indicate that the general direction of groundwater flow in Glasgow is downgradient from areas of high ground towards the lowland valleys and the River Clyde itself, through both bedrock and Quaternary potential aquifers (Turner *et al.* 2015). Groundwater levels in the Quaternary Clyde Valley Aquifer are strongly influenced by the course of the River Clyde and its tributaries. The sand deposits within the Quaternary act as a highly conductive shallow aquifer in the Clyde Valley, and are responsible for regional flow of shallow groundwater. The model also suggests that Quaternary deposits in the Proto Valley of a tributary of the River Clyde constitute a significant aquifer, receiving and promoting groundwater to flow towards the River Clyde, but the dearth of local data make any conclusions conjectural. The conceptual model developed using the 3D modelling as input, has been broadly validated by the numerical modelling. Thus, this study illustrates the value of integrating 3D geological framework models with groundwater flow modelling.

The methodologies and modelling frameworks established for the Glasgow area (Fig. 6) provide a good basis for further modelling using, for example, long-term groundwater observation data, which will soon be available from the monitoring network being set up by the British Geological Survey and the Scottish Environmental Protection Agency (SEPA). In addition to validating the conceptual groundwater model, the numerical model also provides an opportunity to simulate possible future scenarios and to investigate specific urban development issues and proposals in Glasgow (e.g. to understand the groundwater processes that may impact on the remediation of contaminated land).

The future of integrated modelling at the BGS

Future developments in integrated modelling at the BGS are planned along several lines, including

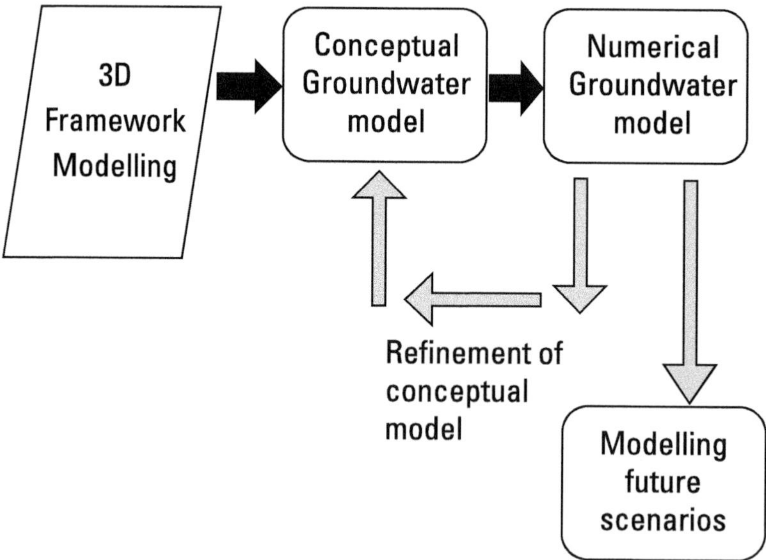

Fig. 6. The modelling process in Glasgow.

further research on parameterizing models with appropriate property values, and the development of voxelated models in order to more accurately predict the distribution of chemical and physical properties within the subsurface. We are also aware of the need to establish a dedicated database of 3D and 4D data to provide storage for these models, support model fusion activities, and to facilitate the delivery of 3D and 4D data to end users. Significant work is also in progress in developing techniques for the visualization and delivery of 3D/4D data. These developments are discussed in more detail below.

The facility to store 3D and, increasingly, 4D models in a manner that allows easy retrieval is an important component of the Environmental Modelling Platform. In developing a modelling workflow, it is imperative that modellers are able to draw on a data store of previous models, and also to incorporate new data that have been assimilated since the last modelling of a particular geographical area. The capacity to be able to validate and, sometimes, update the model source data in the light of modelling results is also an important requirement. The corporate model data store also needs to be software-independent, so that it can be accessed by modellers working with a variety of modelling software tools and, thus, allow the exchange of data in a variety of formats. An important requirement arising out of these model-fusion activities is the increasing need to be able to store potentially large volumes of time-series data in an easily accessible manner, and also store multiple versions of models corresponding to different model scenarios, with appropriate metadata.

In addition to the requirements imposed by the modelling workflow, a geological survey organization also needs to be able to deliver models in a variety of formats, and therefore the BGS data-store architecture must support this. These key requirements arising from modern modelling best practice are currently guiding the creation of a fully functional database and metadata system to support our modelling activities. This work is building on our experience gained in creating systems to store and access models in the Digital Geospatial Model Project (DGSM) undertaken between 2000 and 2005 (Smith 2005).

The development of mechanisms for model delivery is also an important continuing focus. As mentioned earlier, models need to be delivered in a format that is designed to meet the user requirements. The BGS has maintained a strategy to exploit rapidly developing access to mobile technologies for delivery of a number of our datasets, as exemplified by the BGS iGeology and iGeology3D mobile apps for iPhone and Android platforms. Currently, the 'Groundhog' application available via the BGS website (http://www.bgs.ac.uk/research/environmentalModelling/groundhogDesktop.html) and for use in mobile devices provides virtual borehole cross-section and horizontal section viewing functionality; Groundhog is a forerunner for further developments for the delivery of 3D and 4D data.

In addition to technical developments, the importance of close co-operation and collaboration between researchers in different environmental science disciplines has been discussed above, and this continues to be an important area for onwards development in the BGS. The need to integrate various models in order to address a particular problem has been an important factor in influencing the BGS working culture. We are increasingly creating 3D framework and process models as a standard geological survey output rather than 2D maps, and the need to work in a more integrated way is increasingly reflected in the BGS's internal team organization.

A Natural Environment Research Council (NERC) Strategy for integrated environmental modelling (IEM) has been adopted by the NERC Research Centres and is currently being refined (Royse & Hughes 2014). The NERC IEM strategy addresses the challenges of linking environmental models to solve specific science questions by creating a community that prioritizes modelling needs, and a modelling framework to facilitate a greater sharing and linking of data and models. The overall objective is to promote a change in culture towards greater collaborative working, improved accessibility and effective use of existing models and tools developed over many years, and also to encourage a move away from 'silo management', where each problem is addressed in isolation.

This strategy will address a number of key challenges in IEM, including:

- Establishing workable standards that define model input and output parameters, and provide a description of the model and the assumptions on which it is based.
- Understanding and quantifying uncertainty, to understand how uncertainty is propagated within a linked modelling system and to adopt user-specific communication methods.
- Understanding the impact of temporal and spatial scales on model operation and processing, and to design tools that can explore and mitigate the impacts of changing scales on linked model systems.

Final remarks

For the BGS, the crucial issues to be addressed in the coming few years for the successful deployment of an Environmental Modelling Platform for addressing the major questions surrounding energy, and water resource security and sustainability include:

- The successful deployment of 3D geological models in an acceptable, understandable product form, from a dynamically constructed 3D geological framework model, that contains the heart of the BGS geological information.
- The parameterization of this framework model with physical properties and later chemical properties, with error and uncertainty bounds defined for line-work, lithostratigraphy and properties.
- The use of this Environmental Modelling Platform with partners to provide the knowledge base for modelling Earth system processes at all scales. This requires linkages to climate models, surface process models, hydrological and hydraulic models and so on. But to achieve impact and value for society, coupling to social, economic and financial processes and models will also be necessary.

References

Berners-Lee, T., Hendler, J. & Lassila, O. 2001. The Sematic Web: A new form of Web content that is meaningful to computers will unleash a revolution of new possibilities. *Scientific American*, **284**, 34–43, https://www-sop.inria.fr/acacia/cours/essi2006/Scientific%20American_%20Feature%20Article_%20The%20Semantic%20Web_%20May%202001.pdf

Chadwick, A. & Morfett, J. 1986. *Hydraulics in Civil Engineering*. Allen & Unwin, London.

Giles, J.R.A., Hughes, A., Kessler, H., Watson, C. & Peach, D. 2010. *Data, and Research for Applications and Models (DREAM): Scoping Study Report*. OR/10/020. British Geological Survey, Keyworth, Nottingham.

Glynn, P.D. 2015. Integrated Environmental Modelling: human decisions, human challenges. *In*: Riddick, A.T., Kessler, H. & Giles, J.R.A. (eds) *Integrated Environmental Modelling to Solve Real World Problems: Methods, Vision and Challenges*. Geological Society, London, Special Publications, **408**. First published online May 21, 2015, https://doi.org/10.1144/SP408.9

Gregersen, J.B., Gijsbers, P.J.A., Westen, S.J.P. & Blind, M. 2005. OpenMI: the essential concepts and their implications for legacy software. *Advances in Geosciences*, **4**, 37–44.

Gregersen, J.B., Gijsbers, P.J.A. & Westen, S.J.P. 2007. OpenMI: open modelling interface. *Journal of Hydroinformatics*, **9**, 175–191.

Harpham, Q., Cleverly, P. & Kelly, D. 2014. The FluidEarth 2 implementation of OpenMI 2.0. *Journal of Hydroinformatics*, **16**, 890–906.

Hughes, A.G., Mansour, M.M. & Robins, N.S. 2008. Evaluation of distributed recharge in an upland semi-arid karst system: the West Bank Mountain Aquifer, Middle East. *Hydrogeology Journal*, **16**, 845–854, https://doi.org/10.1007/s10040-008-0273-6

ISO 19115-1 2014. *Geographic Information – Metadata – Part 1: Fundamentals*. International Organization for Standardization, Geneva, http://www.iso.org/iso/home/store/catalogue_ics/catalogue_detail_ics.htm?csnumber=53798

Jackson, C.R. & Spink, A.E.F. 2004. *User's Manual for the Groundwater Flow Model ZOOMQ3D*. IR/04/140. British Geological Survey, Keyworth, Nottingham, http://nora.nerc.ac.uk/11829/

Mansour,M., Hughes, A., O Dochartaigh, B. & Graham, M. 2008. Representation of urban recharge processes in the distributed recharge model (ZOODRM) of the Glasgow urban area, Scotland. *In*: Poeter, E., Hill, M.C. & Zheng, C. (eds) *MODFLOW and More 2006: Managing Ground-Water Systems – Conference Proceedings*, http://nora.nerc.ac.uk/9795/

Mansour, M., Mackay, J., Abesser, C., Williams, A., Wang, L., Bricker, S. & Jackson, C. 2013. Integrated Environmental Modeling applied at the basin scale: linking different types of models using the OpenMI standard to improve simulation of groundwater processes in the Thames Basin, UK. *In*: *MODFLOW and More 2013: Translating Science into Practice, 2–5 June 2013, Colorado, USA*, http://nora.nerc.ac.uk/501789/

Mansour, M.M. & Hughes, A.G. 2004. *User's Manual for the Distributed Recharge Model ZOODRM*. IR/04/150. British Geological Survey, Keyworth, Nottingham.

Merritt, J., Monaghan, A., Entwisle, D., Hughes, A., Campbell, D. & Browne, M. 2007. 3D attributed models for addressing environmental and engineering geoscience problems in areas of urban regeneration–a case study in Glasgow, UK. *First Break*, **25**, 79–84.

OGC 2013. *Open Modelling Interface Standard (OpenMI)*. Open Geospatial Consortium, http://www.opengeospatial.org/standards/openmi

Royse, K.R. & Hughes, A.G. (eds). 2014. *Meeting report: NERC Integrated Environmental Modelling Workshop: held at the British Geological Survey*, Keyworth, 4–5th February. OR/14/042. British Geological Survey, Keyworth, Nottingham.

Sen, M. & Duffy, T. 2005. GeoSciML: development of a generic GeoScience markup language. *Computers and Geosciences*, **31**, 1095–1103.

Smith, I.F. (ed.). 2005. *Digital Geoscience Spatial Model Project Final Report*. British Geological Survey, Occasional Publications, **9**.

Turner, R.J., Mansour, M.M., Dearden, R., O Dochartaigh, B.E. & Hughes, A.G. 2015. Improved understanding of groundwater flow in complex superficial deposits using three-dimensional geological-framework and groundwater models: an example from Glasgow, Scotland (UK). *Hydrogeology Journal*, **23**, 493–506, https://doi.org/10.1007/s10040-014-1207-0

Voinov, A., Kolagni, N. *et al.* 2016. Modelling with stakeholders – Next Generation. *Environmental Modelling and Software*, **77**, 1–25.

Integrated environmental modelling: achieving the vision

R. V. MOORE[1] & A. G. HUGHES[2]*

[1]*British Geological Survey, Maclean Building, Crowmarsh Gifford, Wallingford, Oxfordshire OX10 8BB, UK*

[2]*British Geological Survey, Kingsley Dunham Centre, Keyworth, Nottingham NG12 5GG, UK*

**Correspondence: aghug@bgs.ac.uk*

Abstract: Integrated environmental modelling (IEM) is a recent phenomenon that offers the opportunity to solve complex environmental problems. Whilst it has made great strides in recent years, there are still challenges to be met before IEM is universally accepted and used. This paper describes the current state of IEM and sets out a roadmap for achieving its full potential. A multidisciplinary, multi-agency approach will be required, the main goals of which are to: (1) raise awareness and build confidence in IEM; (2) ensure availability and accessibility of IEM techniques, tools and standards; (3) establish a minimum set of standards; (4) build the IEM skills base; (5) establish an underpinning research and development (R&D) programme; (6) co-ordinate and promote collaboration; and (7) foster IEM use by government, industry and the public. Once these goals have been achieved, then IEM can be deployed to help resolve currently intractable environmental issues, and the IEM methodology can be transferred to other fields.

Purpose

The importance of understanding the world, and all the events and activities within it as a set of interconnected interacting processes is now widely recognized. Technology is now in place that facilitates the linking of process simulation models, helping us to better understand and predict how the world will respond to events and management interventions. What are now needed is a strategy, an institutional infrastructure and resources to move that technology out of the research area into the domain of the early adopters. If these things could be put in place, then opportunities open up for finding sustainable solutions to present challenges, and for developing new products and services.

In many spheres, simulation models have proved to be the most effective method of exploring processes, encapsulating our knowledge of them and predicting their response in real or imagined situations. This is true for all the physical sciences, and applies to many social and economic sciences as well. To date, most modelling development has taken place in relatively isolated, discipline-specific groups; there has been little communication across traditional academic disciplines. By comparison with the investment in model development, little work has been undertaken on the complex problem of linking (or coupling) model codes and their application (in this paper, model codes are used to mean the software created to solve the mathematical models (i.e. MODFLOW) and model instance is the application of that software to a particular situation (i.e. the London Basin)) either within or across disciplinary boundaries: an activity generally referred to as 'integrated modelling' or, in the context of this paper, 'integrated environmental modelling (IEM)'. (The terms linking and coupling are synonymous and mean establishing a connection between two numerical models so that they can exchange data and, hence, simulate the interaction of the two processes. Generally, the link allows the model instances to request data from each other. The behaviour of the requested model instance will, therefore, influence the behaviour of the requesting model instance.) Although much scepticism has been expressed about the viability of IEM, recently a consensus on the need for IEM has emerged from a number of meetings between modellers, who either have an immediate need to study interacting processes or have seen the bigger picture and its opportunities (e.g. Moore & Hughes 2012). Consequently, a number of draft roadmaps have been written to spur on progress in integrated modelling. Although these roadmaps have originated from different disciplines, they are remarkably consistent in their conclusions. They all recognize that collaborations within and between disciplines and nation states will be required to bring about the necessary conditions for the new culture and technology to grow and flourish – environmental problems are seldom confined by national or disciplinary boundaries. An example roadmap developed against this background is the UK's *A Strategic Vision for UK*

From: Riddick, A. T., Kessler, H. & Giles, J. R. A. (eds) 2017. *Integrated Environmental Modelling to Solve Real World Problems: Methods, Vision and Challenges*. Geological Society, London, Special Publications, **408**, 17–34.
First published online May 23, 2016, https://doi.org/10.1144/SP408.12

e-Infrastructure (BIS 2011). The report spans the aerospace and automotive industries, health and pharmaceuticals, entertainment and media, the bio-economy, weather and climate and basic research. It outlines the opportunity for scientific and industrial growth, the revolution in e-enabled science and innovation, and proposes a way forward that covers the actions required, a funding model, and the communication channels between government and the participants. Being a master plan, it is inevitably set at a high level. There is, therefore, a need to translate these high-level aspirations into much more specific aims and objectives to be achieved in the environmental sector, and then to propose how they are met and how the benefits are then applied to environmental modelling challenges. This paper sets out to meet that need for both the environmental sciences and their associated industries. It focuses particularly on integrated environmental modelling, identifying challenges and putting forward proposals for addressing them.

Context

The International Council for Science (ICSU) has recently defined the five 'grand challenges' (ICSU 2010) that need to be met if we, as custodians of the Earth, are to manage our resources under increasing pressure from both population and economic growth and environmental change (i.e. land-use change combined with a changing climate). These challenges revolve around the need to increase our capabilities in the fields of forecasting, observing, confining, responding and innovating, and are as follows:

- To improve the usefulness of forecasts of future environmental conditions and their consequences for people.
- To develop, enhance and integrate the observation systems needed to manage global and regional environmental change.
- To determine how to anticipate, recognize, avoid and manage disruptive global environmental change.
- To determine what institutional, economic and behavioural changes can enable effective steps toward global sustainability.
- To encourage innovation (coupled with sound mechanisms for evaluation) in developing technological, policy and social responses to achieve global sustainability.

Table 1 lists some of the more eye-catching issues that are driving the need to understand the world as a system. It is not possible to answer any of them without that system-wide understanding. The need for that understanding becomes even greater when other questions are considered, such as: 'What can be done about it? What will be the impact? Is the solution sustainable? How do we minimize the chance of unintended consequences?'. Similar questions could have been provided from any other sector: for example, health and transport: 'What might be the impact of adding a third runway at London's Heathrow Airport on human health in the surrounding area?'. What all these questions have in common is that often we have models of the individual processes involved but rarely do we have the means to link them together and answer the bigger question of how will they interact. Such complex questions are not the only drivers. Indeed, the real driver to solve the challenges of integrated modelling may come from everyday tasks that also require the same systems understanding of the world. Examples to consider might include determining whether a proposed discharge to a river, or a new groundwater abstraction, both of which could affect the river environment, should go ahead. Although simple by comparison to some of the questions in Table 1, they can still pose significant linkage challenges.

Table 1. *Challenges requiring IEM*

- What is the risk to infrastructure of multiple natural disasters?
- What would be the impact of a 'Carrington' type space weather event on electrical distribution systems and civil society?
- What would be the impact of leakage from an oil and gas well in UK waters on the national economy, coastal and marine biodiversity, and the well-being of the population affected?
- How will climate change affect:
  - the global distribution of malaria?
  - the incidence of road and rail closures due to landslides?
  - the frequency of drought conditions and, hence, the security of water supply and biological diversity?
  - the number of insurance claims for properties lost to inundation and cliff erosion?
- How economically viable will it be to store CO_2 in a geological formation under the North Sea?
- What impact will a volcanic ash cloud from an Icelandic volcano have on civil aviation and subsequent economic losses for a country?
- Rainfall can trigger eruptions . . .?
  . . . eruptions can affect hydrology?
- How does spatial variability of rainfall interact with geology and affect flooding?

Many people can understand that system modelling might enable better decisions. However, many others are understandably sceptical as to whether integrated modelling can contribute meaningful systems understanding, other than in the most trivial cases. One reason for scepticism is that it is currently difficult to simulate even relatively simple natural processes reliably: that is, to answer the question posed using the model instance to a suitable level of accuracy. Linking poorly constrained models simply compounds the problem; the real world is just too complex and chaotic for linking a poorly constrained model instance to work. However, no alternative approach has been proposed. Accepting that the challenges ahead will be extremely demanding, some past experiences suggest that considerable progress can be made by following the integrated modelling path until an alternative can be found. One example is the development process from paper maps via digital maps to geographical information systems and Google maps, satellite navigation and location-based services – all of the latter now available through smart phones. In the 1950s, 1960s and 1970s, to all but the most farsighted, such as David Bickmore, Director of the UK's Experimental Cartography Unit (established in the Royal College of Art in 1969 – see http://www.casa.ucl.ac.uk/gistimeline/original/1960.html) (Coppock & Rhind 1991), these things appeared equally unattainable for similar reasons. The digital-mapping sceptics of the time had the added disadvantage of not having prior knowledge of the materials and IT revolution that we have witnessed. In spite of all the doubts at the beginning, the present mapping technologies have been created and, today, are taken for granted. It is highly likely that the development path ahead for integrated modelling will be very similar to that for mapping: so, it is worth reviewing it briefly to see what it has to teach us and how the IEM development path might be shortened.

During the 1960s, when computing was becoming available to researchers, the first attempts at computerized mapping were concerned with automating the drawing of maps. For example, problems such as capturing the linework from existing hand-drawn paper maps, correcting for distortions in the paper, labelling each line with the pen width and colour for drawing it, and then sending the information to a pen plotter for drawing needed solving. For many years, the resulting maps were far inferior to anything that a cartographer could achieve. For at least a decade or more, the advances represented the triumph of hope over adversity: the benefits, financial or technical, if any were barely discernible. Problems that today seem inconsequential were major challenges. Examples were: how to place names on maps automatically without overwriting; how and when to simplify a motorway junction when drawing at smaller scales; and how to produce maps by computer that were as artistically pleasing to the eye as those drawn by hand. These and many similar problems were gradually solved in the course of thousands of MSc and PhD projects. During this phase, there was a growing awareness that the geometric data being amassed contained vast amounts of information that could be exploited for all sorts of purposes beyond reproducing maps. The problem was that most of the coordinates related to the sheet of paper from which they had been captured: they had not been matched in any way to their position in the real world. Further, none of the data had been structured or labelled – the phrase at the time was 'spaghetti data'. There was no way of telling what a line represented, how it connected to other lines and, if it was a river or road, which way water or cars could pass along it.

A long and frustrating hiatus then followed. Considerable energy was expended on how to structure and label lines, and on developing algorithms to exploit those structures. At first, these algorithms addressed what, today, will seem very simple problems, such as how to find out quickly which of a set of points or polygons, representing, for example, houses, lay within another polygon, representing, say, a district boundary. Against the computing power of the day, this was a difficult problem. As people and then commerce slowly began to see the opportunities, pressure for more sophisticated applications grew, together with the need to merge map data with other data. Among the earliest datasets to be merged were the environmental data required for flood estimation, but these were closely followed by planning, road and utility network characteristics and census data; today's satellite navigation systems depend upon knowing the speeds attainable along each stretch of road throughout the day, now in real time. The initial systems for managing and using these data were large, cumbersome and expensive (Goodchild 1988). There were no standards, and moving data from one system to another generally involved reformatting. Looking back, progress was grindingly slow for most of the time that it has taken to make the transition to today's systems. It is sobering to realize that that period now covers 50 years! The development and uptake of digital mapping followed the classical 'Diffusion of innovation' (Rogers 2003), which results in an exponential curve towards market saturation. However, what is possible now is not just due to the immediate digital-mapping community. It has only become possible because of all the materials science, the physics, the chemical, the communications and the IT advances that have occurred in parallel (e.g. see fig. 2 in Crampton 2000). Also important has been the long hard slog of first transforming world maps into digital form, as well as the development of standards (and standards organizations to support and maintain them) that have allowed digital maps to be brought together to create a global dataset. It was when a critical mass of map data existed that progress began to accelerate and industry to invest. At one point, a real show-stopper was the mistaken belief that enormous income could be generated from the sale of digital map data. This, unfortunately, coincided with the period in the 1980s and 1990s when many public undertakings, the main holders of map data, were privatized. It took several years for prices to fall to levels where user organizations were prepared to start buying and a proper market established itself. Several more years passed for prices to come down to where a mass market could open up, giving everyone access to the benefits of digital data. Here, price was not the only factor. Ease of use was central. Today, it requires no more skill than that needed to rub a finger across a glass screen of a mobile to find the nearest restaurant for lunch. All the complexity involved in making that possible is hidden. This last point highlights a very important lesson. Most of the initial funding, which transformed a wild idea into a viable technology and digitized the base information, came from public sources. When critical mass had been achieved, commerce joined in and created a whole new industry that has already and will continue to find new and innovative ways to exploit digital map data, most of them much simpler and far removed from the weightier applications that the pioneers had in mind. There are many lessons to be learned from this story and many more could have been added had there been space to give a more extended account. These have been incorporated into the proposals that follow later in the paper.

Background

What is IEM?

Integrated environmental modelling (IEM) is becoming an increasingly valuable technique for understanding how processes interact and then using that knowledge to assess the likely outcomes of given scenarios. The detailed nature of IEM is evolving rapidly, but an IEM toolkit can be visualized as a set of tools and modelling components. The modelling components include simulation models, databases, analysis tools and visualization tools, among many others: the key feature of these components is that they are linkable, meaning that they can request data and send data from and to each other as they run. This is achieved by all the components adopting a standard interface for data exchange: for example, the OpenMI (Gregersen *et al.* 2007). See Table 2 for a glossary of terms and acronyms used in this paper. The tools help developers and users undertake such tasks as making pre-existing models linkable, linking components to create an integrated model, perhaps a decision support system, and running it.

It is convenient to think of IEM under four headings (based on the four suggested by Laniak *et al.* 2013):

- IT science;
- modelling science;
- organization;
- applications.

At the 'IT science level', in concept, IEM is simple: an integrated model is created by linking together models of individual or groups of processes (Moore & Hughes 2012). The links allow data to pass between the models and the technical challenge is how best to achieve the transfer. Essentially, it is a pure IT problem, the solution of which should be independent of the data being passed. The problem is challenging because the requirement is to be able to link models based on different concepts, working at different scales, using different spatial and temporal resolutions and representations (including none at all), and sourced from different suppliers. In the simple case where the processes being simulated are sequential, the results of the first model can become the input of the next. All that is needed to affect the transfer is for the second model to be able to read the output of the first. More sophisticated approaches are required to handle situations where the processes are running in parallel and interacting time step by time step. This challenge can become greater if the models are running in different computing environments or in the high-performance or cloud environments. At the time of writing, several approaches exist for solving the linkage problem and, in principle, most are very similar (e.g. OpenMI: Moore *et al.* 2005; OMS: David *et al.* 2013; CSDMS: Peckham *et al.* 2013). They work by providing each model with a standard interface through which the model can request and receive data or respond to requests.

Having developed a number of viable solutions, the IT science focus is now turning to the problem of transforming integrated modelling from a research tool that requires a high degree of skill to use into a tool that anyone can apply with ease and confidence. Can the processes of making models linkable, finding appropriate models, linking and running them be made invisible, as have the multitude of processes that take place when two postcodes are entered into a satellite navigation system, which then responds with maps and a voice to show the way from point A to point B?

Now that it is possible to link models more easily than in the past, it is becoming easier to consider the opportunities that IEM creates for studying interacting processes. Integrated 'modelling science' is concerned with the issues and opportunities

Table 2. *Glossary*

Term	Meaning	Further information
ADCIRC	ADCIRC is a system of computer programs for solving time-dependent, free surface circulation and transport problems in two and three dimensions	adcirc.org
BIS	UK Government's Department of Business, Innovation and Skills	www.gov.uk/government/organisations/department-for-business-innovation-skills
BGS	British Geological Survey	www.bgs.ac.uk
CCA	Common Component Architecture	www.cca-forum.org
CCMP	Chesapeake Community Modeling Program	ches.communitymodeling.org
CHyMP	Community Hydrologic Modeling Platform	www.cuahsi.org/CHYMP
CSDMS	Community Surface Dynamic Modeling System	csdms.colorado.edu
CUASHI	Consortium of Universities for the Advancement of Hydrological Sciences Inc.	https://www.cuahsi.org
EarthCube	EarthCube is a community-led cyberinfrastructure initiative for the geosciences	earthcube.org
e-infrastructure	This refers to the ecosystem of resources that allows distributed collaboration and computation, large-scale simulation and analysis, and fast access to (large) data collections (well organized according to accepted standards and with rich metadata), analytical and visualization services and facilities	https://www.gov.uk/government/uploads/system/uploads/attachment_data/file/249474/bis-13-1178-e-infrastructure-the-ecosystem-for-innovation-one-year-on.pdf
EPSRC	Engineering and Physical Sciences Research Council	www.epsrc.ac.uk
ESMF	Earth System Modeling Framework	www.earthsystemmodeling.org
ESRI	Environmental Systems Research Institute	www.esri.com
FRAMES-3MRA	Framework for Risk Analysis in Multi-media Environmental Systems – Multi-media, Multi-pathway, and Multi-receptor Risk Assessment	www.epa.gov/exposure-assessment-models/3mra
GIS	Geographical Information System	www.esri.com/what-is-gis
GRASS	Geographic Resources Analysis Support System	grass.osgeo.org/grass64
GSFLOW	USGS coupled groundwater and surface-water flow model	water.usgs.gov/ogw/gsflow/
HSPF	USGS Hydrological Simulation Program – Fortran	water.usgs.gov/software/HSPF/
ICSU	International Council of Science	www.icsu.org
ISCMEM	Interagency Steering Committee on Multi-media Environmental Modeling	sites.google.com/a/environmental-modeling.org/environmental-modeling/Home
IHM	Penn State Integrated Hydrologic Model	www.pihm.psu.edu
LOIS	Land Ocean Interaction Study – NERC funded research project	www.bodc.ac.uk/projects/uk/lois/
MapWindow	GIS project that includes a free desktop geographical information system application with an extensible plug-in architecture	www.mapwindow.org
MODFLOW	USGS groundwater model code	water.usgs.gov/ogw/modflow/
NERC	Natural Environment Research Council	www.nerc.ac.uk
NSF	National science Foundation	www.nsf.gov
OGC	Open Geospatial Consortium	www.opengeospatial.org
OpenMI	Open Modelling Interface	www.openmi.org
OMS	Object Modeling System	nrrc.ars.usda.gov/ModelFrameworks/ObjectModelingSystem.aspx
PRMS	USGS Precipitation-Runoff Modeling System	http://www.brr.cr.usgs.gov/projects/SW_MoWS/PRMS.html
SWAT	Soil & Water Assessment Tool	swat.tamu.edu
USACE	United States Army Corps of Engineers	www.usace.army.mil
US EPA	United States Environmental Protection Agency	www.epa.gov
USGS	United States Geological Survey	www.usgs.gov
VPH	Virtual Physiological Human	www.vph-institute.org

that linking models of the same or different processes raises or creates, for instance:

- The validity of integrated modelling – is it valid to link (e.g. Lloyd *et al.* 2011) an economic model to an ecological model; two models at different scales; two models based on different conceptualizations of the world?
- How to approach a problem that requires an integrated approach? Experience shows that it is very easy to focus on the mechanics of model integration and lose sight of the real problem. Voinov & Shugart (2013) discussed how to conceptualize process interactions, and how to calibrate and validate model chains.
- Can the propagation of uncertainty along a model chain be represented and assessed (e.g. Bastin *et al.* 2013)? How can uncertainty be modelled in a multidimensional web of interacting models with complex feedbacks, rather than a chain?
- How to establish and populate: catalogues for both model codes and applications of the model code – descriptions that allow people and machines to find and evaluate appropriate model codes and their instances (e.g. Harpham *et al.* 2014); controlled vocabularies of names of model inputs and outputs to reduce the chance of invalid links and for use in automated linking; ontologies (e.g. van Ittersum *et al.* 2008).
- How to exploit artificial intelligence and ontologies to reduce the chance of 'unintended consequences'. These are model instances that are linked inappropriately: see, for example, Voinov & Shugart (2013) for the concept of 'intergronsters'. There are clearly limits to the human mind's ability to foresee all the possible consequences of a particular course of action. A limitation of current modelling is that is necessary to be able to define the scenario to be modelled before it can be modelled. If we can hold our knowledge of things and of processes in a structured manner, then it might be possible to construct algorithms that can search the knowledge base for potentially disadvantageous situations that we could then model in detail.
- How to reconcile the different objective functions used by different disciplines when evaluating similar interacting processes (e.g. those used to optimize the carrying capacities of sewers and rivers in relation to flooding)?
- How to analyse and present linked model results?
- How to record the details of linked model runs so that the results are reproducible, and the lineage of all input data and model components is traceable?

The 'organizational level' is concerned with creating the conditions for IEM to flourish. At present, much of our knowledge about Earth system processes and their integrated modelling is concentrated in small groups spread across the world, each working on a particular problem or challenge. For some time there has been a shared belief that if some of these groups could be brought together, huge opportunities for technical advancement and innovation could be created. This belief is based partly on observation of the advances over the last 50 years in other disciplines, especially geographic information system (GIS) development, and partly on consideration of the opportunities that would be created if a large array of linkable models spanning diverse processes from many disciplines became widely accessible.

One of the first groups to bring modellers together was the United States Interagency Steering Committee on Multi-media Environmental Modeling (ISCMEM). This group, however, was confined to US federal agencies. Since 2002, there has been growing informal collaboration between the European Community (EC) countries, the United States (USA) and Australia. The relationship has been actively encouraged by the EC and the UK Foreign and Commonwealth Office. It has focused on integrated modelling and, in particular, the standards necessary to enable models to be linked together: for example, the Open Modelling Interface (OpenMI) and several others.

A number of international meetings have been held in Europe and the USA to establish whether the optimistic view of IEM's future is widely shared, and, if so, to raise awareness of the new opportunities among potential users, and to create the conditions for its use to expand and secure the funding to facilitate the change. These meetings arose out contacts between the OpenMI team, the United States Geological Survey (USGS), the US Army Corp of Engineers (USACE), the US Environment Protection Agency (US EPA), the Community of Universities for the Advancement of Hydrological Sciences Inc. (CUAHSI), the Community Surface Dynamic Modeling System (CSDMS), and leaders such as Prof. David Maidment of the University of Texas and Dan Ames of Idaho University. Following US EPA meetings in 2007 and 2008 to discuss a draft white paper on the role of integrated modelling in its regulatory work (Gaber *et al.* 2008), an embryo international Community of Practice for IEM (CIEM) emerged.

To follow this up, a major meeting, the Summit on Integrated Environmental Modelling, was held in Washington, DC in December 2010 (Moore & Hughes 2012). Its chief outcome was an IEM Roadmap (Laniak *et al.* 2013). More recently, the White House sought similar advances in modelling, public communication and transparency (US Government 2012), and this is being advanced in the US water–environment arena by the EarthCube initiative, amongst others (NSF 2011). The environmental sector is not alone in recognizing the importance of integrated approaches for taking forward its science and in finding sustainable solutions to society's challenges. For example, medical researchers have eloquently argued the case in two papers (Viceconti & Clapworthy 2011; VPH 2009). As well as these discipline-specific papers, the bigger picture covering the need to be able to bring together knowledge from all disciplines is covered in *A Strategic Vision for UK e-Infrastructure* (BIS 2011). This work has been monitored to see how successful it has been (BIS 2013) and subsequently implemented by the UK's research councils (e.g. EPSRC 2014).

IEM 'applications' have the potential to deliver what modellers have long sought: 'plug and play' modelling, and decision support system development. In practical terms, this means that the modeller has access to a pool of linkable models and other modelling components, such as graphical user interfaces (GUIs). These components will be located and discoverable from anywhere in the world. Anybody will be able to contribute components and they will be available as open source or commercial code. For a component to usefully join the pool, it must be able to, or be capable of being made able to, expose the variables it accepts or provides through one of the standard interfaces. There are a number of modelling platforms available already that allow users to browse available models, link them together and run the resulting composition. The linking process simply involves dragging each relevant output variable of one model onto the corresponding input of another. When the models compute values of each variable for a number of nodes within a model (e.g. end points of a river reach or cells in a grid), a second step maps the nodes of the two models onto each other. Larger models can involve many thousands of connection points. While the process of identifying and making geographical linkages is largely a manual exercise at present, work is underway to automate it. A water resources example might involve connecting a 1D or 2D hydraulic model of a river network to an underlying groundwater model based on a 2D or 3D grid. In most cases, the linkages can handle differences between models in time steps and units of measurement. Object-orientated programming and open standards have greatly facilitated these linkages. Open standards, in particular, allow modellers working independently to produce components that they can be confident will link to other independently developed components. This has profound implications for the modelling market place, for science and for the innovation community.

The attainment of a critical mass of modelling components, and advances in the ease of use and reliability of modelling platforms (such as have been reached by GIS and office suite containing word processing, presentation and spreadsheet software) will be crucial to the take up of IEM. Once these are realized, then the current levels of time and effort required to link models and reformat data will largely disappear. More time will be available to study actual problems or issues and to find solutions. Shell has not been alone in discovering that its employees spend far more time reformatting data than in analysing them (NERC 2009)!

Although the original reason for a component-based approach to integrated modelling was to increase the understanding of process interactions and to apply this to real-world problems, its first practical application has been by developers. Many of their applications have become large, cumbersome, pieces of code. They need a way of breaking down these systems into more manageable components. If they could do this, not only would maintenance and development become simpler, but components could more easily be reassembled to create product variations or new products.

Science and societal challenges, however, remain the primary drivers. A first quick win, that is already beginning to be achieved, is the relative ease with which one modelling component can now be replaced by another in sensitivity testing. Bigger wins will accrue, however, if IEM can help bring better or timelier answers: for example, to challenges of climate change or to incidents such as the Deepwater Horizon oil spill (Machlis & McNutt 2010). The National Park Service Science Advisor, who had to provide a quick response for the Park Service to the Deepwater Horizon tragedy, lamented that he had no model to help assess the situation immediately after the blowout occurred or to indicate the decisions most likely to mitigate the situation (from the Park Service's perspective). He needed an integrated model that could indicate how the oil was most likely to disperse, how it would behave when it reached the surface, which way it would go, what the environmental and economic damage might be should the oil reach the coast, and the likely effectiveness of different strategies.

From earlier integrated modelling exercises, it is known that modellers should expect to encounter and have to overcome a range of problems as new codes from different disciplines are linked together. For example, when a sewer model was linked to a river model in order to investigate the impact of sewer flows on river flooding and vice versa, it emerged that the criteria by which sewers were designed differed from those used to assess the adequacy of river channels. One was evaluated against a design storm, while the other was assessed according to risk exceedance probabilities. Until these very different objectives were reconciled, meaningful conclusions were not possible.

Integrated modelling makes semantics increasingly important, especially as linkages are made across models/codes incorporating different disciplinary expertise or sourced from countries with different languages. Even within an area of disciplinary expertise, many terms used as variable names and the names of model inputs and outputs are often loosely defined. As numerical modelling is opened up to a broader range of scientists across the world, it will become increasingly important to have clear definitions of what each variable represents, so that modellers can have cognizance of codes and their linkages and simulations, and so that they can make appropriate judgements regarding the validity of their model constructs.

Most models conceptualize their view of the world at a particular scale. Most, if not all, of the current interfaces make no attempt to judge the scientific validity of a connection. This is the correct approach as scientists should be given the freedom to make their own judgement, and the technology does not yet exist to automatically make the links. However, the freedom to join anything to anything also leaves open the possibility of invalid connections (i.e. 'Integronsters': Voinov & Shugart 2013). Linking models at different scales, whilst perfectly valid, can result in misleading results if the aggregation and disaggregation of data at the join is not carefully considered. It is important to establish good practice early.

There are clearly number of challenges to making IEM a useable tool: however, a significant amount work has been already undertaken. The following section reviews the current state of the art.

State of the art

Introduction

Integrated modelling is not new. The analogue flight simulators of the 1950s and 1960s were highly sophisticated integrated models of aeroplanes. It has been argued that some of the early digital models of environmental processes from the 1970s and 1980s could be called integrated if the code representing the different processes was separated out and placed in different subroutines. More recent examples of coupled models linking different processes include: linkages to the MODFLOW (Harbaugh 2005) groundwater simulation code; such as the coupling (Sophocleous & Perkins 2000) with the SWAT (Arnold *et al.* 1998) land surface model; or the coupling with the HSPF rainfall runoff model (Donigian *et al.* 1995) to create IHM (Zhang *et al.* 2009); or the coupling with the PRMS watershed model to create GSFLOW (Markstrom *et al.* 2008; http://www.brr.cr.usgs.gov/projects/SW_MoWS/index.html)

The first large-scale environmental integrated modelling exercise was the UK's £30 million Land Ocean Interaction Study (LOIS), the overall objective of which, set in the late 1980s, was, 'To study and better understand coastal zone processes and their interactions in order to facilitate the development of sustainable management policies' (Wilkinson *et al.* 1997) and more specifically:

- To measure the contemporary fluxes of materials (sediments, nutrients, contaminants) through the coastal zone.
- To characterize key physical and biogeochemical processes that govern coastal morphologies and ecosystem functioning.
- To describe the evolution of coastal ecosystems over the last 10 kyr in relation to climate and sea-level change.
- To develop linked land–ocean models to simulate the transport, change and fate of materials in the coastal zone as a basis for predicting changes over the next 50–100 years.

As will be explained below, the science at this time was severely limited by the available IT. Further, the environmental drivers were only just appearing and being recognized in the political world. During the 1990s and at the start of the present century, public awareness changed dramatically and led to the need for a much better understanding of the Earth as a system, so that politicians and managers could find sustainable solutions to the emerging challenges so admirably summarized in the Belmont Forum Report (ICSU 2010).

Recent developments

There is a growing realization that it is neither practical nor useful to construct a single model encapsulating all the processes needed for decision-making and planning (Argent *et al.* 2006; US EPA 2008). Not only are such large models extremely wasteful of resources, they are rarely reusable and frequently fail to make use of existing process models. These are often referred to as 'legacy models' – the result of a huge, historic investment representing state-of-the-art modelling. Consequently, there are currently attempts to convert existing models into building blocks from which more complex models can be assembled (Barthel *et al.* 2008; Warner *et al.* 2008; Argent *et al.* 2009). In today's IT terminology, these are referred to as 'components'; components that can be linked are called 'linkable components'. The term 'modelling component' now has a wider meaning that is not limited to models: it includes files, databases, analytical tools and visualization tools, and, indeed, any component required to make up a modelling system. At the time of writing, although IEM technology is orders of magnitude simpler to apply than 10 years ago, it is recognized that there is a long way to go before it matches what has been achieved in other technologies. Attempts to streamline model integration are given in the following examples.

The US EPA, in conjunction with the US Nuclear Regulatory Commission, the US Army Corps of Engineers and the US Department of Energy's Pacific Northwest National Laboratory, has been developing the Framework for Risk Analysis in Multi-media Environmental Systems (FRAMES-1) (FRAMES 2009) system to manage the execution and data flow between multiple science modules. It uses a fixed file format system to exchange data between components. The Multi-media, Multi-pathway, Multi-receptor Risk Analysis system (FRAMES-3MRA) (FRAMES 2009) (Babendreier & Castleton 2005) is an extension of FRAMES-1 and is based on an API and dictionary system to exchange data. 3MRA is a collection of 17 modules that describe the release, fate and transport, exposure, and risk (human and ecological) associated with contaminants deposited in various land-based waste management units (e.g. landfills, waste piles). The 17 models in 3MRA cannot be easily replaced. FRAMES-2 (Whelan *et al.* 2010) represents the best attributes of FRAMES-1 and FRAMES-3MRA, and is designed to allow for easier registration and replacement of models and support components.

The Open Modelling Interface and Environment (OpenMI 2009), developed by a consortium of European private companies, research establishments and universities co-funded by the European Commission, is a standard for model linkage (Moore *et al.* 2005). The OpenMI Standard version 1.4 defined an interface that allows time-dependent models to exchange data at runtime: hence, OpenMI-compliant models can be run in parallel and share information at each time step. It is, therefore, particularly appropriate for situations where it is necessary to simulate interacting processes, such as changes in river flow which increase nutrients which affect plant growth which, in turn, affects flow. It can handle feedback loops and iteration. It can link models based on different modelling concepts. OpenMI is a generic standard, and can be used to link models from different domains (e.g. hydraulics, hydrology, ecology, water quality, economics), environments (e.g. atmospheric, freshwater, marine, terrestrial, urban, rural, etc.), scales and resolutions (spatial or temporal), platforms, or suppliers. It is not limited to linking models, but can also link any modelling components. OpenMI version 2.0 paves the way for linking models that run in a super-computing environment and models provided as web services. It can exchange a wider range of data types, and simplifies the exchange process when models have either no spatial and/or temporal dimensions, such as a terrain model. While version 1.4 only provided a 'get values' data option, version 2.0 also provides a 'set values' option to facilitate model optimization. The OpenMI version 2.0 is now an Open Geospatial Consortium (OGC) international standard.

The Common Component Architecture (CCA) is a product developed by the US Department of Energy and the Lawrence Livermore National Laboratory teams (Bernholdt *et al.* 2006), which targets high-performance computers and complex models. The CCA supports parallel and distributed computing, as well as local high-performance connections between components, in a language-independent manner. The design places minimal requirements on components and facilitates integration of existing legacy code into the CCA environment by means of the Babel (2004) language interoperability tool, which currently supports C, C + +, Fortran 77, Fortran 90/95 and Python. The CCA is being applied in a variety of disciplines, including combustion research, global climate simulation and computational chemistry: it has also been adopted as the backbone in the Community Surface Dynamic Modelling System (CSDMS: csdms.colorado.edu: Peckham *et al.* 2013).

The Object Modelling System (OMS) was developed by the US Department of Agriculture (Kralisch *et al.* 2004; Ahuja *et al.* 2005; David *et al.* 2013). In contrast to FRAMES and some other systems, OMS requires modules to be rewritten in Java prior to insertion into the system library. Instead of just linking pre-existing blocks or components, OMS provides the tools and integrated framework to develop the components of an IEM in a coherent way.

Despite the evident need for IEM, it has yet to 'take off' in the way its proponents hoped. The reasons include: (1) a lack of convincing demonstrated added value provided by IEM; (2) a lack of a critical mass of available and accessible linkable modelling components; and (3) a difficulty of use and other barriers to entry. The latter include having users with the right skills, a lack of linkable models or linkable components, poor accountability on uncertainties and a lack of the necessary computer power. However, communities of practice are slowly emerging as a part of a number of IEM initiatives that are attempting to address the challenges of IEM. Currently, the initiatives and their communities are relatively isolated because there is no umbrella organization to bring them together. Examples of these communities and initiatives include:

- OpenMI Association, which makes the OpenMI standard freely available (www.openmi.org);

- CSDMS – Community Surface Dynamic Modelling System, which 'makes earth surface process models available, has computational resources for model simulations, and couples models that bridge critical process domains' (csdms.colorado.edu);
- CCMP – Chesapeake Community Modelling Program, dedicated to advancing the cause of accessible, open-source environmental models of the Chesapeake Bay in support of research and management efforts (ches.communitymodeling.org);
- ESMF – the Earth System Modelling Framework: software for building and coupling weather, climate and related models (www.earthsystemmodeling.org);
- CHyMP – the Community Hydrologic Modelling Platform (www.cuahsi.org/chymp).

There are also communities designed to support individual models and software packages, examples include, but are not limited, to:

- GRASS – free geographical information system (GIS) software used for geospatial data management and analysis, image processing, graphics/maps production, spatial modelling, and visualization (grass.fbk.eu);
- MapWindow – another GIS project that includes a free desktop geographical information system application with an extensible plug-in architecture (www.mapwindow.org);
- ADCIRC – a system of computer programs for solving time-dependent, free surface circulation and transport problems in two and three dimensions (adcirc.org).

There is now a need for an umbrella organization, perhaps in two parts, one academic and one commercial, to coordinate these separate efforts. Collectively, they need to raise awareness, promote interoperability between current de facto standards, lower the barriers to entry, and organize and fund a global research programme to accelerate innovation and the use of IEM.

What are the gaps?

As expected, the new ease of model linking has brought out new challenges, many of which have yet to be addressed. An early discovery in the OpenMI-Life project was that joining models designed to operate separately can result in some or all of the models being pushed outside their design limits (Safiolea *et al.* 2009). For example, a hydraulic model normally used for flood studies was found to become unstable when the simulation was continued into a period of lower flows, as required by the linked model. Hence, when metadata standards are developed for describing models, it will be important that they indicate the assumptions made in developing the models and the circumstances in which they can be safely applied (Harpham & Danovaro 2015).

As mentioned above, it was discovered in an OpenMI-Life case study on the Pinios River catchment (Makropoulos *et al.* 2010) that the design criteria used by different disciplines for same process could differ. Clearly, it will be important for modellers to resolve such differences if they to be are able to draw any useful conclusions from the linked model. When existing models of an area are linked, the geographical extent and representation of the two models are often slightly different. At present, there is little understanding of the impact of these differences on the results. Similarly, the effects of linking models running at different time steps are not fully appreciated. Currently, there are few validation checks ensuring that links between models are valid. This is intentional and leaves the modellers free to make any connection they wish. However, there are many operational contexts where it will be important that the risk of an invalid connection is kept to a minimum. The starting point for addressing this problem will be the introduction of controlled vocabularies and later ontologies for input and output variables; the valuable preliminary work of the SEAMLESS project on ontologies for models will need to be greatly extended (van Ittersum *et al.* 2008).

As the Deepwater Horizon oil spill showed, there is the challenge of having a collection of linked models ready before a situation occurs (Machlis & McNutt 2010). At present, an impact has to have been anticipated before it can be modelled. Can Artificial Intelligence (AI) enable an ontology to be searched for a potential impact and be more successful than a normal human brain. Can AI be better at the identification and linkage of components and, if so, could AI be used to at least semi-automate the present manual model-linking process? IEM skills are spread across different organizations, however, and the communication between these groups is very poor; the resources of groups such as the Integrated Environmental Modelling Software and Systems society need to be increased so that the society can extend its reach. In the UK, HR Wallingford has developed Fluid Earth (Harpham *et al.* 2014) for use internally and to open up the means of model linkage to other organizations. This is beginning to achieve a significant number of downloads, especially in China. In the UK, it has been used on operational problems within the British Geological Survey (BGS) to help it solve advanced groundwater problems, such as linking groundwater models with surface models to provide an integrated understanding of surface-water-driven groundwater flooding on the Oxford floodplain (Macdonald *et al.*

2012). It has also been used to holistically simulate groundwater processes in the Thames Basin (Mansour *et al.* 2013). Other commercial developers are producing similar platforms, but each addressing a slightly different user need. These are the first signs of a market forming and need to be encouraged. However, despite this and other similar projects (e.g. those undertaken for OpenMI-Life: Safiolea *et al.* 2009), there is still a dearth of 'showcase' projects that can be used to give others the confidence to join in. Demonstration projects probably represent the largest gap at the moment and will be a priority task in the strategy to be outlined later in this paper.

One final point is the recognition of the human dimension of IEM and how complexity should be dealt with. Given that linked-model instances will be more complex than the single model approaches of the past, there needs to be a structured way to deal with complexity and, in particular, the 'complexity paradox' (Oreskes 2003). The latter being the innate desire of scientists to make their models more complex as a better understanding of natural systems develops. Along with this, there is an increasing need to involve stakeholders in this process and to undertake participatory modelling (e.g. Voinov *et al.* 2016). Any IEM development should be coupled with a critical analysis of the development of linked models, and Glynn (2015) proposed a 'red team' approach to ensure that IEM is applied appropriately. This uses separate teams that ensure that 'groupthink' does not develop, whereby an idea or approach is not properly challenged.

IEM has now reached the stage where it is ready to move out of the research arena and to be taken up by early adopters. As one modeller put it, 'IEM is a no brainer' (Jackson pers. comm. 2014). Whatever its current imperfections, for the moment there does not appear to be an alternative. Therefore, we have to strive to overcome the imperfections, make it easy and safe to use, and to reduce the barriers to take up. In particular, it needs to be moved out into the market place where it can demonstrate its potential. It can then attract the investment that can enable it to form the basis of new industries comparable and exceeding those spawned by digital mapping and its related technologies.

Roadmap

Vision

The vision for IEM is a world where the user (who can be anyone from a politician to a member of the public) can:

- articulate a problem that involves understanding and/or predicting how many processes will interact in a given scenario, for example:
    - will climate change alter the frequency and magnitude of flood damage across Europe?;
    - are the medical plans for treating X a threat to water supplies?;
    - will a switch to biofuels drive up the demand for water and, hence, energy (large amounts of energy are used in the delivery of water)?;
- be guided to an appropriate set of modelling components (in some cases and for some users, this can be hidden);
- be assisted to assemble the components (in some cases and for some users, this can be hidden);
- run the composition (in some cases and for some users, this can be hidden);
- analyse the results or be presented with an analysis;
- be presented with the conclusion and, if relevant, information about the confidence that the user can place in that conclusion.

Mission

The mission of the IEM community is to:

- raise awareness of, and build confidence in, IEM by showing its utility to solve complex problems, the so-called 'stress test';
- improve the ability to understand and predict how processes, particularly environmental processes, interact;
- create the opportunity for innovation, especially by making it possible to link models of processes from different disciplines and market sectors;
- create new markets.

Strategy

The meetings, formal and informal, of the leaders of the IEM community over the last 4 years have been remarkable for the degree of consensus as to what needs to be done to move IEM from the present state of the art to a technology that is widely available and accessible. The essential elements of the strategy are:

- raise awareness and build confidence in IEM;
- ensure availability and accessibility of IEM techniques, tools and standards;
- establish a minimum set of standards;
- build the skills base;
- establish an underpinning R&D programme;
- co-ordinate and promote collaboration;
- increase IEM use by government, industry and the public.

Implementation plan

Set out below is a plan whose objective is to carry through the main elements of the IEM strategy.

Raise awareness and build confidence in IEM

Situation. The leading IEM R&D organizations have begun the process of raising awareness of the benefits of IEM. The first audience has been their own managements and, interestingly, it has been within the commercial players that they have been most successful: indeed, three are now routinely using the new technology internally. What is needed next is to raise awareness in the following areas:

- Governments:
  - in the departments responsible for business development;
  - in the departments responsible for policy development (i.e. the potential direct or indirect users and beneficiaries of IEM);
  - in the departments responsible for overseeing and funding research.
- Academia:
  - the organizations responsible for developing research programmes;
  - the organizations for developing the teaching curriculum.
- Government agencies and local authorities:
  - those that undertake, provide or commission modelling and purchase modelling components and/or services.
- Consultants:
  - those to whom modelling studies are outsourced.
- Utilities and other commercial users of models and modelling services.
- Major investors.

The major investors are placed last because experience suggests that they will want to see a strong demand that justifies the significant investment that will be required to take IEM from its present state to one comparable to that which now exists for GIS and location-based services.

The audience will be extended later, eventually down to the public at large. It is important that at this early stage such resources as are available are focused on those who can bring about change.

It is anticipated that the various audiences may well require very different information packs ranging from case studies to business plans.

Actions.

- Task 1: Identify and prioritize the information about IEM that each audience requires and the medium through which it will be most effectively communicated. Issues known to be of concern are:
  - integrated environmental modelling: What is it? What will it do for me and my organization? How will it add value?;
  - a wide range of demonstrations of added value illustrating: better science; better management answers (greater certainty; fewer unanticipated outcomes); increased opportunity for innovation; opportunity for wealth creation; reduced costs;
  - will the ease of code reuse, reduced development time and increased efficiency produce a worthwhile return on investment?;
  - will IEM results be accepted in a legal context?;
  - how will QA issues be addressed: transparency; audit trails; reproducibility of results?;
  - the reliability of IEM;
  - support into the future;
  - are the backers of IEM credible?;
  - what is the business plan for developing an IEM market?
- Task 2: Generate the information. It is expected that the highest priority requirement will be for a wide range of case studies/exemplars.
- Task 3: Disseminate the information

Ensure availability and accessibility of IEM techniques, tools and standards

Situation. Currently, only a first generation of IEM tools, standards and a small set of applications can be said to be available. They are only accessible to a small community of modellers with the skills to use them. Development of a supply chain and market is now needed to enable the delivery of IEM technologies to a broader set of end users. These users can be found in organizations faced with challenges whose resolution requires an understanding of many interacting processes. In most cases, these organizations are unaware of the advances in IEM technology and certainly do not have the in-house skills to apply them. The same is true of the consultants who provide them with knowledge and expertise, although they could come up to speed quickly if the necessary incentives were in place. The short-term strategy is, therefore, for the developers of IEM to provide IEM support services to those needing to grow IEM expertise or use. These services range from training to undertaking IEM aspects of particular projects. Some may well follow the Environmental Systems Research Institute (ESRI) model in which ESRI makes their software available to the academic community at very advantageous prices. This pays significant dividends later, when those students with their GIS skills move out into industry and understandably recommend the purchase of ESRI software. Engagement with the academic community is important: only they have the resources to produce graduates in the numbers required. That said, it will be important to consider how remote learning could also be exploited to quicken the pace of growth.

The success of IEM is totally dependent on achieving interoperability and communication between people and between modelling components. In a global development exercise, such as will be needed to get IEM off the ground, standards will be key to success. To ensure that standards do not become a barrier either to development or to entry to the market place, there is a consensus view that all the key IEM standards should be 'Open Standards'. For similar reasons, the OpenMI Association and others have made their IEM tools and platforms open source and the strategy is to encourage others to do the same. In this way, the barriers to entry for newcomers are kept low. All that is needed to join the IEM community is a PC and an Internet connection.

For those who want to apply IEM for themselves, the cost of purchasing IEM applications can be a deterrent. One option to reduce the cost and encourage participation is to provide access through web services or the cloud. This is clearly only an option where the data volumes exchanged between components is low. There are many such instances and, as data transfer speed increases, it will become feasible in more situations. This approach has the advantage of leaving software and data with the people in a position to maintain it.

Whilst ease of use has been transformed in the last decade, there is still a long way to go before the ease of use of IEM matches that of office software such as word processors, spreadsheets or GIS. However, ease of use has been identified as a major barrier to uptake and, therefore, is a priority for action.

Actions.

- Task 1: Increase the availability of IEM by:
  - Creating an open source and commercial IEM market place for:
    - IEM modelling components;
    - IEM modelling tools;
    - IEM modelling expertise;
    - related data;
    - IEM compliance testing;
    - encouraging the development of new and upgraded models, and applications as linkable components;
    - encouraging the development of 'adapters' that will allow modelling components using a different data-exchange interface to be linked and so increase the size of the pool of linkable models.
- Task 2: Increase the accessibility of IEM to users by:
  - Improving the ease of use of IEM tools and techniques by:
    - developing the means of finding linkable models (see the model metadata standard below);
    - automating, where practicable, the process of making a model linkable;
    - providing debugging tools;
    - providing compliance testing tools;
    - minimizing the opportunity for error in the coupling process through vocabularies, ontologies and AI (see also the standards in the following subsection);
    - providing adapters for spatial and temporal aggregation and disaggregation, unit conversions, and interface mapping when data pass between models;
    - hiding processes and procedures irrelevant to end users;
    - providing a wide range of exemplars that can easily be adapted to specific problems by those new to IEM.
  - Developing tools for calibration and validation.
  - Providing GUIs that allow all the results from a linked model to be seen together, rather than individually through the GUI of each component.
  - Providing training (see the subsection on 'Build the skills base' later).

Establish a minimum set of standards

Situation. Standards for data exchange between models fall into two types. File format standards are used where models run sequentially, and data exchange is affected by making the output file of one model the input file of the following model. WaterML (Taylor *et al.* 2012) is an example of such a standard. Where models run in parallel with data exchange happening time step by time step, and the exchange takes place in memory, then interface standards are used.

To date, standards work in IEM has been focused on developing interfaces for run-time data exchange between models. A number of de facto standards have emerged, of which one, the Open Modelling Interface (OpenMI), is approved as an international standard by the Open Geospatial Consortium. These informal standards have emerged to serve different modelling communities (e.g. climate, ocean and water industry modellers). The OpenMI arose from the European Commission's programme of underpinning research in support of the Water Framework Directive and the implementation of its policy of an integrated approach to water management across Europe. None of these standards purports to be the ultimate answer to all modeller's needs and much further work is required. Although, for the moment, it is probably more important to grow the set of linkable models and, for the time being, to create adapters that can link models following the different interface standards – see Task 2 of the earlier subsection on 'Ensure availability and accessibility of IEM techniques, tools and standards'.

No other IEM standards have yet emerged, but the need for them is widely recognized. Now that interface standards are widely used within IEM, the next priority is for a minimum standard for the description of models and modelling components. The need to standardize the descriptions of linkable modelling components arises for several reasons:

- Models are developed across the world by independent groups. If their basic descriptions are published in a common format, it will greatly increase the chances of users, who are also dispersed, finding the components that most closely match their needs.
- The nearest equivalents to a model metadata standard are the metadata standards for describing datasets. However, these were conceived before the needs of model linking emerged. They will need to be built upon and extended to accommodate a description of the model or component process, and descriptions of the the inputs and outputs. It will be important that these descriptions are structured for both human and machine searching.
- The model metadata standards will have a key role in automating the construction of model chains.
- The model metadata standards will have a key role in reducing the chances of making invalid links between models.

A foreseeable problem in model linking is that, even within disciplines, the interpretation of the terms used for input and output variables is often highly context specific. When interpretation occurs across disciplines and across natural language boundaries, the opportunity for misunderstanding and, hence, making invalid connections between models is considerable. For situations where it is critically important that such misunderstanding does not occur, it will be essential that clear definitions of input and output variables are provided, ideally in both human and machine readable forms and that such variables are assigned globally unique identifiers.

Actions.

- Task 1: Define the purpose of, and functions to be supported by, a model metadata standard.
- Task 2: Identify the minimum sets of requirements for the description of a modelling component.
- Task 3: Draft a model metadata standard following the OGC standards template.
- Task 4: Define the purpose of a controlled vocabulary for model component input and output variables.
- Task 5: Adopt, adapt or conceive a design for a controlled vocabulary appropriate to the needs of IEM.
- Task 6: Research the use of ontologies and AI as future options to replace controlled vocabularies and for underpinning improved ways of searching for models and building model chains.

Build the skills base

Situation. When viewed on a world scale and against, for example, GIS, IEM skills are concentrated in a relatively small set of groups, almost all of which are in the research sphere. For IEM to become widely taken up within a reasonable time span, say 5–10 years, all modern channels of communication need to be exploited to disseminate this new knowledge and skills into the wider community. At the time of writing, these could include:

- inclusion in appropriate university courses at undergraduate and postgraduate level as specialist modules;
- e-learning courses;
- summer schools;
- commercial training packages.

The process could be stimulated if the IEM product developers were to make versions of their software available under academic licences. It would be similarly helpful to the course developers if any existing teaching aids could be made available to the universities.

Actions.

- Task 1: Identify a set of universities interested in developing IEM courses.
- Task 2: Identify IEM developers willing to make available software and training material.
- Task 3: Develop and deliver course material.
- Task 4: Develop equivalent e-learning course material and publish it on the web.

Establish an underpinning R&D programme

Situation. While there has been very significant research investment in the design and development of particular models, there has never been a research programme specific to the issues that arise when two or more are brought together. These could be technical, commercial or legal.

Operational trials for IEM are revealing a large number of often small, but important, problems that need to be addressed and handled by IEM tools and applications. A typical example might be instability at the point where the models have been joined. Another might be the validity of linking two models whose geographical representations of the river system might not match exactly. Major issues are how to automate the process of finding and linking modelling components. At present, someone must have thought of a potential impact before it can be modelled and assessed. Could AI improve on our present ability in this field? These problems need to be catalogued and made the subject of a number of research programmes to be carried out at whichever is the appropriate level – MSc, PhD or postgraduate; national, major EC programme within, say, Horizon 2020 or Belmont Forum funded.

The unknowns are not confined to the technical arena. Although the commercial and legal problems that arise when products from different suppliers are brought together are not new, they are new for model developers and vendors. No one is yet sure whether linking models will create a set of problems for which there are, as yet, no precedents for managing.

When the development of IEM began to accelerate around 2000, there was a long period of uncertainty as to who was the 'end user' for the work being undertaken. Although there is a feeling that there is now a better understanding of how an IEM market place could operate, this has never been subjected to professional market research. If the parallels with digital mapping are valid, then the potential future market for IEM products and expertise is very large. It seems logical, therefore, that there should be a series of, first, exploratory and then more detailed research exercises to assess the market and to recommend how it can be grown, and the investment required to realize it then found.

Actions.

- Task 1: Set up a process for cataloguing known IEM problems.
- Task 2: Find suitable funding agencies.
- Task 3: Draft and issue calls for proposals.
- Task 4: Ensure results are published.

Co-ordination and promotion of collaboration

Situation. It has become clear that there are 'islands of excellence' emerging all around the world related to IEM. For example, in Europe, the OpenMI Association is championing the use of the OpenMI standard for linking models at run time. In the USA, FRAMES is being developed by the US Environmental Protection Agency (US EPA), Common Component Architecture (CCA) by the Community Surface Dynamics Modeling System (CSDMS) and Object Modeling System (OMS) by the US Department for Agriculture (USDA). The complete list is too long to enumerate here: however, the shared objective of all these initiatives is to enable us to better understand and predict the wider implications of environmental events and their management. Various meetings have been held to bring these communities together:

- Environmental Software Systems Compatibility and Linkage Workshop (March 2000);
- Integrated Modeling for Integrated Environmental Decision Making (January 2007; US EPA 2008);
- Collaborative Approaches to Integrated Modeling: Better Integration for Better Decision-Making (December 2008);
- iEMSs 2010 Conference Workshop: The Future of Science and Technology of Integrated Modeling (July 2010: Voinov *et al.* 2010)
- Washington Summit (December 2010), which brought these parts of the community together (Moore & Hughes 2012).
- Arlington Workshop (March 2012), developing a roadmap implementation plan.

From these meetings and others, it is clear that there is a nascent community dedicated to using IEM and addressing its challenges.

Action.

- Task 1: Ensure that the community meets regularly by arranging relevant workshops, conferences or special sessions within existing conferences.
- Task 2: Promote a Community of Practice for researchers within IEM.
- Task 3: Utilize existing organizations, such as the OGC, to provide the forum for meetings.
- Task 4: Create a business club to allow commercial organizations to communicate.

Growth take up by government, industry and the public

Situation. The first tranche of models and modelling tools have reached the stage of being 'near market ready' and suitable for first application by early adopters. They are out of the research phase and have been tested under operational conditions: for example, by the OpenMI-Life project (OpenMI 2009). As has been explained, as yet, there is little IEM experience outside the R&D community, therefore the proposed approach to growing the uptake of IEM is as follows:

(1) **Select**, with the aid of end users who have problems that could benefit from the application of IEM, a set of problems.
(2) **Develop** solutions for those problems based on IEM: the solutions may take the form of products (probably software) or services or a combination of the two.
(3) Undertake the **first application** of those products and services with the end users.
(4) **Package** the solutions in ways that will allow them to be sold in different countries or climatic regions or in support of different manifestations of the same underlying problem: for example, an agricultural flood estimation model could be repackaged as an urban flooding estimation model as the underlying equations are essentially the same.
(5) Explore how the IEM products and services can be extended across and **replicated** in other market sectors: for example, energy, transport and health, which have very similar requirements for integrated modelling.

Actions.

- Task 1: Commercialization – identify and fill the gaps in the market.
- Task 2: Solutions – provide 'off the shelf' products that can help decision-makers solve their problems.
- Task 3: Packaging – bundle up the compositions so that they can be more easily used.
- Task 4: Dissemination – promote or market the IEM products and solutions that are available.

Summary and conclusions

IEM is essential to help solve the many problems that require an understanding of multiple interacting processes. Many such problems are described in the Belmont Forum 'grand challenges' and include providing improved forecasts of future conditions within the environment, particularly with respect to global environmental change. IEM can help suggest the best steps to mitigate this change.

IEM technology still requires maturation. This can be considered similar to the path that digital mapping took from its initial hesitant steps in the 1970s. This industry is now a fully-fledged field of commerce with applications that were undreamt of when the first attempts rolled off the printers. To ensure that IEM achieves the same status, the availability of linkable components needs to be improved and their interoperability, along with the ability to link them easily, needs to be addressed. In addition, the description of components, along with their ease of discovery through metadata, needs to be improved.

The main strands of any strategy to solve the problems of IEM are to: raise awareness and build confidence in IEM; ensure the availability and accessibility of IEM techniques, tools and standards; establish a minimum set of standards; build the skills base; establish an underpinning R&D programme; co-ordinate and promote collaboration; and, finally, increase use by government, industry and the public. Without the latter, the demand for IEM will not exist. Once stakeholders and potential users start demanding the use of IEM, then this will provide the 'pull' for the gaps in IEM to be filled.

An important consideration is 'what will success look like?'. This can be imagined as a world where IEM solutions are used almost unknowingly by decision-makers of all types, from householders to governmental level, and where complex decisions are solved with its assistance and resources are much better managed.

There are many and various people to acknowledge who have contributed to the meetings that have fed into this paper. They are almost too numerous to mention. Some individuals have been mentioned in the text, and included in the references below. However, many remain un-named, for which the authors apologise.

The contribution of one anonymous reviewer and Pierre Glynn (USGS) is gratefully acknowledged. Addressing their comments improved the paper. This paper is published with the permission of the Executive Director, BGS (NERC).

References

Ahuja, L.R., Ascough, J.C., II & David, O. 2005. Developing natural resource models using the object Modelling system: feasibility and challenges. *Advances in Geosciences*, **4**, 29–36, http://hal.archives-ouvertes.fr/docs/00/29/68/06/PDF/adgeo-4-29-2005.pdf

Argent, R.M., Voinov, A. *et al.* 2006. Comparing modelling frameworks – a workshop approach. *Environmental Modelling & Software*, **21**, 895–910.

Argent, R.M., Perraud, J.-M., Rahman, J.M., Grayson, R.B. & Podger, G.M. 2009. A new approach to water quality modelling and environmental decision support systems. *Environmental Modelling & Software*, **24**, 809–818.

Arnold, J.G., Srinivasan, R., Muttiah, R.S. & Williams, J.R. 1998. Large area hydrologic modeling and assessment part i: model development. Jawra. *Journal of the American Water Resources Association*, **34**, 73–89.

BABEL. 2004. High-Performance Language Interoperability. Lawrence Livermore National Laboratory, http://www.llnl.gov/CASC/components/babel.html

Babendreier, J.E. & Castleton, K.J. 2005. Investigating uncertainty and sensitivity in integrated, multimedia environmental models: tools for FRAMESe3MRA. *Environmental Modelling & Software*, **20**, 1043–1055.

Barthel, R., Janisch, S., Schwarz, N., Trifkovic, A., Nickel, D., Schulz, C. & Mauser, W. 2008. An integrated modelling framework for simulating regional-scale actor responses to global change in the water domain. *Environmental Modelling & Software*, **23**, 1095–1121.

Bastin, L., Cornford, D. et al. 2013. Managing uncertainty in integrated environmental modelling: the UncertWeb framework. *Environmental Modelling & Software*, **39**, 116–134.

Bernholdt, D.E., Allan, B.A. et al. 2006. A component architecture for high performance scientific computing. *International Journal of High Performance Computing Applications*, **20**, 163–202, https://e-reportsext.llnl.gov/pdf/314847.pdf

BIS 2011. *A Strategic Vision for UK e-Infrastructure: A Roadmap for the Development and Use of Advanced Computing, Data and Networks*. Department for Business Innovation and Skills, London, https://www.gov.uk/government/uploads/system/uploads/attachment_data/file/32499/12-517-strategic-vision-for-uk-e-infrastructure.pdf [last accessed 4 April 2014].

BIS. 2013. *e-Infrastructure: The Ecosystem for Innovation One Year On*. Department for Business Innovation and Skills, London, https://www.gov.uk/government/uploads/system/uploads/attachment_data/file/249474/bis-13-1178-e-infrastructure-the-ecosystem-for-innovation-one-year-on.pdf [last accessed 3 March 2016].

Crampton, J.W. 2000. A history of distributed mapping. *Cartographic Perspectives*, **35**, 48–65.

Coppock, J.T. & Rhind, D.W. 1991. The history of GIS. *In*: Maguire, D.J., Goodchild, M.F. & Rhind, D.W. (eds) *Geographical Information Systems: Principles and Applications, Volume 1*. Longman, Harlow, 21–43.

David, O., Ascough, J.C., II, Lloyd, W., Green, T.R., Rojas, K.W., Leavesley, G.H. & Ahuja, L.R. 2013. A software engineering perspective on environmental modeling framework design: the object modeling system. *Environmental Modelling and Software*, **39**, 201–213.

Donigian, A.S., Jr, Bicknell, B.R. & Imhoff, J.C. 1995. Hydrological simulation program – Fortran (HSPF). *In*: Singh, V.P. (ed.) *Computer Models of Watershed Hydrology*, Water Resources Publications, Littleton, CO, 395–442.

EPSRC. 2014. *e-Infrastructure Roadmap*. Engineering and Physical Sciences Research Council (EPSRC), Swindon, https://www.epsrc.ac.uk/newsevents/pubs/e-infrastructure-roadmap/ [last accessed 3 March 2016].

FRAMES 2009. *Framework for Risk Analysis in Multimedia Environmental Systems – Multimedia, Multipathway, and Multireceptor Risk Assessment (FRAMES-3MRA)*. http://www.epa.gov/ATHENS/research/Modelling/3mra.html

Gaber, N., Laniak, G. & Linker, L. 2008. *Integrated Modeling for Integrated Environmental Decision Making*. EPA100/R-08/010. United States Environmental Protection Agency, Washington, DC.

Glynn, P.D. 2015. Integrated Environmental Modelling: human decisions, human challenges. *In*: Riddick, A.T., Kessler, H. & Giles, J.R.A. (eds) *Integrated Environmental Modelling to Solve Real World Problems: Methods, Vision and Challenges*. Geological Society, London, Special Publications, **408**, first published on May 21, 2015, https://doi.org/10.1144/SP408.9

Goodchild, M.F. 1988. Stepping over the line: technological constraints and the new cartography. *The American Cartographer*, **15**, 311–319.

Gregersen, J.B., Gijsbers, P.J.A. & Westen, S.J.P. 2007. OpenMI: open modelling interface. *Journal of Hydroinformatics*, **9**, 175–191.

Harbaugh, A.W. 2005. *MODFLOW-2005, The U.S. Geological Survey Modular Ground-Water Model – the Ground-Water Flow Process*. United States Geological Survey, Techniques and Methods, **6-A16**.

Harpham, Q. & Danovaro, E. 2015. Towards standard metadata to support models and interfaces in a hydro-meteorological model chain. *Journal of Hydroinformatics*, **17**, 260–274.

Harpham, Q., Cleverly, P. & Kelly, D. 2014. The Fluid Earth 2 implementation of OpenMI 2.0. *Journal of Hydroinformatics*, **16**, 890–906, https://doi.org/10.2166/hydro.2013.190

ICSU. 2010. *Earth System Science for Global Sustainability: The Grand Challenges*. International Council for Science, Paris.

Kralisch, S., Krause, P. & David, O. 2004. Using the Object Modelling System for hydrological model development and application. *In*: *Proceedings of the iEMSs 2004 International Conference*. University of Osnabrück, Germany, http://www.iemss.org/iemss2004/pdf/integratedmodelling/kralusin.pdf

Laniak, G.F., Olchin, G. et al. 2013. Integrated environmental modeling: a vision and roadmap for the future. *Environmental Modelling & Software*, **39**, 3–23, https://doi.org/10.1016/j.envsoft.2012.09.006

Lloyd, W., David, O. et al. 2011. Environmental Modeling Framework Invasiveness: Analysis and Implications. *Environmental Modelling & Software*, **26**, 1240–1250.

Macdonald, D., Dixon, A., Newell, A. & Hallaways, A. 2012. Groundwater flooding within an urbanised flood plain. *Journal of Flood Risk Management*, **5**, 68–80, https://doi.org/10.1111/j.1753-318X.2011.01127.x

Machlis, G.E. & McNutt, M.K. 2010. Scenario-building for the Deepwater Horizon oil spill. *Science*, **329**, 1018–1019.

Makropoulos, C., Safiolea, E., Baki, S., Douka, E., Stamou, A. & Mimikou, M. 2010. An integrated, multi-modelling approach for the assessment of water quality: lessons from the Pinios River case in Greece. *In*: *Proceedings of the International Congress on Environmental Modelling and Software (iEMSs 2010)*, Ottawa, Canada, **1**, 1023–1032, http://www.iemss.org/iemss2010/papers/S10/S.10.06.An%20integrated%20modeling%20approach%20for%20the%20assessment%20of%20water%20quality%20lessons%20from%20the%20Pinios%20river%20case%20in%20Greece%20-%20CHRISTOS%20MAKROPOULOS.pdf

Mansour, M., Mackay, J., Abesser, C., Williams, A., Wang, L., Bricker, S. & Jackson, C. 2013. Integrated Environmental Modeling applied at the basin scale: linking different types of models using the OpenMI standard to improve simulation of groundwater processes in the Thames Basin, UK. *In*: *MODFLOW and More 2013: Translating Science into Practice*, 2–5 June 2013, Colorado, USA (unpublished), http://nora.nerc.ac.uk/501789/

Markstrom, S.L., Niswonger, R.G., Regan, R.S., Prudic, D.E. & Barlow, P.M. 2008. *GSFLOW-Coupled Ground-Water and Surface-Water FLOW model based on the Integration of the Precipitation-Runoff Modeling System (PRMS) and the Modular Ground-Water Flow Model (MODFLOW-2005)*. United States Geological Survey, Techniques and Methods, **6-D1**.

Moore, R. & Hughes, A. (eds). 2012. *International Summit on Integrated Environmental Modeling, December 7–9, 2010, USGS Headquarters, Reston, VA*. United States Environmental Protection Agency, Washington, DC (OR/12/087) (unpublished), http://nora.nerc.ac.uk/21007/

Moore, R., Gijsbers, P., Fortune, D., Gregersen, J. & Blind, M. 2005. *OpenMI Document Series: Part A Scope for the OpenMI (Version 1.0), HarmonIT*. OpenMI Association.

Natural Environment Research Council 2009. *Natural Environmental Research Council Science Information Strategy*. Natural Environment Research Council, Swindon.

NSF 2011. *The 'Earth Cube' – Towards a National Data Infrastructure for Earth System Science*, http://www.nsf.gov/pubs/2011/nsf11065/nsf11065.jsp?org=NSF [last accessed 4th April 2014].

OPENMI 2009. The OpenMI-Life project website, www.openmi.org/archives/openmi-life

Oreskes, N. 2003. The role of quantitative models in science. *In*: Canham, C.D., Cole, J.J. & Lauenroth, W.K. (eds) *Models in Ecosystem Science*. Princeton University Press, Princeton, NJ, 13–31.

Peckham, S.D., Hutton, E.W.H. & Norris, B. 2013. A component-based approach to integrated modeling in the geosciences: the design of CSDMS. *Computers and Geosciences*, **53**, 3–12.

Rogers, E. 2003. *Diffusion of Innovations*, 5th edn. Free Press, New York.

Safiolea, E., Makropoulos, C. & Mimikou, M. 2009. Benefits and challenges in integrated water resources modeling using OpenMI. *In*: *Integrating water systems. Proceedings of the Tenth International Conference on Computing and Control for the Water Industry, CCWI 2009 – 'Integrating Water Systems'*, 1–3 September 2009, Sheffield, UK, CRC Press/Balkema, Boca Raton, 481–484.

Sophocleous, M. & Perkins, S.P. 2000. Methodology and application of combined watershed and groundwater models in Kansas. *Journal of Hydrology*, **236**, 185–201.

Taylor, P., Cox, S. & Walker, G. 2012. Adapting standard information models for water data. *In*: *10th International Conference in Hydroinformatics, HIC2012*, Hamburg, Germany. IWA Publishing, London, 14–18.

US EPA 2008. *Integrated Modeling for Integrated Environmental Decision Making*. EPA100/R-08/010. Office of the Science Advisor, Washington, DC.

US GOVERNMENT 2012. http://www.whitehouse.gov/sites/default/files/omb/egov/digital-government/digital-government.html?goback=.gde_3427896_member_120023144 [last accessed 4th April 2014].

van Ittersum, M.K., Ewert, F. *et al.* 2008. Integrated assessment of agricultural systems – a component-based framework for the European Union (SEAMLESS). *Agricultural Systems*, **96**, 150–165, https://doi.org/10.1016/j.agsy.2007.07.009 (http://www.sciencedirect.com/science/article/pii/S0308521X07000893)

Viceconti, M. & Clapworthy, G. (eds) 2011. *VPH-FET Research Roadmap: Advanced Technologies for the Future of the Virtual Physiological Human*. VPH-FET Consortium.

Voinov, A. & Shugart, H.H. 2013. 'Integronsters', integral and integrated modeling. *Environmental Modelling & Software*, **39**, 149–158.

Voinov, A., Kolagani, N. *et al.* 2016. Modelling with stakeholders–Next generation. *Environmental Modelling & Software*, **77**, 196–220.

Voinov, A.A., DeLuca, C., Hood, R.R., Peckham, S., Sherwood, C.R. & Syvitski, J.P.M. 2010. A community approach to Earth systems modeling. *Eos, Transactions of the American Geophysical Union*, **91**, 117–118, https://doi.org/10.1029/2010EO130001

VPH. 2009. *A Vision and a Strategy for the Virtual Physiological Human in 2010 and Beyond*, http://www.vph-institute.org/upload/vph-vision-strategy-submitted-141209-4_519244d49f91e.pdf [last accessed 4th April 2014].

Warner, J.C., Perlin, N. & Skyllingstad, E.D. 2008. Using the Model Coupling Toolkit to couple earth system models. *Environmental Modelling & Software*, **23**, 1240–1249.

Whelan, G., Tryby, M.E., Pelton, M.A., Soller, J.A. & Castleton, K.J. 2010. Using an integrated, multidisciplinary framework to support quantitative microbial risk assessments. *In*: Swayne, D.A., Yang, W., Voinov, A.A., Rizzoli, A. & Filatova, T. (eds) *International Environmental Modelling and Software Society (iEMSs) 2010 International Congress on Environmental Modelling and Software Modelling for Environment's Sake, Fifth Biennial Meeting*, Ottawa, Canada, **1**, 1223–1230, http://www.iemss.org/iemss2010/papers/S10/S.10.29.Using%20an%20Integrated,%20Multidisciplinary%20Framework%20to%20Support%20Quantitative%20Microbial%20Risk%20Assessments%20-%20GENE%20WHELAN.pdf

Wilkinson, W.B., Leeks, G.J.L., Morris, A. & Walling, D.E. 1997. Rivers and coastal research in the Land Ocean Interaction Study. *Science of the Total Environment*, **194**, 5–14.

Zhang, J., Ross, M.A., Geurink, J.S. & Gong, H. 2009. Modelling vadose zone moisture dynamics in shallow water table settings with the Integrated Hydrologic Model: field-scale testing and application. *Water and Environment Journal*, **24**, 9–20.

From integration to fusion: the challenges ahead

J. SUTHERLAND*, I. H. TOWNEND, Q. K. HARPHAM & G. R. PEARCE

HR Wallingford, Howbery Park, Wallingford, Oxfordshire, OX10 8BA, UK

**Correspondence: j.sutherland@hrwallingford.com*

Abstract: The increasing complexity of numerical modelling systems in environmental sciences has led to the development of different supporting architectures. Integrated environmental modelling can be undertaken by building a 'super model' simulating many processes or by using a generic coupling framework to dynamically link distinct separate models during run-time. The application of systemic knowledge management to integrated environmental modelling indicates that we are at the onset of the norming stage, where gains will be made from consolidation in the range of standards and approaches that have proliferated in recent years. Consolidation is proposed in six topics: metadata for data and models; supporting information; Software-as-a-service; linking (or interface) technologies; diagnostic or reasoning tools; and the portrayal and understanding of integrated modelling. Consolidation in these topics will develop model fusion: the ability to link models, with easy access to information about the models, interface standards such as OpenMI and software tools to make integration easier. For this to happen, an open software architecture will be crucial, the use of open source software is likely to increase and a community must develop that values openness and the sharing of models and data as much as its publications and citation records.

The past few decades have seen the inexorable rise of numerical modelling as a useful tool in hydro-environmental and geomorphological modelling. Models have become more and more detailed, representing more and more processes and, with increasing computer power, being solved using larger and larger geo-spatial structures. These models can be aimed at solving a single set of equations, but have often branched out to include a wider range of processes with the formation of modelling suites.

Albeit a little belatedly, numerical modelling has followed mainstream information technology (IT) in the way that the code is structured and deployed. Programmers began by writing short, self-contained bespoke applications, consisting of sequential lines of procedural code in languages such as FORTRAN. The benefits of callable sub-routines or functions were then quickly realized as applications grew in complexity, leading to a desire for reuse and clean interfaces. Many legacy applications, and those developed by scientific programmers, are still this way today. Object-oriented languages took this trend to its logical conclusion where every code segment has its own attributes and interfaces, and component-driven architectures have increased the scale of such implementations.

Alternatively, a set of environmental phenomena can be simulated using a complex composition of linked models, each of which is considered as a component at this level. In this way the progression of numerical model code structure can be summarized as in Table 1, from sequential programs of procedural code on the left to compositions of linked models on the right.

The structure of model code, and its associated development and execution environments, is also influenced by the growing awareness of the need to model environmental systems as a whole, as one component of a system affects other components and, in turn, is affected by yet more components. For example, 'whole catchment modelling' is required to deliver the objectives of the Water Framework Directive (European Commission 2000) and this requires the linking of a wide variety of models. In the UK, the Foresight Future Flooding project identified a need 'to improve the capability of coastal morphological models to support decision-making by providing accurate predictions of local morphological change and broad-scale morphological responses to coastal engineering and management' (Office of Science and Technology 2004). A framework has been developed (Whitehouse *et al.* 2009) for this based on a technique for mapping coastal systems (from the systems theory of von Bertalanffy 1951; Chorley 1962) which influences the development of an interacting system of reduced complexity morphological models, constrained and guided by sediment pathways derived from detailed coastal area modelling. The implementation of this framework (Nicholls *et al.* 2012) will therefore require the linking of a set of models during run-time with two-way exchange of information to capture feedback effects.

From: Riddick, A. T. Kessler, H. & Giles, J. R. A. (eds) 2017. *Integrated Environmental Modelling to Solve Real World Problems: Methods, Vision and Challenges*. Geological Society, London, Special Publications, **408**, 35–54.
First published online December 1, 2014, https://doi.org/10.1144/SP408.6

Table 1. *Progression of numerical model code structure*

Single code	Multiple sub-routines	Self-contained and reuseable objects	Component-based models
Line 1	Main	Object 1	Model 1
Line 2	Subroutine 1	Object 2	Model 2
Line 3	Subroutine 2	Object 3	Model 3
...	...	...	...

The increasing need to model systems, rather than just single processes, has led to different methods for combining models of different processes. Lu & Piasecki (2012) identified the following four categories:

(1) The use of a range of models of a geographical area. This approach does not attempt to link models, but compares and contrasts the results from different models and approaches to help improve the models themselves and overall understanding.
(2) The construction of a monolithic 'super model' containing multiple processes as multiple sub-routines or sub-models that can be included or excluded as necessary. These models can be constructed by a number of research groups and proposed changes are typically submitted through a version control system for validation and approval before release. Examples include MIKE by DHI (http://mikebydhi.com) and the Weather Research and Forecasting model (Janjic *et al.* 2010).
(3) The use of a generic component-based modelling framework, such as the Community Surface Dynamics Modelling System (http://csdms.colorado.edu), Programme for Integrated Earth System Modelling, PRISM (Valke *et al.* 2006) and the Earth System Modelling Framework (www.earthsystemmodeling.org).
(4) The use of a coupling framework to develop a community modelling system, by providing software to assist with linking different models during run-time. Coupling frameworks ensure that data can be transferred between different component models at known times and places. A good example of this approach is the Open Modelling Interface (OpenMI) standard (as described later in this paper).

All four methods are in use and there remains the issue of whether one should simply build larger monolithic models (one form of 'integration'), or use existing models and link them together – as implied by the concept of model fusion. One could of course argue that linking at the sub-routine level is a logical extension of the concept of integration and, for this reason, many practitioners are still not convinced of the benefits of the dynamic linking of models.

However, most of the established hydraulic and morphodynamic models were developed as proprietary or bespoke packages. These models largely followed the closed architecture philosophy, which emerged in response to the development of a niche market for modelling services but which also posed problems for the clients of these services (Khatibi *et al.* 2004):

- Organizations that use or manage the results of these models need to manage the risk that a software capability becomes obsolete.
- Consultants need to maintain a store of proprietary software products.
- Each of these products contains similar tools that produce similar results but cannot be transferred from one software product to another.
- User-designed systems are not possible, as clients cannot select which tools they want to use.

Khatibi (2003*a*) proposed that 'an understanding of the interfaces between the interacting systems is the key for systemic problem-solving', while Khatibi (2003*b*) promotes a move towards an open architecture which uses standard interfaces to link modules. Khatibi *et al.* (2004) promote the use of both an open architecture and open source software. The benefits of open source software have also been discussed by Harvey & Han (2002) and others.

This paper explores many of the current issues that need to be overcome to promote the concept of model integration by the dynamic linking of models. It starts by introducing two frameworks that have been used to assess the progression in numerical modelling (largely in hydraulics) and considers their application to the modelling of integrated environmental systems. It then discusses the increasingly blurred line between observations (data) and models, before describing the evolution of OpenMI and the OpenMI standard (Gregersen *et al.* 2007). The FluidEarth initiative, which promotes a user community and provides software tools for implementing the OpenMI standard, is then described, before the paper discusses the challenges that must be addressed to move forward from today's integrated models towards model fusion. The *Concise Oxford English Dictionary* defines fusion as the 'fusing, melting, blending of different things into one', so it might be argued that it should involve more than just the ability to link models and this paper looks at what other factors may be involved in this concept.

Frameworks for assessing model development

The development of numerical modelling systems has never occurred in isolation as a purely technical development of code. The development of numerical models is inextricably linked to the development of different types of code, changes in hardware, peripherals and operating systems. Moreover, it is also depends on the increasing number of stakeholders whose decisions may be influenced by the results from modelling and their interactions with the innovators, developers and users of numerical models.

This paper considers the historical development of numerical modelling as described by Abbot (Abbott 1991) and Khatibi and colleagues (Khatibi 2001, 2003*a*, *b*; Khatibi *et al.* 2004).

Abbott's five generations

Abbott (1991) considered there to be five generations of hydraulic models:

(1) Numerical solutions to algebraic equations (1950s)
(2) Project-customized modelling and the development of modelling groups (1960s)
(3) Introduction of modelling systems. Modelling is undertaken as a service by specialized centres (1970s and 1980s).
(4) Development of software tools as packaged products, with help, support, pre- and post-processing tools (1980s to 2000s)
(5) Development of hydroinformatics tools (2010s) aimed at 'raising the level of discourse of its clients' (Abbott 1991), which developed into 'making electronically encapsulated modelling knowledge available over the internet' (Abbott *et al.* 2006) or 'Software-as-a-service' (Abbott & Vojinovic 2009).

The dates in brackets are approximations of when each generation rose to prominence. The level of computational hydraulic knowledge required of the typical user has fallen dramatically between generations (Abbott 1991; Abbott & Vojinovic 2009). This carries risks, which are mitigated by reductions in the freedom given to users by providing compiled code with error checks, help and support. As you move from generation to generation, the roles of scientist, developer, user and supporter have gradually split and become distinct. The role of the common user in each generation is illustrated in Table 2 (Abbott 1991; Abbott & Vojinovic 2009; Khatibi *et al.* 2004).

The re-definition of the fifth generation as 'Software-as-a-service' (Abbott & Vojinovic 2009) represents a logical continuation of the changing role of the user, who no longer needs to be an expert to run a piece of software. As such, it is up to the developer to produce a service that has sufficient help and is sufficiently constrained to be used by a competent professional.

Systemic knowledge management

Khatibi *et al.* (2004) point out that the causes of the changes from one generation of models to another are not readily identifiable in Abbott's account and preferred to apply 'systemic knowledge management' (Khatibi 2003*a*, *b*) to explain the current status and potential future for software tools in hydraulic modelling. Khatibi (2003*a*) developed systemic knowledge management as a methodological tool for assessing the development of a scientific paradigm, in this case hydraulic modelling. Systemic knowledge management was derived from: (i) the concept of paradigm and paradigm shift (Kuhn 1962; Khatibi 2001); (ii) systemic problem solving developed in systems science (von Bertalanffy 1951); and (iii) knowledge management from management science (Nonaka 1998).

In systemic knowledge management, any theory or concept can be viewed as a paradigm that shifts through 'pre-paradigm', 'forming', 'proliferation', 'norming' and 'performing' stages (Khatibi 2003*a*). The different stages were derived from the model developed by Tuckman (1965) of how group behaviour evolves (though Tuckman's 'storming', as in 'brainstorming', has been replaced by the rather more prosaic 'proliferating'). It is interesting that a model of social interaction is used to define the stages a scientific paradigm evolves through. This emphasizes the social nature of science, including software evolution, where scientific paradigms are often conditioned by the social outlook of their time (Prigogine & Stengers 1985). Over development timescales, this is reflected in the many communities of practice that form around different software and between modellers of the same phenomena.

Each stage has different characteristics:

- **Pre-paradigm**. A range of disparate and rudimentary approaches may be developed to begin to tackle an issue. Efforts are often in isolation, may appear random and offer no competitive advantage.

Table 2. *Role of the user in generations of modelling software*

Generation	Date	Description	Role of user
1st	1950s	Numerical solutions to algebraic equations	User is dedicated professional who is also an innovator and developer.
2nd	1960s	Project-customized modelling and development of modelling groups	User is member of dedicated modelling group who innovate and develop model.
3rd	1970s and 1980s	Introduction of modelling systems. Modelling as a service by specialized centres	User need not be an innovator or developer, but is likely to work with experts.
4th	1980s, 2000s	Commercial software tools as packaged products, including help and support	User trained by model owner/ agent and pays for support. User supported by pre- and post-processing tools. Reduction in required knowledge by user.
5th	2010s	Development of hydroinformatics systems (1991), software as a service (2006, 2009)	Limited knowledge of software by user – but freedom constrained by developer.

- **Forming stage**, with the development of a paradigm from the diversity of approaches around. A form of natural selection governs how models develop and determines which models flourish and which die out. As this is a social as well as scientific process, this does not necessarily mean that it is the best scientific models which win out during this stage.
- **Proliferating stage**, with either different variations of the components of the paradigm being developed, or the paradigm spreading to different disciplines. Many people adopt the paradigm. The level of organization increases, while natural selection still governs the fate of individual models. The prevalence of many viable, normally inflexible, options leads to the law of diminishing returns, with many incremental returns being repeated across systems.
- **Norming stage**, where the gains to be made from the hierarchical organization of the competing options are realized. There is a 'conscious process of consolidation', with an active search for synergy between components of the paradigm (that is, within a particular field – 'longitudinal holism') or between the different disciplines sharing a paradigm ('lateral holism'). Partnerships between scientists, practitioners and stakeholders emerge. The principles for the performing stage are developed.
- **Performing stage**, where high performance can be achieved within a flexible environment, where custom solutions can be delivered to meet clients' needs and the clients are aware of the limits of the system. The connections between components and the influences they have on each other are understood and utilized. There is an interplay between technical, economic and social needs. Utility and usability are high.

Application to hydraulic modelling. Khatibi (2003*b*) applied systemic knowledge management to open channel flow modelling, irrigation systems, municipal water supplies and flood drainage systems, while Khatibi *et al.* (2004) applied it to hydraulic modelling software to obtain the following five stages:

- **Pre-paradigm**. Before the invention of computers a number of mathematical and empirical approaches became established in hydraulic engineering. But while this resulted in many elegant analytical solutions or data driven empirical representation, such solutions were hindered by the need for manual computation.
- **Forming stage**, with the development of inflexible project-specific codes for fundamental problems in niche markets. The speed of a numerical calculation gave it a competitive (or selective) advantage over a hand calculation. The user was a dedicated professional, who was also an innovator and developer. A form of natural selection governs how models develop and determines which models continue to be developed and which die out. (This stage has similarities with Abbott's first and second generations.)
- **Proliferating stage,** with the development of many general-purpose, modular (often subroutine based) software products, with pre-processing and post-processing tools. Closed architectures were used, which made it impossible to plug-in innovative components or those from other providers. (This is similar to Abbott's third and fourth generations.)

- **Norming stage**, where open architectures are expected to prevail, allowing interoperability between software systems and thus enabling the user to design bespoke models from a choice of components. The development of published interfaces will play an important role in allowing different models to be used within a modelling system. The current situation represents the onset of this phase, which is still punctuated by many attributes of those previous.
- **Performing stage**, where the open source movement is expected to play a pivotal role in sharing freely and improving software source code, and web services allow increased flexibility, availability and uptake. The last two stages, which concentrate on the development of integrated modelling and data services, complement Abbott's fifth generation (Software-as-a-service).

Khatibi *et al.* (2004) postulated that software architecture, defined as 'the conceptual structure and logical organization of a computer or computer-based system', is the cause of the paradigm shifts observed. This is affected by developments in processors, data storage, peripherals and the user interface, as well as programming environments, standards and the rise of the World Wide Web, as illustrated in Table 3.

Software developers were able to differentiate their products in the market through their user-friendly interface or attractive post-processing graphics (as occurred in Abbott's fourth generation or Khatibi's proliferating stage) only after the software for these had been developed. As developments progressed, web interfaces to software were built on the standards developed for the Web.

Khatibi *et al.* (2004) promoted the use of open architecture and open source code as important concepts for moving hydraulic modelling through the norming stage to the performing stage. Figure 1 (after Khatibi *et al.* 2004) shows how the openness of an environmental modelling system's architecture can be assessed in terms of its openness to models and data. Its openness to 'third-party models' (external model components) is categorized as follows.

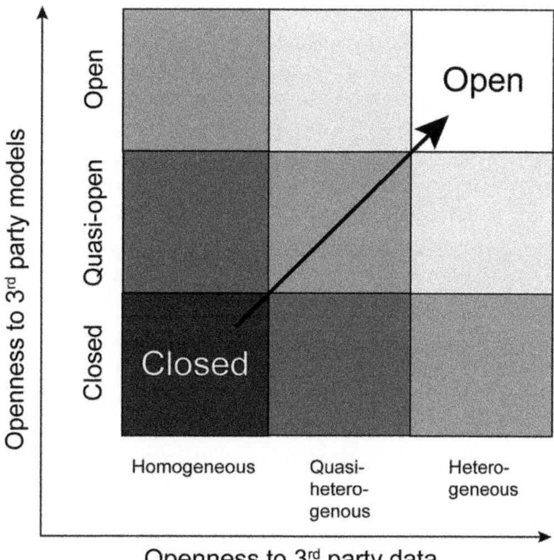

Fig. 1. Assessment of the openness of a model's architecture.

- An open system is one that can link to models from other software providers.
- A quasi-open system can link to a particular sub-set of models.
- A closed system can only link models from within its own system.

Its openness to other data is judged as follows.

- A heterogeneous (open) system can use datasets from third-party products.
- A quasi-heterogeneous system can use only a proprietary sub-set of data products.
- A homogeneous (closed) system accepts only native data.

Table 3. *Developments in computing environments*

Decade	Hardware	UI	Data storage	Software architecture
1970s	Mainframe	Terminals, keyboard	Floppy disk	Closed system and closed code
1980s	PC	Monitor, keyboard and mouse	3.5″ floppy disk	Free software foundation, GNU
1990s	Server	GUI, WWW, Graphical output	CD	Linux, GPL
2000s	HPC	Web-style UI	DVD, USB drive	Standards (including OpenMI) open source Initiative
2010s	Cloud, grid, tablet	Semantic web, SaaS	Cloud/grid	Integrated modelling, web services

Khatibi *et al.* (2004) saw that one of the requirements for the development of open architectures would be the development and application of standards, as this would enable people to work together by establishing common rules and protocols. In recent years there has been a great increase in commonly used standards, approved by bodies such as the International Standards Organization (www.iso.org), the Open Geospatial Consortium (www.opengeospatial.org) or the World Wide Web Consortium (www.w3.org) for web applications. Important topics for standards in integrated modelling include temporal and spatial definitions, phenomenon dictionaries, metadata, time-stepping and interface definitions.

Developers around the world have written tools and applications that use these standards and, as a result, a wide range of software packages has become available to end-users, many of which are open source. Gregersen *et al.* (2007) maintain that a successful standard has three characteristics:

(1) It is technically sound.
(2) Adoption reaches a critical level.
(3) The standard is supported and developed to meet new demands and cope with changes to utilized software packages.

This emphasizes the need for a community to develop around a standard, or software product, for it to become successful.

Application to integrated environmental modelling. Khatibi *et al.* (2004) noted that the different areas where a paradigm is applied may be at different stages. Hydraulic modelling is in the norming stage, with the development of standard approaches. When the stages of systemic knowledge management are applied to the broader field of integrated environmental modelling, the following developments can be identified:

- **Pre-paradigm:** development of single models and modelling systems. The sequential running of different models with data exchange via output files.
- **Forming:** an appreciation developed that separate models can be linked at run-time (Reed *et al.* 1999; van der Wal & van Elswijk 2000; Havnø *et al.* 2001; Gijsbers *et al.* 2002; Whelan & Nicholson 2002). Individual projects set up integrated environmental modelling systems, such as HarmonIT (Gijsbers *et al.* 2002; Blind & Gregersen 2005; Moore & Tindall 2005). Different approaches covered the same ground or separate areas within the subject (Safiolea *et al.* 2011; Bastin *et al.* 2013; Lu & Piasecki 2012).
- **Proliferating:** Many approaches and systems were developed and tested. Bastin *et al.* (2013) divided existing frameworks for model coupling into three classes: (i) standard languages and interfaces; (ii) workflow and integration tools; and (iii) frameworks and provided examples. Lu & Piasecki (2012) used the categories: (iv) geographical area; (v) monolithic super model; (vi) component-based modelling framework; and (vii) coupling framework. Examples from Bastin *et al.* (2013) and Lu & Piasecki (2012) are given in Table 4 and Table 5, respectively.
- **Norming:** The current situation appears to exhibit the onset of the norming stage as there is a clear appreciation of the need to adopt common approaches, which are often manifest as a formal standard. Many standards, candidate standards and implementations of standards have already been developed (see Tables 4 & 5 for a selection). We expect that a limited number of approaches and standards will flourish, covering a complementary set of required functions. The members will be decided by natural selection; practitioners will chose which approach to adopt, so some will flourish and some will become redundant. However, this process will be augmented by the belief that consolidation will be needed. The competitive advantage offered by an approach may depend on a number of things, including the availability of funding to develop and maintain the core software, the development of an active community, the ability to link to other software (for example, the approach can be used over the Web or within different workflow tools). An open architecture will be crucial and the use of open source software is likely to increase. This movement has been boosted by the recent switch of existing closed source codes into open source codes by major players in the hydraulic modelling community such as TELEMAC (www. opentelemac.org) and elements of Delft3D (www.deltaressystems.com). However, given the great resources devoted to legacy code, its track record and the number of instances of its successful use, the ability to include closed source code will remain an advantage for years to come.
- **Performing**: The performing stage will be achieved when high performance can be achieved within a flexible environment, where custom solutions can be delivered to meet clients' needs and the clients are aware of the limits of the system (Khatibi 2003*a*). Although the owners of individual environmental modelling systems may each claim to be at this stage now, the entire community encompasses too many alternative approaches without any single one offering all that is required. The performing stage will be reached after more testing, revision and consolidation of approaches.

Table 4. *Elements in integrated environmental models (Bastin* et al. *2013)*

(i) Standard languages and interfaces	
Business Process Execution Language	docs.oasis-open.org/wsbpel/2.0/wsbpel-v2.0.pdf
Open Modelling Interface	www.openmi.org
Common Component Architecture	www.cca-forum.org
Predictive Model Markup Language	www.dmg.org
Interactive Component Modelling System	www.clw.csiro.au/products/icms/
(ii) Workflow and orchestration tools	
Taverna	www.taverna.org.uk
Kepler	https://kepler-project.org/
VisTrails	www.vistrails.org
Trident	tridentworkflow.codeplex.com
(iii) Frameworks	
Delta Shell	Donchyts & Jagers (2010)
Model Coupling Toolkit	www.mcs.anl.gov/research/projects/mct/
Earth System Modelling Framework	www.earthsystemmodeling.org
Ocean Atmosphere Sea Ice Soil	www.cerfacs.fr/3-26568-OASIS.php
Community Earth System Model	www.cesm.ucar.edu
O-PALM	www.cerfacs.fr/globc/PALM_WEB/
Bespoke Framework Generator	Armstrong *et al.* (2009)
Tarsier	ecoviz.csumb.edu/wiki/index.php/Tarsier
FluidEarth	https://fluidearth.net
Integrated Component Modelling System	www.clw.csiro.au/products/icms/
The Invisible Modelling Environment	www.toolkit.net.au
Spatial Modelling Environment	Maxwell & Costanza (1997)
Framework for Risk Analysis of Multi-media Environmental Systems	mepas.pnnl.gov/framesv1/sum3ug.stm

All web links accessed 4 November 2014.

Tables 4 and 5 indicate that there are many different approaches and solutions available. All have restrictions on their use and are, to a variety of extents, tailored to the issues facing their user communities. The question of how this plethora of approaches might be consolidated will be addressed

Table 5. *Elements in integrated environmental models (Lu & Piasecki 2012)*

(iv) Geospatial context	
Chesapeake Community Modelling Programme	ches.communitymodeling.org
(v) Monolithic code framework	
Weather Research and Forecasting	www.wrf-model.org/index.php
(vi) Generic component-based modelling framework	
Community Surface Dynamics Modelling System	csdms.colorado.edu
Partnership for Research Infrastructures in Earth Systems Modelling	Valke *et al.* (2006)
Earth System Modelling Framework	www.earthsystemmodeling.org
(vii) Coupling frameworks	
Modular Modelling System	Leavesley *et al.* (1996)
Object Modelling System	David *et al.* (2004)
ModCom modular simulation system	Hillyer *et al.* (2003)
Community Hydrologic Modelling Platform	https://www.cuahsi.org/
Dynamic Information Architecture System	Campbell & Hummel (1998)
Tarsier	ecoviz.csumb.edu/wiki/index.php/Tarsier
Open Modelling Interface	www.openmi.org
Interactive Component Modelling System	www.clw.csiro.au/products/icms/
The Invisible Modelling Environment	www.toolkit.net.au
Spatial Modeling Environment	Maxwell & Costanza (1997)
Next-generation Framework for Aquatic Modelling of the Earth System	Lakhankar *et al.* (2008)

All web links accessed 4 November 2014.

later. First, the relationship between data and models is considered in light of the needs of integrated environmental modelling.

Data and models

Traditionally, the boundary between measured and modelled data has been presented as being quite clear: data are direct observations used to set up bathymetry and boundary conditions, and also used for verification and validation. Model runs are simulations that produce outputs (modelled data) that can be compared with other measured data.

However, all data are abstractions of reality; there are merely different levels of abstraction. Possibly the lowest level involves a direct observation of the environment (for example, water level up a marked rule). However, it is impossible to collect much data through direct observation and so instruments are used for repetitive observations. At the next level of abstraction, an instrument will utilize a well-understood mathematical relationship (or conceptual model) to derive the observational value (for example, water level based on pressure measurement). However, a much higher level of abstraction is required to calculate many derived quantities from observational data. For example, the calculation of suspended sediment concentrations from satellite data requires a modelling process, whereby algorithms, assumptions and calibration (normally using a different type of data) are all applied to the captured signal to provide 'measured' data. Hence, as data collection and analysis becomes more complex, the boundaries between model and data become blurred.

Moreover, there are many different types of numerical model, each of which is a representation of one part of reality (Cunge 2003). The following three model types are discussed here:

- deterministic numerical simulations
- data driven modelling
- data mining and assimilation.

Deterministic modelling

The traditional deterministic model is a numerical representation of a physical law or laws such as conservation of mass, energy or momentum. The equations are discretized and solved using a variety of numerical schemes. As such, they encapsulate our knowledge of the physics of a problem. However, some behavioural (or data-driven) representation of processes is needed, such as the use of a roughness length in shallow water flow modelling. This is even more evident in the modelling of more complex, less well-understood phenomena such as sediment transport models. These range from the simulation of scour round an object using computational fluid dynamics code (Dixen *et al.* 2013) through coastal area modelling, to beach plan-shape modelling and models of the coastal tract. These models, even the most detailed, contain behavioural representations of sediment transport at one length-scale or another.

Data driven models

A data driven model is a means of deriving a functional relationship between input and output data, where the parameters and coefficients have been fitted to the data, and so are not based on physical laws (Cunge 2003). Data driven model types include correlations, autoregressive–moving-average (ARMA) methods, artificial neural networks, genetic algorithms and genetic programming. These models are rapid to run, but depend on the number, range and accuracy of the input data. Cunge (2003) warned against possible misuse of these models, but they have become increasingly popular in the past decade and commonly feature in almost any recent issue of popular journals such as the *Journal of Hydroinformatics*.

However, the application of such methods to environmental modelling will result in new transformations and relationships that have not been thought of, but arise out of the formalized exploration of large data sets and the power of recursive algorithms that are made possible by computer programs rather than the physical laws that would underpin a deterministic model. For example, it will be possible to treat deterministic models as data providers (or the sources of the variables to be transferred to another model) and use a data driven model to evolve a sensible overall scenario.

Data mining and data assimilation

Cunge (2003) argued that a theory is both a description and an explanation of physical processes, and that data driven and data mining approaches are therefore not theories (and, by implication, are less worthy than theories and their deterministic models). However, the idea of using computers to augment human intelligence dates back to the memex (Bush 1945), while today some areas of science such as astronomy and biology have collected so much data that data mining techniques are being used to extract information (in the form of statistical models of complex phenomena) that cannot be determined by human intelligence alone. These techniques are not generally used with environmental data (although the volumes of remote sensing data being collected are huge) but they cannot be ignored, even when accepting many of the caveats that Cunge (2003) supplies.

Data assimilation takes measured data and incorporates it into the running of a numerical modelling suite, influencing the final outcome. Data assimilation starts to integrate data and models explicitly and is an area of active research in many modelling disciplines.

Implications for integrated modelling

The increasingly blurred boundaries between measured and modelled data are one driver for the development of common standards for its interoperation. The standards for model linking should, as a corollary, allow for the import of data from more permanent storage media. File-based data transfer is sometimes inefficient but offers a well utilized and simple structure for data from a variety of sources targeted at applications for accessing, reading, writing and analysing.

Modern data standardization is tending to occur at two levels: the structure of the data and its technical implementation. Definitions of data structure are independent of the file encoding. For example, ISO 19115 outlines the data structure of spatial metadata with its XML encoding given in ISO 19139. The supporting (use and discovery) metadata can be given in separate files to the values themselves. This is exhibited in formats such as CSML, NetCDF and XDMF, which offer a binary file type (such as HDF5) for high volumes. Also, directives such as the one establishing an Infrastructure for Spatial Information in the European Community (INSPIRE) (http://inspire.jrc.ec.europa.eu) provide a legal and technical framework for data interoperability. INSPIRE includes specifications for the data, discovery, use and download services and is aimed at making the finding, using and sharing of data easier across the European Union (EU). However, for any practitioner wishing to offer a dataset to the wider community, the set of standards on offer is incomplete, overlapping and highly esoteric.

The Open Modelling Interface

The need for integrated environmental modelling tools led to the development of the Open Modelling Interface (OpenMI) during two European Commission funded projects, HarmonIT (2002–2005) and OpenMI LIFE (2006–2010). The particular driver for these projects was that whole catchment modelling is required to deliver the objectives of the Water Framework Directive (European Commission 2000) which requires Member States to achieve 'good ecological status' of surface waters by 2015. This places significant demands on water managers and requires the linking of a wide variety of models. Integrated modelling was seen as a realistic mechanism for this, which would enable process interactions to be simulated across catchments (Gijsbers *et al.* 2002; Blind & Gregersen 2005; Moore & Tindall 2005). The creation of an open modelling environment was intended to capitalize on the huge prior investment in model development (which was mainly in proprietorial code). It was inspired by a number of national initiatives that demonstrated the feasibility of integrated modelling frameworks and involved work with major commercial players in the water resources software market to ensure that the vast amount of encapsulated knowledge in existing (proprietorial) tools was not abandoned, but rather was modernized, recycled and reused (Gijsbers *et al.* 2002; Blind & Gregersen 2005; Moore & Tindall 2005).

The chosen path to integrated modelling was the development of the OpenMI standard, a set of software interfaces that a compliant component must implement. Version 1.0 (.Net) was released at the end of HarmonIT (Gregersen *et al.* 2005, 2007). Implementation of this standard was tested for a wide range of cases in the OpenMI LIFE project under the European Commission's LIFE Environment programme. OpenMI was applied in the Scheldt basin in Belgium and the Netherlands (Safiolea *et al.* 2011) and in the Pinios basin in Greece (Makropoulos *et al.* 2010; Safiolea *et al.* 2011) to demonstrate that OpenMI can assist competent water authorities in joint model integration to achieve the objectives of the Water Framework Directive. The standard was updated, as was the release procedure, leading to OpenMI version 1.4 (available for both .Net and Java), which became the only official version of the standard. In addition the project set up a legal body, the OpenMI Association, to support, maintain and publicize the OpenMI standard.

Development work continued and, by the end of OpenMI LIFE, a beta release of OpenMI version 2.0 was published for external review. Work continued and version 2.0 of the standard (OpenMI Association 2010*a*) and reference (OpenMI Association 2010*b*) were officially released in December 2010 during a EU–US summit in Washington DC. Following discussions with the Open Geospatial Consortium (OGC), OpenMI has become an OGC standard (http://www.opengeospatial.org/standards).

OpenMI standard

The stated aim in the development of OpenMI was to provide a mechanism for physical and socio-economic models to be linked to each other, other data sources and tools at run-time (Gijsbers *et al.* 2002; Blind & Gregersen 2005). This was achieved through the development of the OpenMI standard

(Gregersen *et al.* 2005, 2007; Gijsbers *et al.* 2010) which is a software component interface (cf. Khatibi *et al.* 2004) that enables OpenMI components to:

- be configured to exchange data during computation (at run-time);
- run simultaneously and share information at each time step making model integration feasible at the operational level.

The OpenMI standard was originally conceived to facilitate the numerical modelling of interacting environmental processes related to whole catchment modelling. However, what was developed is a generic solution to the problem of data exchange between models or software components. For example, it can be applied to link models of different domains and environments such as models of hydraulics, hydrology, ecology, water quality and economics. It can be used to link models of different dimensionality, so that a one-dimensional (1D) river model can be coupled to a two-dimensional (2D) flow model when the river broadens or a 2D flow model could be coupled to a three-dimensional (3D) flow model. OpenMI can also be applied to link models that operate at different time steps, so are running asynchronously. It can link different spatial representations (e.g. networks, grids, polygons) and can cope with different projections, units and categorizations, and with models that have no temporal or spatial representation.

Examples of the use of OpenMI include, but are far from limited to, the following:

- Becker & Schüttrumpf (2011) used the programming interface of the finite element groundwater flow model Feflow to achieve OpenMI compliance without altering the source code by implementing remote procedure calls. This was demonstrated using a model of dike seepage under transient water level boundary conditions.
- Bulatewicz *et al.* (2010) demonstrated the integration of agriculture, groundwater and economic models using OpenMI, while Bulatewicz *et al.* (2013) wrote a Simple Script Wrapper, which simplifies the linking of scripted models (in Matlab, Scilab or python, for example) to other OpenMI components.
- Shrestha *et al.* (2012) linked models of hydrology, hydraulics, water temperature, sediment transport and faecal bacteria using OpenMI and applied this composition to the River Zenne in Belgium, demonstrating interaction between the sediments and the bacteria.
- Becker *et al.* (2012) linked the software package RTC Tools (for the control of hydraulic structures) to the hydraulic model SOBEK using OpenMI to provide a real time control package for hydraulic structures on the River Rhine.

OpenMI compliant components may come from any suppliers and can be based on legacy software or a new model. The standard supports two-way links (Gregersen *et al.* 2007) where involved models mutually depend on calculation results from each other. Castronova & Goodall (2013) found that computational overhead imposed by OpenMI's run-time exchange of data was not significant when applied on a semi-distributed watershed model. Developers have also found it practical to use OpenMI in conjunction with other software tools, as listed in Table 6.

The new features in OpenMI 2.0 centre on making the standard more flexible and extensible. They include the following.

(1) The concept of adaptors which allow component outputs to be transformed (adapted) before inputting into other components. This is to allow situations such as transformation between differing spatial structures held by different components (for example: a triangular model grid passing data to a rectangular grid; and a 2D grid passing data to 1D). It is also possible to chain adaptors together if multiple transformations are required in series.
(2) Easier incorporation of data from other sources such as files, databases and web services.
(3) A more flexible overall structure with a core set of mandatory interfaces and optional extension sets. The extension governing space- and time-dependent components is included in the current edition as this is the most common, current requirement, but is not part of the mandatory interface set. This allows the core standard to be easily applied by other model types.
(4) A representation of geographical data structures which is closer to common, modern implementations. This points towards OpenMI dovetailing with specific geospatial data standards in the future.

Roles and responsibilities

The OpenMI standard allows the passing of many different types of data between models, which is a great strength. The more widely applicable an interface is, the closer it gets to 'plug and play' interoperability. However, this is also a weakness, as when the description of the passed data becomes less prescriptive, the onus passes to the modeller to understand what is being offered by one model and required by another, as this is not specified by the standard.

There is already an emerging issue with large complex models in that many assumptions used in the construction of suitable algorithms and code

Table 6. *Software tools that have been linked to OpenMI*

Software linked to	Reference
Simple Script Wrapper	Bulatewicz *et al.* (2013)
CUAHSI Hydrologic Information System (HIS)	Castronova *et al.* (2013)
Simple Model Wrapper	Castronova & Goodall (2010)
Real Time Control (RTC) tools	Becker *et al.* (2012)
UncertWeb	Gupta *et al.* (2012), Bastin *et al.* (2013)
Pyxis workflow manager	San Roman Blanco *et al.* (2012)

are not explicit. When the user was also the developer, an understanding of these assumptions was retained. Increasingly users do not have this knowledge and model developers have been slow to help make the underlying assumptions explicit, for example, by incorporating tests that check whether an assumption is being violated and warning the user. As we begin to link models that deploy different sets of assumptions this problem is compounded.

So, for example, consider the scenario where there are two models passing data between themselves during run-time (see Fig. 2). A person constructing a composition by linking the models needs to know:

- the assumptions that underpin each model, so their suitability for the target application can be judged;
- that both models are OpenMI compliant (or how to achieve this);
- that they can obtain the model under suitable licence terms;
- details of the data being offered by the first model;
- details of the data expected by the second model;
- how to adapt the output from the source model to suit the target.

It is not always clear where this information will come from. Ideally a model should be developed with the following accompanying elements:

- strong version control (especially if open source);
- documentation and training material;
- an application for verification and validation (ideally one that is linked to version control, such as that by Farrell *et al.* 2011);
- improved 'use' metadata describing the model itself, in particular the underlying assumptions and interfaces;
- a track record including examples of use;
- increased compliance to recognized versions of standards;
- a standard licence;
- supporting web portals with standardized cataloguing and documentation.

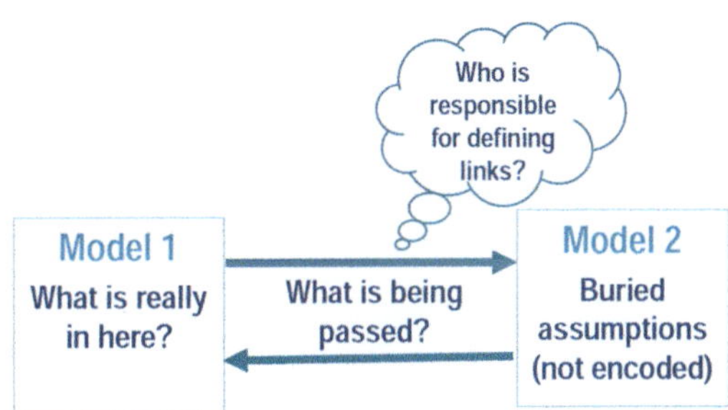

Fig. 2. The perils of plug and play.

The scientists who write component models are not normally software developers familiar with standards and may not themselves have any need to link their models. Neither scientist nor software developer has the wrapping of models as a primary goal, so the wrapping of models falls between two stools (Knapen *et al.* 2013). This situation could be assisted by the provision of software tools and guidance to simplify the model wrapping process.

Moreover, unless the model wrapping is documented along with the underlying component model, it may not be clear which variables are exposed (available to be exchanged) or exactly how they are defined. As an example of the latter, radiation stress is defined differently in the wave model SWAN (Booij *et al.* 1999) and the shallow water flow model TELEMAC-2D (Hervouet 2007) so when they are linked in an OpenMI composition, an adaptor must be used to translate the output from SWAN for TELEMAC-2D.

To address the challenges of integrated modelling, the community will have to start to provide more information about their models, as suggested above. It will only be by providing this information, ideally in a standardized way, which would be assisted by the development of common ontologies, that people will trust models enough to take up and use third-party models to create new model compositions. Moreover, the community will have to address questions such as 'who owns the intellectual property rights to a new composition?' and 'what is the quality of the modelled data?'.

FluidEarth

FluidEarth, formerly known as OpenWEB, is a collaborative initiative between the academic community and users with the aim of researching and implementing integrated computer modelling approaches to environmental systems (Pearce *et al.* 2010). One of the main problems with the OpenMI standard is that it requires application development skills above those of a typical scientific programmer. In response, HR Wallingford has developed the FluidEarth implementation of the OpenMI standard (Harpham *et al.* 2014) making its use easier for the scientific community through the provision of a graphical user interface, 'Pipistrelle', and a software development kit. Both are open source and available on SourceForge (http://sourceforge.net).

FluidEarth is developing a dialogue between key academic partners, providing the tools needed to reduce duplication of effort within the research community, making the task of translating research into applications easier and increasing the commercial potential of research outputs by creating a large community of active users. To facilitate these activities, FluidEarth has developed a web portal (http://fluidearth.net) where people can find out about the FluidEarth implementation of the OpenMI standard, post discussion questions and replies, present their studies, view e-learning tutorials, read community announcements and gain access to the following key software and repositories:

(1) Pipistrelle, the FluidEarth open source user interface, provides a run-time environment for linking OpenMI compliant components. This gives modellers the ability to create and run compositions of linked components (Fig. 3).
(2) The FluidEarth Software Development Kit (SDK) is an open source tool allowing model developers to more easily adapt their models and other components to be compliant with the OpenMI standard, thus reducing the requirement for specialist programming.
(3) A repository is provided for research partners to upload software that can be tried and tested by the community.
(4) The FluidEarth catalogue of model engines and instances of their application is an implementation of the open source GeoNetwork web interface for searching geospatial data (http://geonetwork-opensource.org).

A SourceForge project (http://sourceforge.net/projects/fluidearth) gives access to all source code, feature requests and amendments of the FluidEarth 2.0 tools for version 2.0 of OpenMI.

FluidEarth communities

FluidEarth recognizes the split in responsibilities and skills that has occurred in the development of

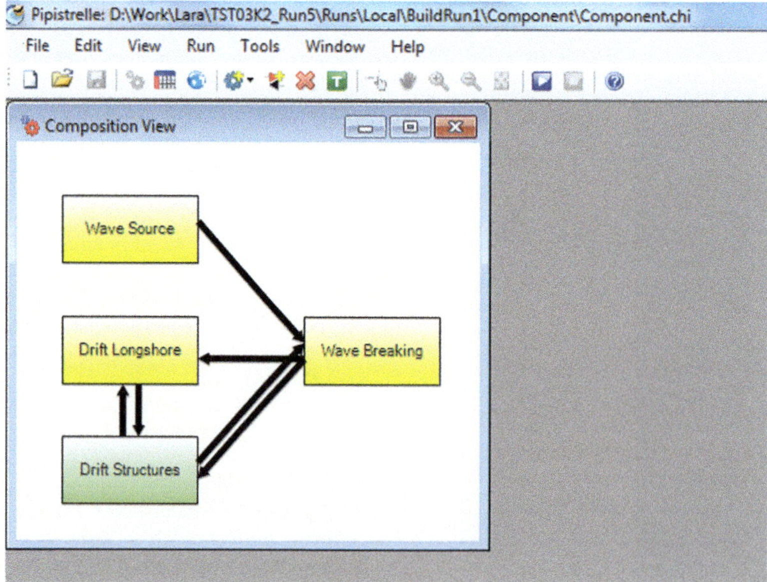

Fig. 3. Screenshot of Pipistrelle editor with simple composition.

modelling tools (from Abbott's first to fifth generations and from Khatibi's forming to performing stages) and seeks to facilitate the activities of the following five main groups and to act as a conduit between them.

(1) **Software architects and developers** who develop and test software tools for common tasks. They have a strong background in software development, but may have no expertise in environmental modelling.
(2) **Model integrators**, whose role is to take individual models and form integrated models (or compositions) from them. The role of model integrator is a new one and requires a particular set of skills in using integrating software (such as the FluidEarth SDK and Pipistrelle) and checking the metadata to ensure compatibility.
(3) **Researchers** wishing to develop new techniques for integrated modelling to explore more complex feedback mechanisms, or simply test new algorithms within existing models. They are commonly scientists with a knowledge of coding but with limited knowledge of accepted software development practices, who are used to working on the development of individual models.
(4) **Users** who apply existing models, or even model compositions, to real world problems. They are likely to be scientists or engineers, with a background in environmental or physical sciences or engineering.
(5) **End-users**, such as the developers of schemes, regulatory authorities, local and national governments who use the results supplied by users to influence their decision making (for example in developing policy or planning).

Members of the FluidEarth community may be in more than one group. For example, it is not unknown for a researcher to contribute to software development as well as the writing of individual models. It is more common perhaps for a user to also be a researcher. End users may well not have any background in science or engineering, or any experience of coding models or software. The groups are interdependent and share the FluidEarth resources, as shown in Figure 4.

Towards model fusion

Previous sections have shown how the demand for modelling across disciplines has led to the development of a large number of solutions to the problem of achieving integrated modelling, which may lead ultimately to model fusion. As noted in the introduction, fusion involves the 'melting [or] blending of different things into one' (Oxford English Dictionary), so it involves not just the ability to link models, but also easy access to information about the models and linking technologies and access to software tools to make the process easier.

Using Khatibi (2003*a*), we propose that model fusion will occur: when problems can be addressed

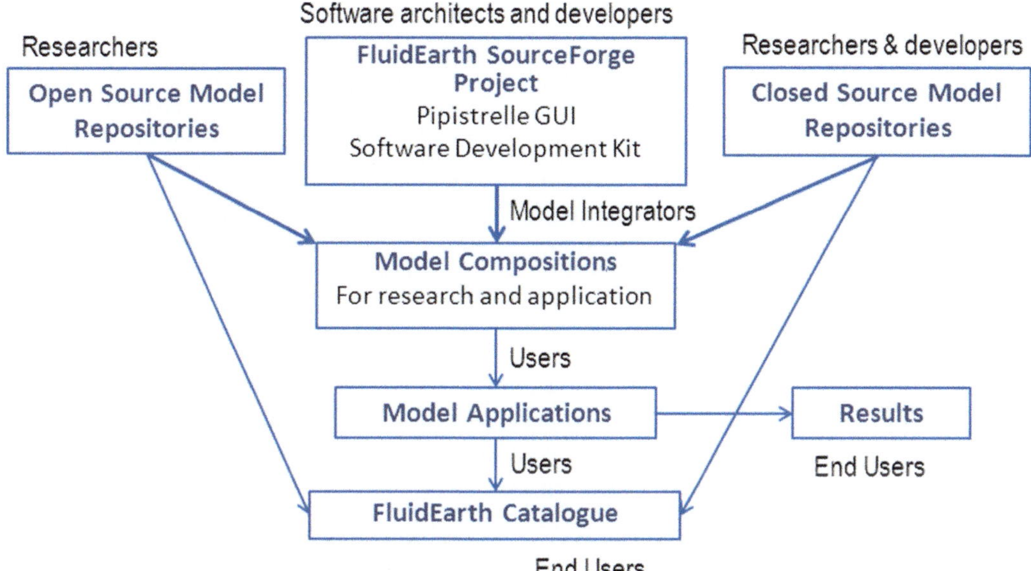

Fig. 4. FluidEarth#software, repositories and communities.

efficiently within a flexible environment; where custom solutions can be delivered to meet clients' needs and the clients are aware of the limits of the system; and where the connections between components and the influences they have on each other are understood and utilized. There is an interplay between technical, economic and social needs. Model fusion will occur when integrated environmental modelling has reached the performing stage (Khatibi 2003*a*).

Many approaches to integrated modelling already exist (Tables 4 & 5) but the norming phase will involve conscious attempts at consolidation, as part of a Darwinian struggle for survival, and the emergence of a few key sets of standards. Note that it is highly improbable that a single approach will evolve from this process, as no approach is likely to be optimal for all situations. It is much more probable that a reduced number of approaches will co-exist, some in niche markets and some in open competition. This will help to drive continued innovation.

We put forward in this paper an approach to achieving model fusion, not just considering technical issues but also the involvement of the communities of model developers, scientists, users and end-users which develop around an integration method. This approach recognizes the emergence of an era of more open science. The concept of integrated environmental modelling benefits from open architecture and is commonly associated with open source software. Meanwhile the importance of open access to public sector research has been recognized by the Organisation for Economic Co-operation and Development (OECD 2004) the European Commission (European Commission 2011) and the UK (HM Government 2012). The European Commission has launched an Open Data Strategy for Europe, which is expected to deliver a €40 billion boost to the EU's economy each year (European Commission 2011) and there is a commitment to increasing access to data and publications within the Commission's Horizon2020 research programme. Open access to research publications is becoming increasingly common and open access journals are proliferating. Moreover, we are in an era of increasingly open scientific collaboration. Nielsen (2011) presents a number of case studies of networked science, from citizen science to mathematical problem solving, which have occurred by open collaboration over the Web and advocates this as the future of science.

The move towards open science, comprising open architecture, open source, open data, open access publication and open collaboration (or networked science) is ongoing but faces many difficulties. Nielsen (2011) predicts that it will only take place when we learn to value openness and the sharing of models and data as much as our publication and citation records. However, openness will only work efficiently through the widespread application of standards, as these assist people in working together. This paper is not concerned with how to make open science happen, but how to move integrated modelling towards model fusion against this backdrop.

We have seen in this paper the importance of standards, software architecture, model metadata and documentation, and tools to assist with linking models and collaboration. These ideas form the basis for the following six topics:

(1) Enhanced metadata for data and models
(2) Provision of supporting information
(3) Software-as-a-service
(4) Consolidation of linking technologies
(5) Diagnostic and reasoning tools
(6) Verification, validation and explanation

Each of the six topics is described in more details below.

Enhanced metadata for data and models

There will be benefits to enhancing the richness of information about shared data and models and in standard forms. This extra information will enable any interested party to judge the suitability and quality of the data or model. In order to do this we need to develop (and ideally consolidate) metadata standards and ontologies. A primary requirement will be to extend from an established standard set to reduce the likelihood of independent bespoke implementations of the same standard being developed. Scientists and engineers will have to get used to the routine generation and use of metadata. Tools that map between different standard ontologies are beginning to arrive in the marketplace and will become increasingly useful.

We will also need to improve the standards of model documentation. There are already forms of automatic code documentation, such as Doxygen (www.doxygen.nl/features.html) which show flow charts of information between sub-routines, but these are themselves generally not sufficient to understand a code. A poorly documented open source code is a considerable barrier to understanding and hence take-up, while a model without documented verification or validation is impossible to judge.

A community of modellers will also be more effective when there are improved tools to search for, discover and link to data and models. Indeed, tools that address many semantic issues between communities will become more fundamental as integrated modelling increasingly crosses disciplines.

Provision of supporting information

The development of shared knowledge bases will help to create a free market in information, which every member of the community can access equally. This is more manageable within an organization than between organizations, but the concept is similar to a distributed database, where the data owner maintains the data, with the ability for these data to be integrated with other data held in the distributed system. This will require protocols and standards to be developed, published, accepted and used.

Harvesting data or models from external knowledge bases is a much bigger challenge, not so much due to the technical problem of searching for information over the Web or having common formats, but due to the need to establish ownership, licence conditions and the quality of both data and models; hence the need for enhanced richness of information about data and models. The knowledge bases also have to contain information on input/output protocols, the principles underpinning each model, implicit assumptions in the models, and information on data calibration and processing.

Software-as-a-service

The evolution of modelling systems and the changing role played by the user (Table 2) combined with the changes in the computing environment (Table 3) indicate a move towards offering Software-as-a-service (SaaS) over the internet. Although this is common in other fields and has been trialled in academic circles, it is only starting to be offered as a commercial proposition in environmental modelling (Bourban *et al.* 2012). This has been enabled by rapid progress in the development of the Cloud, the Grid and emerging standards for web services.

Consolidation of linking technologies

There are a considerable number of candidate approaches to model integration (Tables 4 & 5). These come with significant overlap in functionality, but also many distinct features which usually arise from the specific needs of the community that created them. Some consolidation is sensible and inevitable as the community comes to fully appreciate the benefits to be gained from adopting common standards. The particular needs of each community and the degree of overlap will severely hamper this process and, of course, nobody wants to abandon the time and effort they have put into developing their own system. Moreover, anyone changing systems will have to make an investment of time and effort in learning the new system and there are limited incentives to make this happen. In the short term, consolidation may occur as researchers follow the money and/or join in with active communities. In the longer term, we propose that one or both of two outcomes are possible:

(1) A set of mutually compatible standards will be adopted to solve particular issues, such as standards for metadata (e.g. ISO 19115) web services (e.g. WPS) and memory-based model coupling (e.g. OpenMI) file-based model coupling (with a variety of standard file formats). These standards will be universally adopted by virtue of their utility and technical credibility. Existing frameworks will adapt to incorporate these standards into the appropriate components.

(2) A large technology company will produce a product which covers a large percentage of modelling requirements. Irrespective of the quality of the underlying product, it becomes 'standard' because it exists, is common and works with minimal technical help.

Diagnostic and reasoning tools

The development of integrated environmental models, composed of a number of component models (each of which may have many sub-routines) calls for the development of a new set of tools for testing integrated models, analysing the results and synthesizing outputs.

Each model should have its own published verification and validation tests, preferably including measures of model skill (Sutherland *et al.* 2004) rather than just qualitative assessments. Models should be developed under version control and there should be regular, preferably automated (Farrell *et al.* 2011) testing of new code, ranging from unit tests of functions and sub-routines to validation of the entire model.

However, we also need to test integrated models to ensure that the data exchange and feedbacks mechanisms have been adequately captured. This should also involve skill scores, if at all possible, and it may be possible to adapt techniques such as variance based sensitivity analysis (VBSA) (Saltelli *et al.* 2008) to isolate which model parameters contribute the most to the variance in the output result.

As multidisciplinary model compositions are often time-consuming to run, the use of model emulators may be developed and the mixed use of a full model composition and a quicker emulator has the potential to speed up VBSA or optimization techniques with integrated models. Where a process or link is poorly understood, a simple behavioural representation may be developed, with the coefficients or even the form of the behavioural representation optimized using data-driven techniques. Another concept involves replacing a poorly defined or uncertain model link with data in some simulations and either comparing the results from the full model and the model/data mix, or allowing the linked model data to influence but not entirely define the data passed (Voinov & Cerco 2010).

The development of diagnostic and reasoning tools involves bringing in the tools of professional software development and is already changing the skill sets needed to develop software. The modelling community will continue to need people with a computational sciences background as well as people with a background in applied sciences.

Verification, validation and understanding

Models are only useful if they are used in a way that acknowledges their limitations and serves to help establish and inform understanding. There is always a danger that with increasing complexity, clarity of understanding is lost and users become confused under the welter of, often conflicting, information being provided. What is worse, models, like statistics, can be used to obfuscate or mislead. This is already giving rise to public scepticism about the use and value of models. Too often model outputs have been 'over sold', or conversely have been shown to have missed critical outcomes (for example, the infamous 'hurricane' that was not going to hit the south of England in 1987 (Met Office 2013)). As a result there is a preference among non-specialists to prefer the simple over the complex and this can lead to 'rules of thumb' being preferred over sophisticated model results (no matter how good the representation of the physics, chemistry and biology). Given the inherent uncertainty that abounds in environmental modelling, results have to be communicated with care. A careful balance is needed between not undermining the value of the model outputs while at the same time recognizing the limitations of the model. Often the biggest gains are when a model, or models, leads to a better understanding of how a system behaves, rather than specific predictions.

This complexity also reinforces the need for better audit trails of the modelling that has been undertaken. All the various steps in the modelling process need to be documented, ideally in an openly accessible form that allows the whole process to be reproduced by others. This is rarely possible at present but the move to greater open access to data and models should enable the possible for in-depth and independent reviews of model applications. However, there will also be a need for improved interrogation tools that allow the model user to generate reports on particular aspects of the model. This is not dissimilar to the process of querying a complex database. The requirement is to be able to drill down from high level, perhaps summary outputs, to the underlying information. This is likely to include detailed outputs, data inputs, resolution in space and time, assumptions made in each of the component models, details of the exchanges between models, and the underlying science that is being represented in the model. Whilst metadata together with model documentation and manuals are a good start, they are a long way short of providing what is needed for a fully interactive set of interrogation tools.

Discussion

Many models are already very large, with hundreds of sub-routines and a host of assumptions buried deep in the formulation and coding of the model. Individually they can be a challenge to maintain. If such models are merged, maintenance becomes even more difficult. One of the biggest advantages of fusion through linking is that individual model developers retain responsibility for their model, while at the same time enabling others to use the model with other models to emulate the 'larger' model. This is not dissimilar to the evolution of databases. Initially it was thought that a central database was essential for the efficient use of data in an organization. After a few years it quickly became apparent that such databases were extremely difficult to maintain and that it was far better to have a distributed system, where the data owner maintained the data, with the ability for these data to be integrated with other data held in the distributed system.

This comparison with databases is also relevant in the context of the future development of model fusion. As models increasingly use data assimilation techniques and data monitoring uses models to aid interpolation, the difference between model and data becomes increasingly blurred. We are moving towards a new era, in a world that is data rich and with the ability to undertake large computational activities. This creates new opportunities for modellers in terms of the model complexity, model detail, and perhaps most importantly, the ability to run model ensembles and so start to better represent the uncertainty inherent in the analysis of real world systems.

However, this comes at a price. The new role of model integrator requires a different set of skills from that of researcher or user (although in practice they may be the same person) in order to ensure the validity of what is being represented by the collection of models. This requires a much clearer appreciation of the fact that all models are wrong but some are useful (Box 1979), not only amongst modellers but also those who use the outputs.

Looking further ahead one might envisage the ability for models to 'self-assemble' compositions to represent a particular problem. This requires some substantial advances from where we are now, not least making explicit the knowledge and assumptions that are currently buried deep in existing models.

Conclusion

The past decade has seen the development of the new field of integrated environmental modelling where compositions of linked models exchange data at run-time. The application of systemic knowledge management to integrated environmental modelling indicates that we are at the onset of the norming stage, where gains will be made from the hierarchical organization of the competing options. This implies that there will be consolidation in the range of approaches that have proliferated in recent years, which is likely to become manifest in the predominance of a limited number of standards (covering ontologies, metadata, model interfaces, data formats and so on). An open software architecture (consisting of a user interface with published interfaces to a range of models, data and data processing routines) will be crucial and the use of open source software is likely to increase.

We propose six topics that will help integrated modelling to move through the norming stage towards the performing stage, where problems can be addressed efficiently within a flexible environment and custom solutions can be delivered to meet customers' needs. The six topics are:

(1) Enhancing metadata for data and models to enable any interested party to find and then judge the suitability and quality of the data or model.
(2) Provision of supporting information to create a free market in information including the ownership of models, their licence conditions, underlying principles, input/output protocols, calibration, verification and validation.
(3) Software-as-a-service, where models are offered for use through a web-interface.
(4) Consolidation of linking technologies, such as OpenMI, where one or both of two outcomes are possible: (i) a set of mutually compatible standards will be adopted to solve particular issues; and/or (ii) a large technology company will produce a commonly-available product that covers most needs and works with minimal technical help.
(5) Development of diagnostic or reasoning tools for testing integrated models, analysing the outputs and synthesizing the results.
(6) Verification, validation and understanding of integrated modelling should improve. Results have to be communicated with care, due to the inherent uncertainty in integrated modelling. Ways must be developed to acknowledge limitations and to help establish and inform understanding. This will require improved audit trails, tools for the propagation and assessment of uncertainty, and improved tools for the interrogation of integrated models.

When these six topics are developed, integrated modelling will have developed into model fusion, which involves the ability to link models, but also easy access to information about the models and interface standards (such as OpenMI) and access to software tools to make the process easier. In order for this to happen, a community must develop that is prepared to openly share information about models and data. This will be assisted by the adoption of standards, but will require the community to value openness and the sharing of models and data as much as it loves its publication and citation records. The development of a more open culture and the adoption of standards therefore go hand-in-hand.

This work was funded by HR Wallingford's internal R & D programme as part of the FluidEarth initiative, by the *iCOASST – integrating coastal sediment systems* project, which is funded by UK Natural Environment Research Council under grant NE/J00541X/1 with support from the Environment Agency and by the European Commission's 7th Framework Programme through the DRIHM project (grant number 283568) and DRIHM2US project (grant number 313122). The topics for model fusion were influenced by a presentation given by Professor Rizzolli at the final meeting of the OpenMI-LIFE project.

References

Abbott, M. B. 1991. *Hydroinformatics: Information Technology and the Aquatic Environment.* Avebury Technical, Aldershot, UK.

Abbott, M. B. & Vojinovic, Z. 2009. Applications of numerical modelling in hydroinformatics. *Journal of Hydroinformatics*, **11**, 308–319.

Abbott, M. B., Tumwesigye, B. E. & Vojinovic, Z. 2006. The fifth generation of modelling in hydroinformatics. *In*: Gourbesville, P., Cunge, J., Guinot, V. & & Liong, S.-Y. (eds) *Proceedings of the 7th International Conference on Hydroinformatics*. Research Publishing, Singapore, 2091–2098.

Armstrong, C. W., Ford, R. W. & Riley, G. D. 2009. Coupling integrated Earth system model components with BFG2. *Concurrency and Computation: Practice and Experience*, **21**, 767–791.

Bastin, L., Cornford, D. *et al.* 2013. Managing uncertainty in integrated environmental modelling: the UncertWeb framework. *Environmental Modelling &*

Software, **39**, 116–134, https://doi.org/10.1016/j.envsoft.2012.02.008

Becker, B. P. & Schüttrumpf, H. 2011. An OpenMI module for the groundwater flow simulation programme Feflow. *Journal of Hydroinformatics*, **13**, 1–12.

Becker, B. P. J., Schwanenberg, D., Schruff, T. & Hatz, M. 2012. Conjunctive real time control and hydrodynamic modelling in application to Rhine river. *10th International Conference on Hydroinformatics, HIC 2012*, Hamburg, Germany.

Blind, M. & Gregersen, J. B. 2005. Towards an Open Modelling Interface (OpenMI) the HarmonIT project. *Advances in Geosciences*, **4**, 69–74.

Booij, N., Ris, N. R. & Holthuijsen, L. H. 1999. A third-generation wave model for coastal regions, 1, model description and validation. *Journal of Geophysical Research*, **104**, 7649–7666.

Bourban, S., Durand, N., Turnbull, M., Wilson, S. & Cheesman, S. 2012. Coastal shelf model of northern European waters to inform tidal power industry decisions. *In*: Bourban, S., Durand, N. & Hervouet, J.-M. (eds) *Proceedings of XIXth TELEMAC-MASCARET User Conference*, 18–19 October 2012, Oxford, HR Wallingford, Wallingford, UK, 143–150.

Box, G. E. P. 1979. Robustness in the strategy of scientific model building. *In*: Launer, R. L. & Wilkinson, G. N. (eds) *Robustness in Statistics: Proceedings of a Workshop*, 11–12 April 1978, Durham, NC, USA, Academic Press, New York, 201–236.

Bulatewicz, T., Yang, X., Peterson, J. M., Staggenborg, S., Welch, S. M. & Steward, D. R. 2010. Accessible integration of agriculture, groundwater, and economic models using the Open Modeling Interface (OpenMI): methodology and initial results. *Hydrology and Earth System Sciences*, **14**, 521–534.

Bulatewicz, T., Allen, A., Peterson, J. M., Staggenborg, S., Welch, S. M. & Steward, D. R. 2013. The simple script wrapper for OpenMI: enabling interdisciplinary modeling studies. *Environmental Modelling & Software*, **39**, 283–294, https://doi.org/10.1016/j.envsoft.2012.07.006

Bush, V. 1945. As we may think. The Atlantic. http://www.theatlantic.com/magazine/archive/1945/07/as-we-may-think/303881/ [accessed 4 November 2014]

Campbell, A. P. & Hummel, J. R. 1998. The dynamic information architecture system: an advanced simulation framework for military and civilian applications. *In*: *Proceedings of the Advanced Simulation Technologies Conference*, Boston, MA, USA.

Castronova, A. M. & Goodall, J. L. 2010. A generic approach for developing process-level hydrologic modeling components. *Environmental Modelling & Software*, **25**, 819–825, https://doi.org/10.1016/j.envsoft.2010.01.003

Castronova, A. M. & Goodall, J. L. 2013. Simulating watersheds using loosely integrated model components: evaluation of computational scaling using OpenMI. *Environmental Modelling & Software*, **39**, 304–313, https://doi.org/10.1016/j.envsoft.2012.01.020

Castronova, A. M., Goodall, J. L. & Ercan, M. B. 2013. Integrated modeling within a Hydrologic Information System: an OpenMI based approach. *Environmental Modelling & Software*, **39**, 263–273, https://doi.org/10.1016/j.envsoft.2012.02.011

Chorley, R. J. 1962. *Geomorphology and General Systems Theory*. United States Geological Survey Professional Paper 500B, 10p.

Cunge, J. A. 2003. On data and models. *Journal of Hydroinformatics*, **5**, 75–98.

David, O., Schneider, I. W. & Leavesley, G. H. 2004. Metadata and modelling frameworks: the object modelling system example. *In*: *Transactions of the 2nd Biennial Meeting of the International Environmental Modelling and Software Society*, 14–17 June 2004, Osnabruck, Germany, International Environmental Modelling and Software Society, Manno, Switzerland, 439–443.

Dixen, M., Sumer, B. M. & Fredsøe, J. 2013. Numerical and experimental investigation of flow and scour around a half-buried sphere. *Coastal Engineering*, **73**, 84–105. https://doi.org/10.1016/j.coastaleng.2012.10.006

Donchyts, G. & Jagers, B. 2010. DeltaShell – an open modelling environment. *In*: Swayne, D. A., Yang, W., Voinov, A. A., Rizzoli, A. & Filatova, T. (eds) *IEMSS 5th International Congress on Environmental Modelling and Software*, 5–8 July 2010, Ottawa, Canada, International Environmental Modelling and Software Society, Manno, Switzerland, 1058–1065.

European Commission 2000. Directive 2000/60/EC of the European Parliament and of the Council of 23 October 2000 establishing a framework for Community action in the field of water policy. *Official Journal of the European Communities L*, **327**, 1–170.

European Commission 2011. Open Data, an engine for innovation, growth and transparent governance. Communication from the Commission to the European Parliament, the Council, the European Economic and Social Committee and the Committee of the Regions, COM(2011) 882 final, European Commission, Brussels.

Farrell, P. E., Piggott, M. D., Gorman, G. J., Ham, D. A., Wilson, C. R. & Bond, T. M. 2011. Automated continuous verification for numerical simulation. *Geoscientific Model Development*, **4**, 435–449, https://doi.org/10.5194/gmd-4-435-2011

Gijsbers, P. J. A., Moore, R. V. & Tindall, C. I. 2002. HarmonIT: towards OMI, an Open Modelling Interface and environment to harmonise European developments in water related simulation software. *In*: Falconer, R. A., Lin, B., Harris, E. L., Wilson, C., Cluckie, I. D., Han, D., Davis, J. P. & Heslop, S. (eds) *Hydroinformatics 2002. Proceedings of the 5th International Conference on Hydroinformatics*, Cardiff, UK. IWA Publishing, London, 1628–1275.

Gijsbers, P., Hummel, S. *et al.* 2010. From OpenMI 1.4 to 2.0. *In*: Swayne, D. A., Yang, W., Voinov, A. A., Rizzoli, A. & Filatova, T. (eds) *IEMSS 5th International Congress on Environmental Modelling and Software*, 5–8 July 2010, Ottawa, Canada, International Environmental Modelling and Software Society, Manno, Switzerland, 1081–1088.

Gregersen, J. B., Gijsbers, P. J. A., Westen, S. J. P. & Blind, M. 2005. OpenMI: the essential concepts and their implications for legacy software. *Advances in Geosciences*, **4**, 37–44.

Gregersen, J. P., Gijsbers, P. J. A. & Westen, S. J. P. 2007. OpenMI: open modellingInterface. *Journal of Hydroinformatics*, **9**, 175–191, https://doi.org/10.2166/hydro.2007.023

Gupta, T., Jones, R., Bastin, L. & Cornford, D. 2012. Integrating OpenMI and UncertWeb: managing uncertainty in OpenMI models. *In*: Seppelt, R., Voinov, A. A., Lange, S. & Bankampa, D. (eds) *Managing Resources of a Limited Planet: Pathways and Visions under Uncertainty, IEMSS Sixth International Congress on Environmental Modelling and Software*, 1–5 July 2012, Leipzig, Germany, International Environmental Modelling and Software Society, Manno, Switzerland, 1151–1158.

Harpham, Q., Cleverley, P. & Kelly, D. 2014. The FluidEarth 2 implementation of OpenMI 2.0. *Journal of Hydroinformatics*, **16**, 890–906, https://doi.org/10.2166/hydro.2013.190

Harvey, H. & Han, D. 2002. The relevance of open source to hydroinformatics. *Journal of Hydroinformatics*, **4**, 219–234.

Havnø, K., Sørensen, H. R. & Gregersen, J. B. 2001. Integrated water resources modelling and object oriented code architecture. *In*: *Proceedings of Water Resources Modelling and Management*. Japan Academic Society of Hydraulics. Chuo University, Japan.

Hervouet, J.-M. 2007. *Hydrodynamics of Free Surface Flows: Modelling with the Finite Element Method*. John Wiley & Sons, Chichester, UK.

Hillyer, C., Bolte, J. & Lamaker, A. 2003. The ModCom modular simulation system. *European Journal of Agronomy*, **18**, 333–343.

HM Government 2012. *Open Data White Paper. Unleashing the Potential*. The Stationary Office, Norwich, UK.

Janjic, Z., Gall, R. & Pyle, M. E. 2010. *Scientific documentation for the NMM solver*. NCAR Technical Note NCAR/TN-477 + STR. National Center for Atmospheric Research, Boulder, CO, USA.

Khatibi, R. H. 2001. Some issues in computational hydraulic practices. *Journal of Hydraulic Engineering*, **127**, 438–442.

Khatibi, R. H. 2003*a*. Systematic knowledge management in hydraulic systems: I, a postulate on paradigm shifts as a methodological tool. *Journal of Hydroinformatics*, **5**, 127–140.

Khatibi, R. H. 2003*b*. Systematic knowledge management in hydraulic systems: II, application to hydraulic systems. *Journal of Hydroinformatics*, **5**, 141–153.

Khatibi, R., Jackson, D., Curtin, J., Whitlow, C., Verwey, A. & Samuels, P. 2004. Vision statement on open architecture for hydraulic modelling software tools. *Journal of Hydroinformatics*, **6**, 57–74.

Knapen, R., Janssen, S., Roosenschoon, O., Verweij, P., De Winter, W., Uiterwijk, M. & Wien, J.-E. 2013. Evaluating OpenMI as a model integration platform across disciplines. *Environmental Modelling & Software*, **39**, 274–282, https://doi.org/10.1016/j.envsoft.2012.06.011

Kuhn, T. S. 1962. *The Structure of Scientific Revolution*. 3rd edn. University of Chicago Press, Chicago.

Lakhankar, T., Fekete, B. M. & Vörösmarty, C. J. 2008. Semantic web infrastructure supporting NextFrAMES modelling platform. *American Geophyiscal Union Fall Meeting 2008*, San Francisco.

Leavesley, G. H., Markstrom, S. L., Brewer, M. S. & Viger, R. J. 1996. The modular modelling system (MMS): the physical process modelling component of a database-centred decision support system for water and power management. *Water, Air and Soil Pollution*, **90**, 303–311.

Lu, B. & Piasecki, M. 2012. Community modeling systems: classification and relevance to hydrologic modelling. *Journal of Hydroinformatics*, **14**, 840–856, https://doi.org/10.2166/hydro.2012.060

Makropoulos, C., Safiolea, E., Baki, S., Douka, E., Stamou, A. & Mimikou, M. 2010. An integrated, multi-modelling approach for the assessment of water quality: lessons from the Pinios River case in Greece. *In*: Swayne, D. A., Yang, W., Voinov, A. A., Rizzoli, A. & Filatova, T. (eds) *IEMSS 5th International Congress on Environmental Modelling and Software*, 5–8 July 2010, Ottawa, Canada, International Environmental Modelling and Software Society, Manno, Switzerland, 1023–1032.

Maxwell, T. & Costanza, R. 1997. A language for modular spatio-temporal simulation. *Ecological Modelling*, **103**, 105–114.

MET Office 2013. The great storm of 1987. http://www.metoffice.gov.uk/education/teens/case-studies/great-storm [accessed 4 November 2014]

Moore, R. V. & Tindall, C. I. 2005. An overview of the Open Modelling interface and environment (the OpenMI). *Environmental Science & Policy*, **8**, 279–286.

Nicholls, R. J., Bradbury, A. *et al.* 2012. iCOASST – integrating coastal sediment systems. *In*: Lynett, P. & Mckee Smith, J. (eds) *Proceedings of the 33rd International Conference on Coastal Engineering*, 1–6 July 2012, Santander, Spain. International Conference on Coastal Engineering, Los Angeles, 3685–3699.

Nielsen, M. 2011. *Reinventing Discovery: The New Era of Networked Science*. Princeton University Press, Princeton, NJ, USA.

Nonaka, I. 1998. The knowledge-creating company. *In*: Drucker, P. F. (ed.) *Harvard Business Review on Business Management*, Harvard College, Cambridge, MA, USA, 21–45.

OECD 2004. *Science, Technology and Innovation for the 21st Century. Meeting of the OECD Committee for Scientific and Technological Policy at Ministerial Level, 29–30 January 2004, Final Communique*. Organisation for Economic Co-operation and Development, Paris.

Office of Science and Technology 2004. Foresight Future Flooding. Office of Science and Technology, London.

OpenMI Association 2010*a*. OpenMI Standard 2 Specification for the OpenMI (Version 2.0). Part of the OpenMI Document Series. http://www.openmi.org [accessed 22 November 2012].

OpenMI Association 2010*b*. OpenMI Standard 2 Reference for the OpenMI (Version 2.0). Part of the OpenMI Document Series. http://www.openmi.org [accessed 22 November 2012]

Pearce, G., Millard, K. & Harper, A. 2010. Application of OpenMI interfacing to promote integrated modelling across the water/environment sector, the FluidEarth platform. *Proceedings of the 1st European IAHR Congress*, 4–6 May 2012, Edinburgh, UK.

Prigogine, I. & Stengers, I. 1985. *Order out of Chaos: Man's Dialogue with Nature*. Harper Collins, London.

Reed, M., Cuddy, S. M. & Rizolli, A. E. 1999. A framework for modelling multiple resource management issues – an open modelling approach. *Environmental Modelling & Software*, **14**, 503–509.

San Roman Blanco, B. L., Sutherland, J. & Harper, A. 2012. Pyxis: a new workflow manager for integrated modelling. *In*: Hinkelmann, R., Nasermoaddeli, M. H., Liong, S. Y., Savic, D., Fröhle, P. & Daemrich, K. F. (eds) *Understanding Changing Climate and Environment and Finding Solutions, Proceedings of 10th International Conference on Hydroinformatics, HIC 2012*, 14–18 July 2012, Hamburg, Germany.

Safiolea, E., Baki, S. *et al.* 2011. Integrated modelling for river basin management planning, Proceedings of the ICE. *Water Management*, **164**, 405–419.

Saltelli, A., Ratto, M. *et al.* 2008. *Global Sensitivity Analysis. The Primer*. John Wiley & Sons, Chichester, UK.

Shrestha, N. K., Leta, O. T. *et al.* 2012. Integrated modelling of river Zenne using OpenMI. In: Hinkelmann, R., Nasermoaddeli, M. H., Liong, S. Y., Savic, D., Fröhle, P. & Daemrich, K. F. (eds) Understanding Changing Climate and Environment and Finding Solutions, *Proceedings of 10th International Conference on Hydroinformatics, HIC 2012*, 14–18 July 2012, Hamburg, Germany.

Sutherland, J., Peet, A. H. & Soulsby, R. L. 2004. Evaluating the performance of morphodynamic modelling. *Coastal Engineering*, **51**, 917–939.

Tuckman, B. W. 1965. Development sequence in small groups. *Psychological Bulletin*, **63**, 383–399.

Valke, S., Guilvardi, E. & Larsson, C. 2006. PRISM and EMES: a European approach to earth system modelling. *Concurrency and Computation – Practice and Experience*, **18**, 247–262.

Van Der Wal, T. & Van Elswijk, M. J. B. 2000. Generic framework for hydro-environmental modelling. *Proceedings of the 4th International Conference on Hydroinformatics*, 23–27 July 2000, Cedar Rapids, IA, USA. IWA Publishing, London.

Voinov, A. & Cerco, C. 2010. Model integration and the role of data. *Environmental Modelling & Software*, **25**, 965–969.

Von Bertalanffy, L. 1951. General systems theory – a new approach to unity of science. *Human Biology*, **23**, 303–361.

Whelan, G. & Nicholson, T. J. (ed.) 2002. Proceedings of the Environmental Software Systems Compatibility and Linkage Workshop. NUREG/CP-0177, PNNL-13654. US Nuclear Regulatory Commission, Rockville, MD, USA. http://www.nrc.gov/reading-rm/doc-collections/nuregs/conference/cp0177/ [accessed 31 May 2013].

Whitehouse, R., Balson, P. *et al.* 2009. *Characterisation and prediction of large scale long-term change of coastal geomorphological behaviours: final science report*. Science Report SC060074/SR1. Environment Agency, Bristol, UK.

Splicing recharge and groundwater flow models in the Environment Agency National Groundwater Modelling System

ROLF FARRELL[1]*, MARCEL VERVERS[2], PAUL DAVISON[3], PAUL HOWLETT[4] & MARK WHITEMAN[5]

[1]*Environment Agency, Lateral House, 8 City Walk, Leeds LS11 9AT, UK*

[2]*Deltares, Rotterdamseweg 185, 2629 HD Delft, Postbus 177, 2600 MH Delft, The Netherlands*

[3]*AMEC Foster Wheeler, Canon Court, Abbey Lawn, Abbey Foregate, Shrewsbury SY2 5DE, UK*

[4]*Haskoning DHV UK Ltd, 36 Park Row, Leeds LS1 5JL, UK*

[5]*Coverdale House, Aviator Court, Amy Johnson Way, Clifton Moor, York YO30 4GZ, UK*

**Correspondence: rolf.farrell@environment-agency.gov.uk*

Abstract: This paper explains the background and development of the Environment Agency National Groundwater Modelling System (NGMS) to integrate recharge functionality with the existing groundwater modelling (Modflow functionality).

The Environment Agency groundwater models were originally developed primarily as a tool for making high-level strategic decisions but their use for short-term extreme event scenarios, such as drought or flood, has been relatively limited. This functionality has been constrained by the format of the rainfall/recharge input datasets. Undertaking scenario runs based on change in climate and weather has only been possible by direct manual alteration of those input datasets, which is not always practical on a day-to-day basis.

Full implementation of recharge models into NGMS changes this, allowing the recharge models to be run in the NGMS environment and output to be generated. The fusion aspect of the process involves that output being processed in such a way that it can be used by NGMS as input into a Modflow scenario run. This process is explained using the example of a recent drought scenario in the Wessex Basin groundwater model.

The Environment Agency is the environmental regulator for England. A wide variety of models are used by the Environment Agency for the specific purpose of making a broad range of environmental decisions: for example, relating to hydrology, air, land contamination, and groundwater resources and quality.

Technical specialists within the Environment Agency have historically developed models relating to their own specialism, and ownership of models tends to belong within these functions. As such, the content of the models is orientated towards the function that developed the models. This leads to a legitimate criticism that models may not fully contain the overall environmental knowledge that the Environment Agency holds; river flow models may simplify groundwater contributions, and groundwater models may calibrate poorly against surface water flow. At present, these deficiencies are acceptable on the basis that the existing models are fit for purpose and reasonable in cost, which are the over-riding priorities of the Environment Agency (Whiteman *et al.* 2012*a*), but we should be aware that technological development may change the balance.

This implies that more fully integrated models may be a positive future to aim for and model handling within the Environment Agency is changing in ways that will make this more attainable. The Data, Modelling, Mapping and Information team (DMMI) of the Environment Agency has developed a centralized modelling platform (CMP) to which most Agency models have either been migrated or will be moved onto as funding and scheduling allow. This has been developed for a number of reasons. For example, it is not possible to place models onto the Environment Agency internal computer network without expensive and time-consuming user-acceptance testing: this has meant that there has been a tendency for them to be stored on

From: Riddick, A. T., Kessler, H. & Giles, J. R. A. (eds) 2017. *Integrated Environmental Modelling to Solve Real World Problems: Methods, Vision and Challenges*. Geological Society, London, Special Publications, **408**, 55–69.
First published online September 30, 2016, https://doi.org/10.1144/SP408.14

standalone computers, which clearly poses a risk as these are not automatically backed up. Another benefit concerns hardware: Environment Agency staff use laptop and desktop PCs of differing age and performance, as well as Ultra Thin Clients. This does mean that different users may expect different performance, particularly where non-standard applications, such as models, are concerned. Central hosting of models effectively means that performance is dependent on the servers, ensuring a more consistent performance across a substantial network.

Whilst the above are examples of clear and immediate benefits of central hosting of models, it also means that integration of models is theoretically simpler both in terms of the physical location of the models and a move towards standardization of model formats. An example of this is the National Groundwater Modelling System (NGMS: Whiteman *et al.* 2012*b*), which uses the Delft-FEWS software published by Deltares (Gijsbers *et al.* 2008) – this also provides a platform for the Environment Agency National Flood Forecasting System (NFFS). NFFS takes rainfall and river flow inputs (e.g. observed, forecast, radar, physically monitored), and passes that data through rainfall runoff, hydraulic and tidal models to produce forecasts of river and tidal flow and level. Input data are processed automatically for the NFFS models, unlike that for NGMS which needs to be heavily processed before use. However, the presence of rainfall input files (either rain gauge data as already used in the groundwater models or, more interestingly, rainfall radar data) in NFFS means that the same data could be processed through NGMS without incurring excessive development costs – whether that is desirable in practical terms rather than purely theoretically has yet to be tested.

A further obstruction to model fusion is the effectiveness of the existing software and methodologies. The construction of recharge models (Heathcote *et al.* 2004; Quinn *et al.* 2012), which provide input to separate groundwater models built using Modflow (Harbaugh & McDonald 1996), is well tested and minimizes the risk that the completed model is not fit for purpose. A standardized approach has been used for many years in the Environment Agency's groundwater modelling programme (Shepley *et al.* 2012), using the USGS Modflow software and 4R recharge models.

The experience of having built many regional-scale groundwater models across the country has shown that the traditional techniques can result in very effective and fit for purpose models with risks minimized. The models and the monitoring networks used to calibrate them are well balanced, and the benefits of risking limited funds trying different novel approaches are uncertain.

On this basis, the decision to move to an unfamiliar and more complex code on which to model a system would add significant risk to the project without delivering a guaranteed sufficient benefit to justify that risk.

This paper demonstrates how the integration of 4R and Modflow in NGMS enables operational hydrogeologists, as opposed to groundwater modellers, to be provided with a predictive scenario tool that has a great deal more flexibility and user friendliness than they have otherwise had access to. In an organization where time pressures are high, models need to be capable of generating meaningful output quickly and easily.

Methods and tools

Groundwater modelling

Modelling in the Environment Agency is undertaken on a hierarchical basis according to the nature of the problem to be assessed and its impacts (Farrell & Whiteman 2008; Whiteman *et al.* 2012*b*). Simple issues can be conceptualized and tested using simple analytical models such as IGARF (Environment Agency 2001), with greater investment and more complex models justified, for example, according to the scale of impact on the water environment (Fig. 1) (Whiteman *et al.* 2012*b*).

The most familiar numerical groundwater modelling tool in the Environment Agency remains Modflow. It processes recharge data (see below) to model throughflow, groundwater baseflow, spring discharge, surface water flow and more. The models built tend to represent spatially significant geographical areas – in fact, the Environment Agency has developed (or is in the process of developing) models that cover virtually the entire extent of the

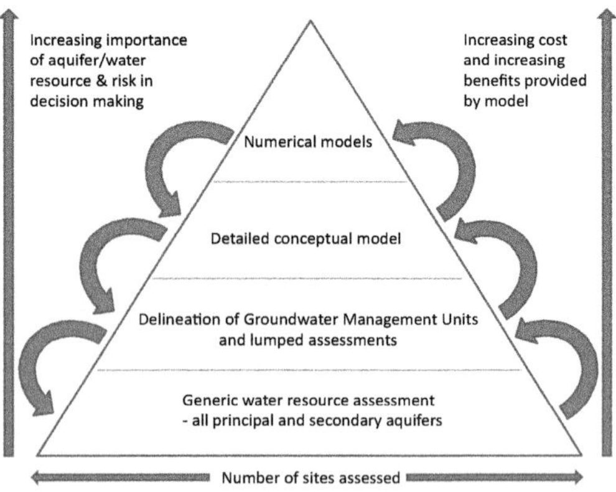

Fig. 1. Tiered approach to modelling for groundwater-resource assessment.

Cretaceous Chalk and much of the Triassic Sandstone principal aquifers in England. These models are split where operationally or hydrogeologically convenient and can be reasonably described as catchment-scale water-resource models (Fig. 2).

Modflow is a finite-difference Fortran code. Input data are in the form of arrays of parameter values defined in space and time at various resolutions. In the Environment Agency models, the former are commonly 200 m grids and the latter typically one–three time steps per month. This resolution is a reflection of the scale of spatial variation of aquifer properties and data available with which to parameterize the models, and the temporal variation of external stresses such as abstraction. The aim is to produce output of a resolution that is realistic and in proportion to the knowledge contained within the model.

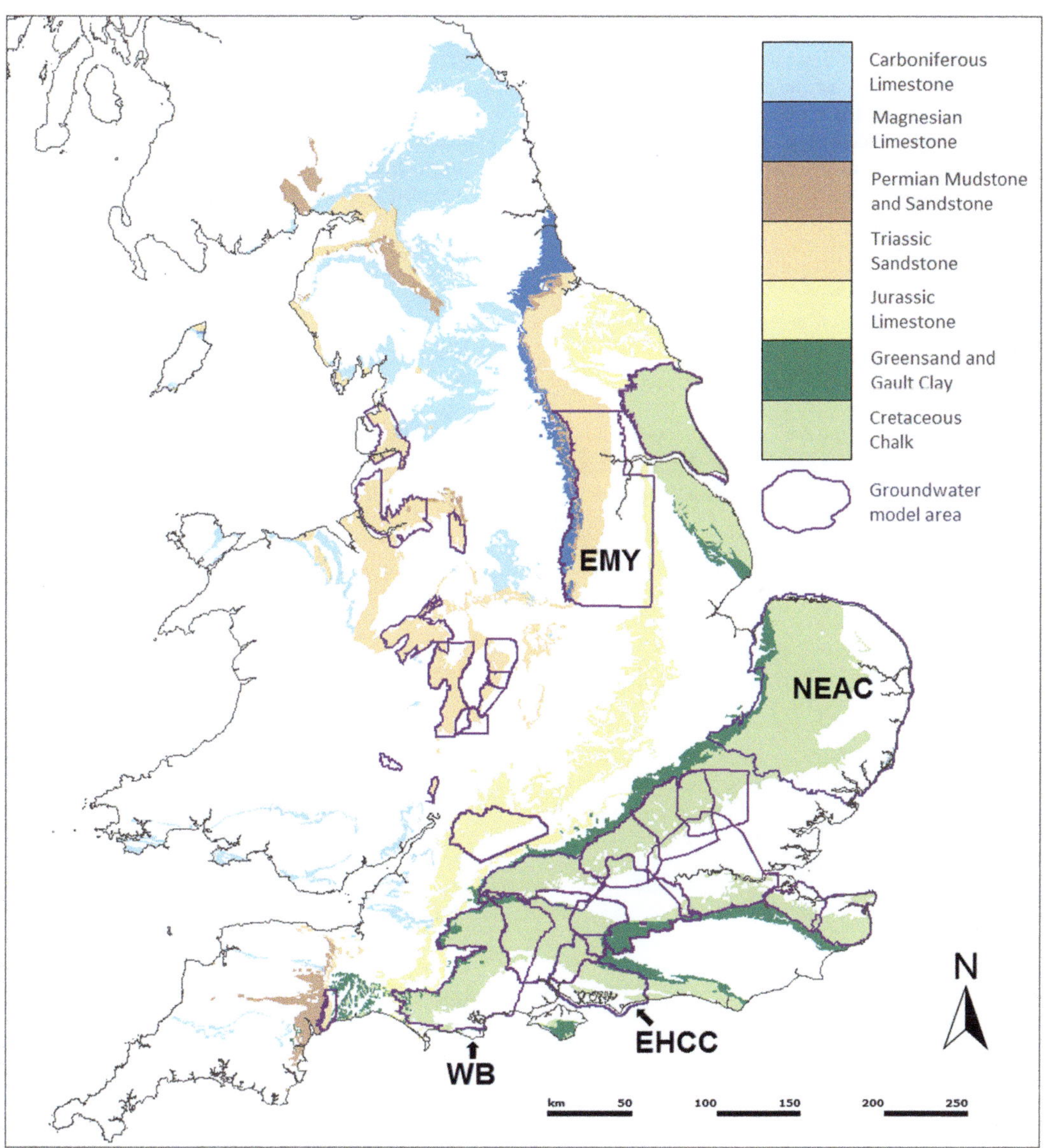

Fig. 2. Map of Environment Agency regional groundwater models on NGMS overlain onto principal aquifers. EMY, East Midlands Yorkshire; NEAC, Northern East Anglian Chalk; EHCC, East Hampshire and Chichester; WB, Wessex Basin.

For example, the East Midlands Yorkshire sandstone groundwater model (EMY in Fig. 2) covers an area of approximately 6171 km^2, and is calibrated against 97 observation boreholes and 15 river gauging stations. Grid spacing is 200 m. In contrast, the East Hampshire and Chichester chalk groundwater model (EHCC: Fig. 2) covers only about 1760 km^2, and has 159 observation boreholes and 34 gauging stations with a 250 m grid spacing (Fig. 3). These discrepancies reflect differences between the high-storage, slow-reacting sandstone catchments and the lower-storage, faster-reacting chalk catchments, but are usually very roughly consistent between models of given aquifer types. Refining the grid can be accomplished relatively easily but, unless the number of observation points is increased accordingly, a finer grid gives an impression of a greater resolution than is meaningfully the case.

How the groundwater models are used in the Environment Agency

The groundwater models are hosted on NGMS. This enables the models to be used widely without the risk of the models being inadvertently altered by the end user.

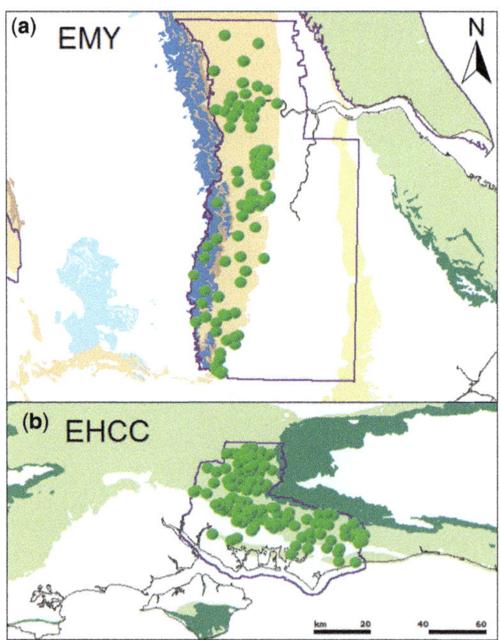

Fig. 3. NGMS screenshots. Maps of (**a**) the East Midlands Yorkshire sandstone model and (**b**) the East Hampshire Chichester Chalk model with principal aquifers both at the same scale showing differing observation borehole densities.

NGMS makes available to the user a number of standard default scenarios:

- The Naturalized Scenario – this is a scenario in which all abstractions and discharges are removed. It is not a true 'natural' scenario in that no change is made to land-use or riverbed conductances or any other non-abstraction-related anthropogenic impact. What it represents is the water environment minus the impact of the licensing system that the Environment Agency regulates.
- The Historic Scenario – this is the standard scenario that represents what actually happened and is the scenario used to calibrate the model.
- The Fully Licensed Scenario – most abstraction licences allow more abstraction than actually takes place (i.e. that modelled in the Historic Scenario). The Fully Licensed Scenario represents what might occur if all abstractions operated at their maximum allowable volumes (they commonly do not abstract total licensed volumes as headroom is required for operational flexibility). This scenario is used as part of the process for determining catchment licensing strategies: that is, what the Environment Agency's response would be to requests for new or changed licences;
- Recent Actual Scenario – this is a scenario that applies the average abstraction regime for the last few (typically 5 or 6) years to the entire model run. It is more commonly used in high-storage aquifers where significant changes to the abstraction regime in a catchment take a long time before the impacts reach a steady state (Shepley & Taylor 2003). By providing a more steady abstraction regime, it is easier to determine what the impacts of new changed existing abstractions would be. This scenario is also increasingly used to help determine future licensing policy;
- The 'What-if' scenarios – The default scenarios are supplemented by user-defined 'what-if' scenarios. Users are able to define changes to the existing groundwater-abstraction regime (e.g. changes to abstraction rates of existing sources, the addition of new boreholes or a combination of the two) to find out what impact such changes might have on water resources – and, particularly, on river flows.

New functionality recently implemented enables users to run sub-models where appropriate. This means that, rather than running the whole model (which can be very large: e.g. see the North East Anglian Chalk (NEAC) model in Fig. 2), only the parts of a model likely to be affected by a specific change to the abstraction need to be run. Given that some models take over a day for a scenario to run, this makes routine use of such models much more manageable on a day-to-day basis. Also available are water balance outputs for user-defined areas using USGS Zonebudget and Flowsource – a tool used to assess where the water abstracted by each borehole comes from (Foley & Black 2013).

Each of these 'what-if' interventions can be run individually or in combination with each other, thus providing a very powerful assessment tool.

What the system does not do is allow the model parameters to be changed; NGMS is a tool that enables the user to interrogate the model to aid in making operational water-management decisions – not to respond to developmental changes to the conceptualization of the system.

Recharge modelling

Groundwater model input datasets include recharge, groundwater abstractions, river-flow data, aquifer properties and various boundary conditions to develop a water table that should calibrate to observed river flow and groundwater level data to an agreed level of accuracy. The process of creating the recharge dataset is dependent on the development of a separate and distinct model from the groundwater model: the recharge model. Quinn *et al.* (2012) reviewed the recharge models currently in use in the UK.

Recharge models take rainfall and reference evapotranspiration (ET) input datasets and carry out soil moisture balance calculations to predict potential recharge to groundwater, taking into account the effects of surface-water abstractions and discharges, land use, and geology. The output from the recharge models provides one of the primary inputs into groundwater models. Normally, scenarios are developed representing historical, naturalized and fully licensed conditions (matching the default groundwater model scenarios), each of which incorporates different scenarios of surface-water abstraction and discharge, but all using the same historical rainfall and ET records.

In recent times, an increasing focus has developed on consideration of the recharge scenarios for the purpose of assessing the impacts of extreme events (i.e. drought and, more controversially, flood), as well as the topical subject of climate change. This is becoming an operational requirement, as the UK Government expects the Environment Agency to provide meaningful drought and climate-change predictions (HR Wallingford 2012; Calder & Faherty 2014; Soley *et al.* 2015).

This then highlights a significant gap in the capabilities of our modelling systems to deliver required outcomes, in that available model functionality has not kept up with the changing needs of the business. NGMS enables operational staff to gain access to the groundwater models to assist in strategic water-resource management decisions but they are still unable to alter the rainfall or evapotranspiration inputs into the model to support, for example, the management of extreme events.

The natural solution to this situation is the development of Delft-FEWS functionality to undertake 'what-if' scenarios on the rainfall data (i.e. fusing recharge models and NGMS). This was not a straightforward process, as the recharge and Modflow models tend to use data in different ways.

How NGMS Modflow works

The process of developing the model is iterative. The initial conceptualization based on examination of the observation data, understanding of the processes and initial water balances is developed into an initial Modflow model. Output is analysed and performance assessed at each of the observation locations. Where calibration fails to achieve target levels of accuracy, the conceptual model is reviewed and any revisions incorporated back into the model. The process of developing the model can be aided by using commercial pre- and post-processing tools such as Groundwater Vistas or GMS, which enable the parameters to be manipulated using a graphical user interface (GUI). These programs allow an easy assessment of model performance: for example, by the graphical display of residuals between modelled and observed data.

NGMS takes the input datasets, in the form of simple arrays, and using them runs the Modflow executable (Gijsbers *et al.* 2008). The input files consist of header information followed by the data: for example, geographical arrays of aquifer parameters or time-series data given for each location and time step. The output files are produced by Modflow and then post-processed and written to a central database, and the raw output files are then purged; this is because the output files can total tens of gigabytes of data.

Once written to the database, the data can only easily be viewed through NGMS. However, it is possible to edit an NGMS configuration file so that the output data are not purged, enabling the raw output to be viewed and compared with the output from the non-NGMS version of the model.

The Delft-FEWS software extracts the data from the database and displays output in a variety of different ways. Delft-FEWS displays are dependent on how a particular version of the software has been configured, and there is a great deal of flexibility in terms of how it can be developed. Within NGMS, data display formats include time-series plots, river flow accretion profiles, spatial plots of river flow data, groundwater level and flow data, and differences between different scenarios (Whiteman *et al.* 2012*b*). Data can be displayed spatially and temporally, and the data can be output easily in the form of spreadsheet data, graphical images and animations.

How NGMS recharge works

The recharge models take information such as distributed rainfall grids, evapotranspiration data, unsaturated zone properties and information from land-use maps. The spatial resolution of these grids tends to be coarser than the Modflow grids but the temporal resolution is much higher (e.g. daily rather than weekly). Also, different rainfall datasets can be used. Whilst commonly rainfall data are based on observed data from rain gauges, the spatial coarseness of their distribution means that they can miss highly intensive but localized rainfall events that can be shown to have an impact on water resources (Quinn *et al.* 2012). These events can be picked up by rainfall radar systems but here, although the spatial detail may be greater, the absolute accuracy can be prone to significant errors. The combination of the absolute values of rainfall derived from the Environment Agency's rain-gauge network and the relative distribution of rainfall derived from radar observation does, however, provide a potentially significant improvement in our ability to model recharge.

These issues have yet to be resolved, but it highlights the fact that there is considerable potential to be had in experimentation with recharge models which, until now, has been difficult to undertake operationally within the Environment Agency.

Configuration of the recharge models into NGMS is similar to that for Modflow, whereby a command within Delft-FEWS instructs the recharge model to be run, output is produced and purged after being post-processed into NGMS. Output is relatively simple – spatial data show the distribution of recharge across the model area, and time-series plots of each active cell can be viewed and exported. Difference plots show the variation between scenarios.

Fusing Modflow and 4R recharge

The processes are summed up in Figure 4. The flowchart (4R and Modflow: Fig. 4) illustrates the relatively complex process of generating recharge model files, running them through the model, producing the output, adding them to the input files required for Modflow, running Modflow and then post-processing the output. In its original format, NGMS holds formatted Modflow input files and the whole process of running the Modflow model is contained within NGMS, with the recharge model process unchanged (4R and NGMS Modflow: Fig. 4).

The revised NGMS, with the recharge model, can handle the whole process from start to finish (NGMS 4R and NGMS Modflow: Fig. 4). The configuration of the recharge models to NGMS (i.e. the development of adapters to enable NGMS to communicate with the recharge model and the software configuration enhancements to allow the new tasks to be undertaken) is relatively straightforward – reflecting that the recharge models themselves are comparatively simple compared to the Modflow models. The difficult step is the one that connects the output of the recharge model with the input to the groundwater model.

Normally, the output from a recharge model is processed to form a fixed input to Modflow. Therefore, theoretically, all that needs to be done to run a recharge scenario is to apply a change to the recharge model output and to use that to replace the Modflow recharge input file in NGMS.

The rainfall and evaporation data used to feed the 4R recharge model (Heathcote *et al.* 2004) can be chosen from various sources (as described above). The first processing steps to get from this rainfall and evaporation gridded input data to the same spatial resolution used in the recharge model are to spatially interpolate the values to a new grid and to cut out the area of interest (to avoid unnecessary data storage). The data interpolation is done once during the 4R model build, to provide the default rainfall and evapotranspiration data that are used in the default recharge scenarios. In NGMS, these data are used as the basis for a modified time series of rainfall and ET that represent what might happen in a drought year or under climate change.

Land use in Environment Agency recharge scenarios currently does not change over time. The option to run recharge models with different inputs for land use is dealt with differently. Since this input is fixed over time, it is possible to choose an alternative dataset to be used by the recharge model. This dataset will have to replace the default dataset used.

Allowing users of the models to adjust rainfall and evaporation series to create climate scenarios is a prerequisite of using recharge models in combination with groundwater models. To allow users to create recharge scenarios, it is necessary to be able to change the default rainfall or evaporation data. This could be achieved through the application of a fixed multiplication factor, or a different multiplication factor for each month or each time step, or the selection of historical years of rainfall data to create rainfall series for the future based on historical rainfall series and multiplication

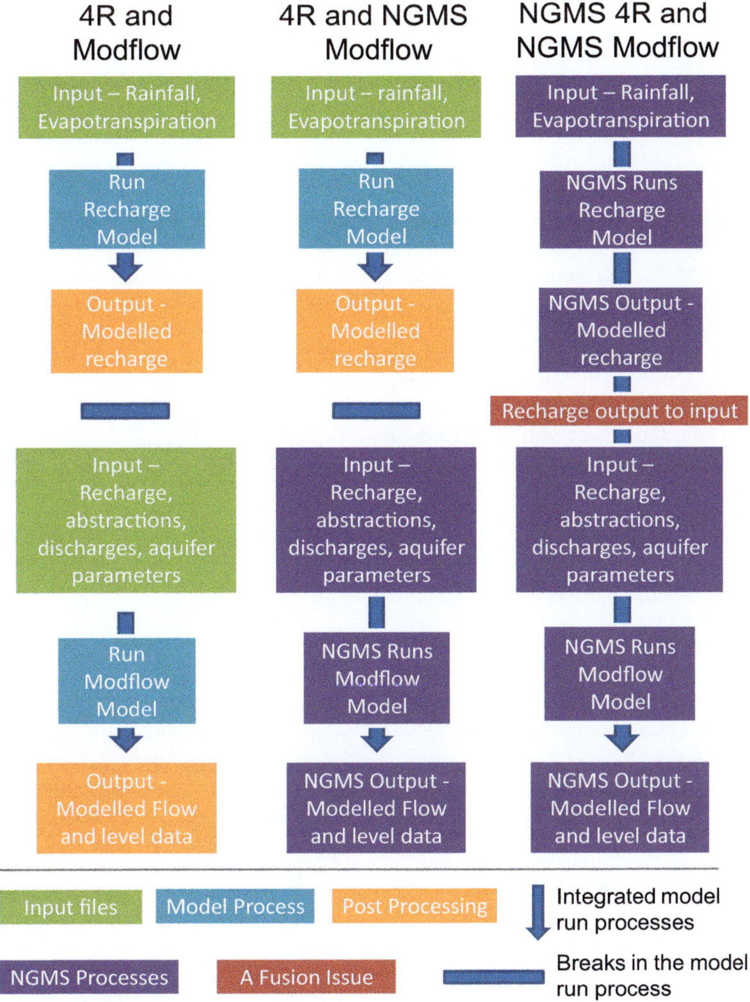

Fig. 4. Processes in running recharge and groundwater models outside of NGMS (left), partly within NGMS (centre) and fully within NGMS (right).

factors. The substitution of historical rainfall series and corresponding ET series is particularly useful as it can be used to represent plausible weather patterns.

Once the recharge model has produced its output (typically on a daily time step), this is used as input for the Modflow model. The resulting recharge grids are converted by 4R to Modflow input format and replace the default recharge input files.

What you get

The result is the ability to run rainfall scenarios to feed into the corresponding NGMS Modflow and to produce appropriate output.

The process is as follows. First, a scenario rainfall sequence is selected, typically a short-term scenario (several months to 1–2 years to assess potential summer droughts).

The rainfall sequence can be based on an entirely new scenario or factored from existing patterns: for example, based on factors of long average term recharge (LTA). A typical technique is to identify individual months in the record that correspond to a proportion of the long-term average (e.g. 60, 80, 100, 120 and 140% LTA) and to build those into a monthly rainfall input. Another option would be to select a historical period as a scenario of concern: for example, the 1976 drought, which is commonly regarded as the most severe single season drought within the period normally covered by Environment Agency groundwater models and often used as a reference benchmark during drought events (Marsh & Cole 2006). In this case then, those months would be selected in the recharge scenario editor and the model run for each scenario.

Next, the abstraction regime is considered. During a drought, it is likely that some boreholes will be shut off, as abstraction may be restricted during periods of low flow. Therefore, a recent historical drought year is used to represent the abstraction during the forecast scenario period.

Finally, initial conditions for the scenario are chosen based on a period where groundwater levels and flows are similar to current conditions. As long as current water levels are within the extremes of the calibrated model run, then there will be periods in the past history of the model that can be used as a basis for the scenario.

So, the above enables a scenario to be set up whereby one historical period in the past is used as a basis to represent current groundwater levels and surface water flow, another period is used to represent the rainfall and PE, and another period again used to represent abstraction.

That scenario is then set up and processed through NGMS in a seamless single workflow, with the recharge scenario being run and the Modflow scenario following on.

This process is described in more detail below in the following section on 'The methodology'.

In the case of the percentage of the long-term average approach, the workflow would be run for each selected percentage of LTA.

To test the functionality, a real-life example was chosen as a comparison. In early 2012, low winter groundwater levels led to some concern that a groundwater drought could occur in the summer. In the end, very high spring rainfall solved this problem quickly but, before this occurred, some attempts had been made to run various rainfall/recharge scenarios through the Environment Agency's Wessex Basin groundwater model (WB: Fig. 2) and to determine likely outcomes. The NGMS/4R configuration was used to replicate this work and to consider whether this approach improved our ability to manage drought events.

The methodology

The approach taken in 2012 was first to identify a period in the historical record where groundwater levels were similar to those at the start point of the scenario run. In this case, the groundwater levels from July 1984 were used to represent groundwater levels in April 2012 For rainfall/recharge, the data from 1976 were used (for the reasons outlined in the previous section) and factors applied to that to give a range of scenarios. Finally, the abstractions from 1995 were applied. 1995 represents the last significant groundwater drought in the UK, prior to 2012, and therefore the most recent period where something approaching the current abstraction regime was affected by drought.

The model was run for 21 months from July 1984 – this effectively represented a 21 month forecast from April 2012 to December 2013. Figure 5 shows how the different periods of different datasets were applied for the forecasts.

The above scenario is carried out in NGMS as follows:

(1) A scenario 4R run is undertaken using rainfall and PE data from 1976 applied to April 1984–December 1985;
(2) On completion of the 4R run, a Modflow scenario run is started. This takes the 1995 abstraction data for all abstractions and applies it to April 1984–December 1985.

The result is the same as the default scenarios but, once the time step representing 1 July 1984 is reached, the rainfall/recharge and PE from April 1976 and onwards is applied along with groundwater abstractions for April 1995 onwards (repeating again until the 21 months are complete). Thus, effectively, April–July happens twice in the scenario run – this is not a problem as Modflow does not have any date awareness. Seasons/years are effectively defined by the input data and therefore change as the input data are changed.

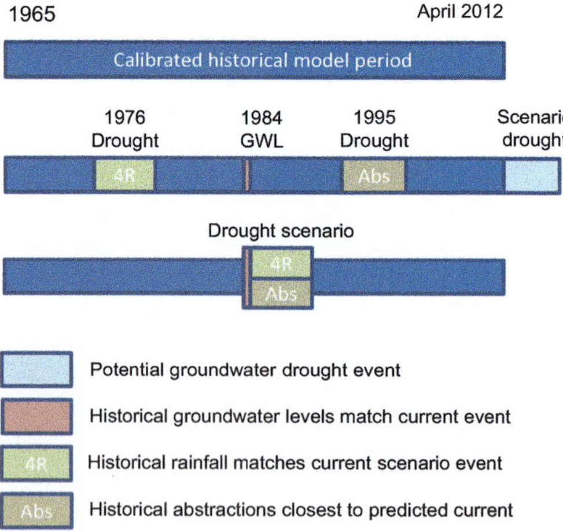

Fig. 5. Components of the Wessex Basin drought scenario indicating which historical years were chosen to represent recharge, groundwater level and abstractions in the scenario. 4R, recharge (rainfall, evaporation); GWL, groundwater levels; Abs, abstractions.

After the modified months complete, the scenario run continues to the normal end point with the same input data as used in the default scenario, and the output from July 1984 is used as the 'forecast' from April 2012.

Figure 6 shows the change in recharge during the scenario drought phase.

In practice, to make this scenario work, the following steps were taken:

(1) A new workflow file (xml file) was created in the NGMS configuration to enable the modified 4R scenario to run, followed by the modified historical Modflow scenario.

(2) Modifications were made to some 4R data input files to enable selection of the rainfall and PE data. Selection of that data depended on editing the input data files (by running a utility to substitute the required rainfall months/years, and editing the PE input data) rather than using the existing scenario functionality within NGMS. This functionality has now been implemented into NGMS and the process will be simpler in future. Although this process is, in fact, fairly simple for the experienced modeller, the ethos of NGMS is that it should be straightforward enough to be used by any competent hydrogeologist – editing and uploading configuration and data files adds complication and risk.

(3) The abstractions are a more complicated issue that has been addressed with newly implemented NGMS scenario functionality. This enables, for example, factors to be easily applied to all abstractions in the model. It also allows an historical scenario to be set up for an abstraction: for example, by inserting a specific year of data into a different period of the model run (see Fig. 7).

Figure 8 shows the resulting scenario abstraction from the Belhuish abstraction. The 1995 historical

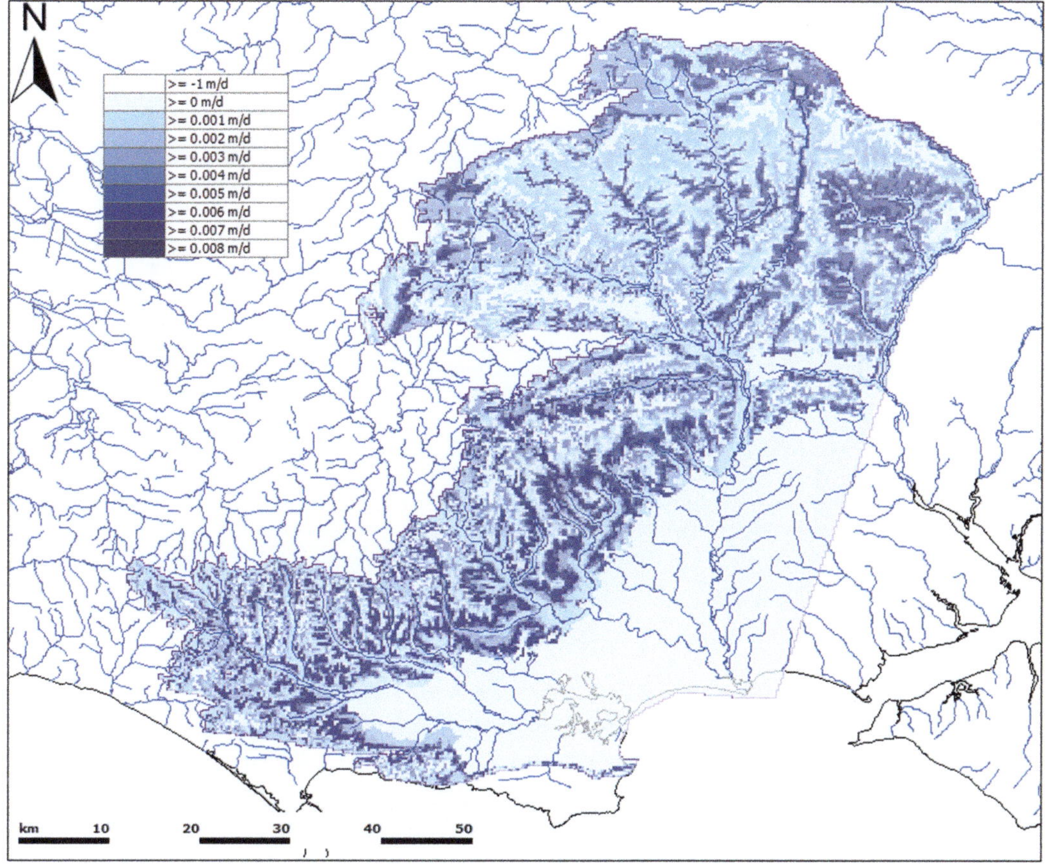

Fig. 6. NGMS screenshot: Wessex Basin model – difference in modelled recharge due the to 1975–76 historical recharge being applied to the 1984–85 historical recharge.

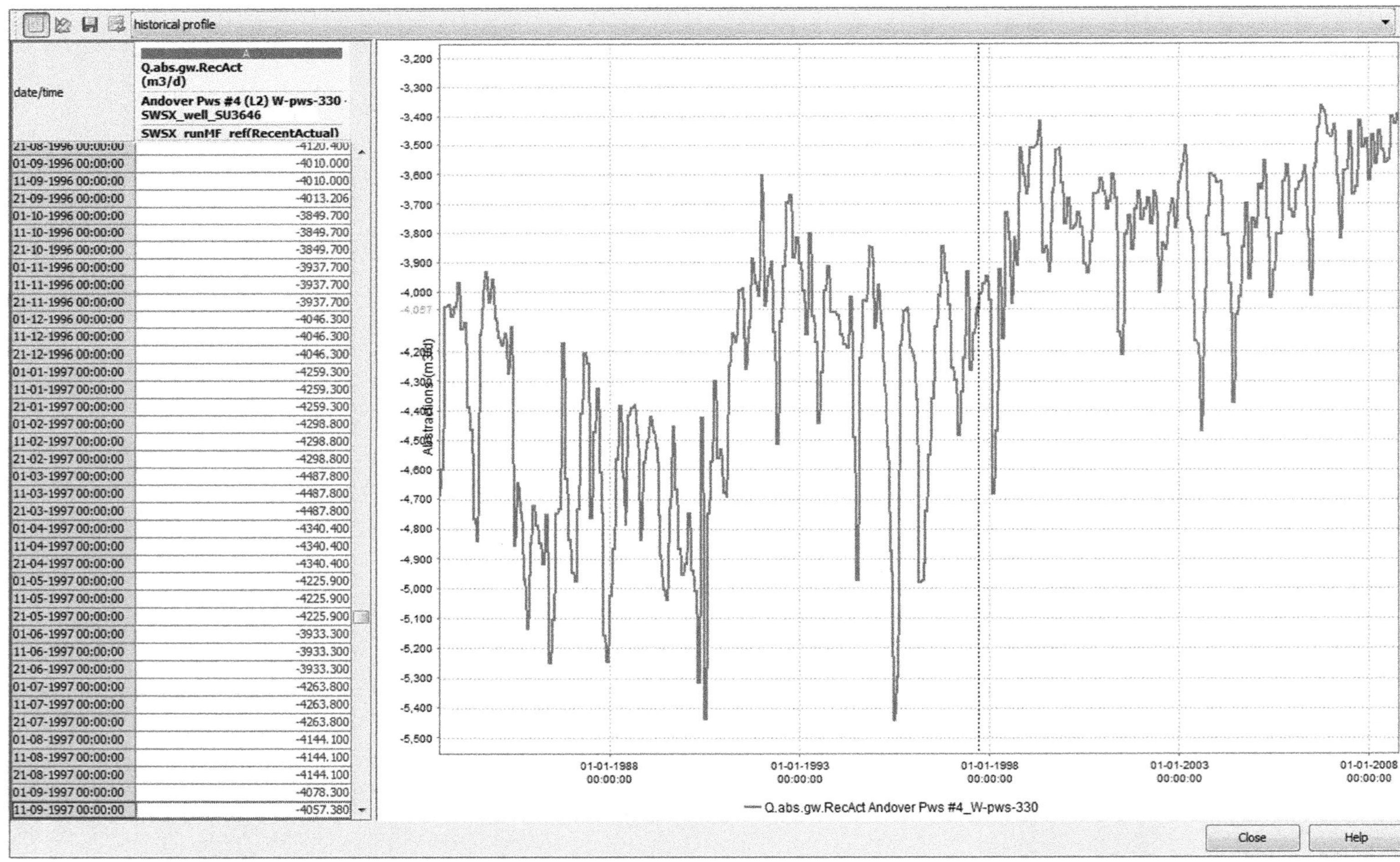

Fig. 7. NGMS screenshot: Scenario Editor – editing abstraction data for a single location using the tabular data editor, Wessex Basin model. m^3/d, m^3/day.

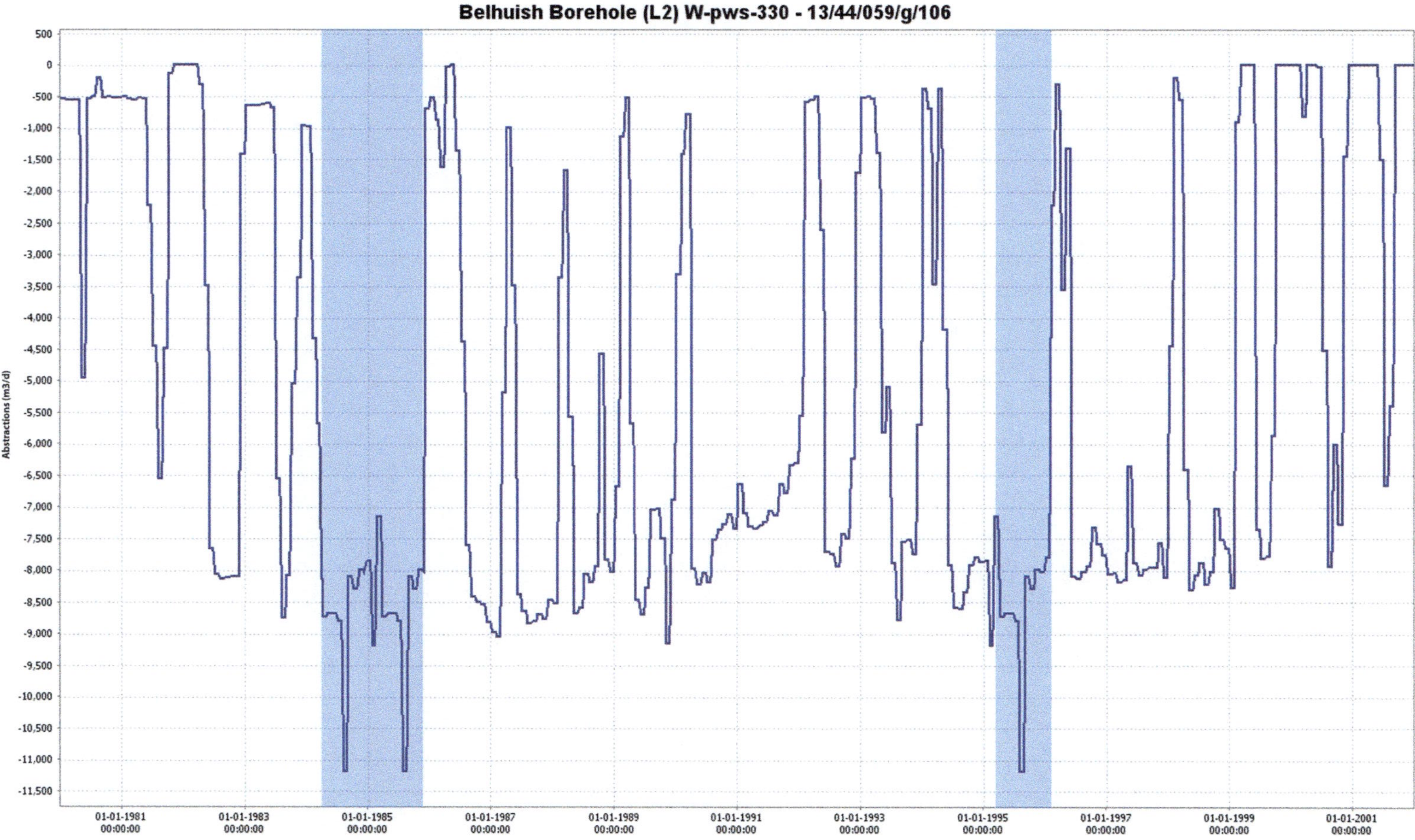

Fig. 8. NGMS screenshot: Modflow output showing the 1995 abstraction data applied to 1984–85. m^3/d, m^3/day.

abstraction can be seen to be repeated during 1984–85 (shown as shaded). The abstraction from 1995 is that from the most recent drought. 1984–85 is the period where the historical conditions match those of the current developing drought most closely. Therefore, the 1995 drought abstractions are applied to 1984–85 to get the most realistic mix of initial conditions and abstractions

This process illustrated a likely common problem with fusing models. In this case, the expectation was that NGMS would run the 4R recharge model and then combine the output of the 4R run with a Modflow scenario run that used the modified wel file. To run a default scenario, NGMS unzips all of the input files from a specific folder and applies them to Modflow. In theory, all that is needed is to edit the wel file and to run the scenario. Where this process initially failed is that the wel file is not used directly by NGMS – rather the input data are copied into a local datastore during the default scenario run and those data are then used directly during the default and 'what-if' scenario runs. The 'what-if' scenario runs do not use the actual wel file at all, so any modifications to it do not end up in the model output.

The solution was relatively simple if untidy: the default scenarios were re-run using the scenario recharge input files. This then activates the new wel file and places a copy of that into the local datastore that can be used by the 'what-if' scenario run. This served as a proof of concept that enabled the final development of the functionality to be justified.

Results

The final NGMS output (Fig. 9) is illustrated by the time series of river flows at the Amesbury gauging station.

The modified historical scenario (i.e. historical but using the modified wel file) is shown in blue, with the drought scenario shown in red. It can be seen that the scenarios only differ from late 1984 to late 1986 – the period representing the forecast period (April 2012–December 2013).

All of the above highlights that, whilst a solution has been found, the method of achieving it was initially not in keeping with the ethos of NGMS. The completion of the run still relies on some simple manual editing of input data files (for potential evapotranspiration and abstractions). The final configuration has a set of default scenarios and a wel file that relates only to the scenario run and cannot be used for routine use of the model. Therefore, a completely new standalone copy of the Wessex Basin NGMS is needed, and that has to be hosted on a drive that can handle over 100 Gb of data being created as part of the model run process.

Discussion

The purpose of this test is to demonstrate that the same scenario model runs can be undertaken using NGMS with 4R and Modflow as were used by operational staff to assess a potential extreme event. The test is not intended to duplicate all the scenarios undertaken nor to prove that specific output matches, but simply to show that the outputs from NGMS are capable of being used in the same way as those generated from the original method. The primary benefit of the NGMS approach would be:

- reduced time taken to undertake the scenario runs and to visualize the outputs;
- a reduced degree of expert knowledge required to access and process output.

The purpose of the above is, of course, to use the combined Modflow/4R NGMS to provide predictive output more quickly than would otherwise be achieved. At the time the above model runs were undertaken, the process was not fully implemented into NGMS and, as such, it was not really possible to quantify accurately how much time benefit was achieved with this process due to the number of manual steps currently required. The functionality has, at the time of writing, just been fully integrated into NGMS: that is, the wel file can be manipulated using the Scenario Editor recharge model functionality and used by the following Modflow scenario, and the PE file can be altered in a similar manner to the rainfall file. The next step is to repeat the process with the new functionality and to quantify the potential time savings.

General issues related to model fusion as a whole

This example is, perhaps, fairly simple but illustrates clear benefits of model fusion. Beyond this, the benefits within a regulatory context are less clear. For example, complex whole water-system models have a clear appeal; groundwater models in the UK are, strictly speaking, water-resource models covering the whole of the resource cycle (Rushton & Skinner 2012) – albeit simplistically in some areas. Developing sophistication (e.g. incorporating river flow models) has an obvious appeal, but has resource implications. The case for development of fused ('complex') models in a regulatory context will depend on the costs, benefits and environmental risks – the 'so-what' test referred to by Whiteman *et al.* (2012*a*). The ability to develop fused models is also likely to be limited by data availability (Quinn *et al.* 2012).

This example using Delft-FEWS illustrates the problems of combining three different applications (i.e. Delft-FEWS, Modflow and 4R recharge).

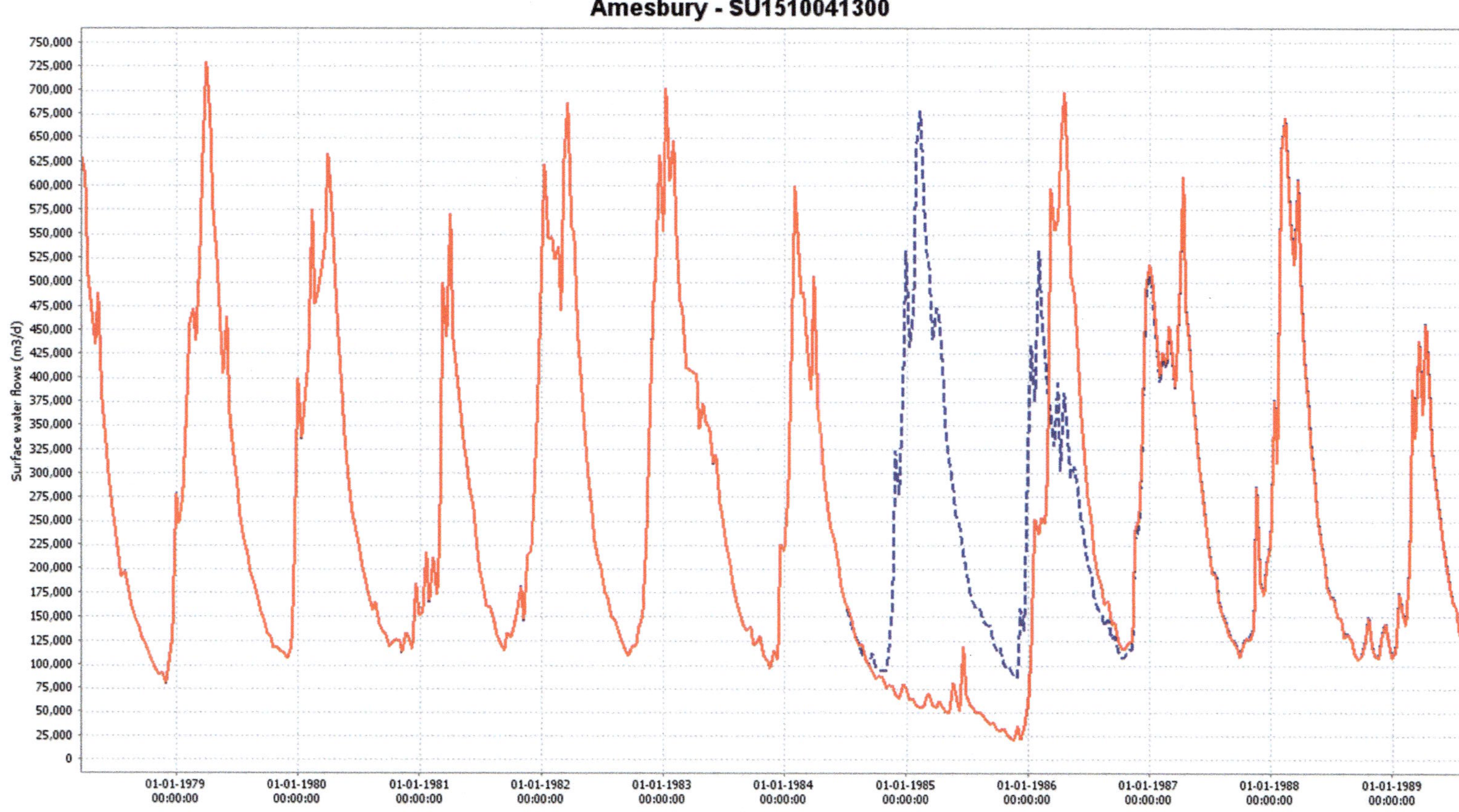

Fig. 9. NGMS screenshot: Amesbury gauging station: dashed line, original default scenario flow; solid line, 'what-if' drought scenario. m^3/d, m^3/day.

Solutions to applying Modflow to NGMS have often ended up being obstructions to the implementation of 4R to NGMS.

The future may be to focus on ownership of datasets and agreement on specific standards, such as OpenMI. The models can then remain distinct, but the use of common datasets would provide a greater degree of confidence in the models. This should lead naturally to further model fusion without the need to force-fit models in the manner described in this chapter.

Further developments

A key weakness of NGMS is its handling of surface-water abstractions. Whilst NGMS has always allowed users to undertake scenario runs on groundwater abstractions, the same functionality has not been available for surface-water abstractions. This can be particularly useful for assessing and modifying river augmentation schemes, where groundwater is discharged to surface water to support flows. The additional flow may cause changes in surface water–groundwater interaction that may not be easily assessed without modelling the discharge.

This limitation is caused by the nature of Modflow. The groundwater-abstraction scenario functionality works by making on-the-fly changes to the Modflow input wel file, which contains the abstraction information. There is no such equivalent Modflow file for surface-water abstractions that are included within the input str file. In this file, specific inflows and outflows (e.g. abstractions, discharges and lateral flow from runoff) are lumped together, so applying changes to surface-water abstractions is less straightforward. Where the surface-water abstractions and discharges are explicitly represented is in the recharge model. To enable the same flexibility for the user to make changes to surface-water abstractions and discharges, as is present for groundwater abstractions, would require NGMS to be able to make on-the-fly changes to the recharge model input file that contains this information. The user would then need to carry out a new recharge simulation, followed by a Modflow model run using the output from the modified recharge scenario.

In the longer term, consistent standards for hydrological datasets do offer the potential for the direct or indirect linkage of models. The feeding of water flux between rivers and groundwater could be a useful improvement to the performance of surface water models – particularly, for example, where flood forecasting is concerned.

The authors would like to express their thanks to the groundwater modelling technical specialists within the Environment Agency who have given their skills and enthusiasm to the development of the models and techniques for management and regulation of groundwater described in this paper. The views expressed in this paper are those of the authors and not necessarily the views of the Environment Agency.

References

CALDER, I. & FAHERTY, J. 2014. *Otter Valley Groundwater and River Flow Model: Future Flows Climate Change Scenarios.* AMEC report to the Environment Agency of England, Devon and Cornwall Area.

ENVIRONMENT AGENCY 2001. *Impact of Groundwater Abstractions on River Flows.* Environment Agency Project Report NC/00/28 Environment Agency.

FARRELL, R.P. & WHITEMAN, M.I. 2008. The role of groundwater models in integrated catchment management. *In*: SÀNCHEZ-MARRÈ, M., BÉJAR, J., COMAS, J., RIZZOLI, A. & GUARISO, G. (eds) *iEMSs 2008: International Congress on Environmental Modelling and Software. Integrating Sciences and Information Technology for Environmental Assessment and Decision Making. Proceedings of 4th Biennial Meeting of iEMSs.* International Environmental Modelling and Software Society, Barcelona, 631–638, http://www.iemss.org/iemss2008/index.php? n¼Main.Proceedings

FOLEY, C. & BLACK, A. 2013. Efficiently delineating volumetric capture areas and flow pathways using directed a cyclic graphs and MODFLOW; description of the algorithms within FlowSource. *In*: *MODFLOW and More 2013: Translating Science into Practice*, June 2–5, Golden, Colorado – Conference Proceedings, Maxwell, Hill, Zheng and Tonkin, 571–581, http://igwmc.mines.edu

GIJSBERS, P.J.A., WERNER, M.G.F. & SCHELLEKENS, J. 2008. Delft FEWS: A proven infrastructure to bring data, sensors and models together. *In*: SÀNCHEZ-MARRÈ, J., BÉJAR, M., COMAS, J., RIZZOLI, A.E. & GUARISO, G. (eds) *Proceedings of the iEMSs Fourth Biennial Meeting: International Congress on Environmental Modelling and Software, Barcelona, Catalonia, Volume 1.* International Environmental Modelling and Software Society (iEMSs), Manno, Switzerland, 28–36.

HARBAUGH, A.W. & MCDONALD, M.G. 1996. *User's Documentation for MODFLOW-96, an Update to the US Geological Survey Modular Finite-Difference Ground-Water Flow Model.* United States Geological Survey Open-File Report 96-485. United States Geological Survey, Reston, VA.

HEATHCOTE, J.A., LEWIS, R.T. & SOLEY, R.W.N. 2004. Rainfall routing to runoff and recharge for regional groundwater resource models. *Quarterly Journal of Engineering Geology and Hydrogeology*, **37**, 113–130.

HR WALLINGFORD 2012 *The UK Climate Change Risk Assessment 2012 Evidence Report.* Presented to Parliament pursuant to Section 56 of the Climate Change Act 2008. Defra Project Code GA0204. Department for Environment, Food and Rural Affairs (Defra), London.

MARSH, T.J. & COLE, G.A. 2006. *A Review of the GLA Drought Severity Assessments Presented at the Beckton Gateway WTW Public Enquiry.* CEH Project

Report Number C03095. Centre for Ecology and Hydrology, Wallingford, Oxfordshire.

Quinn, S.A., Liss, D., Johnson, D., Van Wonderen, J.J. & Power, T. 2012. Recharge estimation methodologies employed by the Environment Agency of England and Wales for the purposes of regional groundwater resource modelling. *In*: Shepley, M.G. & Whiteman, M.I., Hulme, P.J. & Grout, M.W. (eds) *Groundwater Resources Modelling: A Case Study from the UK*. Geological Society, London, Special Publications, **364**, 65–83, https://doi.org/10.1144/SP364.6

Rushton, K.R. & Skinner, A.C. 2012. A national approach to groundwater modelling: developing a programme and establishing technical standards. *In*: Shepley, M.G., Whiteman, M.I., Hulme, P.J. & Grout, M.W. (eds) *Groundwater Resources Modelling: A Case Study from the UK*. Geological Society, London, Special Publications, **364**, 7–17, https://doi.org/10.1144/SP364.2

Shepley, M.G. & Taylor, A. 2003. Exploration of aquifer management options using a groundwater model. *Water and Environment Journal*, **17**, 176–180.

Shepley, M.G., Whiteman, M.I., Hulme, P.J. & Grout, M.W. (eds). 2012. *Groundwater Resources Modelling: A Case Study from the UK*. Geological Society, London, Special Publications, **364**.

Soley, R., Sadowski, J., Faherty, J. & Cook, C. 2015. *Climate Change and Abstraction Reform Modelling of the Lincolnshire Chalk and Spilsby Sandstone Groundwater and River Flow Systems*. AMEC Foster Wheeler Report for the Environment Agency of England, Lincolnshire and Northamptonshire Area.

Whiteman, M.I., Seymour, K.J., van Wonderen, J.J., Maginness, C.H., Hulme, P.J., Grout, M.W. & Farrell, R.P. 2012*a*. Start, development and status of the regulator-led national groundwater resources modelling programme in England and Wales. *In*: Shepley, M.G., Whiteman, M.I., Hulme, P.J. & Grout, M.W. (eds) *Groundwater Resources Modelling: A Case Study from the UK*. Geological Society, London, Special Publications, **364**, 19–37, https://doi.org/10.1144/SP364.3

Whiteman, M.I., Maginness, C.H., Farrell,. R.P., Gijsbers,. P.J.A. & Ververs, M. 2012*b*. The national groundwater modelling system: providing wider access to groundwater models. *In*: Shepley, M.G., Whiteman, M.I., Hulme, P.J. & Grout, M.W. (eds) *Groundwater Resources Modelling: A Case Study from the UK*. Geological Society, London, Special Publications, **364**, 49–63, https://doi.org/10.1144/SP364.5

3D geological models to groundwater flow models: data integration between GSI3D and groundwater flow modelling software GMS and FeFlow®

A. H. PASANEN[1]* & J. S. OKKONEN[2]

[1]*Geological Survey of Finland, Eastern Finland Office, PO Box 1237, 70211 Kuopio, Finland*

[2]*Geological Survey of Finland, Western Finland Office, PO Box 97, 67101 Kokkola, Finland*

**Correspondence: antti.pasanen@gtk.fi*

Abstract: Data integration between different software is routinely needed in order to create suitable data formats or necessary data manipulation prior to importing the data. The procedures and workflows are not usually published. This paper presents the data integration between GSI3D (Geological Surveying and Investigation in 3 Dimensions) and groundwater flow modelling software GMS version 7.0 (Groundwater Modelling Systems) and FeFlow®. Geological models for two sites in Finland, an esker aquifer at Patamäki and a mine site in Luikonlahti, were constructed using dedicated 3D geological modelling software GSI3D.

The data from the GSI3D model in Patamäki was exported as ASCII grid files directly to GMS in order to delineate hydrogeological features prior to groundwater flow modelling. The data from the GSI3D model in Luikonlahti was first manipulated in ArcGIS to make it amenable in FeFlow®.

In both modelling locations, the detailed geological modelling greatly helped to discern different hydraulic conductivity zones that are based on different geological materials. This, in addition, eases the development of conceptual groundwater flow models, the model calibration process and potentially improves the simulation results.

Groundwater flow modelling is an important method of studying the movement of groundwater. It can be used for investigating the impacts of pumping rates on, for example, groundwater level, discharge rates, groundwater–surface water interaction, flow in bedrock fractures. In recent years, numerical distributed models have been actively used in many fields and applications: for example, predicting climate change impacts on groundwater (Scibek & Allen 2006), studying the interaction between surface water and groundwater (Okkonen & Kløve 2011), investigating the impacts of mine drainage (Molson *et al.* 2012), and geothermal applications (Axelsson 2010). Groundwater flow modelling is thus becoming an increasingly valuable tool for managing aquifers in a sustainable manner, and for environmental studies of mining activities and other applications where groundwater plays an import role. In addition, legislation, licensing and public awareness call for the use of novel methods when estimating the impacts on groundwater.

Software that are commonly used for creating groundwater flow models (e.g. GMS (Brigham Young University 2005), FEFLOW® (DHI-WASY GmbH 2012) and MIKE-SHE (DHI 2007)), usually provide an easy to use methodology and user interface for conceptualizing the problem, creation of the grid and meshes, parameterization and setting of boundary conditions. These different versions of software are, in general, highly efficient in calculating groundwater flow and solute transports, and the visualization of the outcome of numerical data. In comparison, however, the modelling of the geological strata can be difficult or is not performed in a practical manner, and the visualization and data input of geological strata may not be as efficient as in software that is specifically developed for handling it. The accurate geological interpretations used alongside the geological model create the backbone of every groundwater flow model and, thus, are an essential part of every successful modelling project.

The geological modelling can be performed using several types of software. The most commonly used software in Finland are ArcGIS (surface models), Surpac (volume/solid models, surface models), Groundwater Modeling System (GMS, volume/solid models, surface models) and the GSI3D (volume/solid models, surface models), which is replaced by the Subsurface Viewer (Insight GmbH 2014). When integrating geological models into groundwater flow models, the modelling software that can calculate volume/solid models and takes into account the stratigraphy makes the modelling more straightforward and faster. The time taken to create the geological model can be a critical factor, for example, for commercial consultancies.

From: Riddick, A. T., Kessler, H. & Giles, J. R. A. (eds) 2017. *Integrated Environmental Modelling to Solve Real World Problems: Methods, Vision and Challenges*. Geological Society, London, Special Publications, **408**, 71–87.
First published online September 28, 2016, https://doi.org/10.1144/SP408.15

Integrating data between different software is most probably a task that is performed almost on a daily basis in many research institutes and companies, but the methodologies are transmitted orally between the users or the written documents are for internal use only. For example, in the Geological Survey of Finland, a typical data-integration task consists of exporting interpreted geophysical data to ArcGIS's personal geodatabase and from there to modelling software using tools in either ArcGIS or in external software, such as the Feature Manipulation Engine (FME).

In this paper, the data integration between geological 3D-modelling software and two different groundwater flow modelling software tools is described, and the methodologies are evaluated. The paper describes two groundwater flow modelling case studies but it concentrates on the data integration rather than modelling. At the Patamäki site in Kokkola, western Finland, the groundwater flow model was created to study different groundwater pumping schemes, and to study the groundwater flow paths from potential risk sites to groundwater wells. In the Luikonlahti site in Kaavi, eastern Finland, the groundwater flow model was created to study the effect of a mining site on groundwater in sediments and in crystalline bedrock. In both cases, the geological modelling was performed using GSI3D modelling software (Mathers *et al.* 2011). In the Patamäki case, the groundwater flow modelling was performed with MODFLOW (Harbaugh 1990) using the Groundwater Modeling System version 7.0 (GMS), and the Luikonlahti case was performed using the FeFlow® software (DHI-WASY GmbH 2012).

Study sites

Patamäki

The Patamäki groundwater reservoir is an unconfined esker aquifer located in western Finland, and is the main groundwater source for the town of Kokkola (Fig. 1). The average abstraction rate is 8229 m^3/day. The studied groundwater area is 94 km^2 and it is partly bordered by the Bay of Bothnia in the north. The eastern side of the aquifer is bordered by a groundwater divide (Fig. 1).

The Patamäki was most probably formed during the last deglaciation approximately 10.6 kyr ago (Boulton *et al.* 2001) and it lies partly on the 1900 myr old Palaeoproterozoic shale formation. The bedrock consists mainly of gneiss and schist, but granites and granodiorites have also been found (Luukas 2014).

The methods used to study the Quaternary strata, the top of the bedrock and the groundwater conditions consisted of sediment borings and the installation of groundwater observation pipes, and geophysical measurements, such as ground-penetrating radar (GPR) profiling, refraction seismic data and gravity measurements. Based on the data, the sediment thickness varies, on average, between 10 and 20 m (Paalijärvi *et al.* 2009). The deposit consists mainly of sands formed in different palaeoenvironmental settings (e.g. glaciofluvial and littoral), and a 200–300 m-wide glaciofluvial gravel zone runs all the way from the south of the aquifer to the north of the modelling area and possibly continues to the seafloor of the Bothnian Bay. The ground surface varies between −1 and 33 m asl (metres above sea level), and the bedrock reaches the ground surface only in the middle of the area.

Luikonlahti

Luikonlahti is a closed mine with an active concentration plant in Kaavi, eastern Finland, where the quarries, both open and underground, are filled with water (Fig. 2). Currently, the concentration plant, including the crushers and conveyors, is in use and the ore is transported from the Kylylahti mine in Polvijärvi, approximately 40 km from the Luikonlahti mine. In recent studies from the Luikonlahti area, it was observed that the mining activities have affected the quality of the surface water and groundwater in the vicinity of the mining area (e.g. Räisänen & Juntunen 2004; Kauppila *et al.* 2006; Heikkinen *et al.* 2009; Kihlman & Kauppila 2009). The tailings area and the Heinälampi settling pond were identified as the main sources of pollution to groundwater and surface water. The flooded underground mine and the open pit were also noted as possible sources of contaminants, but their effects are unknown. During the operation of the concentration plant, two different kinds of tailings material – first, since 1968, sulphide-rich tailings, then magnetite-rich tailings, from 1979 to 2006, and again, since 2012, sulphide-rich tailings – have been deposited in the tailings impoundment (Heikkinen *et al.* 2009). The main contaminants coming from the mining site are Ni, Co, Cu, Zn and S (Heikkinen *et al.* 2009). In addition, in the Heinälampi settling pond, arsenic concentrations are elevated and they are transported to surface waters through the Kylmäpuro ditch.

The bedrock of the study area consists of Palaeoproterozoic rocks, biotite paraschists in the northern part of the study area, granites in the eastern part and granodiorites in the southern part of the area (Luukas 2014). The ore was mined from sulphide-rich, polymetallic ore deposits located inside the biotite paraschist formation. The ore and the host rock belong to the Outokumpu Assemblage type rocks, The ore is hosted by metamorphosed quartz rocks, which are surrounded by skarns and

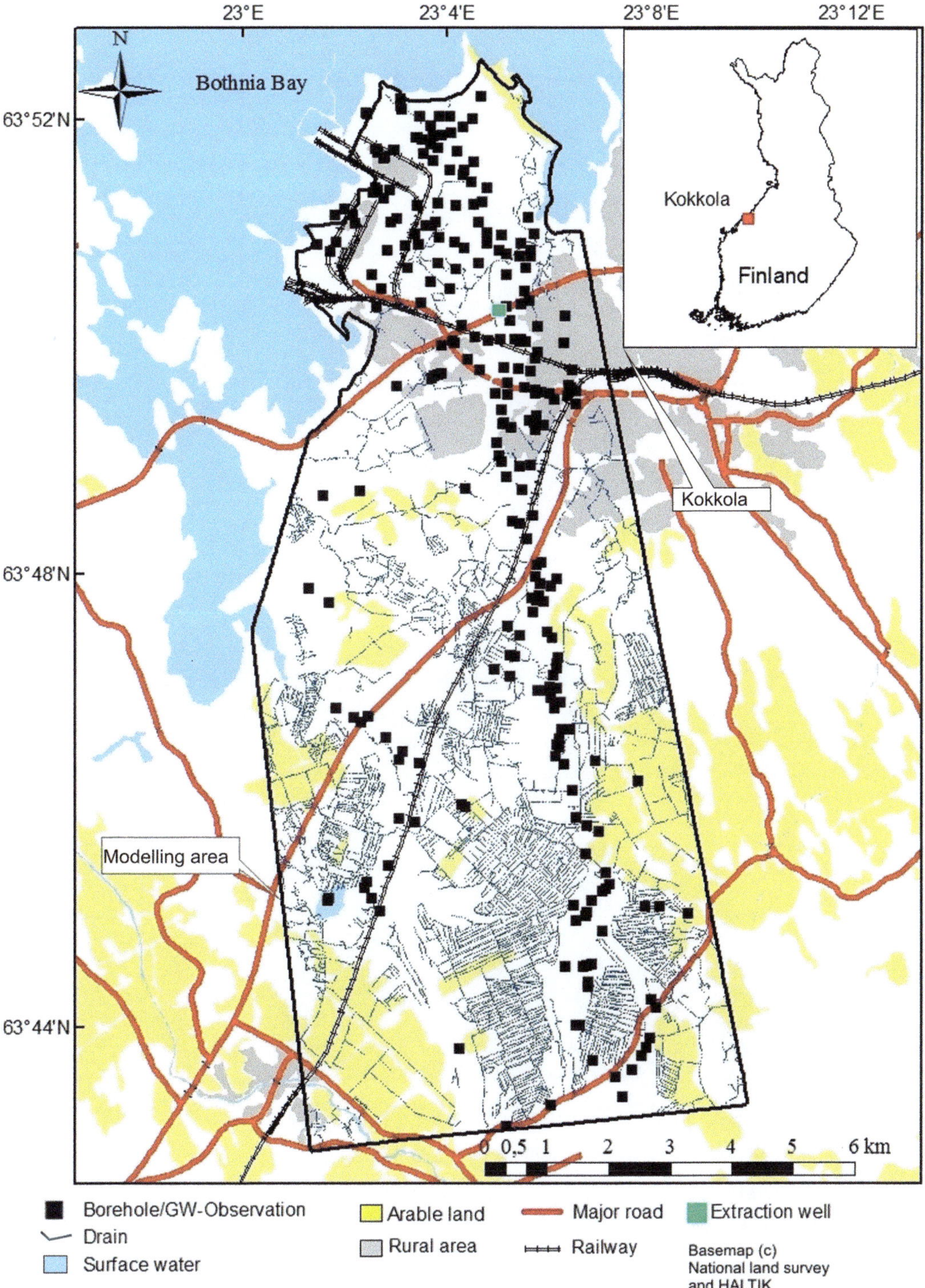

Fig. 1. Map of the Patamäki study site. The inset map shows the location of the city of Kokkola, which lies in the northern part of the study area.

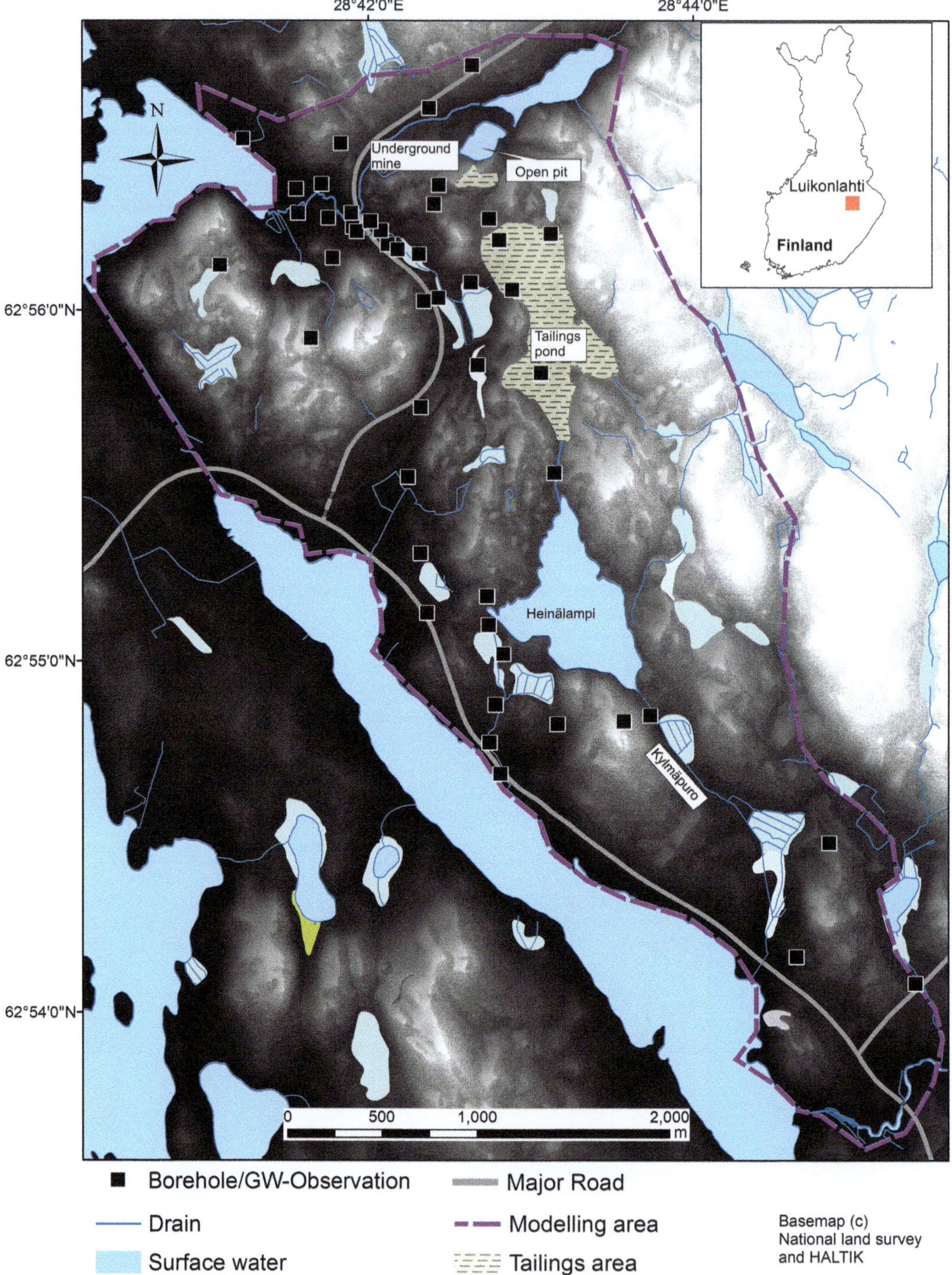

Fig. 2. Map of the Luikonlahti study site. The map shows the hill shaded DEM and the main features relevant to the study. The inset map shows the location of the study area.

carbonate rocks, and associated with serpentinites and black schists (Luukas 2014).

The studies in the Luikonlahti area concentrated on the Quaternary strata and bedrock fracture zones, which were thought to be important transport channels for the contaminants outside the mining area. The methods used to study the Quaternary strata, top of the bedrock and groundwater conditions included borings, installation of groundwater pipes and geophysical measurements, such as GPR, refraction seismic and gravimetric measurements. The bedrock fracture zones were mapped using aeromagnetic data alongside the digital elevation model (DEM), and the interpretation was enhanced with land-based geophysics (refraction seismic, gravimetric measurements, electrical resistivity tomography and Terra-TEM measurements), and boring and installing groundwater observation pipes to the fracture zones. Despite the extensive geological and geophysical studies, the internal structure of the fracture zones still remains largely unknown. The geophysical methods used do not have a sufficient vertical resolution for the detailed analysis of the internal structure of the fracture zones. Further steps will be taken in the future for better definition of the internal structure of the fracture zones: for example, by testing the use of magnetic resonance sounding and three-component high-resolution reflection seismic, especially the Poisson ratio (Poisson 1829; Barton 2007) for rock-quality estimation.

The sediments in the study area are Pleistocene and Holocene glacial and post-glacial sediments. They are composed of glacial tills, glaciofluvial gravel and sands, littoral sand and silts, and biogenic peat deposits (Huttunen 1990). Most of the study area is covered with till, where other sediments are less common.

The modelling area was selected mainly on an environmental basis, because the modelling was part of a larger environmental risk assessment project (Fig. 2). The area includes the surface runoff routes to nearby lakes and suspected groundwater transport routes. This area selection was sufficient for the geological modelling of the Quaternary strata, but the modelling area should have been larger for the bedrock fracture zones. The selection of the sufficient areal extent for bedrock fracture modelling is difficult, because it should be carried out according to, for example, hydraulic and geochemical parameters. Drilling in fracture zones always needs geophysical studies in advance to minimize the risk of drilling into unfractured bedrock. The available resources for using a vast array of geophysical methods to extend the study area were not sufficient in this study. For the groundwater flow modelling, the area limitation was usable because the western part of the model was delimited by two lakes and a channel between them, and the eastern part of the model was mainly delimited by a ditch in a small valley. The northern border is a slope where the water flow is towards the modelling area, whereas the southern border is an arbitrary line taken based on the trade-off between the large modelling area and the available resources available for modelling. It is thought that the mine does not have a significant effect on the groundwater at the southern border.

Methodology and discussion

GSI3D

The 3D modelling in this study was performed with GSI3D software, which is no longer available. It has been replaced by the Subsurface Viewer software, which uses much of the same programming code as GSI3D

GSI3D includes a graphical user interface, and methodology for modelling or mapping and visualizing the geological strata in three dimensions (Mathers *et al.* 2011). The software is based on manually digitizing geological cross-sections from several different types of data, such as DEMs, maps of geological formations, cross-section drawings, boreholes and geophysical data. The software calculates the 3D volume or solid model by triangulating digitized base surfaces of the geological units from the cross-sections. For the triangulation, the stratigraphy of the model has to be defined as a General Vertical Section (GVS), which defines the order of the geological units in a volume/solid model. The lateral extent of the geological units is defined by 2D envelopes, which are digitized from map data and the digitized base surfaces of the cross-sections.

Similar geological models can be produced, for example, in GMS, but GSI3D was selected instead to test the capabilities of the software. In the Patamäki case, GSI3D was tested for the first time by the Geological Survey of Finland, when an inexpensive and simple way to use geological 3D modelling and mapping software for everyday use by several geologists was being sought. The test showed that GSI3D is suitable 3D modelling software for use in complex Quaternary environments. In the Luikonlahti case, the software was in routine use, but its capabilities were further tested in the modelling of combined Quaternary sequence and bedrock fracture zones.

At both study sites, the 3D modelling in GSI3D followed the same methodology, but the original data and how it was used in the software differ to some extent. Both models are geological models where the hydrogeological units are defined by their lithology. In addition, in the Patamäki area, the genesis of the unit was also used in the GVS and in modelling if it could be interpreted.

The GVS at Patamäki consists of an upwards-fining sequence, typical of a retreating glacier and later isostatic uplift (Table 1). All of the data obtained from geological and geophysical methods were interpreted and exported to GSI3D as borehole data. Interpreted geological and geophysical data were used to calculate the top surface of the bedrock in ArcGIS using the 'Topo to Raster' tool. This surface was then exported to GSI3D as a grid. The ground surface, essential for every GSI3D modelling project, was obtained from LiDAR DEM, originally 2 m horizontal and 0.3 m vertical resolution. Because of the large modelling area (18.5 × 6.5 km) and the limitations in GSI3D internal memory handling, the LiDAR DEM was resampled to 20 m horizontal resolution. In addition, a map of Quaternary deposits (Kukkonen 1983*a*, *b*) was simplified to match the geological units and colouring defined in the GVS and legend files, and used to assist in the digitizing of the sections and envelopes. The model consisted of 93 cross-sections that were generated from 111 boreholes, more than 3000 GPR locations converted to boreholes and the bedrock top grid. The final model is shown in Figure 3.

The Quaternary strata at the Luikonlahti study site showed a similar succession to that at the Patamäki site, but the reworking and redeposition by the littoral processes could not be identified on a large scale. This redeposition can be identified in places, but in the GVS littoral reworking and deposition was not differentiated from glaciofluvial deposition as it was at Patamäki (Table 2).

The GSI3D modelling process was similar to that used at Patamäki, but the data were used differently inside GSI3D. The main reasons for this were the advancement of the GSI3D user interface allowing the use of raster backdrops, the unavailability of the LiDAR data from the study area and improved internal memory handling allowing the use of higher-resolution DEMs.

The DEM used was calculated from altitude contours (2.5 m vertical interval) to a 5 m horizontal resolution grid. A simplified map of Quaternary deposits (Huttunen 1990) was used in the digitizing of cross-sections and the envelopes. Boring and GPR data were interpreted and imported as boreholes. It was impractical to import the GPR data as raster backdrops because of the huge amount of data that would mask other data, and the limitations of the GPR data-processing software that does not allow the GPR profiles to be vertically scaled to represent the interpreted interface depths in multilayered strata. This was solved by exporting the interpretation data to *XYZ* points for each interface at an interval of every 10 m and converting the data to GSI3D borehole data in spreadsheet software. This approach gives the true interpreted depths of the interfaces. Another option would have been to import the interpretations without the GPR profiles as raster backdrops, but it was deemed impractical compared to the borehole approach. The rest of the geophysical data were imported as raster backdrops of the geophysical cross-sections with interpretations. This approach helped in a combined interpretation between geophysical methods, which is necessary for the reliable interpretation of the bedrock fracture zones.

The GSI3D model in the Luikonlahti study area was 6.5 km long and 2.5 km wide, and it was constructed using 34 digitized cross-sections consisting of data from 47 borings and 37 km of geophysical sections (Fig. 4).

Table 1. *GSI3D general vertical section used in the modelling of the Patamäki aquifer*

Name	id	Stratigraphy	Lithology	Genesis	Free text
Asfaltti	5	Asphalt	Asphalt		
Ta	20	Earth fill	Earth fill	Anthropogenic	
Ranta1	30	Littoral 1	Fine sand	Littoral deposit	
Ranta2	40	Littoral 2	Sand	Littoral deposit	
Tu	50	Peat	Peat	Biogenic	
Sa	60	Clay	Clay	Waterlain	
Si	70	Silt	Silt	Waterlain	
HHk	80	Fine sand	Fine sand	Glaciofluvial	
Hk	90	Sand	Sand	Glaciofluvial	
Si2	91	Silt2	Silt	Glaciofluvial	Bottomset facies
KaHk	100	Coarse sand	Coarse sand	Glaciofluvial	Tunnel facies
SrHk	110	Gravelly sand	Gravelly sand	Glaciofluvial	Tunnel facies
Sr	120	Gravel	Gravel	Glaciofluvial	Tunnel facies
Mr	130	Till	Till	Glacial	
Ka	140	Bedrock	Bedrock	Crystalline bedrock	Impermeable

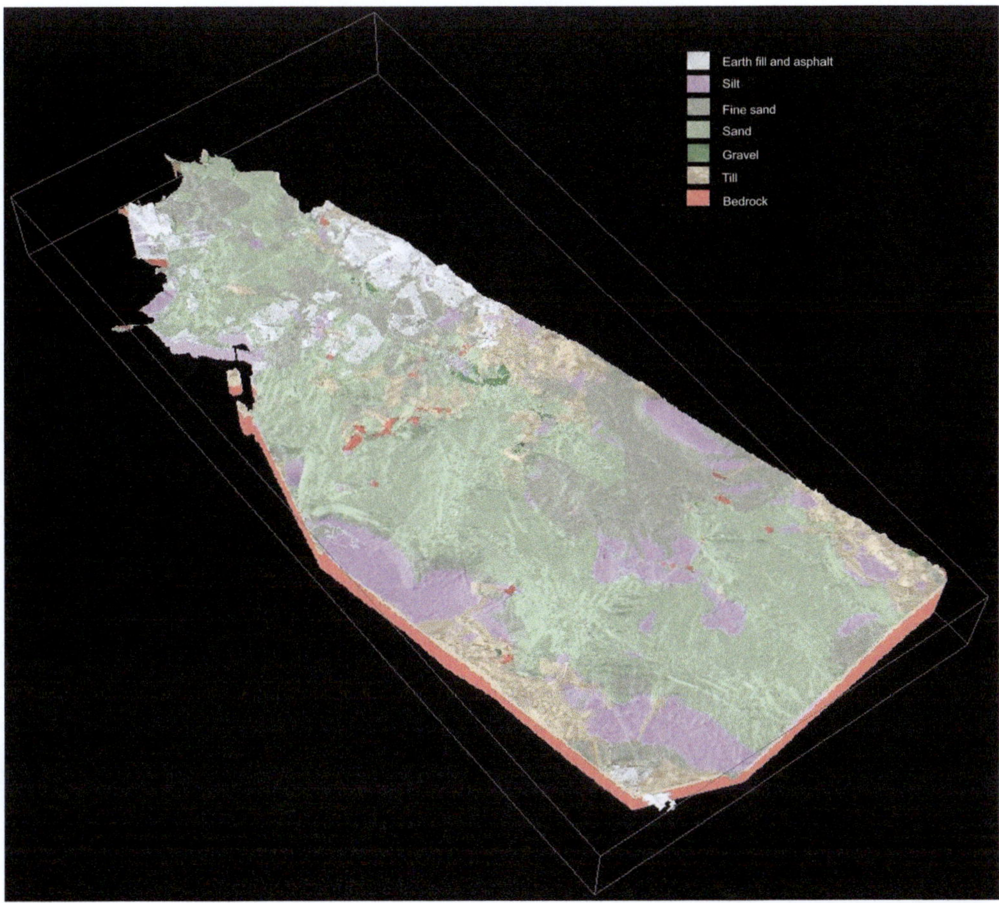

Fig. 3. GSI3D model of hydrogeological units at the Patamäki study site. The location of the model is represented in Figure 1.

Data export from GSI3D

After the geological model was refined and finished, the data needed to be exported from GSI3D. The data export was performed in GSI3D for modelled geological units using the 'Export all as grids' tool. The data were exported as ASCII grids of the top, base and thickness of each layer. The cell size

Table 2. *GSI3D general vertical section used in the modelling of the Luikonlahti study site*

Name	id	Stratigraphy	Lithology	Genesis	Free text
Vesi	10	Water	Water	Water	
Ta	20	Earth fill	Earth fill	Anthropogenic	
Tu	50	Peat	Peat	Biogenic	
Si	70	Silt	Silt	Waterlain/littoral	
HHk	80	Fine sand	Fine sand	Glaciofluvial/littoral	
Hk	90	Sand	Sand	Glaciofluvial/littoral	
Sr	120	Gravel	Gravel	Glaciofluvial	Tunnel facies
Mr	130	Till	Till	Glacial	
Ruhje	0	Fracture zone	Fractured bedrock	Bedrock movement	
Ka2	0	Bedrock	Slightly fractured		Slightly fractured and unfractured

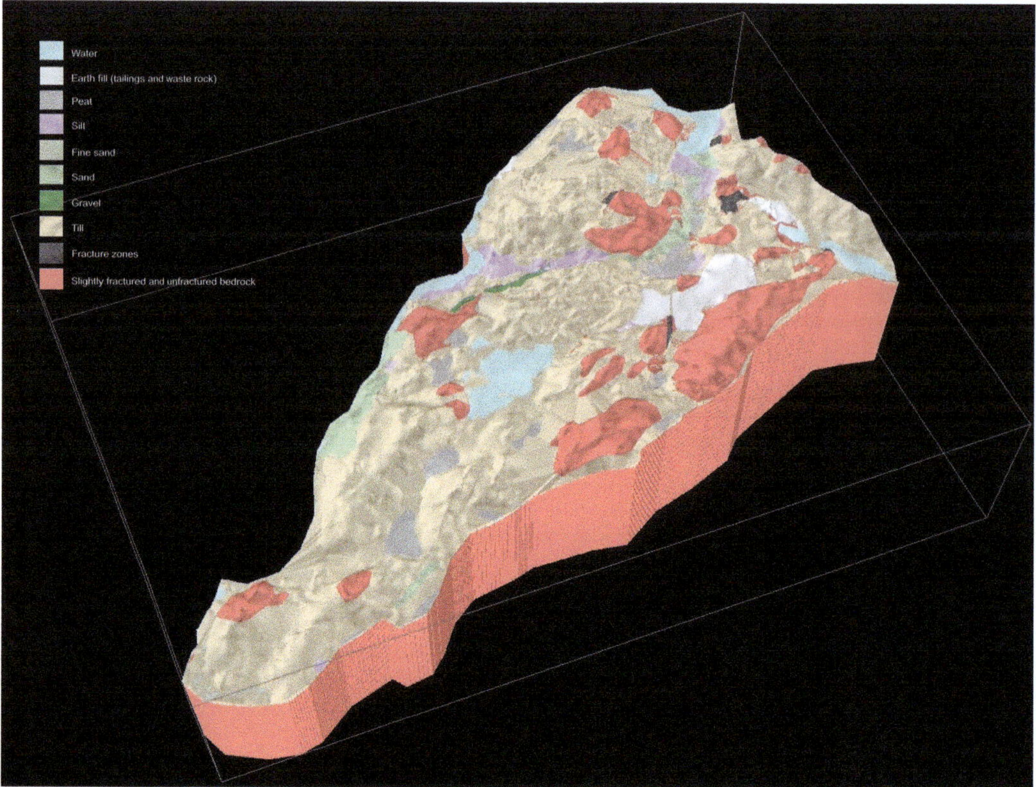

Fig. 4. GSI3D model of the hydrogeological units at the Luikonlahti study site. The location of the model is represented in Figure 2.

for the grids in the Patamäki area was 15 × 15 m, whereas, in the Luikonlahti area, the cell size was 25 × 25 m. In the Patamäki case, the grids were used directly in GMS software and the methodology is described in the following subsection. In the Luikonlahti case, the data were first imported to ArcGIS. The methodology and the data manipulation are also described below.

GMS

GMS (Groundwater Modeling Systems) is a graphical user interface for creating and visualizing of geological structures, groundwater flow and contaminant transport. It also provides many mathematical tools for handling the data in 2D and 3D. The software used in the study was GMS version 7.

The stratigraphy within the GMS software can be represented by creating boreholes. The boreholes can be directly created within the GMS using borehole tools or by importing text files that contain the borehole data. The borehole data can then be used to create cross-sections and 3D geological models (i.e. solid models). Within the GMS, the borehole data can be converted to other types of objects such as scatter points, triangulated irregular networks (TINs) and 3D meshes. The advantages of using the GMS borehole and solid model packages are that solid model data can be used directly for creating groundwater flow simulation.

In GMS, the governing groundwater flow equations can be solved with a finite-element method, FEMWATER (Lin *et al.* 1997), or with a finite-difference method, MODFLOW (Harbaugh 1990). In addition, an analytical element method, MOD-AEM (Analytic Element Model: Aquaveo 2016), can be used in 2D flow simulations. In this study, MODFLOW was used to simulate groundwater flow in the Patamäki. In GMS, four different flow packages are possible to use: block centred flow, BCF6 (Harbaugh *et al.* 2000), layer property flow, LPF (Harbaugh *et al.* 2000), upstream weighting flow, UPW (Niswonger *et al.* 2011), and hydrogeological unit flow, HUF (Anderman & Hill 2000). In the flow packages BCF6 and LPF, layer types and cell attributes, such as hydraulic conductivity (including anisotropy) and storage coefficients, are included. The UPW package is modified from the LPF package in order to use a Newton solver algorithm within the GMS software. The main difference between UPW and LPF (and also between BCF6 and HUF) is for modelling drying and rewetting of the grid cells. When the discrete approaches in the LPF, BCF6 and HUF packages are used for treating rewetting and drying of cells, the UPW package deals with non-linearities of cell drying and rewetting by use of a continuous function of groundwater head (Niswonger *et al.* 2011). The Newton solver used in UPW package, however, requires more memory due to an asymmetrical matrix instead of a symmetrical matrix produced by the LPF, BCF6 and HUF packages. Processing an asymmetrical matrix increases computational time. In the Patamäki unconfined aquifer, the grid was not discretized vertically to avoid thin cells, which can go dry during the simulation. Dry cells in MODFLOW are considered inactive. Inactive cells are no longer part of the model output and may result in mass-balance errors. Drying and rewetting of cells was not considered an issue in Patamäki and we wanted to keep computational time to the minimum: hence, the LPF package was chosen. The UPW package could be used to model groundwater flow in an unconfined aquifer because when using it cells should not go inactive during the simulations. This can be advantageous in keeping cells active and giving more realistic mass-balance calculations in unconfined aquifers (Niswonger *et al.* 2011). With the HUF, the groundwater flow can be modelled in a 'grid-independent' fashion: that is, the vertical elevations of the materials are independent of the finite-difference grid. The parameters, such as hydraulic conductivity and storage coefficient, are assigned to each material and not to grid cells *per se* as in other packages. The HUF package could be a good alternative, especially in the case where hydrogeological properties of the soil change vertically. This was not the case in Patamäki, where the vertical section with respect to hydraulic properties was considered somewhat homogenous. The LPF package is similar to the BCF6 package, and it should be viewed as an alternative to the BCF6 package. The main difference is the user-specified input data. In the LPF package, hydraulic properties are independent of cell dimensions, which is not always the case in the BCF6 package (Harbaugh *et al.* 2000). In the LPF package, the user always assigns hydraulic conductivities and transmissivity is then calculated based on the global elevation data. In the Patamäki, the global elevation data (top of the model area = ground surface elevation) and bottom of the model (non-permeable bedrock) was known, thus we only needed to assess different hydraulic conductivity zones horizontally independently of the grid cell dimensions. The LPF package then calculated automatically transmissivities. This procedure greatly helped in setting up the hydraulic conductivities compared to the BCF6 package, the use of which can be time-consuming.

From the GSI3D modelling, six different hydrogeological units were imported to GMS as ASCII files. The ASCII files can then be saved as point cloud or 2D cell-centred grid files (Fig. 5).

Both the point cloud and the grid files contained the coordinate and the elevation, and either top or the base of the unit. The data were imported in as point clouds and grid files. When saved as grid files, the data needed to be converted immediately to TINs or point clouds because only one 2D grid could be created at a time in GMS. 2D grids were converted to TINs because TINs could be used to create a solid model that corresponds to hydrogeological properties. The solid model data could be converted directly to MODFLOW-HUF to execute groundwater flow simulation: however, this option was not tested in this study, but will be considered in future studies. The ASCII files were also saved as point clouds. When point clouds were converted to TINs, the triangulation did not work as well as when triangulated using 2D grid files, as seen in Figure 6. The issue was that triangulation from point cloud to TIN created strings outside of the modelling domain (Fig. 6a) whereas triangulation directly from 2D grid files produced strings inside of the modelling domain (Fig. 6b) and represented exactly the units as modelled in GSI3D.

In the next phase, the 3D finite-difference grid was created (Fig. 7), one layer design was made and the MODFLOW-LPF package was chosen to simulate groundwater flow. To simplify the computation, it was assumed that each material extended from the ground surface to bedrock, which was the base of the grid. The bedrock was considered impermeable and thus was the bottom of the modelling domain. TINs were then used to define different hydrogeological boundaries horizontally that correspond to different hydraulic conductivity zones This can be done by plotting (see Fig. 7) TINs on

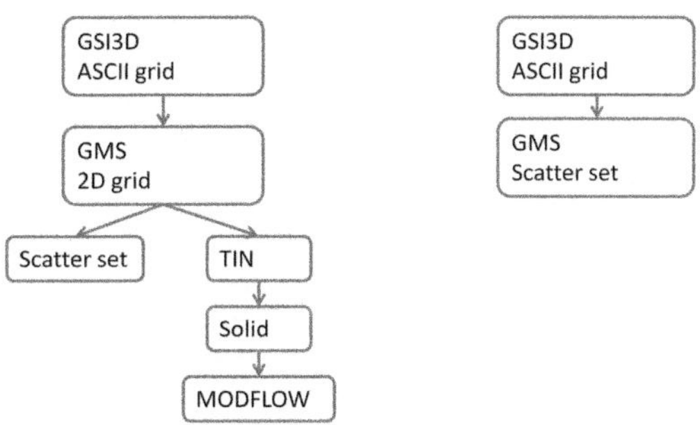

Fig. 5. Framework for importing hydrogeological units from GSI3D to GMS.

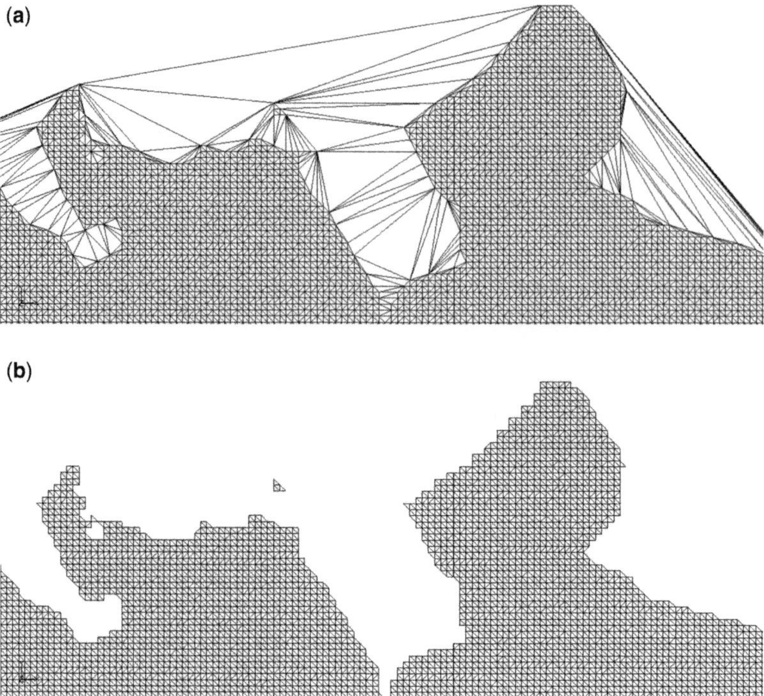

Fig. 6. Triangulation based on the point cloud (**a**) and 2D grid (**b**).

the grid, and using vertices to delineate the boundaries of different units. Six different horizontal hydraulic conductivity zones were determined. Because the simulation was at steady state, storage coefficients (or specific yield) were not needed. From these zones, the attributes were then converted to cells by one command and MODFLOW was ready to run.

Figure 8 presents the simulated hydraulic heads and the goodness of the fit. The groundwater model was calibrated by trial and error. The initial hydraulic conductivities were based on grain-size distribution for different units. The automated calibration was also tested using PEST – Model-Independent Parameter Estimation and Uncertainty Analysis (Watermark Numerical Computing 2005) but it did not improve the results. The final calibrated hydraulic conductivities varied from 0.1×10^{-4} m s^{-1} in the sand to 0.4×10^{-3} m s^{-1} in the gravel (Fig. 9). The hydraulic conductivities fell within observed ranges found in the literature (e.g. Airaksinen 1978). The groundwater recharge was also distributed according to land use and soil cover into different zones, and was part of the model simulation. The simulated groundwater levels match very well to 229 observations. The bias calculated between simulated and observed groundwater levels was -0.004 m, indicating that the errors are basically random. The bias was calculated with the following equation:

$$\text{Bias} = \frac{1}{n}\sum_{i=1}^{n}(y_i - f_i) \qquad (1)$$

where y is the simulated value and f is the observed value and n is the number of observations.

One of the key factors affecting the goodness of the groundwater flow simulation result was the accurate construction of a conceptual model that was based on the accuracy of geological model.

FeFlow®

FeFlow® is a groundwater and transport modelling system that uses a finite-element method to solve the governing groundwater flow and transport equations (DHI-WASY GmbH 2012). The creation of geometry using a 3D calculation mesh in FeFlow® is performed by giving the desired number of slices (a slice is a layer defined by its top and base surfaces) in the 3D Layer configuration tool and assigning the elevations from GSI3D ASCII grids to each slice (DHI-WASY GmbH 2012). The slices need to

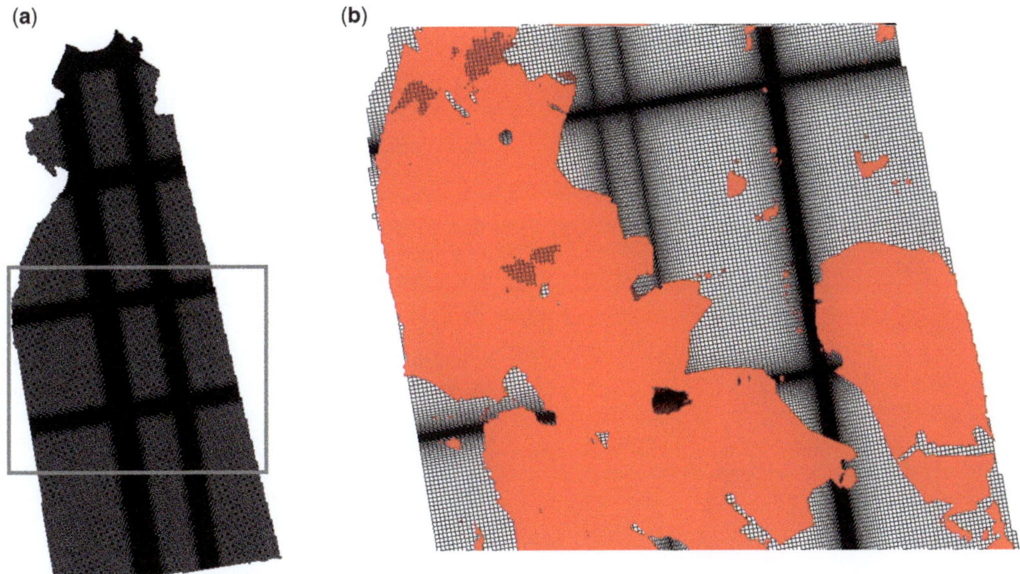

Fig. 7. (**a**) Finite-difference grid of the Patamäki modelling area. The grey rectangle shows the area in (b). (**b**) The 'Silt-TIN' plotted on the finite-difference grid that represents the zone of the silty material.

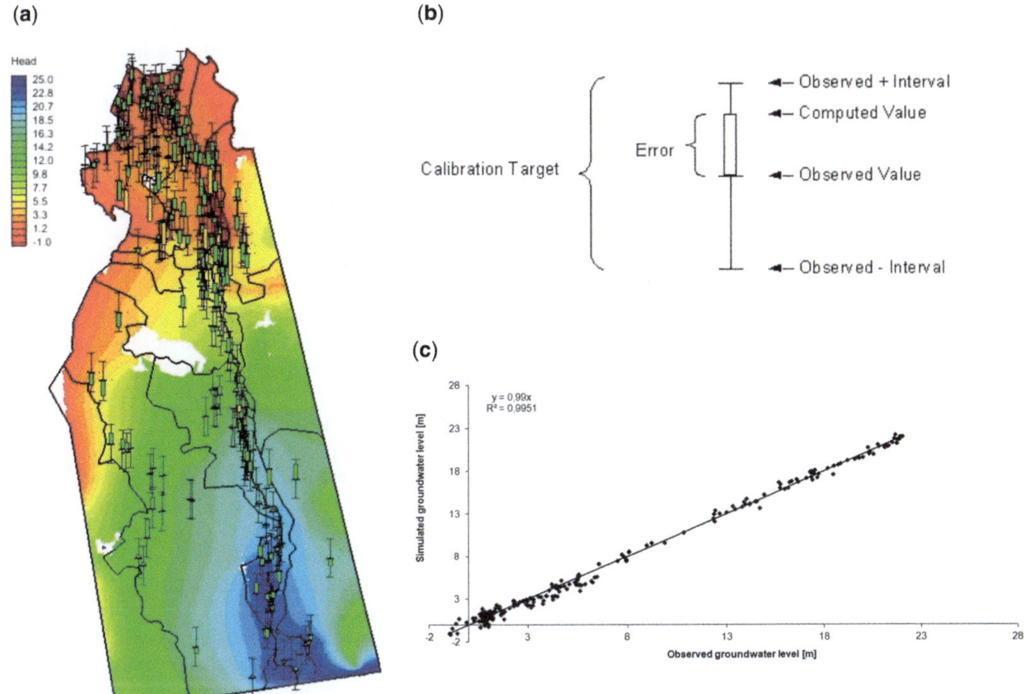

Fig. 8. Simulated heads (**a**), calibration target (**b**) and the goodness of fit (R^2) (**c**) in the Patamäki study site. Black lines in (a) represents the hydraulic conductivity zones. In the calibration target, the coloured bar represents the error. If the computed value lies within the pre-assigned calibration target, the colour bar is drawn in green. The bar is drawn in yellow or red if the error is less or larger than 200%, respectively.

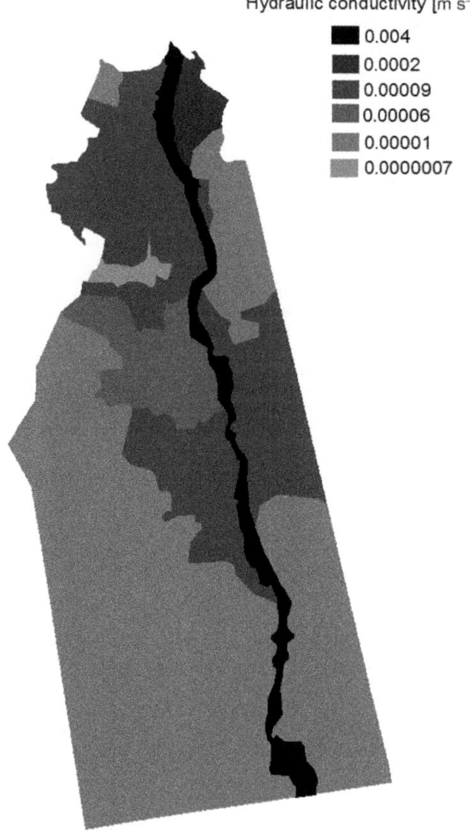

Fig. 9. Hydraulic conductivity zones of the Patamäki site. The location of the map is shown in Figure 1.

be continuous throughout the modelling area and this poses a problem when exporting the data from GSI3D.

Two different models were created for the Luikonlahti area. The initial approach on integrating the data between GSI3D and FeFlow® was performed by creating a 10-layer model where the top of each hydrogeological unit from GSI3D was used to define tops of individual slices. This approach was selected first to import the detailed 3D geological model to FeFlow®, but it proved to be unsuccessful. The revised and more promising approach was to create a three-layer model where the hydrogeological units used were combined Quaternary strata, bedrock fracture zones and bedrock (slightly fractured and unfractured). The groundwater flow in bedrock and bedrock fracture zones were solved using equations for porous media. This was selected because of the lack of precise data of fractures and cracks in bedrock. Using equations for pipe flow would have needed more detailed information on the size of the fractures and their connectivity. This information was not available at the Luikonlahti site.

FeFlow® data integration: initial approach

This initial approach involved importing ASCII grids from GSI3D to ArcGIS using the 'ASCII to Raster' tool and the integer option for numerical data. This integer option must be used because the attribute table is built only for integer data during the import and further modification of the data is impossible without the attribute table. The integer option truncates the data and, in this case, the horizontal resolution is approximately 1 m, which was thought to be sufficient for modelling. The data can also be imported as floating point numbers, but the change to integer must be done separately by using the 'Int' tool in ArcGIS. If more accuracy is needed, the floating-point elevations need to be multiplied by a factor so that the effects of the truncation can be minimized. The attribute table can then be built using 'Build Raster Attribute Table' tool. The attribute table is essential when compiling the data for FeFlow®.

In exported ASCII grids, the null values are set to 10e32. ArcGIS reads this value as 10, which cannot be differentiated from real 10 m elevations. Therefore, it was converted to value −9999 prior to import. ArcGIS reads −9999 as a null value automatically. As mentioned earlier, GSI3D creates grids of the top, base and thickness of each unit. FeFlow® only needs the top elevation of each slice and, therefore, the top surfaces of each unit were imported to ArcGIS. The bottom surface of the bedrock unit is also needed as a base of the model or it can be given an arbitrary elevation, which is on a lower level than the base of the bedrock fracture zones.

The problem when modelling recently glaciated areas, such as in Finland, is that the Quaternary strata are not laterally continuous. GSI3D can easily handle discontinuous strata, but FeFlow® requires that all the slices and layers are continuous throughout the modelling area. The only surfaces that are continuous in the GSI3D model are the top and base of the bedrock unit. The elevations of these surfaces could be used as is and no raster manipulation was needed. All the other surfaces had to be made continuous throughout the modelling area and this was done in ArcGIS using the 'Mosaic to New Raster' tool. The tool was run using the elevations from a top elevation of a selected raster layer and top elevations of all the rasters below in the order of the GVS (Table 2). In the 'Mosaic Method', the option 'First' was used. This option uses the elevations from the first raster in areas where the rasters overlap each other. The result is a continuous raster layer where the elevation of the selected unit is found

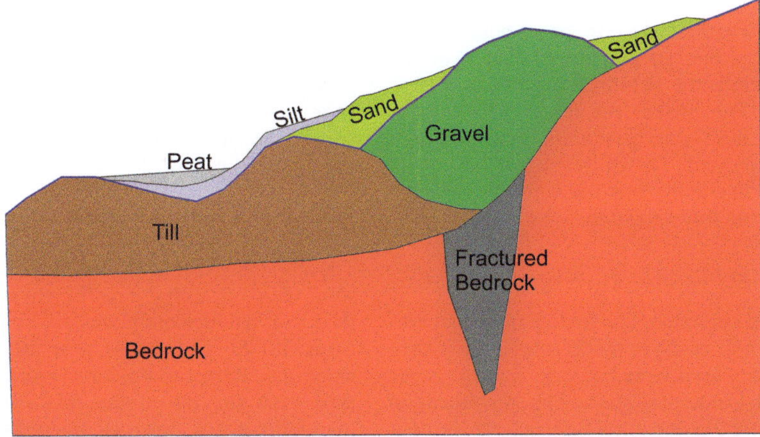

Fig. 10. Conceptual model of the top surfaces needed to create a continuous raster layer from discontinuous data. For the continuous top surface of gravel (blue line) for the entire modelling area, the top surfaces of till and bedrock are also used. The top surface of fractured bedrock is not needed in this case because it is overlain by till and gravel, which are in a higher stratigraphical position.

where the unit is present, and the elevation of the units below it where the selected unit is not present (Fig. 10). The routine presented here is based on the experience gained on the earlier studies by the authors, and other options for creating continuous rasters were not tested. The DEM could not be used for the ground surface because it includes the top surface of the water bodies. Therefore, the ground-surface slice used in FeFlow® was a compiled top surface of all the hydrogeological units, except the water unit, showing the ground surface and the bottom of the lakes. FeFlow® reads the elevation of the slices from a tab-separated text file where the columns are *X* (Easting), *Y* (Northing), Ele (Elevation) and slice (slice number). The *X* and *Y* coordinates were added to the attribute tables of the continuous rasters by first converting them to ArcGIS shape files using the 'Raster to Point' tool and writing the coordinates to the attribute table using the ArcGIS extension XTools Pro's 'Add *XYZ* Coordinates' tool. The attribute tables were then exported as .dbf files and edited in spreadsheet software to match the format read by FeFlow®. The procedure for importing the file to FeFlow® and adding the elevations to the slices is explained in the FeFlow® manual (DHI-WASY GmbH 2012). The flow chart of the methodology is presented in Figure 11.

This first attempt at data integration proved to be unsuccessful. Therefore, the modelling is not described but only the reasons why this method does not work. The main problem in compiling the GSI3D data like this is that it produces very thin layers in FeFlow®. For example, in the areas where the bedrock is exposed to ground surface, there are nine other slices above the bedrock slice because all the slices need to be continuous and uniform throughout the modelling area. FeFlow® does not permit slices within the model to have an elevation that is lower than or equal to slices below. It automatically

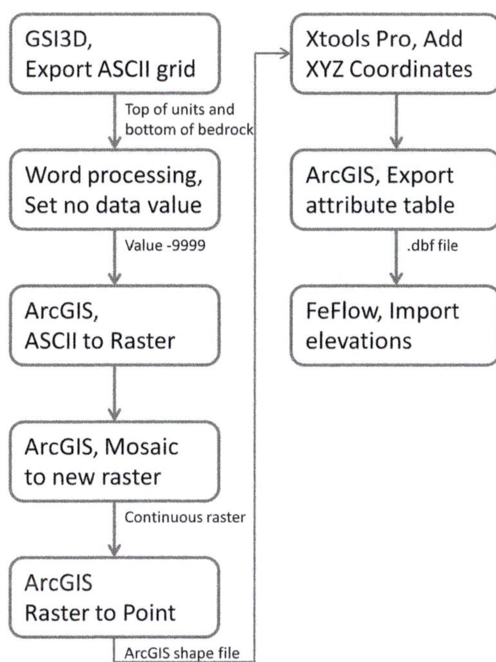

Fig. 11. Framework for importing hydrogeological units from GSI3D to FeFlow.

suggests a correction value that is added to the elevation at that location. The correction value applied was minimal, 1 cm, so that the elevation of the ground surface in the FeFlow® would not change significantly. This makes the layers very thin in places where the Quaternary units above the bedrock are not present. The calibration of the model can be done manually, but the software was not able to solve the flow equation.

FeFlow® data integration: revised approach

In this revised approach, the geological model was simplified to a three-layer model where the hydrogeological units used were bedrock, fracture zones and combined Quaternary strata. This simplification was done in ArcGIS and it gives three hydrogeological layers with distinct hydraulic parameters. Even though the hydraulic parameters in Quaternary strata differ between the hydrogeological units, using similar parameters for the Quaternary strata was thought to be sufficient. Earlier hydrogeological and hydrogeochemical observations from the study area strongly suggests that the surface runoff is the main transport path for the contaminants from the mine site. It is suspected that bedrock fracture zones are also important transport paths, but the transport in Quaternary strata is minimal compared to these two pathways and, therefore, the similar parameters for Quaternary units and the three-layer model is justified.

The continuous elevation rasters to create the FeFlow® slices were created using the same methodology as in the initial approach. The elevation rasters were created for the base of the bedrock, top of the bedrock, top of fracture zones and top of the ground surface, excluding the water. These units were then imported to FeFlow® to create a 3D mesh.

Groundwater flow modelling with FeFlow®

The parameterization zones in FeFlow® (Fig. 12) were derived using ArcGIS shape files created from original (discontinuous) hydrogeological units from GSI3D. The shapes were created in ArcGIS using the 'Raster to Polygon' tool. The zones created were the top of the fractured bedrock and inverted bedrock outcrops where the bedrock outcrops are removed from the ground surface raster. The fractured bedrock raster could be used as it is. The inverted bedrock outcrops were compiled with the 'Mosaic to New Raster' tool from all the other rasters, except top of the bedrock and surface of the water. The compiling can also be done from the GSI3D envelopes exporting them as a shape file using the GSI3D 'Export unit map as Shapefile' tool. The shapefile can then be manipulated to show the required parameterization zones using the definition query action in ArcGIS. The raster manipulation method was used in this study and the envelope method was only briefly tested. Based on earlier experience, the raster manipulation method was

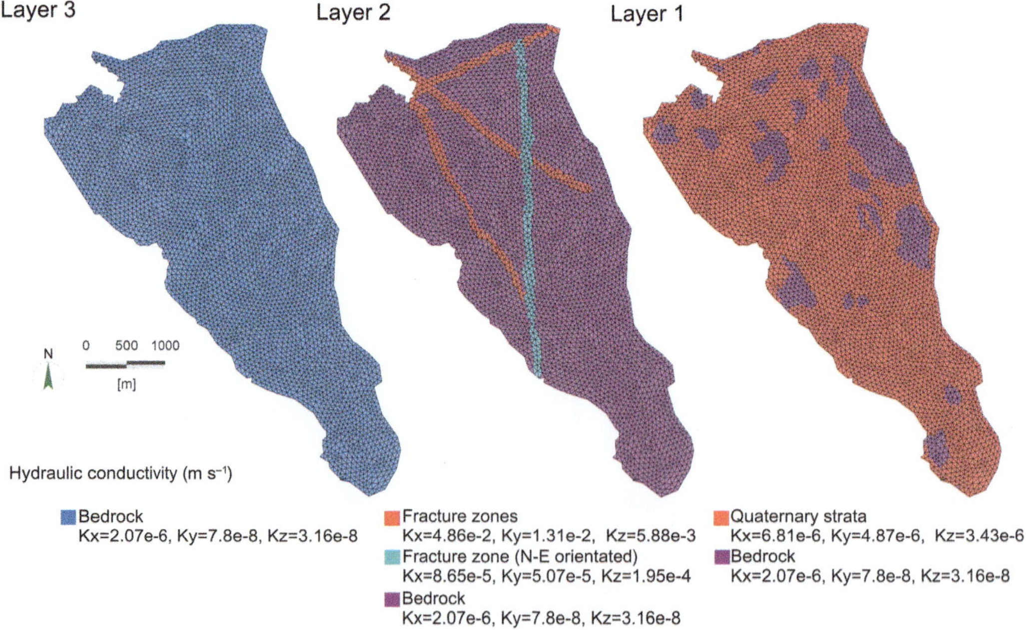

Fig. 12. Hydraulic conductivity zones and the mesh used in the modelling of the Luikonlahti area.

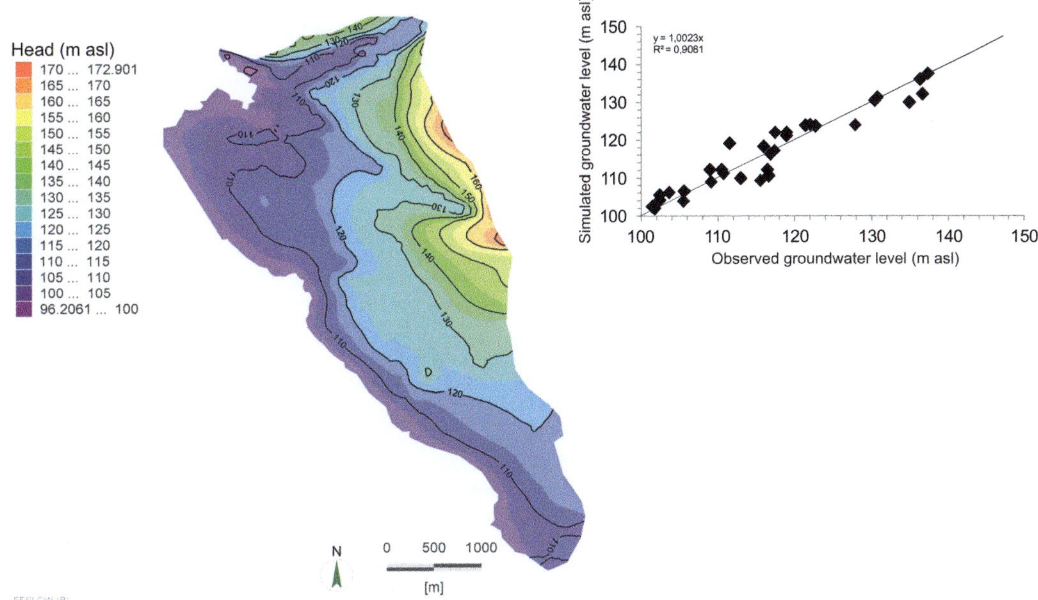

Fig. 13. (**a**) Simulated heads and (**b**) goodness of fit (R^2) in the Luikonlahti study site.

thought to be more straightforward compared to the envelope method. The parameterization zones were then imported to FeFlow® and used in populating the mesh with different parameters, such as hydraulic conductivity and porosity.

The lowermost layer (Layer 3, Fig. 12) was configured with the same parameters throughout the layer. This layer represented the slightly fractured and unfractured bedrock. The middle layer (Layer 2, Fig. 12) represented the fracture zones, and slightly fractured and unfractured bedrock. Within Layer 2, a parameterization zone was configured to represent the fracture zone area. Outside of this fracture zone area within layer 2, the parameters were set to represent slightly fractured and unfractured bedrock. In the topmost layer (Layer 1, Fig. 12), the inverted bedrock outcrop parameter zone was used to set the parameters for the Quaternary strata. Outside this parameter zone, the parameters for slightly fractured and unfractured bedrock were used. The parameterization was first performed manually. During the manual parameterization and model calibration, it was noticed that the fracture zones need to have separate parameter values and the north–south-trending fracture zone was separated (Fig. 12). The parameterization was then calculated using the PEST-software (Watermark Numerical Computing 2005), which slightly improved the results between simulated and observed groundwater levels. The initial hydraulic conductivities were partly calculated from grain-size analyses and partly taken from the literature (Airaksinen 1978), and the calibrated values are within these limits. The simulated hydraulic heads and the goodness of fit are presented in Figure 13. The relatively large differences in the measured hydraulic head and calculated hydraulic head are thought to be the result of rough topography and mining activities, and relatively coarse calculation mesh. Also, the lack of data from the fracture zones most probably causes error in the model.

Conclusions

Data integration is an essential, but a somewhat underestimated, part of many geological procedures. This paper presents, for the first time, two workflows for integrating data from geological 3D software GSI3D to groundwater flow modelling software GMS and FeFlow®. Even though, the integration may be routine in many tasks, the workflows are often not published or are only for internal use. The workflows are designed to be used in data integration between specific software, but the principles presented can be useful with other 3D geological modelling and groundwater flow modelling software.

The use of dedicated 3D geological modelling software for the conceptual geological model improves the groundwater flow modelling. This is achieved through the more versatile modelling procedures and more accurate models compared to the conceptual modelling in groundwater flow modelling software without proper geological modelling capabilities. The more accurate geological model allows the necessary simplification for the parameterization and ease of calculation in groundwater flow modelling. The less convoluted workflow, especially for FeFlow®, presented in this paper will save modellers both time and effort.

The data integration between GSI3D and the groundwater modelling software is fairly straightforward, but a good knowledge of software, data types, and import and export procedures are needed. The data manipulation and use in groundwater flow modelling software sometimes needs lots of trial-and-error testing before it is usable.

The main problem in developing the workflows came from the inadequate simplification of the geological model, which caused the unsuccessful solution in the initial FeFlow® approach. Also, the triangulation from point clouds in GMS did not work as well as from 2D grids. Working with grids needed a lot of trial and error to produce suitable data for groundwater flow modelling software. Some parts of the workflow could be automated, but generating continuous slices, for example, needs very complicated automation. The experience gained by the authors in their earlier work guided the design of the workflow and data manipulation within it.

The following conclusions can be drawn from the study presented in this paper:

- The groundwater flow modelling can be enhanced by first creating the geological model in dedicated software. A better simplification of the conceptual groundwater flow model can be achieved by using dedicated geological modelling software
- The GSI3D ASCII grids can be used directly in GMS 7.0, but data manipulation inside GMS 7.0 is needed. GSI3D ASCII grid files, which represent different soil layers, are imported to GMS 7.0 and are treated as 2D grid files. In GMS, only one 2D grid can be created at one time: thus, this 2D grid file needs to be converted immediately to a 2D point cloud or TINs, which represent the layer obtained from GSI3D. In GMS, we used TINs to assess different hydraulic conductivity zones.
- The GSI3D ASCII grids need to be manipulated in ArcGIS prior to import to FeFlow®. The workflow consists of exporting the ASCII grid from the GSI3D, setting the null value, converting ASCII grids to raster grids in ArcGIS, mosaicing the raster grids to produce a continuous raster grids, and converting the raster grids to point data and *XYZ* coordinates, which are then exported to a spreadsheet for modification of the template, and importing the elevations from modified spreadsheets to the FeFlow® slices.
- The main problems in the study were encountered with too little simplification in the FeFlow® initial approach and with ArcGIS trial and error when testing the suitable workflow.
- The main issue in GMS was creating TINs from the point clouds. When the 2D grid file in GMS was converted to TINs, the shape of the TIN closely approximated to the model created in GSI3D. When point clouds were converted to TINs, the strings were created outside the modelling domain, and did not exactly represent the layers as modelled in GSI3D.

The authors would like to acknowledge the Geological Survey of Finland for support in compiling this work. The city of Kokkola and Altona Mining Ltd (currently Boliden Kylylahti Oy Finland Ltd) are thanked for submitting the data used in the original projects. The original project at Patamäki was partly funded by the city of Kokkola, and in Luikonlahti by Altona Mining Ltd and European Area Development Funding 2008–14 programme, alongside the Geological Survey of Finland.

References

Airaksinen, J.U. 1978. *Maa- ja pohjavesihydrologia [Soil and groundwater hydrology]*. Pohjoinen, Oulu, Finland.

Anderman, E.R. & Hill, M.C. 2000. *MODFLOW-2000, the U.S. Geological Survey Modular Ground-Water Model – Documentation of the Hydrogeologic-Unit Flow (HUF) Package*. United States Geological Survey Open-File Report 00-342. United States Geological Survey, Reston, VA.

Aquaveo 2016. *MODAEM – Analytic element modeling with GMS*. http://www.aquaveo.com/software/gms-modaem [last accessed 28 April 2016].

Axelsson, G. 2010. Sustainable geothermal utilization – Case histories; definitions; research issues and modelling. *Geothermics*, **39**, 283–291.

Barton, N. 2007. *Rock Quality, Seismic Velocity, Attenuation and Anisotropy*. Taylor & Francis Group, London.

Boulton, G.S., Donglemans, P., Punkari, M. & Broadgate, M. 2001. Palaeoglaciology of an ice sheet through a glacial cycle: the European Ice Sheet through the Weichselian. *Quaternary Science Reviews*, **20**, 591–625.

Brigham Young University 2005. *Groundwater Modeling System (GMS) Version 6.5*. Brigham Young University, Provo, UT.

DHI 2007. *MIKE SHE User Manual*. Danish Hydraulic Insitute, Hørsholm.

DHI-WASY GmbH 2012. *FeFlow® 6.1. Finite Element Subsurface Flow & Transport Simulation System*. DHI-WASY GmbH, Berlin.

Harbaugh, A.W. 1990. *A Computer Program for Calculating Subregional Water Budgets using Results from USGS MODFLOW Model*. United States Geological

Survey Open File Report 90-392. United States Geological Survey, Reston, VA.

HARBAUGH, A.W., BANTA, E.R., HILL, M.C. & MCDONALD, M.G. 2000. *MODFLOW-2000, the U.S. Geological Survey Modular Ground-Water Model – User Guide to Modularization Concepts and the Ground-Water Flow Process*. United States Geological Survey Open-File Report 00-92. United States Geological Survey, Reston, VA.

HEIKKINEN, P.M., RÄISÄNEN, M.L. & JOHNSON, R.H. 2009. Geochemical characterisation of seepage and drainage water quality from two sulphide mine tailings impoundments: acid mine drainage v. neutral mine drainage. *Mine Water and the Environment*, **28**, 30–49.

HUTTUNEN, T. 1990. *Luikonlahti. Map of Quaternary Deposits 1: 20 000, 43 1104*. Geological Survey of Finland, Kuopio.

INSIGHT GMBH 2014. *Manual, Subsurface Viewer XL and Subsurface Viewer MX, 6.0*. INSIGHT Geologische Softwaresysteme GmbH, Cologne.

KAUPPILA, T., KIHLMAN, S. & MÄKINEN, J. 2006. Distribution of arcellaceans (testate amoebae) in the sediments of a mine water impacted bay of Lake Retunen, Finland. *Water, Air and Soil Pollution*, **172**, 337–358.

KIHLMAN, S. & KAUPPILA, T. 2009. Mine water-induced gradients in sediment metals and arcellacean assemblages in a boreal freshwater bay (Petkellahti, Finland). *Journal of Paleolimnology*, **42**, 533–550, https://doi.org/10.1007/S10933-008-9303-6

KUKKONEN, E. 1983*a*. *Ytterbråtö. Map of Quaternary Deposits 1: 20 000, 23 2210*. Geological Survey of Finland, Espoo.

KUKKONEN, E. 1983*b*. *Kokkola. Map of Quaternary Deposits 1: 20 000, 23 2211*. Geological Survey of Finland, Espoo.

LIN, H.-C.J., RICHARDS, D.R. & TALBOT, C.A. 1997. *FEMWATER: A Three Dimensional Finite Element Computer Model for Simulating Density-Dependent Flow and Transport in Variably Saturated Media*. Technical Report CHL-97-12. United States Army Corps of Engineers, Vicksburg, MS.

LUUKAS, J. 2014. *Bedrock of Finland – DigiKP. Digital map database*. Geological Survey of Finland, Espoo, http://en.gtk.fi/export/sites/en/informationservices/explorationnews/stakeholderseminar_2014/presentations/GTKxs_bedrock_map_databases.pdf

MATHERS, S.J., WOOD, B. & KESSLER, H. 2011. *GSI3D 2011 Software Manual and Methodology*. British Geological Survey Internal Report OR/11/020. British Geological Survey, Nottingham, http://nora.nerc.ac.uk/13841/1/OR11020.pdf

MOLSON, J., AUBERTIN, M. & BUSSIÈRE, B. 2012. Reactive transport modelling of acid mine drainage within discretely fractured porous media: plume evolution from a surface source zone. *Environmental Modelling & Software*, **38**, 259–270.

NISWONGER, R.G., PANDAY, S. & IBARAKI, M. 2011. *MODFLOW-NWT, A Newton Formulation for MODFLOW-2005*. United States Geological Survey Techniques and Methods, **6-A37**.

OKKONEN, J. & KLØVE, B. 2011. A sequential modelling approach to assess groundwater–surface water resources in a snow dominated region of Finland. *Journal of Hydrology*, **411**, 91–107.

PAALIJÄRVI, M., LEHTIMÄKI, J. & VALJUS, T. 2009. *Patamäen pohjavesialueen geologisen rakenteen selvitys 2007–09*. Tutkimusraportti 17.7.2009 [*Hydrogeology of the Patamäki Groundwater Area, Studies During the Period 2007–09*. Research paper 17.7.2009] Geological Survey of Finland, Western Finland office, Kokkola [in Finnish].

POISSON, S.D. 1829. *Mémoire sur l'équilibre et le movement des corps élastiques* [*Thesis on the quality and movement of the elastic bodies*]. Mémoires de l'Académie Royale des Sciences de l'Institut de France, **8**.

RÄISÄNEN, M.L. & JUNTUNEN, P. 2004. Decommissioning of the old pyritic tailings facility previously used in a talc operation, eastern Finland. *In*: JARVIS, A.P., DUDGEON, B.A. & YOUNGER, P.L. (eds) *International Mine Water Association Symposium*. Proceedings of the Symposium: Mine Water 2004 – Process, Policy, and Progress, 19–23 September 2004, University of Newcastle, Newcastle upon Tyne, UK, Volume **I**, International Mine Water Association, 91–99, http://www.imwa.info/imwaconferencesandcongresses/imwa-symposia/189-proceedings-2004.html

SCIBEK, J. & ALLEN, D.M. 2006. Comparing modelled responses of two high-permeability, unconfined aquifers to predicted climate change. *Global and Planetary Change*, **50**, 50–62.

WATERMARK NUMERICAL COMPUTING 2005. *PEST. Model-Independent Parameter Estimation. User Manual*. 5th Edition, http://www.pesthomepage.org/getfiles.php?file=pestman.pdf [last accessed 19 December 2012].

The potential for the use of model fusion techniques in building and developing catastrophe models

K. R. ROYSE[1]*, J. K. HILLIER[2], A. HUGHES[1], A. KINGDON[1], A. SINGH[1] & L. WANG[1]

[1]*British Geological Survey, Keyworth, NG12 5GG, UK*

[2]*Loughborough University, Loughborough, LE11 3TU, UK*

**Correspondence: krro@bgs.ac.uk*

Abstract: Global economic losses related to natural hazards are large and increasing, peaking at US$380 billion in 2011 driven by earthquakes in Japan and New Zealand and flooding in Thailand. Catastrophe models are stochastic event-set based computer models, first created 25 years ago, that are now vital to risk assessment within the insurance and reinsurance industry. They estimate likely losses from extreme events, whether natural or man-made. Most catastrophe models limit the level of user interaction, stereotyped as 'black boxes'. In this paper we investigate how model fusion techniques could be used to develop 'plug and play' catastrophe models and discuss the impact of open access modelling on the insurance industry and other stakeholders (e.g. local government).

Natural hazards such as earthquakes, tsunamis and hurricanes can be extremely damaging, with a clear majority of countries at risk (e.g. Munich RE 2011*a*, *b*). Even small advances in understanding the distribution of hazards and our susceptibility to them have a huge potential to ultimately translate into substantial and highly visible benefits. Specifically, target benefits include reductions in both mortality and financial losses incurred by individuals, businesses, states and insurance companies. This relationship between natural hazards, geology and economic loss has long been recognized (e.g. Doornkamp 1995; Woo 1999; Grossi *et al.* 2005). Between 1960 and 2000, events defined as disasters caused by natural hazards worldwide more than tripled in number and, in real terms, economic losses increased more than eight fold (Munich RE 2000). The rise in economic loss over this period is due to multiple factors, including rapid and poorly controlled urbanization, ineffective public policy, increasing construction in hazard-prone areas and variations in climate (e.g. Pelling 2003; Walter 2004; Strömberg 2007; Tseng 2012). Hurricane Georges (September 1998) is a demonstration of where poor planning policy decisions increase an event's economic impact. Georges affected five Caribbean countries and in particular the city of Santo Domingo in the Dominican Republic. In this case, it has been estimated that 50% of the economic impact was a result of poor land use practices, with around 35% of these losses due to the erroneous location, design and construction of infrastructure (Mora 2009).

Insurance is a mechanism for managing financial risks. The objective of purchasing insurance is to increase resilience to avoid failure by spreading the cost of a large loss. Destruction of your house by flooding could, for example, bankrupt you. So, you pay an insurer an agreed premium for the year ahead. The insurer groups or 'aggregates' many such risks with the intention that, on average, it will then pay out a steady, predictable amount each year. However, natural hazards or 'perils' can affect many insured properties across a wide area (e.g. hundreds of kilometres across) in a limited time window (e.g. <72 h). Limited diversification, perhaps within a country, may then be inadequate. For example, losses to property caused by Hurricane Andrew in 1992 were around $15.5 billion, which caused 13 insurance companies to fail (AIR 2002; Cummins 2007). To protect against this, insurers buy reinsurance from reinsurers, spreading risk further. Quantifying (for insurers this means costing) the risk of severe events is critical to insurers, reinsurers and reinsurance brokers, and is of interest to the regulators. Claims information held by insurers about previous events is often of little use as such extreme events are seldom similar enough and insured property 'exposure' changes rapidly. For example inflation, real growth in property values, varied insurance penetration and changes in the locations of properties insured within a portfolio must be accounted for to estimate likely insured losses. Therefore if we take the case of Hurricane Andrew, losses today would be in the order of $31.3 billion which is comparable with those for Hurricane Katrina in 2005 of between $40 and 66 billion (Towers Perrin 2005; Swiss Re 2007). Catastrophe models, developed over the last ~25 years (Grossi *et al.* 2005) are one solution to this problem. The failure of only four insurance companies after Katrina indicates that they may have facilitated better financial risk management within the insurance sector (Cummins 2007).

From: Riddick, A. T. Kessler, H. & Giles, J. R. A. (eds) 2017. *Integrated Environmental Modelling to Solve Real World Problems: Methods, Vision and Challenges*. Geological Society, London, Special Publications, **408**, 89–99.
First published online December 3, 2014, https://doi.org/10.1144/SP408.7

There is a wealth of natural hazard models simulating physical processes available to academic hazard and risk modellers (Frankel *et al.* 1996; Mastin *et al.* 2009; Sparks *et al.* 2012; Pappenberger *et al.* 2012; Wang & Rosowsky 2012). Coupled with the availability of a rapidly increasing range of data products, one might hope that these resources would be accessible and interoperable (Nature Geoscience 2008). A key problem, however, is that model code is often highly specialized or setup to accept very specific data formats (Foresight 2012). The available data resources may be constructed according to different standards, making the integration of new models with existing datasets problematic (van Westen *et al.* 2008). Trying to fuse together multiple models, all of which might require input data in different formats or units, seems a challenging mission. These problems typically result in a compartmentalized approach to model development. If Model A depends upon Model B, and the developers of Model A have previously spent a good deal of effort in linking the two models, they are understandably disinclined to use more effort to link with the new, high-tech Model C, even if the resulting combination is expected to lead to a better understanding of the system under study (Rougier *et al.* 2010).

Such discrepancies between software/data standards can form barriers to the investigation of any differences between predictions of different models, or combinations of different models. Easing the process of fusing models together could facilitate assessment of this model uncertainty with the intention of reducing it.

The NERC PURE (Natural Environment Research Council Probability Uncertainty and Risk in the Environment) Research and Knowledge Exchange programme (co-sponsored by the Engineering and Physical Sciences Research Council, Environment Agency and Technology Strategy Board) is developing an 'experimental zone'. The aim of this web-enabled zone is to give unrestricted access to publicly available risk and uncertainty tools. The key objective of the PURE Experimental Zone is to tackle the problem of model fusion by creating an environment in which users can link models and data together via an easy-to-use web interface. Furthermore, the portal will allow models and data to be catalogued, making new advances more visible and more accessible to model users and consultants. It is hoped that this will not only encourage the takeup of the latest datasets or modelling techniques, but that it also make it simple to swap in different models to investigate the overall effect on the linked system.

In this paper we give an overview of how model fusion techniques can help illustrated through the development of a 'plug and play' catastrophe model (Royse *et al.* 2014) and discuss the impact of open-access modelling could have on the insurance and reinsurance industry.

Catastrophe models

Catastrophe models allow insurers to test their current exposure such as property assets against all probable events for a particular hazard within an area such as flooding in the UK. These models provide insurers with a better understanding of their likely liability to events in the next year. Events used are extrapolated from observed historical ones by either a statistical model or a numerical simulation. Figure 1 illustrates a framework for a catastrophe model. The contents, definitions and names of the pieces of such frameworks are not yet standardized and vary (e.g. Grossi *et al.* 2005; Qu *et al.* 2010) depending on the purpose, personal preference and implementation of the model; however, the broad work-flow remains the same. The framework consists of three major 'modules', where processing occurs, and four inputs (Fig. 1). Indeed, this component-based approach reflects how catastrophe models are typically created by linking various elements. Consequently, it is easy to envisage how

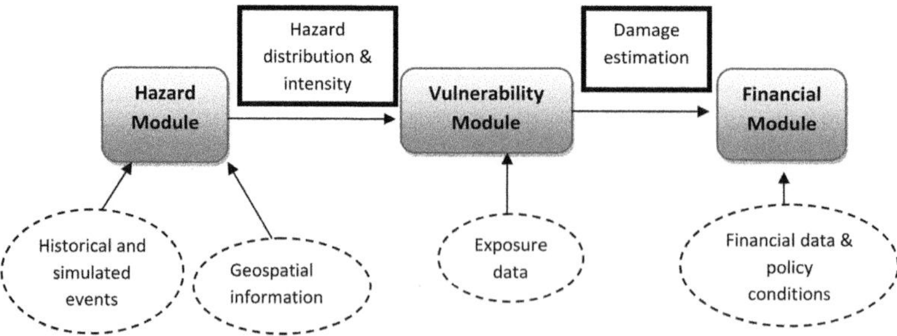

Fig. 1. One possible conceptual framework for a component based catastrophe model. Grey rectangles are modules, dashed ovals are inputs, black outlined squares are the outputs from each module and the arrows indicate the flow of information. Adapted from Royse *et al.* (2014).

a catastrophe model could be broken down into, or compiled from, a series of linked or fused models that perform calculations and exchange data.

Catastrophe models are not perfect and different models can give very different results dependent on the data and assumptions used (e.g. Willis 2007). If, for simplicity, we consider losses resulting from property damage, first-order errors (e.g. Grossi *et al.* 2005) could include:

(a) Hazard estimation because the event sets (which include the number, frequency and size of possible hazard events) used may be imperfect. For instance the 2011 magnitude 9.0 earthquake to hit Japan was larger than suggested by science and so not included in catastrophe models (e.g. Avouac 2011; Lay & Kanamori 2011; Ozawa *et al.* 2011; AIR 2011).
(b) The severity v. damage (i.e. 'vulnerability') relationships may be systematically biased or out of date.
(c) Incomplete or incorrect details about assets on insurers' books such as errors in their location, type or insured value will lead to erroneous estimates.

The scale of the difference that different data or approaches can make is demonstrated by initial estimates of losses caused in Europe by windstorm Kyrill in 2007. Whilst remarkably good considering the speed of their production and the complexity of the problem, the losses reported from the five main models catastrophe models were reported as between \$3 and 5 billion (RMS), \$3.6–8.8 billion (AIR), \$2.5–5 billion (EQECAT), \$4–7 billion (Hanover Re), \$5–7 billion (Munich RE) and \$3.5 billion (Swiss Re) (Willis 2007). The actual property losses were around \$5.8 billion (Munich RE 2011*b*). Note that actual losses lie in or close to all but one of the ranges given, making these estimates reliable, demonstrating that these modellers appreciate the levels of uncertainty present. The potential for over or underestimation of likely losses therefore obviously exists.

Where models exist for perils and large gaps in the coverage of models remain, there are also a number of other issues known to affect the accuracy of loss estimates. Some of the more notable issues reflecting limitations in the models and an approach originating in random and uncorrelated Poisson events are:

(1) **Multi-perils** (Woo 1999): tsunami and fire following earthquake or flood and landslides following hurricanes. These types of scenarios may be included but often only at a basic level if at all.
(2) **Hazard clustering** (Lennartz *et al.* 2008; Rybski *et al.* 2008; Vitolo *et al.* 2009): This is where events are not independent of each other. Normally events are considered to be autonomous, but recent research is showing that this is not always the case. A good example of this is the clustering of European windstorms, for example, Lothar and Martin, Christmas 1999, and the eight storms in December 1989 and January 1990 costing 10.5 billion Euros in insured losses (Mailier *et al.* 2005).
(3) **Unmodelled exposures**: material (e.g. cars or oil rigs) or intangible assets. Although business interruption is normally included, wider economic losses such as supply chain disruption are not. A good example of this was in the Thailand 2011 floods where, out of a total of \$40 billion of insured losses, \$10 billion were due supply chain interruptions (Munich RE 2012; Willis 2012).
(4) **Demand surge**: This is where the cost of repairs goes up because there is a scarcity in materials and/or labour.

Catastrophe models therefore need to be used with due caution (e.g. ABI 2011; Guy Carpenter 2011). These models provide a guide to the risk not answers or precise calculations of risk (e.g. ABI 2011). Most catastrophe models limit the level of user interaction and are often viewed as 'black boxes'. New regulations in Europe (Solvency II) (e.g. Eling *et al.* 2007; EU 2009; ABI 2011) will require insurers to understand the assumptions their solvency calculations are based on. This will provide insurers with the ability to better constrain where the uncertainty originates within their calculations and hence better understand the catastrophe models upon which these calculations are based. Essentially going forward, insurers need to be able to critique similar catastrophe models and understand the impact that changing a single component will have on the outputs.

Drivers for change

Within the insurance and reinsurance industry, decisions are made and millions are gained or lost based on the results of catastrophe modelling. Catastrophe models are limited by our incomplete scientific knowledge and are based on simplifying assumptions. Modelled loss estimates can therefore vary significantly, as they are based on different yet scientifically valid assumptions.

Solvency II is a new harmonized EU-wide regulatory regime which replaces 14 previous directives. Its objectives are that:

- there will be a greater awareness of risk within both the governance and operational side of the insurance sector;

- the EU insurance market will become more integrated and there will be an increase in the competitiveness of EU insurers;
- there will be enhanced protection to policyholders.

The Solvency II (Eling *et al.* 2007; EU 2009; ABI 2011) regulations will require insurance companies to better understand the assumptions on which their solvency calculations ultimately rest. They will be required to hold capital against a range of risks up to a 99.5% confidence level, namely 1 in 200 year financial worst case. In terms of catastrophe modelling, these new regulations require the insurer to understand the strengths and weaknesses of the catastrophe models it uses, to be aware of potential gaps and quality differences in the company's catastrophic risk modelling landscape, to actively seek the levels of information and detail they need to feel comfortable with taking decisions and finally to ensure that the proper policies and procedures for doing so are in place (ABI 2011).This is translating into a call for a greater level of transparency in how catastrophe models are generated.

A second driver is the potential for the wider application of the catastrophe modelling methodology. Complex catastrophe models require significant resource to produce and so are not often developed in areas where insurance penetration is low (Munich RE 2010; Guha-Sapir *et al.* 2011). However, it is widely recognized that the creation of these models for new areas will open up markets for the insurance industry and also raise awareness amongst the population of the possible risks they face (Yucemen 2005; Gurenko *et al.* 2006; Cummins & Mahul 2009; Chavez-Lopez & Zolfaghari 2010). One of the major limiting factors is often that of data. Firstly with low frequency events such as earthquakes in regions where these are infrequent but likely to be highly destructive, this means that there is little data available on hazard or damages. Secondly as these are new markets there will be little insurance claims data to develop reliable vulnerability curves (Chavez-Lopez & Zolfaghari 2010; Foresight 2012). However, even with these limitations, the development of models in these areas can have huge societal and economic benefits, especially in hazard prone areas, where populations are growing, where the replacement value of the properties is increasing, and where changes in building practices, codes and their enforcement can result in time in reducing the lives lost due to natural disasters.

Application of model fusion techniques to catastrophe modelling

In choosing what natural hazard to focus on in our worked example below, we looked at where the biggest potential insurance losses are likely to be derived from in the UK in the future. Dailey *et al.* (2009) suggest that river (fluvial) flooding and flash (pluvial) floods have the potential to impact the UK the most. The Association of British Insurers (ABI) predicts that, with a 4 °C rise in global temperatures, the average insured losses in the UK from river flooding and flash floods could rise by 14% to £633 million each year (Dailey *et al.* 2009). The 2007 floods across Yorkshire, the Midlands and the west of England demonstrated that flooding in the UK can have major economic as well as social impacts. For example the 2007 floods cost insurers over £3 billion (Pitt 2008). The 2009 floods in Cumbria, west Wales and Dumfries and Galloway, although on a smaller scale, still cost insurers over £1.5 million (Munich RE 2010). The financial effects of flooding are therefore significant; the average cost per home affected by flood damage in the UK has been put as between £20 000 and £40 000 by the ABI (Dailey *et al.* 2009). The Pitt review (Pitt 2008) made recommendations for how the UK could improve its flood management in light of the lessons learnt from the 2007 floods. One of the key recommendations was that when managing flood risk in the UK a whole system approach should be taken. This means that groundwater flood risk should be included along with pluvial and fluvial flooding in flood risk planning process. It is therefore appropriate that in our case study we chose to look at groundwater flooding; for a fuller description of the case study see Royse *et al.* (2014).

Integrated environmental modelling methods can be used to make existing numerical models into components or 'building blocks' that can be assembled together to make more complex models (Warner *et al.* 2008). In this approach, models become substitutable and the linking (or fusion) mechanism is transparent (Knapen *et al.* 2013). A model framework can be used to link models together. Several frameworks are available such as FRAMES (Babendreier & Castleton 2005), CSDMS (Overeem *et al.* 2013), OMS (David *et al.* 2002) and OpenMI (Moore & Tindall 2005). Where these frameworks are based on open standards such as the case with OpenMI (Gregersen *et al.* 2005; Moore & Tindall 2005), the framework can be made widely assessable (Knapen *et al.* 2013). The choice between modelling frameworks is largely dependent on the project being undertaken (Knapen *et al.* 2013). This paper focuses on the use of the OpenMI model framework because it uses an open framework based on well-defined standards and was initially developed to facilitate an integrated approach to environmental management as specified in the Water Framework Directive (OpenMI 2009).

OpenMI is an open source model linking standard. OpenMI's development has been funded through two EU projects HarmonIT (EVK1-CT-2001-00090) and OpenMI-Life (LIFE06 ENV/UK/000409). When computational entities (each individual model or dataset) are made OpenMI compliant, they are turned into objects or 'components', which enables them to exchange data by linking the variables exposed in the interface. A key feature is that the component's owner selects which model variables are available for use by another model. It is also possible to choose the number of variables that users can manipulate. This feature will help to address any security issues that linked catastrophe models may have as it restricts access to variables within any OpenMI compliant component.

Fluid Earth (2011), which is available for free on SourceForge (2011), was developed by HR Wallingford to make computational entities easily OpenMI compliant and to link them together. It includes both a software implementation of OpenMI comprising a software development kit (SDK) and an interface to aid the linking of different components called 'Pipistrelle'. The SDK allows models to be adapted into OpenMI compatible 'components' in preparation for linking to other models in OpenMI compositions. Fluid Earth's Pipistrelle is a graphical 'point and click' interface that allows users to link components into 'compositions' and run them.

To demonstrate the use of model fusion techniques in the development of a catastrophe model, we created an illustrative linked model for modelling groundwater catastrophe which can be downloaded from http://pureexperimentalzone.org/. This model of the Marlborough and Berkshire downs, which are located 70 km west of London, covers about 2600 km^2 (Fig. 2). To create a basic catastrophe model some simplifying assumptions were made; the climate time series used was limited to 33 years, the vulnerability curves are extrapolated to groundwater flooding and different financial perspectives are not considered. However, even with these assumptions the losses calculated were similar to recent historical events, that is, within the same time period.

The workflow for the model broadly follows that set out in Figure 1, constructed of seven linked components within three modules (Fig. 3). In should be noted that, unlike catastrophe models currently available to users, in this example the event set generation has been included within the hazard module (Qu *et al.* 2010. Also, the financial loss model combines the vulnerability and loss since the freely available vulnerability curves (Penning-Rowsell *et al.* 2010) are presented in terms of 'pounds per house'. This highlights the ability of a linked model composition to be highly flexible, adaptable to

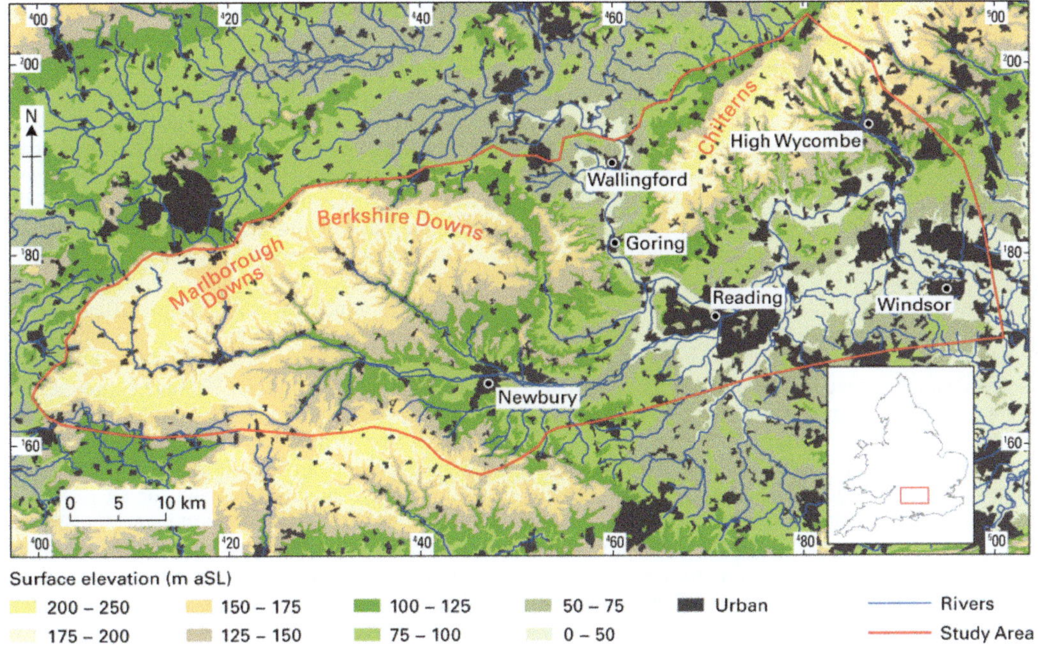

Fig. 2. Map displaying area covered by the groundwater catastrophe model. Red outline depicts the limits of the Marlborough and Berkshire Downs and South West Chilterns groundwater model.

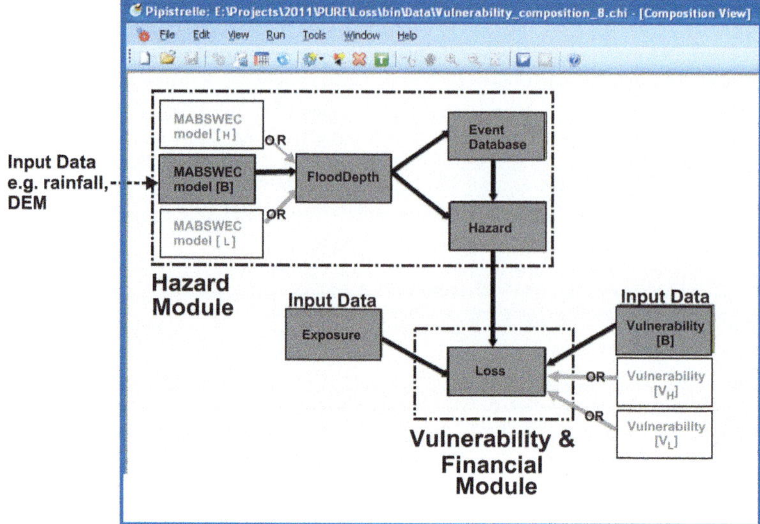

Fig. 3. Screenshot of a Pipistrelle composition for the linked models comprising the catastrophe model. The grey boxes are the seven components of the 'baseline' [B] model. Arrows show data flow. Grey text and arrows have been added to the screenshot to show how alternative components can be swopped in and out of the composition. Adapted from Royse *et al.* (2014).

user requirements and data availability, and create opportunities such as end-to-end uncertainty analysis.

The hazard module has been constructed using four components. It is based on the 'MABSWEC model' component, a pre-existing groundwater recharge model called ZOODRM (Mansour & Hughes 2004), which is driven by daily rainfall (1971–2003) from the UK Met Office via the Centre for Ecology and Hydrology and potential evapotranspiration datasets provided directly by the Met Office. The model calculates the surface runoff based on a 50 m resolution Digital Elevation Model (DEM; Morris & Flavin 1990) and applies the soil moisture deficit (SMD) method (Penman 1948; Grindley 1967) to simulate the balance between evapotranspiration, soil moisture changes and potential groundwater recharge. Using a daily time step produces the lowest practical error in the soil moisture balance calculation (see Howard & Lloyd 1979). A regional groundwater flow model developed using ZOOMQ3D then simulates transient fluctuations in groundwater head, river baseflow and spring discharge. The rivers are approximated using an interconnected river network that exchanges water with the underlying aquifer according to a Darcian type flux equation (Hughes *et al.* 2008; Campbell *et al.* 2010; Guardiola-Albert & Jackson 2011; Mansour *et al.* 2011).

The 'FloodDepth' component compares calculated groundwater heads to the DEM (e.g. CEH-DTM; Morris & Flavin 1990) to create daily maps of flooding and its depth whenever water height breaches the land surface. It is now possible for the 'EventDatabase' component to identify discrete flood events, assigning them a number and probability up to $1/33$ (i.e. once in 33 years). Events are defined as discrete periods of time where the flood depths on average exceed 0 m. To mitigate model artefacts (e.g. near the boundary of the model domain), the average was taken only over depths between -5.6 and 10 m, and implausible groundwater floods in areas capped with surface clay were masked. The final *'hazard'* component of the hazard module creates hazard maps of the maximum flood depths attained during each event.

The 'Loss' component combines information provided, upon request, by the 'exposure' and 'vulnerability' components with hazard maps to estimate losses for each event. Here, the components supplying data simply read text files, but could equally be SQL databases. We have simplified the exposure by assuming that urban areas have a constant housing density of a 100 houses per km^2, with no exposure outside. Vulnerability was calculated using a residential average for all exposure. Losses for all properties affected by each event are first summed to obtain an 'event loss table' of losses per event and then occurrence exceedance probability (OEP) curves (e.g. Grossi *et al.* 2005) are generated, see Figure 4. Appendix 1 contains further details of the components used in the model. It should be noted that there is no barrier to generating a more complex composition being created. This could include a physical or statistical climate model at the beginning, a component to assign financial losses to different parties in the context of insurance at the end, or a full database of vulnerability curves.

Development of the experimental zone

To facilitate the sharing and fusion of models and data from a variety of different sources, we are developing an experimental zone (http://pureexperimentalzone.org). The experimental zone platform will enable users to share and link data, models, and analytical and visualization tools (in real time) across the internet using OpenMI. The OpenMI standard has already been applied successfully to several models in the water sector and the Fluid Earth initiative has been created to provide a community based around the use of OpenMI in environmental modelling. The objective of the experimental zone is to create a community incorporating both natural hazard scientists and risk modellers. The aim is to facilitate collaboration between research groups thus enabling cross-disciplinary models to be linked together and encourage the integration of commercial and open source data and model components. The portal will make use of the Web processing services being developed by the Open Geospatial consortium (http://www.opengeospatial.org) which is providing rules for standardizing how inputs and outputs for geospatial processing need to be formulated as well as defining how clients request executions of a process and how the output from the process is handled. Significant time will also be given to ontologies and how they are defined so that an unambiguous exchange of information between different models can be achieved.

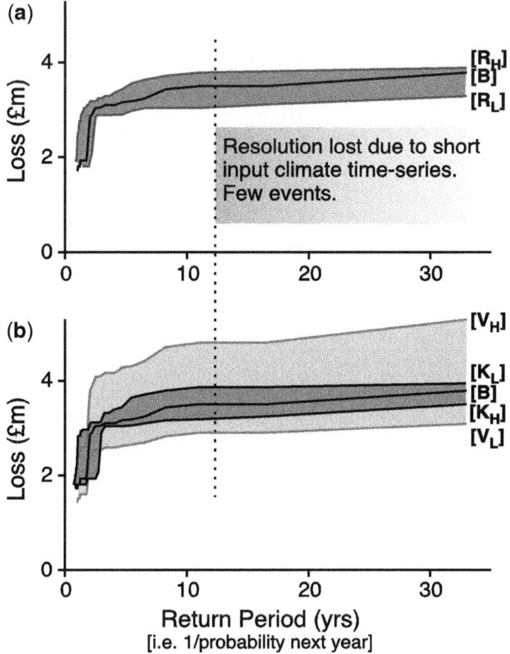

Fig. 4. Occurrence exceedance probability (OEP) curves produced by the model. (**a**) Baseline model [B] is represented by a black line. The dark grey lines are for high [R_H] and low [R_L] recharge scenarios, with shading representing a plausible range. (**b**) Baseline as (a), but dark grey lines represent low [K_L] and high [K_H] hydraulic conductivity scenarios.

The experimental zone will provide an environment where users will be able to test model compositions, demonstrate the sensitivity of model component choice on results and support the construction of alternative models for existing as well as unmodelled natural hazards. Crucially it will also lower the barrier to entry permitting researchers, whether in industry or academia, to offer up new modules rather than complete catastrophe models. In essence, the experimental portal will provide users with the ability to generate 'plug and play' catastrophe models (e.g. Royse *et al.* 2014).

Our vision is that the experimental zone will contain a 'library' of both models and data sources that can be linked together into a fused model (or 'composition') via a version of the Pipistrelle interface. It is hoped that by operating in this way not only will the insurance community have access to the latest research in hazards and vulnerability assessments, but it will also lower the boundaries to linking models by developing a web-based interface. This will make it easy to add, remove or swap model components and run compositions remotely. In time, the experimental portal will help to open up the insurance and reinsurance market to new models and data resulting in better catastrophe models and improved risk assessments. An additional aim for the experimental zone is that it will encourage the flow of information between scientists and end users in a much more interactive way, thereby speeding up the dissemination of information to a larger user community, widening the opportunity to evaluate simulations and model outputs from a variety of different community perspectives.

Discussion and conclusion

In this paper we have looked at how model fusion techniques can be used to develop linked catastrophe models with 'plug and play' capability and how access to model fusion software on the Web will be able to lower the entry barriers to probabilistic loss modelling. Ultimately, this will create an opportunity for both commercial and academic providers and developers to be able to supply the insurance industry with specialist modelling components. The illustrative groundwater flooding catastrophe model of Royse *et al.* (2014), described in overview here, took around a 100 h to prepare. Although the groundwater component was already available, this time included coding six additional components, making all of them OpenMI compliant, and linking the models and other components together. So this is relatively time-efficient and would become more so with standard variants of the six new components available. If this method were taken up, even quick-and-dirty catastrophe modelling would have the potential to improve on rough scenario-based estimates for eventualities such as a dam collapsing. Another use is in generating catastrophe models for areas where none exist currently. For example, recent flooding in Thailand (2011) highlighted the need for flood risk models for south-east Asia (AON Benfield 2012; Willroth *et al.* 2012). Catastrophe models for a particular area are often lacking because either the cost of collecting adequate data is prohibitively expensive and/or there is a perceived low level of financial risk. Using model fusion techniques as described in this paper would provide a way of producing quick and cheap catastrophe models using available open source data drastically improving financial risk management in those areas affected.

Furthermore advances made in the understanding of hazard, vulnerability and exposure will bring considerable societal as well as economic benefits. This will translate, not only in a reduction in financial losses incurred, but will also save lives. Our global populations are becoming increasingly susceptible to the impacts of natural hazards as they move progressively to more urban centred living. Nations will need to develop a better understanding of where their vulnerabilities lie so that they can adapt, monitor and mitigate against major natural hazard events. It is therefore essential that all the relevant information related to the distribution and severity of hazards, and a region's vulnerability to particular natural hazards, are presented in an accessible, reliable and understandable form so that those that need it are able to make use of the information being provided. One way of achieving this is through the use of model fusion technology to produce open accessed 'plug and play' catastrophe models for use not only within the insurance sector but for all those that have to model and manage risk.

This paper is published with the permission of the Executive Director of the British Geological Survey. The work was funded under the NERC PURE research consortium and forms part of K. R. Royse's NERC KE Fellowship.

Appendix 1

The table below provides information on all the datasets exchanged by the OpenMI compatible components in the catastrophe model composition discussed in this paper. See Gregersen *et al.* (2005) and Moore & Tindall (2005) for more information on how OpenMI exchanges data and Royse *et al.* (2014) for more information on the groundwater catastrophe model generation. The full model can be downloaded from http://pureexperimentalzone.org

Model component	Input	Output
MABSWEC model	Three files: 1. Recharge scenario: gridded for each daily time step. Calculated from rainfall, potential evaporation, land use and topography 2. Boundary conditions: rivers and springs locations; groundwater abstraction locations and rates 3. Hydraulic parameters: hydraulic conductivity, storage coefficients	Gridded map of groundwater head at each time step.
Flood Depth	Two files: 1. Gridded groundwater heads for each time step 2. DEM (Digital Elevation Model) CEH-DTM (Morris & Flavin 1990)	Depth of groundwater flooding in X, Y, Z format for each day.
Event Database	Three files: 1. Depth of groundwater flooding in X, Y, Z format per day	For each flood event: • Event number • Start time • End time

(Continued)

Model component	Input	Output
	2. Lower and upper limits: When ingdepth a range is used. This is because groundwater model calibration is imperfect and so limits are necessary to screen out anomalous values. 3. Threshold value: height, on average, of groundwater during flooding event	• Depth of groundwater flooding in X, Y, Z format for each day within the event
Hazard	For each flood event three pieces of information are required: 1. Event number 2. Start and finish time 3. Flood depth maps for each day within the flood	For each event: • Event number • Hazard intensity maps (the map gives the maximum flood depth attained during the each event)
Vulnerability	Vulnerability curves	Flood damage costs
Exposure	Exposure data	Exposure information in X, Y, Z format
Loss	Three files: 1. Hazard intensity maps 2. flood damage costs 3. Exposure information	Occurrence exceedance probability (OEP) curve

References

ABI (Association of British Insurers) 2011. *Industry Good Practice for Catastrophe Modelling: A Guide to Managing Catastrophe Models as a Part of an Internal Model under Solvency II.*

AIR 2002. *Ten Years After Andrew: What Should We be Preparing for Now?* AIR Special Report, Technical Document HASR_0208. AIR Worldwide Corporation, Boston, MA, USA.

AIR 2011. *Understanding Earthquake Risk in Japan following the Tohoku-Oki Earthquake of March 11, 2011.* AIR Worldwide Corporation, Boston, MA, USA.

AON Benfield 2012. *2011 Thailand Floods Event Recap Report. Impact Forecasting March 2012*, Impact Forecasting, Chicago, IL, USA.

Avouac, J. P. 2011. The lessons of Tohoku-Oki. *Nature*, **475**, 300–301.

Babendreier, J. E. & Castleton, K. J. 2005. Investigating uncertainty and sensitivity in integrated, multimedia environmental models: tools for FRAMES-3MRA. *Environmental Modelling & Software*, **20**, 1043–1055.

Campbell, S. D. G., Merritt, J. E. O. *et al.* 2010. 3D geological models and their hydrogeological applications: supporting urban development: a case study in Glasgow-Clyde, UK. *Zeitschrift der Deutschen Gesellschaft fur Geowissenschaften*, **161**, 251–262.

Chavez-Lopez, G. & Zolfaghari, M. 2010. Natural catastrophe loss modelling: the value of knowing how little you know. *14th ECEE 2010*, 30 August to 3 September 2010, Ohrid, Macedonia.

Cummins, J. D. 2007. Reinsurance for natural and man-made catastrophes in the United States: Current state of the market and regulatory reforms. *Risk Management and Insurance Review*, **10**, 179–220.

Cummins, J. D. & Mahul, O. 2009. *Catastrophe Risk Financing in Developing Countries. Principles for Public Intervention.* The World Bank, Washington DC, USA

Dailey, P., Huddleston, M., Brown, S. & Fasking, D. 2009. *The Financial Risks of Climate Change: Examining the Financial Implications of Climate Change using Climate Models and Insurance Catastrophe Risk Models.* ABI Research Paper no 19, Association of British Insurers, London.

David, O., Markstrom, S. L., Rojas, K. W., Ahuja, L. R. & Schneider, I. W. 2002. The object modelling system. *In*: Ahuja, L., Ma, L. & Howell, T. (eds) *Agricultural System Models in Field Research and Technology Transfer.* CRC Press, Boca Raton, FL, USA, 317–330.

Doornkamp, J. 1995. Perception and reality in the provision of insurance against natural perils in the UK. *Transactions of the institute of British Geographers*, **20**, 68–80.

Eling, M., Schmeiser, H. & Schmit, J. T. 2007. The Solveny II process: Overview and critical analysis. *Risk Management and Insurance Review*, **10**, 69–85.

EU 2009. Directive 2009/138/EC of the European parliament and of the council of 25 November 2009 on the taking-up and pursuit of the business of insurance and reinsurance (Solvency II). *Official Journal of the European Union*, **L335**, 1–155.

Foresight 2012. *Reducing Risk of Future Disasters: Priorities for Decision Makers.* Final report. Government Office for Science, London.

Frankel, A., Mueller, C. *et al.* 1996. *National Seismic Hazard Maps Documentation.* US Geological Survey Open-File Report 96–532. US Government Printing Office, Washington, DC.

Gregersen, J. B., Gijsbers, P. J. A., Westen, S. J. P. & Blind, M. 2005. OpenMI: The essential concepts and their implications for legacy software. *Advances in Geosciences*, **4**, 37–44.

Grindley, J. 1967. The estimation of soil moisture deficits. *Meteorological Magazine*, **96**, 97–108.

Grossi, P., Kunreuther, H. & Windeler, D. 2005. An introduction to catastrophe models and insurance. *In*: Grossi, P. & Kunreuther, H. (eds) *Catastrophe Modelling: A New Approach to Risk Management*. Springer, New York, 23–42.

Guardiola-Albert, C. & Jackson, C. R. 2011. Potential impacts of climate change on groundwater supplies to the Doñana Wetland, Spain. *Wetlands*, **31**, 907–920, https://doi.org/10.1007/s13157-011-0205-4

Guha-Sapir, D., Vos, F., Below, R. & Ponserre, S. 2011. *Annual Disaster Statistical Review 2010: The Numbers and Trends*. CRED, Brussels.

Gurenko, E., Lester, R. & Mahul, O. 2006. *Earthquake Insurance in Turkey: History of the Turkish Catastrophe Insurance Pool*. World Bank Publications, Washington, DC.

Guy Carpenter 2011. *Managing Catastrophe Model Uncertainty: Issues and Challenges*. Guy Carpenter & Company, London and New York.

Howard, K. W. F. & Lloyd, J. W. 1979. The sensitivity of parameters in the Penman evaporation equations and direct recharge balance. *Journal of Hydrology*, **41**, 329–344.

Hughes, A. G., Mansour, M. M. & Robins, N. S. 2008. Evaluation of distributed recharge in an upland semi-arid karst system: the West Bank Mountain Aquifer, Middle East. *Hydrogeology Journal*, **16**, 845–854, https://doi.org/10.1007/s10040-008-0273-6

Knapen, R., Janssen, S., Roosenschoon, O., Verweij, P., De Winter, W., Uiterwijk, M. & Wien, J.-E. 2013. Evaluating OpenMI as a model integration platform across disciplines. *Environmental Modelling & Software*, **39**, 274–282.

Lay, T. & Kanamori, H. 2011. Insights from the great 2011 Japan earthquake. *Physics Today*, **64**, 33–39.

Lennartz, S., Livinia, V. N., Bunde, A. & Havlin, S. 2008. Long-term memory in earthquakes and the distribution of interoccurrence times. *Earth and Planetary Letters*, **81**, https://doi.org/10.1209/0295-5075/81/69001

Mansour, M. M. & Hughes, A. G. 2004. *User's Manual for the Recharge Model ZOODRM*. British Geological Survey Internal Report CR/04/151N, British Geological Survey, Keyworth Nottingham, UK.

Mansour, M. M., Barkwith, A. & Hughes, A. G. 2011. A simple overland flow calculation method for distributed groundwater recharge models. *Hydrological Processes*, **25**, 3462–3471, https://doi.org/10.1002/hyp.8074

Mastin, M., Guffanti, R. *et al.* 2009. Multidisciplinary effort to assign realistic source parameters to models of volcanic ash-cloud transport and dispersion during eruptions. *Journal of Volcanology and Geothermal Research*, **186**, 10–21, https://doi.org/10.1016/j.jvolgeores.2009.01.008

Moore, R. V. & Tindall, I. 2005. An overview of the open modelling interface and environment (the OpenMI). *Environmental Science & Policy*, **8**, 279–286.

Mora, S. 2009. Disasters are not natural. *In*: Culshaw, M. G., Reeves, H., Jefferson, I. & Spink, T. (eds) *Engineering Geology for Tomorrow's Cities*. Geological Society, London, Engineering Geology Special Publications, **22**, 101–112.

Morris, D. G. & Flavin, R. W. 1990. A digital terrain model for hydrology. *In*: *Proceedings of 4th International Symposium on Spatial Data Handling*, 23–27 July, Zurich, **1**, 250–262.

Munich RE 2000. *Topics Geo: Annual Review. Natural Catastrophes*. Munich RE, Munich, Germany.

Munich RE 2010. *Topics Geo: Natural Catastrophes 2009. Analyses, Assessments, Positions*. Munich RE, Munich, Germany.

Munich RE 2011*a*. *Nathan World Map of Natural Hazards*. Munich RE, Munich, Germany.

Munich RE 2011*b*. Significant natural catastrophes 1980–2011: 10 costliest winter storms in Europe ordered by insured loss. http://www.munichre.com/en/reinsurance/business/non-life/georisks/natcatservice/significant_natural_catastrophes.aspx [last accessed December 2012].

Munich RE 2012. NatCatSERVICE. Figures accompany press release on 4 January 2012 'Review of natural catastrophes in 2011: Earthquakes result in record loss year'. http://www.munichre.com/app_pages/www/@res/pdf/media_relations/press_releases/2012/2012_01_04_munich_re_natural-catastrophes-2011-overview_en.pdf [last accessed December 2012].

Mailier, P. J., Stephenson, D. B., Ferro, C. A. T. & Hodges, K. I. 2005. Serial clustering of extratropical cyclones. *Monthly Weather Review*, **134**, 2224–2240.

Nature Geoscience 2008. Globalising quake information. *Nature Geoscience*, **1**, 803. http://www.nature.com/ngeo/journal/v1/n12/full/ngeo368.html

OpenMI 2009. The Open-MI life project website. http://www.openmi-life.org/

Overeem, I., Berlin, M. M. & Syvitski, J. P. M. 2013. Strategies for integrated modelling: the community surface dynamics modelling system example. *Environmental Modelling & Software*, **39**, 314–321.

Ozawa, S., Nishimura, T., Suito, H., Kobayashi, T., Tobita, M. & Imakiire, T. 2011. Coseismic and post-seismic slip of the 2011 magnitude-9 Tohoku-Oki earthquake. *Nature*, **475**, 373–377.

Pappenberger, F., Dutra, E., Wetterhall, F. & Cloke, H. 2012. Deriving global flood hazard maps of fluvial floods through a physical model cascade. *Hydrology and Earth System Sciences Discussions*, **9**, 6615–6647, https://doi.org/10.5194/hessd-9-6615-2012

Pelling, M. 2003. *The Vulnerability of Cities: Natural Disaster and Social Resilience*. Earthscan Publications, London.

Penman, H. L. 1948. Natural evaporation from open water, bare soil and grass. *Proceedings of the Royal Society London A*, **193**, 120–145.

Penning-Rowsell, E., Viavattene, C., Parode, J., Chatterton, J., Parker, D. & Morris, J. 2010. *The Benefits of Flood and Coastal Risk Management: A Handbook of Assessment Techniques-2010 (Multi-Coloured Manual)*. Flood Hazard Research Centre, London.

Pitt, M. 2008. *Learning Lessons from the 2007 Floods*. Cabinet Office, London.

QU, Y., DODOV, B., JAIN, V. & HAUTANIEMI, T. 2010. An inland flood loss estimation model for Great Britain. *British Hydrological Society 3rd International Symposium, Managing Consequences of a Changing Global Environment*, 19–23 July, Newcastle.

ROUGIER, J., SPARKS, S. *ET AL.* 2010. *SAPPUR: NERC Scoping Study on Uncertainty and Risk in Natural Hazards. Summary and Recommendations,* http://www.nerc.ac.uk/research/funded/programmes/pure/sappur-summary-report.pdf

ROYSE, K. R., HILLIER, J. A., WANG, L. A., LEE, T. F., O'NEILL, J. A., KINGDON, A. & HUGHES, A. 2014. The application of componentised modelling techniques to catastrophe model generation. *Environmental Modelling and Software*, **61**, 65–77.

RYBSKI, D., BUNDLE, A. & VON STORCH, H. 2008. Long-term memory in 1000-year simulated temperature records. *Journal of Geophysical Research*, **113**, D02106, https://doi.org/10.1029/2007JD008568

SPARKS, R. S. J., LOUGHLIN, S. C. *ET AL.* 2012. Global volcano model. *Geophysical Research Abstracts*, **14**, EGU2012-13299.

STRÖMBERG, D. 2007. Natural disasters, economic development, and humanitarian aid. *Journal of Economic Perspectives*, **21**, 199–222.

SWISS RE 2007. *Natural Catastrophes and Man-Made Disasters In 2006: Low Insured Losses*, Sigma No. 2/2007. Swiss RE, Zurich.

TOWERS PERRIN 2005. *Hurricane Katrina: Analysis of the Impact on the Insurance Industry. White Paper.* Towers Perrin, Stamford, CN, USA.

TSENG, C.-P. 2012. Natural disaster management mechanisms for probabilistic earthquake loss. *Natural Hazards*, **60**, 1055–1063.

VAN WESTEN, C. J., CASTELLANOS, E. & KURIAKOSE, S. L. 2008. Spatial data for landslide susceptibility, hazard, and vulnerability assessment: an overview. *Engineering Geology*, **102**, 112–131.

VITOLO, R., STEPHENSON, D. B., COOK, I. & MITCHELL-WALLACE, K. 2009. Serial clustering of intense European windstorms. *Meterologische Zeitschrift*, **18**, 411–424.

WALTER, J. 2004. *World Disasters Report 2004. Focus on Community Resilience.* Kumarian Press, Bloomfield, CT, USA.

WANG, Y. & ROSOWSKY, D. V. 2012. Joint distribution model for prediction of hurricane wind speed and size. *Structural Safety*, **35**, 40–51.

WARNER, J. C., PERLIN, N. & SKYLLINGSTAD, E. D. 2008. Using the model coupling toolkit to couple earth system models. *Environmental Modelling & Software*, **23**, 1240–1249.

WILLIS 2007. *Windstrom Kyrill, 18 January 2007.* Willis Analytics, London. http://www.willisresearchnetwork.com/assets/templates/wrn/files/Willis%20Windstorm%20Kyrill%20Report%2001%20Feb%202007.pdf

WILLIS, 2012. *2012 Property Insurance Market Update – January 2012.* Willis North America, New York. http://www.willis.com/Documents/Publications/Services/Property/2012/Property_Market_Update_2012.pdf

WILLROTH, P., MASSMANN, F., WEHRHAHN, R. & REVILLA DIEZ, J. 2012. Socio-economic vulnerability of coastal communities in southern Thailand: the development of adaptation strategies. *Natural Hazards and Earth System Sciences*, **12**, 2647–2658, https://doi.org/10.5194/nhess-12-2647-2012

WOO, G. 1999. *The Mathematics of Natural Catastrophes.* Imperial College Press, London.

YUCEMEN, M. S. 2005. Probabilistic assessment of earthquake insurance rates for Turkey. *Natural Hazards*, **35**, 291–313.

A stochastic bioenergetics model-based approach to translating large river flow and temperature into fish population responses: the pallid sturgeon example

MARK L. WILDHABER[1]*, RIMA DEY[2], CHRISTOPHER K. WIKLE[2], EDWARD H. MORAN[1], CHRISTOPHER J. ANDERSON[3] & KRISTIE J. FRANZ[4]

[1]*US Geological Survey, Columbia Environmental Research Center, 4200 New Haven Road, Columbia, MO 65201-8709, USA*

[2]*Department of Statistics, University of Missouri–Columbia, 146 Middlebush Hall, Columbia, MO 65211- 6100, USA*

[3]*Climate Science Initiative, Iowa State University, 2021 Agronomy Hall, Ames, IA 50011, USA*

[4]*Geological and Atmospheric Sciences, Iowa State University, 3023 Agronomy Hall, Ames, IA 50011, USA*

**Correspondence: mwildhaber@usgs.gov*

Abstract: In managing fish populations, especially at-risk species, realistic mathematical models are needed to help predict population response to potential management actions in the context of environmental conditions and changing climate while effectively incorporating the stochastic nature of real world conditions. We provide a key component of such a model for the endangered pallid sturgeon (*Scaphirhynchus albus*) in the form of an individual-based bioenergetics model influenced not only by temperature but also by flow. This component is based on modification of a known individual-based bioenergetics model through incorporation of: the observed ontogenetic shift in pallid sturgeon diet from marcroinvertebrates to fish; the energetic costs of swimming under flowing-water conditions; and stochasticity. We provide an assessment of how differences in environmental conditions could potentially alter pallid sturgeon growth estimates, using observed temperature and velocity from channelized portions of the Lower Missouri River mainstem. We do this using separate relationships between the proportion of maximum consumption and fork length and swimming cost standard error estimates for fish captured above and below the Kansas River in the Lower Missouri River. Critical to our matching observed growth in the field with predicted growth based on observed environmental conditions was a two-step shift in diet from macroinvertebrates to fish.

Bioenergetics models can be used in fish life-history models to help partition energy intake based on the laws of thermodynamics, where the energy consumed by a fish must balance the energy required by physiological processes and growth (Enders & Scruton 2006). A bioenergetics model provides an approach of estimating the consumed energy that is partitioned into three basic components: metabolism, waste loss and growth. This modelling framework accounts for the energy cost experienced by the fish and is used to solve for the level of consumption consistent with the observed growth, integrating the array of environmental conditions experienced by the fish (Moss 2001). Essentially, bioenergetics models are used to understand the relationship between growth and feeding rates under different environmental conditions. Over the last few decades, bioenergetics models have been used widely as a tool in fisheries management and the agricultural industry to address issues related to management of sport fish populations (Chipps & Wahl 2008).

One well-known bioenergetics modelling approach used in fisheries is known as the Wisconsin model (Kitchell *et al.* 1977); its application has been reviewed by Hanson *et al.* (1997). The Wisconsin model refers to an approach that incorporates maximum feeding rates ($C_{\max}$) and P values (proportion of $C_{\max}$) as a way to explore consumption patterns. The P values are meant to put consumption estimates into context (i.e. a fish feeding at 40% or 95% of $C_{\max}$). As part of the Wisconsin approach, consumption is equated to the sum of metabolic cost, waste loss and net gain in weight, with metabolic cost being a function of temperature and body mass. Previous analyses have shown that the model output is sensitive to consumption and

From: RIDDICK, A. T., KESSLER, H. & GILES, J. R. A. (eds) 2017. *Integrated Environmental Modelling to Solve Real World Problems: Methods, Vision and Challenges*. Geological Society, London, Special Publications, **408**, 101–118.
First published online May 28, 2015, updated October 12, 2015 and March 10, 2016, https://doi.org/10.1144/SP408.10

respiration parameters (Bartell *et al.* 1986). In practice, estimates of bioenergetics model parameters are obtained through field studies and carefully designed laboratory experiments in which fish of various sizes are tested at different water temperatures and food-ration levels. However model parameters are often borrowed from related species or, due to difficulty in obtaining the parameter estimates for different sized fish, by extrapolating parameters to other size classes. This borrowing and/or extrapolation of parameter estimates consequently results in a source of uncertainty in the bioenergetics model. Therefore, it is of paramount importance to develop model parameter estimates in a way that accounts for such uncertainty and bias to increase the utility of the bioenergetics model.

Bioenergetics models have been developed for a wide variety of fish species and ontogenetic stages that include freshwater (e.g. mosquitofish (*Gambusia affinis*); Chipps & Wahl 2004) to salt water species (e.g. Pacific herring (*Clupea pallasi*); Megrey *et al.* 2007), larvae (walleye (*Sander vitreus*); Madon & Culver 1993), small species (e.g. minnows; Schindler *et al.* 1993) and large species (e.g. tunas; Boggs 1984). Although the bioenergetics literature is vast and dates back over 50 years, research on sturgeon *Acipenser* spp. populations began fairly recently for white sturgeon (*A. transmontanus*; Bevelhimer 2002), green sturgeon (*A. medirostris*; Mayfield & Cech 2004), Atlantic sturgeon (*A. oxyrinchus*; Niklitschek 2001; Niklitschek & Secor 2009), and gulf sturgeon (*A. oxyrinchus*; Flowers *et al.* 2010). However, a bioenergetics model for the endangered pallid sturgeon (Dryer & Sandvol 1993) has only recently been developed (Chipps *et al.* 2010).

Wildhaber *et al.* (2007, 2011) introduced a conceptual life-history model for pallid sturgeon (*Scaphirhynchus albus*). The pallid sturgeon is an endangered fish endemic to the turbid waters of the Missouri and Lower Mississippi rivers (Dryer & Sandvol 1993). The model was developed to delineate how *Scaphirhynchus* sturgeon ecology relates to river management. It provided the framework for expanding the Bajer & Wildhaber (2007) population forecasting model to include environmental variables for prediction of future population size and distribution of Missouri River pallid and shovelnose (*S. platorynchus*) sturgeon. Because sturgeon in large rivers may move long distances (DeLonay *et al.* 2007), a life-history model should incorporate various environmental conditions related to different parts of the river and tributary system. For greatest utility in assessing habitat effects on population processes, the life-history model should accommodate fine-scale, three-dimensional models of habitat use and availability, and fish behaviour nested within a broader geographic extent. One way to incorporate spatial and temporal environmental and habitat conditions into such a life-history model for large river fishes is through the use of bioenergetics models.

Chipps *et al.* (2010) developed a bioenergetics model for juvenile pallid sturgeon using laboratory measurements of consumption and respiration and field-derived information for growth, diet and water temperature from the inter-reservoir sector of the Missouri River above Gavins Point Dam in South Dakota. Their model was developed on two size classes of juvenile pallid sturgeon: 200–400 mm fork length (FL) and 500–700 mm FL. The model estimates of food consumption agreed well with those from laboratory experiments over a wide range of water temperatures. Although uncertainty quantification was not incorporated in individual parameter estimates, uncertainty was accommodated in the model by developing a corrective equation that reduced mean error and hence increased reliability of the model output.

In applying bioenergetics models like that of Chipps *et al.* (2010) for pallid sturgeon, parameter estimates are assumed true, and held constant among individuals and across time and space. Without incorporating stochasticity into parameter estimates, the results of such models provide an unrealistic perception of certainty in their predictions. Furthermore, inclusion of parametric stochasticity in the bioenergetics model can be used to help account for differences in observed responses among individual fish. The deterministic bioenergetics model as developed by Chipps *et al.* (2010) produces a point estimate for the individual parameters that could potentially vary among individual fish due to physiological differences. The stochastic bioenergetics model we develop here includes not only point estimates of individual model parameters but also their associated distributions, and thus reflects current uncertainty about their true values.

This paper focuses on development of an individual-based bioenergetics model for pallid sturgeon that extends Chipps *et al.*'s (2010) pallid sturgeon bioenergetics model with swimming costs (Geist *et al.* 2005), diet shift (Grohs *et al.* 2009; Chipps *et al.* 2010), and parameter uncertainty as key components within a framework for informing management decisions. The goal is a model that comes closer to providing more realistic predictions of pallid sturgeon growth potential within the Lower Missouri River mainstem (the Missouri River mainstem below Gavins Point Dam to its confluence with the Mississippi River). To inform and validate the model we used on-going US Geological Survey (USGS) Comprehensive Sturgeon and Research Program work on migration, physiology, habitat choice and spawning success of pallid sturgeon (DeLonay *et al.* 2009) and US Army Corps of Engineers (USACE) Pallid Sturgeon Population Assessment Program (PSPAP) (Drobish 2008) data. Temperature and discharge data obtained from the USGS (USGS 2011) for gauging stations at Nebraska City, Nebraska (USGS 06 807 000 Missouri River at Nebraska City, NE; referred to here as Nebraska) and Boonville, Missouri (USGS 06 909 000 Missouri River at Boonville, MO; referred to here as Missouri) (Fig. 1) were used for input into the pallid sturgeon bioenergetics model to produce simulated pallid sturgeon growth trajectories. Ultimately, this model provides the basis of a forecasting tool that could be used to quantify the relative importance of various environmental factors in the Lower Missouri River on individual growth and consumption of juvenile and adult pallid sturgeon.

Methods

Study area

Using a modified version of Chipps *et al.*'s (2010) bioenergetic model for pallid sturgeon, we developed growth/consumption relationships for two locations along the mainstem of the Lower Missouri River (Fig. 1). These two locations lie along the Lower Missouri River: Above Kansas River (AKR) (Gavins Point Dam to Kansas River confluence) and Below Kansas River (BKR) (Kansas River confluence to Missouri River mouth), with the AKR segment tending to have higher velocities than the BKR segment (Galat *et al.* 2001).

Model development

The bioenergetics model developed here uses parameter estimates from Chipps *et al.* (2010) for pallid sturgeon and Geist *et al.* (2005) for white sturgeon with additional information on parameter distributions from various other sources (Table 1). This model, along with daily river velocity (calculated from discharge) and temperature data, were used to predict pallid sturgeon growth potential under different river conditions both in time (season – described later) and space (location – AKR and BKR).

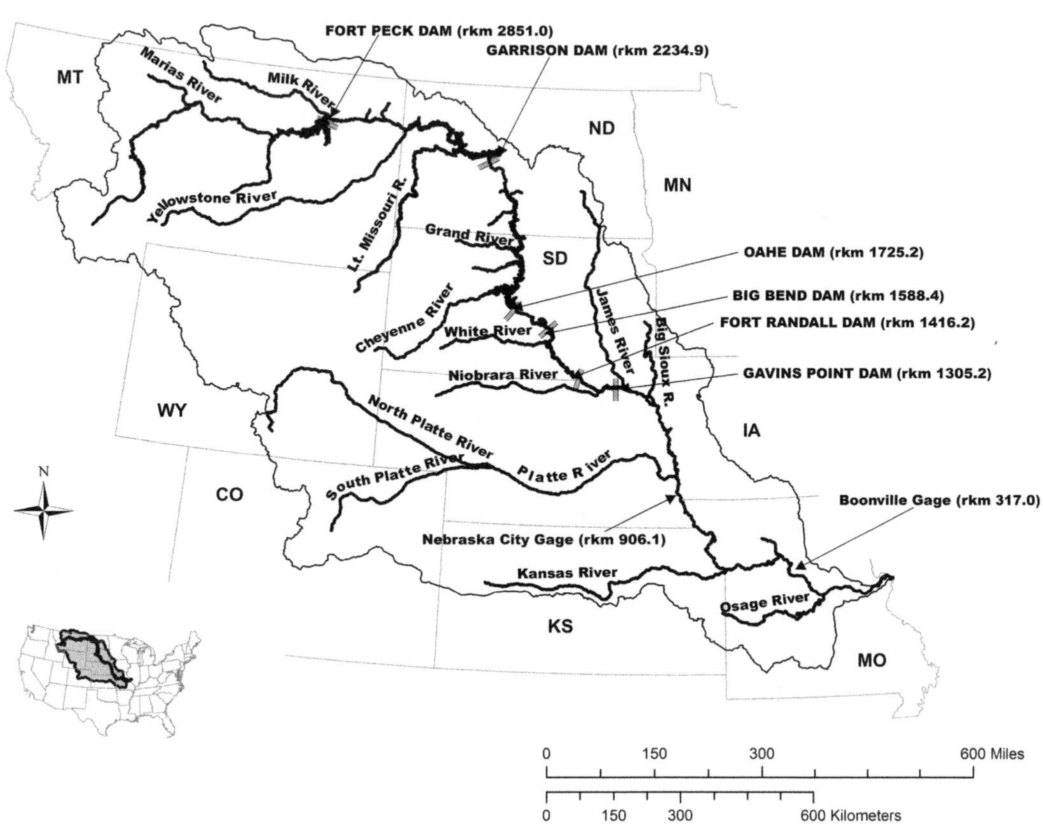

Fig. 1. Missouri River study area (rkm, river kilometre).

Table 1. *Distributional assumptions of bioenergetics model parameters*

Parameter	Description	Value	Distribution	Source
Consumption ($g\ g^{-1}\ day^{-1}$)	$C = a_c W^{b_c} f(T) P$			
a_c	Intercept	0.553	exp(trunc-N(-2.6698, 0.135) + ($\ln(Q_C)/10$)28), truncated 2 SE	Chipps *et al.* (2010).
b_c	Slope	−0.326	trunc-N(0.674, 0.0767) - 1, truncated 2 SE	Chipps *et al.* (2010)
Q_C	Q_{10} rate at low temperature	2.1	exp(10*trunc-N(0.074, 0.00286)), truncated 2 SE	Chipps *et al.* (2010), distribution Dowd *et al.* (2006*b*).
T_{o_C} (°C)	Optimum feeding temp.	28		Chipps *et al.* (2010)
T_{m_C} (°C)	Maximum feeding temp.	33		Chipps *et al.* (2010)
P	Proportion of maximum consumption for small fish		trunc-N(0.544, 0.028) AKR trunc-N(0.528, 0.024) BKR truncated 2 SE	Model fitting Pallid Sturgeon Population Assessment (PSPAP) data (Drobish 2008)
P	Proportion of maximum consumption for large fish		$=1-\exp[-(FL/\theta)]I[FL \geq x]$	
θ			trunc-N(764.98, 6.22), truncated 2 SE, Boonville trunc-N(697.43, 5.56), truncated 2 SE, Nebraska	Model fitting PSPAP data
Respiration ($J\ g^{-1}\ day^{-1}$)	$R = 13562 a_r W^{b_r} g(T) \exp(a_s p_u(0.58V + S_u) + (1 - p_u)S_d$			
a_r	Intercept	0.017	exp(trunc-N(-6.041, 0.147) + ($\ln(Q_R)/10$)30), truncated 2 SE	Chipps *et al.* (2010), distribution Dowd *et al.* (2006*a*)
b_r	Slope	−0.15	trunc-N(0.85255, 0.01198) − 1, truncated 2 SE	Chipps *et al.* (2010), distribution Dowd *et al.* (2006*a*)
Q_R	Q_{10} rate at low temp.	1.92	exp(10*trunc-N(0.06554, 0.00912)), truncated 2 SE	Chipps *et al.* (2010), distribution Dowd *et al.* (2006*a*)
T_{o_R} (°C)	Optimum temp. for respiration	30		Chipps *et al.* (2010).
T_{m_R} (°C)	Lethal water temp.	35		Chipps *et al.* (2010).
a_s	Coefficient of swimming speed	0.00914	trunc-N(0.00914, 0.008), truncated left, 0	Geist *et al.* (2005).

p_u	Proportion of time swimming upstream		trunc N(0.0097,0.031), with prob p_1 trunc N(0.471,0.178), with prop p_2, Mar–Jun trunc-N(0.4469, 0.255), Jul–Oct trunc N(0.0167,0.047), with prob p_1 trunc N(0.4275,0.149), with prop p_2 trunc N(0.97,0.064), with prob p_3, Nov–Feb truncated at 0 and 1	Model fitting non-spawning pallid sturgeon tracking data (Delonay *et al.* 2009). Based on a combination of Dirichlet(1,2) for Mar–Jun and Dirichlet(3,5,2) for Nov–Feb and multinomial(1,1,p) with p based on appropriate Dirichlet distribution.
S_u (cm s^{-1})	Upstream swimming speed		exp(0.556), Mar–Jun exp(0.923), Jul–Oct exp(1.616), Nov–Feb	Model fitting Delonay *et al.* (2009) tracking data.
S_d (cm s^{-1})	Downstream swimming speed		exp(0.527), Mar–Jun exp(0.685), Jul–Oct exp(2.300), Nov–Feb	Model fitting Delonay *et al.* (2009) tracking data.
Fa	Proportion of energy egested	0.1	triangular(0.1, 0.05, 0.2)	Chipps *et al.* (2010), distribution Wetherbee & Gruber (1993)
Ua	Proportion of energy excreted	0.04	triangular(0.04, 0.02, 0.08)	Chipps *et al.* (2010), distribution Brett & Groves (1979), Duffy (1999)
D	Proportion of C adjusted for egestion (F), used to calculate specific dynamic action (SDA)	0.13	triangular(0.13, 0.065, 0.26)	Chipps *et al.* (2010), distribution DuPreez *et al.* (1988), Sims & Davies (1994), Duffy (1999), Ferry-Graham & Gibb (2001)
W (g)	Fish weight		trunc-N(603.647, 270.93), truncated left, 0	Model fitting PSPAP data
FL (mm)	Fish fork length = $a_w W^{b_w}$			
a_w	Intercept		trunc-N(82.339, 2.46), truncated 2 SE	Model fitting PSPAP data
b_w	Slope		trunc-N(0.301924, 0.005), truncated 2 SE	Model fitting PSPAP data
m_{11} (J g^{-1})	Prey (fish) energy density	4473.533		Hansen *et al.* (1993), Venturelli & William (2006), Grohs *et al.* (2009).
m_{12} (J g^{-1})	Prey (mayfly) energy density	3065.83		Sitaramaiah (1967).
m_2 (J g^{-1})	Predator energy density	2698	trunc-N(2698, 254), truncated 2 SE	Chipps *et al.* (2010).

Basic bioenergetics model

The bioenergetics model (starting with Chipps *et al.* 2010) was developed using a general mass balance equation:

$$C = (R + \text{SDA}) + (F + U) + G,$$

where C is food consumption, R is standard metabolism, SDA is specific dynamic action, F is egestion, U is excretion and G is gonadal or somatic growth.

The rate of feeding (C) is defined as a function of the maximum consumption rate ($C_{\max}$) achievable at the optimal temperature for consumption for an individual fish of a given size, as follows,

$$C = C_{\max} P r_c, \tag{1}$$

where C is obtained by simply balancing the model as in the previous equation, and through an iterative process, a P value is fitted until $C_{\max} P r_c$ is equal to C with r_c being a temperature dependent proportional adjustment of consumption rate that lies between 0 and 1 with

$$r_C = (V_C{}^{X_C})(e^{X_C(1-V_C)}),$$

$$V_C = \frac{T_{m_C} - T}{T_{m_C} - T_{o_C}},$$

$$X_C = \frac{Z_C{}^2(1 + (1 + 40/Y_C)^{1/2})^2}{400},$$

$$Z_C = \ln(Q_C)(T_{m_C} - T_{o_C}),$$

$$Y_C = \ln(Q_C)(T_{m_C} - T_{o_C} + 2).$$

where T is ambient water temperature in degree Celsius, T_{o_C} optimum feeding temperature = 28 °C, T_{m_C} maximum feeding temperature = 33 °C, Q_C approximates a Q_{10} rate over the low range of temperatures = 2.1. The proportionality constant (P) is used to adjust the actual consumption as a proportion of maximum consumption in fitting the growth curve, having values in [0,1]. $C_{\max} = 0.553W^{-0.326}$ is the maximum specific feeding rate as estimated from feeding experiments conducted at the optimum temperature for the particular fish species, where W is fish weight (g) and 0.553 and −0.326 are the intercept and slope values for a 1-g fish, respectively.

The SDA value is estimated based on the metabolic cost of digestion, deposition and absorption of consumed energy and is assumed to be a proportion of the food consumption rate as follows (Kitchell *et al.* 1977):

$$\text{SDA} = D(C - F), \tag{2}$$

where D is a proportion. Waste losses, based on F and U, are modelled as constant proportions of consumed energy and are given by Chipps *et al.* (2010) as

$$F = FaC, \tag{3}$$

$$U = Ua(C - F), \tag{4}$$

where Fa is the proportion of energy egested and Ua is the proportion of energy excreted.

Proportional consumption and length

The proportion of maximum consumption (P) is one of the key parameters in determining the growth of fish. In practice, the true P associated with a pallid sturgeon is unknown. Based on the fixed parameter model, we estimated P using observed beginning and ending weights over corresponding time periods for pallid sturgeon recorded in PSPAP data. Observed water temperature and velocity (computed from observed discharge) at the USGS gauging station central to the river location in which the fish were collected were used as input to the bioenergetics model. Preliminary analyses determined that a constant P across all size classes of pallid sturgeon could not account for observed growth in the field. We found that a function that related P to FL was necessary for the model to successfully mimic observed growths for fish larger than 500 mm FL. However, for smaller fish (FL less than 500 mm), there was no significant relationship between P and FL.

In order to correctly parameterize the P to FL relationship, we trained the bioenergetics model for the gauges at Boonville, MO and Nebraska City, NE separately as representative of BKR and AKR, respectively. By 'training' the model, we refer to developing a separate P to FL relationship for each of the two segments of the Missouri River based on their specific water velocities and temperatures. To accomplish this, we used mark-recapture data from the PSPAP; Drobish 2008) and Pallid Sturgeon Stocking Program of the US Fish and Wildlife Service (USFWS 2008). Only beginning and ending weights for pallid sturgeon captured and recaptured in the same river location (i.e. AKR, BKR) were used to estimate growth and to develop a separate relationship between FL and P for each of these river segment. To develop these relationships, we ran the model keeping all physiological parameters of the bioenergetics model constant. To estimate P, we used observed water temperature and flow velocity from the corresponding river segment where the fish had been captured and recaptured for the period of time in which the fish growth occurred. Finally, given the initial and final weight of the fish, we solved for the proportion of maximum food necessary under the model to attain the observed final weight at recapture. Based on our preliminary assessment mentioned above, once the P values were estimated, we fitted the non-linear regression model (Table 1) to a set of P values for fish having FL larger than 500 mm (referred to as large fish) and a truncated normal distribution to the set of P values estimated for fish 500 mm or smaller (referred to as small fish).

Diet

Diet composition plays an important role in determining growth of pallid sturgeon (Grohs *et al.* 2009). Grohs *et al.* (2009) found that mayflies (Ephemeroptera), particularly the family Isonychiidae, are an important component of the juvenile pallid sturgeon diet. In addition, the diet of pallid sturgeon changes from macroinvertebrates to fish as their body length increases, with fish between 350 and 500 mm FL consuming 57% fish by wet weight and those more than 500 mm FL consuming 90% fish by wet weight (Gerrity *et al.* 2006; Grohs *et al.* 2009). We used these size ranges and their associated percentages of fish and macroinvertebrates to determine the relationship between FL and P. Based on previous work (Sitaramaiah 1967; Hansen *et al.* 1993; Venturelli & William 2006; Grohs *et al.* 2009), the caloric densities used for prey fish and macroinvertebrates were 4473.5 J g^{-1} wet mass and 3065.8 J g^{-1} wet mass, respectively, with caloric density assumed to be constant over time. We used a mean caloric density of 2698 J g^{-1} wet mass for pallid sturgeon (Table 1).

Respiration and swimming speed

Respiration rate (R) as given in Chipps *et al.* (2010) is measured by oxygen consumption, which is dependent on water temperature, fish mass and activity cost:

$$R = R_{\max} r_R A, \quad (5)$$

where $R_{\max} = a_r W^{b_r}$, $R_{\max}$ is the maximum weight-specific standard respiration rate at the optimum temperature, a_r and b_r are the intercept and slope for a 1-g fish, respectively, in the relationship between $R_{\max}$ and W, and A is an activity parameter to specify respiration rates. Furthermore, r_R is a temperature dependent proportional adjustment of respiration rate that lies between 0 and 1 with

$$r_R = (V_R{}^{X_R}) e^{X_R(1-V_R)}$$

$$V_R = \frac{T_{m_R} - T}{T_{m_R} - T_{O_R}},$$

$$X_R = \frac{Z_R{}^2(1 + (1 + 40/Y_R)^{1/2})^2}{400},$$

$$Z_R = \ln(Q_R)(T_{m_R} - T_{O_R}),$$

$$Y_R = \ln(Q_R)(T_{m_R} - T_{O_R} + 2),$$

where T is ambient temperature in degree Celsius (°C), T_{O_R} optimum temperature for respiration = 30 °C, T_{m_R} maximum (lethal) water temperature = 35 °C, and Q_R approximates a Q_{10} rate over the low range of temperatures = 1.92.

The costs associated with swimming under varying water velocity conditions were incorporated into the model by replacing the constant value most often used for activity costs with a function relating costs to swimming speed. The swimming speed function was taken from work done on white sturgeon (Geist *et al.* 2005). To increase the accuracy of swimming cost estimates, seasonality in swimming was introduced into the model based on observed swimming patterns, and associated variability, from tracking data on pallid sturgeon in the Lower Missouri River (DeLonay *et al.* 2009). Using tracking data for non-reproductive pallid sturgeon, we estimated the average upstream and downstream swimming speeds and the proportion of time the fish moved in either direction during spawning season or spring (March through June), summer/fall (July through October), and winter (November through February) (Fig. 2). Fish expend more energy moving upstream than downstream as a result of swimming against the flow. We incorporated this additional cost by assuming that a fish's swimming speed would be equivalent to the observed upstream swimming speed plus the water velocity through which the fish was swimming. This upstream estimate of swimming speed was further modified based on work by McElroy *et al.* (2012) on pallid sturgeon upstream migration paths. In an initial attempt to estimate energetic costs of swimming from observed telemetered fish, McElroy *et al.* (2012) observed that the chosen paths of upstream travel of pallid sturgeon were energetically less costly than the average possible path. The actual path chosen had an estimated energetic equivalent of an average velocity of 1.18 m s^{-1} compared to the average available estimated energetic equivalent of a random set of paths (i.e. velocity of 2.03 m s^{-1}) (B. McElroy, pers. comm., 2013). It is important to realize that this value is provisional and may change as more fish are evaluated. Based on McElroy *et al.* (2012), we used 58% (i.e. $(1.18/2.03) \times 100$) of the estimated average velocity found in a cross section of the Lower Missouri River as the estimate of the velocity experienced by

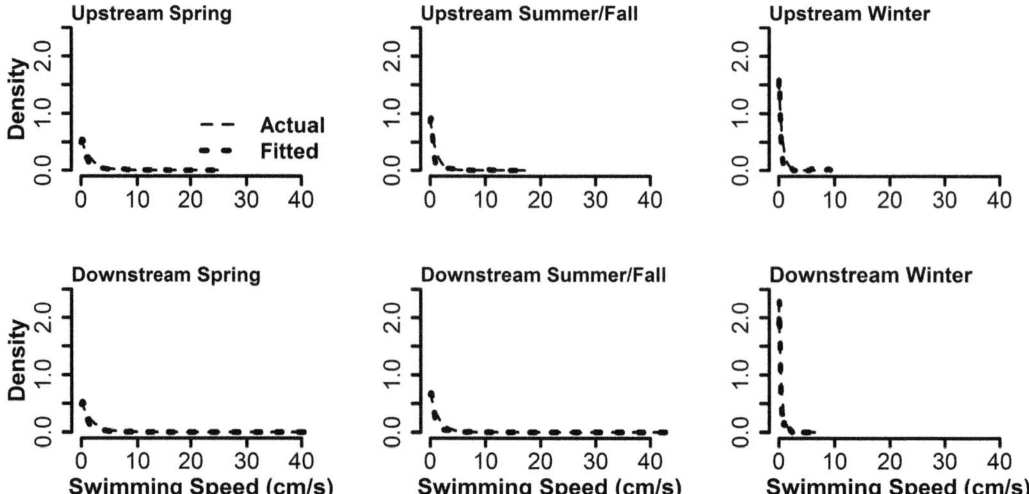

Fig. 2. Actual (dashed) and fitted (dotted) exponential distribution of upstream (upper row) and downstream (lower row) swimming speed of fish based on tracking data and separated on the basis of three seasons (from left to right columns: spring, summer/fall, and winter).

the fish. The logic of using a proportion is that when the average velocity changes in a channel section the relative distribution of high and low velocities does not (i.e. fastest in channel and slowest along the edges). Using a proportion also provides the ability to apply the model to both migrating and non-migrating pallid sturgeon through the use of actual, observed differences in swimming speeds (see below). Therefore, upstream swimming speed based on velocity was adjusted using a factor of 0.58 (see equation for *S* below).

The resulting model for daily oxygen consumption (*M* in g O_2/day) based on the above described modifications becomes:

$$\log(M_1) = -6.041 + 0.852 \log W + 0.065\,T + 0.00914\,S,$$

where

$$S = p_u(0.58V + S_u) + (1 - p_u)S_d.$$

Here, *S* is relative swimming speed (cm sec^{-1}), *V* is the flow velocity (cm s^{-1}), p_u is proportion of time the fish swims upstream, S_u is upstream swimming speed relative to a stationary point (cm s^{-1}) and S_d downstream swimming speed relative to a stationary point (cm s^{-1}).

We rewrite the above equations as follows:

$$M_1 = e^{-6.041+0.065(T)+0.00914(S)}\,W^{0.852}.$$

At $T = 30$ °C (optimum temperature for respiration) maximum specific respiration rate in turn, is estimated as

$$R_{max} = \frac{M_1}{W} = 0.017\ W^{-0.15}\ e^{0.00914(S)}.$$

Intermittent data collected in the field and reported for each gauging station by the USGS (USGS 2011) were used to calculate velocity based on methods discussed by Leopold & Maddock (1953). Discharge and velocity are related by the equation,

$$V = aD^b,$$

where *V* is the velocity (cm s^{-1}) and *D* is the discharge (m^3 s^{-1}).

In this case, a and b are the empirical coefficients of the non-linear equation. The estimated a and b for AKR (Nebraska) were 0.173 and 0.302, and for BKR (Missouri), 0.117 and 0.303, respectively. Knowing that these relationships change with changes in channel form, we used the longest sequence of field measured data available, between 1950 and 2009, to get long-term-average estimates for these relationships. Once estimated, these parameter values were used to convert daily discharge data as reported at each station to daily velocity data. It must be noted that using mean cross-sectional velocities at gauging stations is a simplification since gauging stations are preferentially located to measure in straight, uniform reaches and tend to have higher velocities compared to adjacent reaches. The computed velocity data were rounded to 0.5 dm s^{-1} (5 cm s^{-1}) to account for the precision of the reported discharge data owing to the number of significant digits (USGS 2002, 2012).

Model parameter distribution

Distributions of the physiological parameters are based on literature, statistical analysis and pilot studies (Table 1). Most of the parameters, as found in previous literature (Bevelhimer *et al.* 1985), were given a truncated normal distribution. Parameters involved in waste losses and SDA were assumed to follow a triangular distribution (Dowd *et al.* 2006*a*). A truncated normal distribution was found from model fitting to be the best fit for computed *P* of small fish. For large fish, the distribution of the parameters of the non-linear regression model fitted between the *P* value and the FL was determined based on the Normality test (Shapiro-Wilk's test). The distributions for upstream and downstream swimming speeds were determined by model fitting of tracking data for non-spawning fish. All truncated normal distributions were truncated at $\pm$ two standard errors (SE) to maintain the nature of the relationships (i.e. preserve the sign associated with each parameter) and control for unrealistically extreme relationships.

The distribution of swimming speed was obtained from the tracking data of pallid sturgeon captured between 7 April 2006 and 5 May 2011 in the Lower Missouri River between 0 and 1303.5 rkm (0 and 810 river miles). Swimming speed was computed based on 109 unique non-reproductive pallid sturgeon individuals with average swimming for an individual based on 1 to 101 relocation points, discarding three non-spawning fish from the analysis since they had been recaptured and taken to the hatchery for potential broodstock. As previously mentioned, swimming speed was divided into spring, summer/fall and winter and, based on direction of travel, upstream and downstream. We found that an exponential distribution fitted seasonal upstream and downstream swimming speeds (Fig. 2). We fitted a mixture of truncated normal distributions to the different seasonal proportions of upstream swimming times.

As mentioned in 'Proportional consumption and length', we estimated the proportion of maximum consumption or *P* based on the PSPAP data (Drobish 2008) available for AKR and BKR. A total of 353 fish of unknown sex or reproductive stage were captured between November 2002 and October 2010 in various parts of the Lower Missouri River. We discarded those fish that were recaptured in less than 4 months to assure the fish had grown for one full season. A total of 265 fish (i.e. 181 for Nebraska, 84 for Missouri) were used to inform the model for initial and final weight and FL. For fish captured at a river kilometre less than 593.8 km (369 miles) or BKR, the Missouri (i.e. Boonville) gauge was used as the representative gauge with *P* computed using temperature and flow velocity from November 2002 to May 2010 when, essentially, a complete record was available. Consistent with the selection process applied for BKR, for fish captured at river kilometres above 593.8–1303.6 (369–810 miles) or AKR, the Nebraska (i.e. Nebraska City) gauge was used as the representative gauge with temperature and flow velocity data from March 2005 to May 2010. The recorded temperature for Nebraska had numerous missing data points, some of which were supplemented by the temperature recorded at Omaha (DeLonay *et al.* 2009). Missing values in the observed temperature data for AKR from 22 April to 18 November 2008 and 10 March to 7 December 2010 were linearly interpolated. For each fish, the analysis period for the temperature and flow velocity data overlapped with the time period between consecutive capture and recapture events to assure that the calculation of *P* was based on environmental conditions present during the time fish growth was measured. A regression model was fitted to the relationship between *P* and FL for big fish as a result of the clear indication of two separate relationships corresponding to the diet shift from macroinvertebrates to fish *c.* 500 mm FL (described in 'Results'). To incorporate stochasticity into the diet shift, we assumed that the pallid sturgeon diet starts to contain 57% fish between 250 and 350 mm FL, whereas fish comprise up to 90% of fish diet at 500–600 mm FL (Gerrity *et al.* 2006; Grohs *et al.* 2009). These shifts were incorporated into the model by assuming a uniform distribution for the two diet shifts over the ranges 250–350 mm and 500–600 mm, respectively.

For the swimming speed energetic cost coefficient (*S*), an SE was not available from the literature. Hence, to set SE for *S*, we used a separate, iterative procedure for AKR and BKR. In this procedure, all parameters of the model and associated SEs, except *S*, were set. To determine the SE for *S*, we set the SE and then simulated 1000 fish (see 'Model evaluation' below) for each of the observed PSPAP fish initial weights. We compared the median of the 1000 fish with the observed final weight. We increased or decreased SE and reran the simulations until an equal number of final-weight medians of the simulated fish ended up above and below the observed final fish weights.

Model evaluation

To assess the effectiveness of the model at forecasting pallid sturgeon growth, we applied it to a new set of 18 non-reproductive pallid sturgeon captured and subsequently recaptured in BKR between January 2008 and October 2011 for the BKR growth/consumption relationship, and 14 non-reproductive fish initially captured and later recaptured in AKR between May 2006 and April 2011 for the AKR growth/consumption relationship. Extensive movement and infrequent relocations of these fish precluded identification of actual movement pathways and locations within the river between consecutive recaptures.

To demonstrate the importance of location on the growth/consumption relationship, we applied both the AKR and BKR modelled relationships to water temperature and flow velocity from Nebraska from 1 January 1972 to 31 December 1977. This location and the time period were chosen for the length and completeness of the record of water temperature that occurred after Missouri River mainstem reservoirs were filled and regulation began. These water temperature and flow velocity data were used as input in the model. For this location comparison, we used the relationship of P and FL obtained using inputs from Nebraska.

For each simulation, all parameters and variables of the bioenergetics model were sampled from their respective distributions daily. The model was run on a daily timeframe with no birth and mortality in the population. We started with a 1000-fish cohort of pallid sturgeon sampled from a truncated normal distribution with mean weight 571.21 g (SD = 317.36 g, truncated from left at 0) based on PSPAP data (1 February 2010 to 22 October 2010).

Results

On average we found that non-reproductive, tracked pallid sturgeon in the Lower Missouri River swam upstream 36% of the time during spring, 45% during summer/fall, and 39% during winter. Also, these pallid sturgeon travelled an average of 1.85 cm sec^{-1} (SD = 3.9 cm sec^{-1}) in spring, 1.27 cm sec^{-1} (SD = 3.27 cm sec^{-1}) in summer/fall, and 0.52 cm sec^{-1} (SD = 1.27 cm sec^{-1}) in winter (combining upstream and downstream movements) (Fig. 2). Average upstream movement of these pallid sturgeon in spring was 1.80 cm sec^{-1} (SD = 3.20 cm sec^{-1}), in summer/fall 1.08 cm sec^{-1} (SD = 2.18 cm sec^{-1}) and in winter 0.62 cm sec^{-1} (SD = 1.54 cm sec^{-1}), whereas that of downstream movement was 1.90 cm sec^{-1} (SD = 4.47 cm sec^{-1}), 1.46 cm sec^{-1} (SD = 4.09 cm sec^{-1}), and 0.43 cm sec^{-1} (SD = 1.00 cm sec^{-1}), respectively. Overall upstream and downstream average swimming speeds were 1.34 cm sec^{-1} (SD = 2.64 cm sec^{-1}) and 1.52 cm sec^{-1} (SD = 4.04 cm sec^{-1}), respectively.

As previously described, we separated PSPAP collected pallid sturgeon from the Lower Missouri River based on whether they were captured BKR or AKR. There were 126 big ($\geq$500 mm FL) and 55 small (<500 mm FL) fish for AKR and 72 big and 12 small fish for BKR available for training the model. For AKR, we had to discard 6 fish since their capture dates fell outside the range of dates when temperature data were available in Nebraska. It was evident from the plot of P to FL (Fig. 3) that for big fish P increases as FL increases in a non-linear fashion.

The P for small fish captured BKR, having average FL 430.8 mm (SD = 59.2 mm) and computed based on the water temperatures and velocities in Missouri, ranged from 0.48 to 0.57 (Fig. 4), whereas big fish, having average FL 638.4 mm (SD = 88.3 mm), ranged from 0.49 to 0.70. For small fish captured AKR, with FL 426.8 mm (SD = 38.2 mm), P ranged from 0.49 to 0.94 (Fig. 4), whereas, for big fish with FL 671.7 mm (SD = 140.7 mm), P ranged from 0.48 to 0.80.

The slope of the non-linear regression fit for large fish for both AKR and BKR are similar (Fig. 3; Table 1). The difference in the two relationships lies in the intercept of the corresponding regression fit, suggesting that P increases with FL in a similar fashion irrespective of location. The reason behind the higher value of intercept of the regression fit for AKR can be attributed to the higher swimming speeds as estimated by recorded velocities (Fig. 7) from that location that translated into higher energy costs.

Validation of the model using independent fish showed that the model for BKR performed well; all of the observed weights of the 18 fish fall in the 95% quantile interval (Fig. 5). The model for AKR was less effective than the BKR model; however, it still correctly predicted 8 of the 14 final weights while underestimating or overestimating the final weights of the other 6.

Simulation results for Nebraska indicate that the 1000 fish started with an initial average weight of 597.15 g (SD = 285.70 g) and ended up with mean weight of 2164.78 g (SD = 2183.95 g) at the end of the 5-year simulation period, which is evidenced by the shift in fish weights to the right (Fig. 6). The final distribution of fish weight was wider (i.e. it had higher variance) than was the distribution for initial weights. Simulation results using the growth/consumption relationship for Missouri based on Nebraska temperatures and velocities predicted much less fish growth than that predicted using the Missouri temperatures and velocities. When conditioned on the Missouri growth/consumption relationship, a mean weight of 336.18 g (SD = 240.65 g) resulted (Fig. 6).

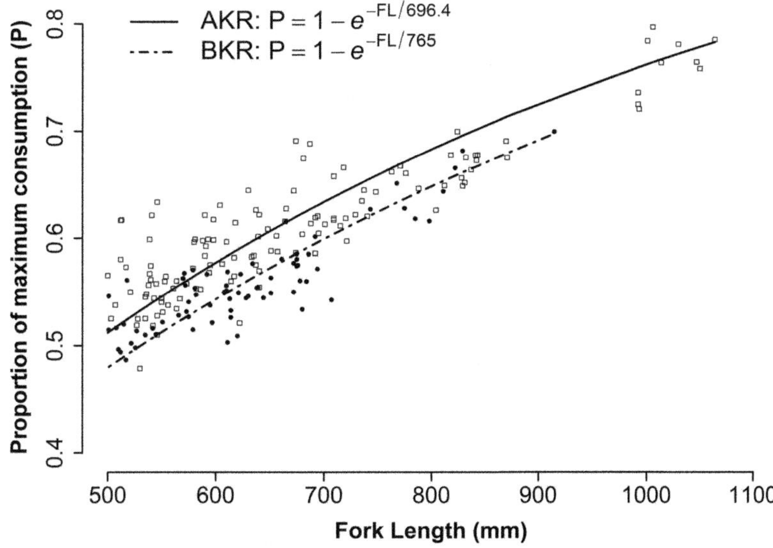

Fig. 3. Proportion of maximum consumption (*P*) v. fork length (FL) computed based on fish captured AKR trained on Nebraska City, Nebraska and BKR trained on Boonville, Missouri. Non-linear regression models were fitted on big fish having FL greater than 500 mm. The rectangles correspond to the fish captured AKR and circles to BKR. There was not a significant relationship between *P* and FL for fish less than 500 mm FL.

Discussion

Here we developed an individual-based, stochastic bioenergetics model for pallid sturgeon. The results indicate that pallid sturgeon growth estimates are dependent on the environmental conditions of the location where the model is trained. Assuming the observed velocities are representative of average velocities experienced by fish in the river, observed velocity of the narrower channelized river AKR tends to be higher resulting in higher estimated metabolic costs associated with swimming, and thus slower estimated growth rates. The varying flow velocity of the river and its effect on growth played an important role in determining consumption–length relationships for fish in different parts of the Lower Missouri River.

Fish growth rate is affected by genetic, environmental and nutritional factors (Very & Sheridan 2002). Studies by Harvey *et al.* (2006) on salmonids suggest that changes in streamflow with only small changes in other aspects of physical habitat can result in substantial changes in individual growth. These authors demonstrated that fish allocate energy to growth only after meeting maintenance costs; they also showed that relatively modest differences

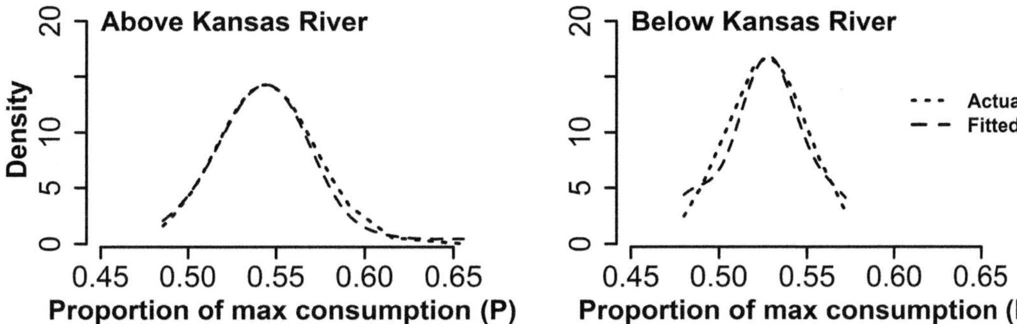

Fig. 4. Actual (dashed) and fitted (dotted) density plot of proportion of maximum consumption (*P*) for fish with FL less than 500 mm computed on fish captured in two different segments of the Lower Missouri River and trained using associated USGS gauge data. Left panel, AKR trained using Nebraska City, Nebraska data; right panel, BKR trained using Boonville, Missouri data.

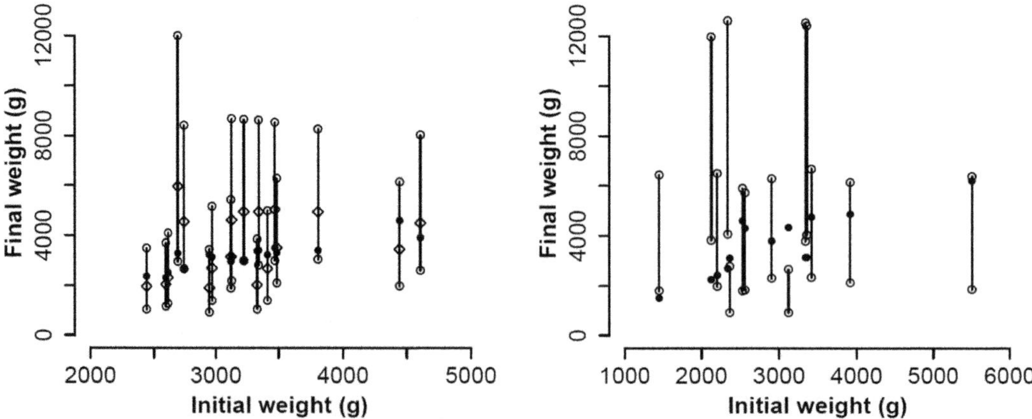

Fig. 5. Predicted 95% quantile interval and observed final fish weight. The left figure exhibits 18 fish captured BKR between January 2008 and October 2011. The right figure displays 14 fish captured AKR between May 2006 and April 2011. The *x* and *y*-axes indicate the observed initial and final weights respectively. The open diamonds indicate the median, circles indicate the 97.5 percent quantile, and the 2.5 percent quantile. The filled circles indicate the observed final weight of the corresponding fish. The 95% quantile intervals in both the plots are based on the data simulated from the bioenergetics model trained for BKR and AKR, respectively.

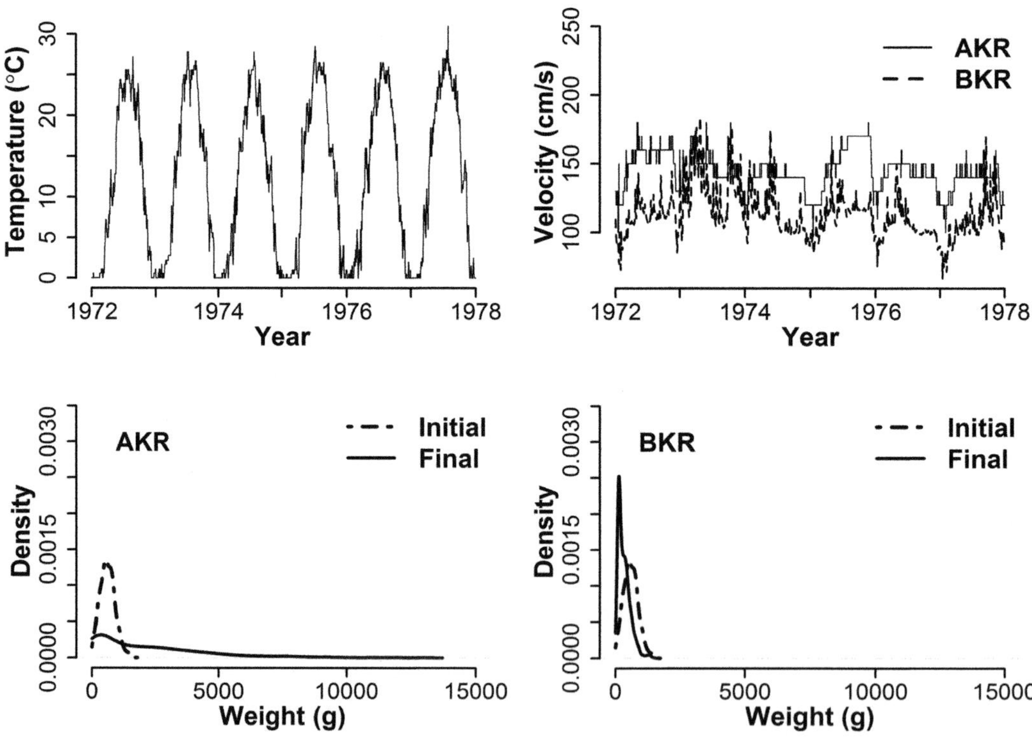

Fig. 6. Water temperature at Nebraska City, Nebraska (upper left, AKR); water velocity at Nebraska City, Nebraska and Boonville, Missouri (upper right, BKR); and simulated weight density plots for model trained AKR (lower left) and BKR (lower right) using the temperature and flow velocity of AKR from 1 January 1972 to 31 December 1977.

in total energy expenditure could potentially result in large differences in individual growth. As with salmonids, our observations from modelling pallid sturgeon growth suggest that water velocity may play an important role in their growth, and hence on population dynamics.

We found that shifting the diet of pallid sturgeon in the model was key to correctly estimating growth. Smaller fish diets are made up almost exclusively of macroinvertebrates that have lower energy densities compared to fish prey. Without shifting the diet to fish (higher energy density) as pallid sturgeon become larger, pallid sturgeon growth rate estimates were lower than the rates necessary to obtain the sizes observed in the field. Consistent with observed life-history attributes, as pallid sturgeon developed from juvenile to adult life stages, our modelling efforts indicated that a shift from aquatic invertebrates to piscine prey was necessary to attain growth rates observed in the field. It is important to note, however, that all things being equal, if a fish switches to a higher energy density food source, then its consumption rate (i.e. P value) should decrease because it would require less consumption per unit body weight to grow by the same amount. Hence, our model predictions, displaying increased P values in larger pallid sturgeon for growth estimates to approximate observed growths, were surprising. We hypothesize that once pallid sturgeon switch to fish, they become more efficient foragers. We did not have any measure of available prey. Though a shift in prey was necessary in our model, the hypothesis that a shift in prey is necessary for pallid sturgeon in the field requires further testing.

A primary controlling factor of fish growth is diet during different life stages; this is true for pallid sturgeon (Wildhaber *et al.* 2007, 2011). Grohs *et al.* (2009) found that the Isonychiidae family was an important component of juvenile pallid sturgeon diet. Duffy *et al.* (2010) found that the growth and survival of juvenile Chinook salmon (*Oncorhynchus tshawytscha*) during the marine stage of their life depended on the quality and quantity of prey consumed. These authors found that variation in diet composition and quality of prey could translate into strong annual growth patterns for this marine stage of life. Daly *et al.* (2009) suggested that a successful shift to a more piscivorous diet could be an important factor in the growth and survival of juvenile coho salmon (*O. kisutch*) and Chinook salmon. Piscivorous fishes that consume other fishes early in development generally experience a dramatic increase in growth after the ontogenic diet shift to piscine prey (Juanes *et al.* 1994).

To improve the accuracy of metabolic cost estimates, seasonality in swimming speed was included in the model along with its associated variability. The inclusion of seasonality in the observed swimming pattern was important as high swimming speed translates into higher metabolic cost and consequently slower rate of growth. As evidenced from Figure 2, the observed swimming speed of pallid sturgeon when swimming upstream is highest in summer/fall and lowest in winter suggesting the need to account for seasonal variability in the model. We found that accurately reflecting the actual swimming costs to fish required inclusion of seasonal variation in swimming behaviour. Most previous work on swimming behaviour and associated energetics costs were conducted using laboratory acclimated fish without any consideration of season. Adams & Parsons (1998) found that swimming performance of field acclimated smallmouth buffalo (*Ictiobus bubalus*) varied significantly among seasons suggesting the need to account for seasonal variation. Examination of environmental change effects on seasonally acclimated white crappie (*Poximus annularis*) critical swimming speed revealed that both size and swimming speed were instrumental in determining swimming ability (Parsons & Smiley 2003). Similar to our observations, Parsons & Smiley (2003) found that performance of a fish was lowest during winter and highest in summer/fall. Similarly, we found that the mean swimming speed of pallid sturgeon was highest in summer and decreased with temperature. This seasonal effect on fish swimming speed was also seen by Facey & Grossman (1990).

Any bioenergetics model must be evaluated to ensure the accuracy of model predictions and to identify potential sources of error (Ney 1993). Here we evaluated our model by comparing our predicted growth of pallid sturgeon to the observed growth of an independent set of pallid sturgeon collected from the Lower Missouri River. The effectiveness of the model to predict pallid sturgeon growth was demonstrated by final observed fish weights occurring within the predicted distribution of final weights for fish captured BKR and 8 of the 14 fish captured AKR with a combination of underestimation and overestimation for the other 6. For whitefish (*Coregonus clupeaformis*), Madenjian *et al.* (2006) found that their model on average initially underestimated growth by 16%. Their model provided much more accurate predictions of growth after adjusting the estimated cost of respiration. These results are similar to ours in that respiration seemed to drive overestimation and underestimation through a limited certainty of the estimate of work in the form of swimming.

In general, estimated P for fish captured in Nebraska tended to be higher than those captured in Missouri as evidenced by the shift in the intercept of the regression lines (Fig. 3). This outcome reflects flow velocity being greater in Nebraska than in

Missouri (Fig. 7), forcing modelled fish to consume more food as a consequence of higher metabolic cost associated with swimming. Swimming speed and *P* played important roles in estimating fish growth. Small *P* and high flow velocity, as an estimate of swimming speed, can result in an estimated weight loss, thus the need to separate fish in the Lower Missouri River based on their river kilometres. The model, trained using temperature and velocity data obtained from Missouri but run with temperature and velocity inputs from Nebraska (Fig. 6), indicates the potential impact of using an inaccurate consumption rate to estimate fish growth. The change in the population growth response is linked to *P* and estimated swimming speed based on flow velocities in Missouri that are less than those in Nebraska; hence, the importance of training the model using the environmental conditions to which the fish was exposed.

There are several possible reasons for the tendency of the AKR model to underestimate or overestimate the growth. First, several model parameters were borrowed from different species. For example, the activity parameters were borrowed from the white sturgeon (Geist *et al.* 2005). The high metabolic cost of swimming, as estimated based on water velocity, was very influential on the estimates of fish growth. It may be that swimming costs for white sturgeon are different from those of pallid sturgeon. This would result in overestimation or underestimation of growth, depending on whether the cost was higher or lower, respectively, of actual metabolic cost allocated by our model to swimming for pallid sturgeon when exposed to the same water velocities. Though not possible from current data, more intensive tracking with location-specific velocities and temperatures could be used to compare swimming speeds at different velocities. Additionally, Geist *et al.* (2005) found no relationship between slopes for swimming speed and fish weight, but they used a very narrow range of fish in their lab studies (600–800 mm). It may be possible that if the size range of the fish studied included a larger sized fish (as our study did), such a relationship might have occurred and yielded different regression outcomes.

Second, the model assumed the upstream pathway taken by a pallid sturgeon required that the

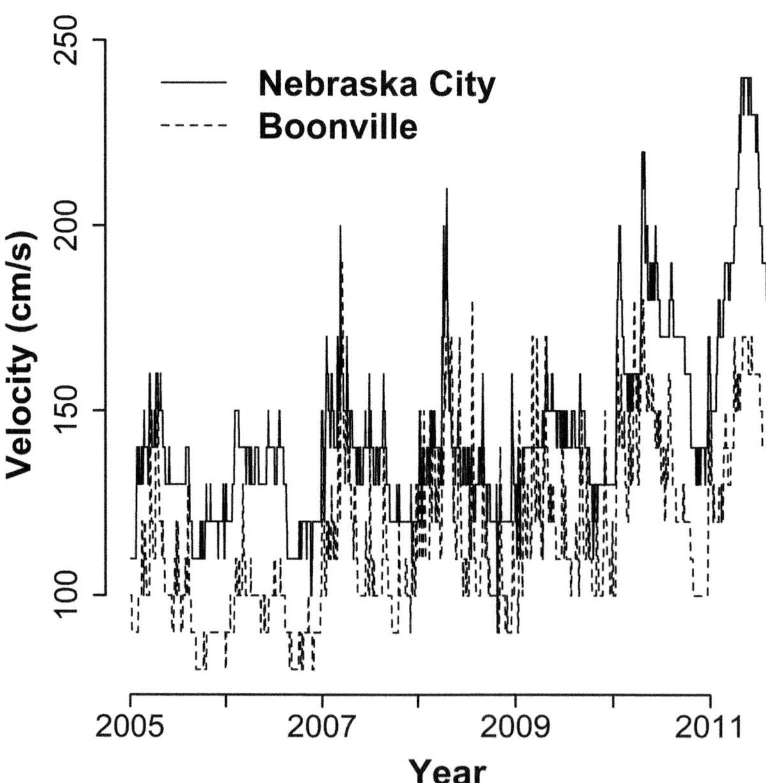

Fig. 7. Daily water velocity of Lower Missouri River at Nebraska City, Nebraska (solid line) and Boonville, Missouri (dashed line) from March 2005 to September 2011.

fish experiences 58% of the average flow velocity found in the segment of the river through which it is travelling (McElroy *et al.* 2012). Yet, the pathway taken by pallid sturgeon may expose them to lower or higher flow velocities.

Lastly, we were not able to model growth of the individual pallid sturgeon using the actual river conditions they experienced during their observed growth due to very sparse location data. We had to use velocities and temperatures from an available USGS gauge found in the portion of the river the fish was assumed to inhabit during its growing period based on first and last locations as representative of the conditions experienced by the fish. Using mean cross-sectional velocities at gauging stations is a simplification since gauging stations are preferentially located to measure in straight, uniform reaches and tend to have higher velocities compared to adjacent reaches. Further, empirical data from adjacent gauging stations can differ because discharge–velocity relationships depend on local hydraulics; hence, model outputs dependent on these values will be variously affected spatiotemporally. These field conditions and related observations suggest that more work is needed to further develop a precise model for pallid sturgeon, in particular: direct estimation of metabolic costs associated with swimming; finer scale estimation of actual swimming done in the field; and additional model validation using simultaneous measurements of water temperature and velocity actually experienced to estimate growth for individuals.

Even with inclusion of stochasticity, there are limitations to our model. First, the bioenergetics model, as originally developed by Chipps *et al.* (2010), is based on juvenile pallid sturgeon. Parametric values derived in Chipps *et al.* (2010) are all based on either larvae or juveniles (0.03–617.3 g), which were much smaller in weight than the upper range of weights used in this study. Because parameter estimates for large pallid sturgeon are not available in the existing literature, we extrapolated the model to adult life stages; hence, parameters used in our bioenergetics model are based on experimental study using juvenile pallid sturgeon and may not represent comparable values in adults.

Second, all parameter estimates come from laboratory studies whereas we implemented the model for pallid sturgeon in the wild. Fish behaviour and performance in the laboratory may be artificial and not truly representative of fish in the wild (Hansen *et al.* 1993; Ney 1993).

Third, the distributional assumptions obtained for the physiological parameters are based on literature found on other species, which may poorly represent pallid sturgeon physiology. Species-borrowing is common in scientific literature, yet a recent study found that a model based on Pacific salmon species was unable to accurately predict the metabolic cost for another Pacific salmon species (Trudel *et al.* 2004), suggesting the need to avoid the practice of species-borrowing whenever possible. In our case, we borrowed parameter estimates for metabolic cost from the white sturgeon, but our model validation test based on a new set of fish indicated that the metabolic costs of swimming for white sturgeon and/or the estimated work done by pallid sturgeon based on path analyses failed to fully represent the metabolic costs due to swimming for pallid sturgeon. The current model, trained for Nebraska, either overestimated or underestimated metabolic cost due to swimming in just under half the fish. Despite limitations linked to species-borrowing, the stochastic bioenergetics model we developed is capable of capturing the embedded uncertainty of the parameters and emerges as a promising mechanism to predict the observed growth of pallid sturgeon. Indeed, as a starting point in bioenergetics research, initial developments in model building inevitably identify shortcomings associated with existing data derived from different species. However, these initial efforts point us toward species-specific studies to garner empirical values that offset these downsides to data extracted from existing literature.

The purpose of this work was to develop an individual-based bioenergetics model for the endangered pallid sturgeon to serve as the basis of a tool for predicting fish growth under changing environmental conditions. This model is being used as part of a larger effort to develop a tool that could be used to provide a better understanding of the effects of climate change on the Missouri River ecosystem (Wildhaber *et al.*, this volume, in prep). Such a tool would help managers identify strategies for mitigating potential negative impacts of climate change on endangered sturgeon populations. In doing this, the model we developed provides a methodological framework for development of a management tool that would have the ability to integrate various environmental factors along with accommodating their uncertainties. Ultimately, such a tool may serve as a means to assess the growth dynamics of the pallid sturgeon under changing environmental conditions. It could potentially be applied to help predict the response of pallid sturgeon populations to changes in environmental conditions as a result of management actions and/or changes in climate, thus providing information that could be used to guide recovery efforts.

Future research should be done to help develop a better understanding of the various components of the model and how their stochasticity affects model outcomes. For example, model components could be grouped as physiological parameters (e.g.

respiration), ecological parameters (e.g. diet shift), and environmental inputs (e.g. river flow and velocity, and temperature) to assess the relative influence of stochasticity of each of these groups, with the other two groups held constant, on model outcomes. The physiological (bioenergetics) parameters would seem to be conceptually different from the other two groups. For instance, these bioenergetics values have some 'correct' value, but we may not be sure of that value. Therefore, they may be (approximately) constant, but the value is uncertain. However, ecological and environmental values are 'inherently' stochastic, varying daily.

The model could be further improved in several ways. First, caloric density of fish and invertebrate prey and prey items consumed could be made stochastic. We did not include these components of stochasticity because, essentially, no information on energy density distributions of potential invertebrate and fish prey of pallid sturgeon were available. Here such stochasticity was implicitly incorporated to some degree through the daily variation in pallid sturgeon caloric density. Second, with the addition of stochasticity to prey energy density and makeup, a different form of stochasticity, such as Brownian motion where today's value is dependent on yesterday's value plus some change, could be used to model pallid sturgeon caloric density. Third, estimates of caloric density of adult pallid sturgeon should be obtained to ensure the representativeness of juvenile pallid sturgeon caloric density for all pallid sturgeon. If differences do exist, they could impact the estimates of P and the energy balance of the model.

Funding for this work was provided by the USGS through the National Climate Change Wildlife Science Center. We would like to thank the PSPAP Team, USACE, USFWS and USGS for providing access to PSPAP, Pallid Sturgeon Propagation Program, and USGS Comprehensive Sturgeon Research Program data, respectively, and S. Chipps of USGS for use of the pallid sturgeon basic bioenergetics model parameterization to help parameterize the initial bioenergetics model. We would also like thank the following University of Missouri undergraduate students (S. Childers, L. Crawford, Q. Hall, J. Kastman, T. Schrautemeier, R. Scott, A. Wettlaufer), USGS staff (J. Albers, E. Beahan, N. Green, and L. Johnson) for their assistance in compiling data and running model simulations. Thanks to R. Jacobson and S. Chipps of USGS for reviewing earlier versions of this manuscript. Any use of trade, product or firm names is for descriptive purposes only and does not imply endorsement by the US Government.

Correction notice: The previous version was incorrect. This was due to an error in Table 1.

References

Adams, S. R. & Parsons, G. R. 1998. Laboratory measurements of swimming performance and related metabolic rates of field sampled smallmouth buffalo (*Ictiobus bubalus*): a study of seasonal changes. *Physiological and Biochemical Ecology*, **71**, 350–358.

Bajer, P. G. & Wildhaber, M. L. 2007. Population viability analysis of Lower Missouri River shovelnose sturgeon with initial application to the pallid sturgeon. *Journal of Applied Ichthyology*, **23**, 457–464.

Bartell, S. M., Breck, J. E., Gardner, R. H. & Brenkert, A. L. 1986. Individual parameter perturbation and error analysis of fish bioenergetics models. *Canadian Journal of Fisheries and Sciences*, **43**, 160–168.

Bevelhimer, M. S. 2002. A bioenergetics model for white sturgeon *Acipenser transmontanus*: assessing differences in growth and reproduction among Snake River reaches. *Journal of Applied Ichtyology*, **18**, 550–556.

Bevelhimer, M. S., Stein, R. A. & Carline, R. F. 1985. Assessing significance of physiological differences among three esocids with a bioenergetics model. *Journal of Fisheries and Aquatic Sciences*, **42**, 57–69.

Boggs, C. H. 1984. *Tuna Bioenergetics and Hydrodynamics*. PhD Thesis, University of Wisconsin-Madison, WI.

Brett, J. R. & Groves, T. D. D. 1979. Physiological energetics. *In*: Hoar, W. S., Randall, D. J. & Brett, J. R. (eds) *Fish Physiology*. Academic Press, New York, NY, 279–352.

Chipps, S. R. & Wahl, D. H. 2004. Development and evaluation of a western mosquitofish bioenergetics model. *Transaction of the American Fisheries Society*, **133**, 1150–1162.

Chipps, S. R. & Wahl, D. H. 2008. Bioenergetics modeling in the 21st century: reviewing new insights and revisiting old constraints. *Transactions of the American Fisheries Society*, **137**, 298–313.

Chipps, S. R., Klumb, R. A. & Wright, E. B. 2010. *Development and Application of a Juvenile Pallid Sturgeon Bioenergetics Model.* Final report to South Dakota Department of Game, Fish and Parks, Federal Aid Project Number T-24-R, Study 2424, Pierre, SD.

Daly, E. A., Brodeur, R. D. & Weitkamp, L. A. 2009. Ontogenic shifts in diets of juvenile and subadult coho and Chinook salmon in coastal marine waters: important for marine survival? *Transactions of the American Fisheries Society*, **138**, 1420–1438.

DeLonay, A. J., Papoulias, D. M. *et al.* 2007. Use of behavioral and physiological indicators to evaluate *Scaphirhynchus* sturgeon spawning success. *Journal of Applied Ichthyology*, **23**, 428–435.

DeLonay, A. J., Papoulias, D. M. et al. 2009. *Ecological Requirements of Pallid Sturgeon Reproduction and Recruitment in the Lower Missouri River – a Research Synthesis 2005–08*. USGS, Scientific Investigation Report 2009-5201, 91.

Dowd, W. W., Brill, R. W., Bushnell, P. G. & Musick, J. A. 2006*a*. Standard and routine metabolic rates for juvenile sandbar sharks (*Carcharhinus plumbeus*), including the effects of body mass and acute temperature change. *Fisheries Bulletin*, **104**, 323–331.

Dowd, W. W., Brill, R. W., Bushnell, P. G. & Musick, J. A. 2006*b*. Estimating consumption rates of juvenile sandbar sharks, (*Carcharhinus plumbeus*), in Chesapeake Bay, Virginia, using a bioenergetics model. *Fisheries Bulletin*, **104**, 332–342.

Drobish, M. R. 2008. *Pallid Sturgeon Population Assessment Project, Volume 1.3*. U.S. Army Corps of Engineers, Omaha District, Yankton, South Dakota.

Dryer, M. P. & Sandvol, A. J. 1993. *Recovery Plan for the Pallid Sturgeon (Scaphirhynchus albus): Bismarck, North Dakota*. US Fish & Wildlife Service, 55.

Duffy, K. A. 1999. *Feeding, Growth and Bioenergetics of the Chain Dogfish, Scyliorhinus Retifer*. PhD thesis, University of Rhode Island, Kingston, RI.

Duffy, E. J., Beauchamp, D. A., Sweeting, R. M., Beamish, R. J. & Brennan, J. S. 2010. Ontogenic diet shifts of juvenile Chinook salmon in nearshore and offshore habitats of Puget Sound. *Transactions of American Fisheries Society*, **139**, 803–823.

DuPreez, H. H., McLachlan, A. & Marais, J. F. K. 1988. Oxygen consumption of two nearshore marine elasmobranchs, *Rhinobatus annulatus* (Muller and Henle 1841) and *Myliobatus aquila* (Linnaeus 1758). *Comparative Biochemistry and Physiology*, **89A**, 283–294.

Enders, E. C. & Scruton, D. A. 2006. *Potential Application of Bioenergetics Model to Habitat Modeling and Importance of Appropriate Metabolic Rate Estimates with Special Consideration for Atlantic Salmon*. Canadian Technical Report of Fisheries and Aquatic Science **2641**, Fisheries and Oceans Canada, St. John's, Newfoundland.

Facey, D. E. & Grossman, G. D. 1990. The metabolic cost of maintaining position for four North American stream fishes: effects of season and velocity. *Physiological Zoology*, **63**, 757–776.

Ferry-Graham, L. A. & Gibb, A. C. 2001. A comparison of fasting and post-feeding metabolic rates in a sedentary shark *Cephaloscyllium ventriosum*. *Copeia*, **2001**, 1108–1113.

Flowers, H. J., van Poorten, B. T., Tetzlaff, J. C. & Pine, III, W. E. 2010. Bioenergetics approach to describing Gulf sturgeon (*Acipenser oxyrinchus*) growth in two Florida rivers. *The Open Fish Science Journal*, **3**, 80–86.

Galat, D. L., Wildhaber, M. L. & Dieterman, D. J. 2001. *Population Structure and Habitat use of Benthic Fishes Along the Missouri and Lower Yellowstone Rivers, volume 2: Spatial Patterns of Physical Habitat Variables Along the Missouri and Lower Yellowstone Rivers*. Final Report of Missouri River Benthic Fish Study PD-95-5832 to USACOE and US Bureau of Reclamation, U.S. Army Corps of Engineers, Omaha District, Omaha, NE.

Geist, D. R., Brown, R. S., Cullimam, V., Brink, S. R., Lepla, K., Bates, P. & Chandler, J. A. 2005. Movement, swimming speed, and oxygen consumption of juvenile white sturgeon in response to changing flow, water temperature, and light level in the Snake River, Idaho. *Transactions of the American Fisheries Society*, **134**, 803–816.

Gerrity, P. C., Guy, C. S. & Gardner, W. M. 2006. Juvenile pallid sturgeon are piscivorous: a call for conserving native cyprinids. *Transactions of the American Fisheries Society*, **135**, 604–609.

Grohs, K. L., Klumb, R. A., Chipps, S. R. & Wanner, G. A. 2009. Ontogenetic patterns in prey use by pallid sturgeon in the Missouri River, South Dakota and Nebraska. *Journal of Applied Ichthyology*, **25**, 48–53.

Hansen, M. J., Boisclair, D., Brandt, S. B., Hewett, S. W., Kitchell, J. F., Lucas, M. C. & Ney, J. J. 1993. Applications of bioenergetics models to fish ecology and management: where do we go from here? *Transactions of the American Fisheries Society*, **122**, 1019–1030.

Hanson, P. C., Johnson, T. B., Schindler, D. E. & Kitchell, J. F. 1997. *Fish Bioenergetics 3.0*. University of Wisconsin Sea Grant Institute, Madi-son, WI.

Harvey, B. C., Nakamoto, R. J. & White, J. L. 2006. Reduced streamflow lowers dry-season growth of rainbow trout in a small stream. *Transaction of the American Fisheries Society*, **135**, 998–1005.

Juanes, F., Buckel, J. A. & Conover, D. O. 1994. Accelerating the onset of piscivory: intersection of predator and prey phenologies. *Journal of Fish Biology*, **45**, (Supplement A) 41–54.

Kitchell, J. F., Stewart, D. J. & Weininger, D. 1977. Applications of a bioenergetics model to yellow perch (*Perca flavescens*) and walleye (*Stizostedion vitreum vitreum*). *Journal of the Fisheries Research Board of Canada*, **34**, 1922–1935.

Leopold, L. B. & Maddock, T. 1953. *The Hydraulic Geometry of Stream Channels and Some Physiographic Implications*. USGS Professional Paper, Washington, DC.

Madenjian, C. P., O'Connor, D. V. et al. 2006. Evaluation of a lake whitefish bioenergetics model. *Transactions of American Fisheries Society*, **135**, 61–75.

Madon, S. P. & Culver, D. A. 1993. Bioenergetics model for larval and juvenile walleyes: an in situ approach with experimental ponds. *Transactions of the American Fisheries Society*, **122**, 797–813.

Mayfield, R. B. & Cech, Jr., J. J. 2004. Temperature effects on green sturgeon bioenergetics. *Transactions of the American Fisheries Society*, **133**, 961–970.

McElroy, B., DeLonay, A. & Jacobson, R. 2012. Optimum swimming pathways of fish spawning migrations in rivers. *Ecology*, **93**, 29–34.

Megrey, B. A., Rose, K. A., Klumb, R. A., Hay, D. E., Werner, F. E., Eslinger, D. L. & Smith, S. L. 2007. A bioenergetics based population dynamics model of Pacific herring coupled to a lower trophic level nutrient-phytoplankton-zooplankton model: description, calibration, and sensitivity analysis. *Ecological Modeling*, **202**, 144–164.

Moss, J. H. H. 2001. *Development and Application of a Bioenergetics Model for Lake Washington Prickly*

Sculpin (Cottus asper). MSc thesis, University of Washington, Seattle, WA.

NEY, J. F. 1993. Bioenergetics modelling today: growing pains on the cutting edge. *Transactions of the American Fisheries Society*, **122**, 736–748.

NIKLITSCHEK, E. J. 2001. *Bioenergetics Modeling and Assessment of Suitable Habitat for Juvenile Atlantic and Shortnose Sturgeons (Acipenser Oxyrinchus and A. Brevirostrum) in the Chesapeake Bay*. PhD Thesis, University of Maryland at College Park, College Park, MD.

NIKLITSCHEK, E. J. & SECOR, D. H. 2009. Dissolved oxygen, temperature and salinity effects on the ecophysiology and survival of juvenile Atlantic sturgeon in estuarine waters: II. Model development and testing. *Journal of Experimental Marine Biology and Ecology*, **381**, S161–S172.

PARSONS, G. R. & SMILEY, P. 2003. The effect of environmental changes on swimming performance of the white crappie. *Journal of Freshwater Ecology*, **18**, 89–96.

SCHINDLER, D. E., KITCHELL, J. F., HE, K., CARPENTER, S. R., HODGSON, J. R. & COTTINGHAM, K. L. 1993. Food web structure and phosphorus cycling in lakes. *Transactions of the American Fisheries Society*, **122**, 756–772.

SIMS, D. W. & DAVIES, S. J. 1994. Does specific dynamic action (SDA) regulate return of appetite in the lesser spotted dogfish, *Scyliorhinus canicula*? *Journal of Fish Biology*, **45**, 341–348.

SITARAMAIAH, P. 1967. Water, nitrogen, and calorific values of freshwater organisms. *Journal Du Conseil*, **31**, 27–30.

TRUDEL, M., GEIST, D. R. & WELCH, D. W. 2004. Modeling the oxygen consumption rates in Pacific salmon and steelhead trout: an assessment of current models and practices. *Transactions of the American Fisheries Society*, **133**, 326–348.

USFWS 2008. *Pallid Sturgeon (Scaphirhynchus Albus) Range-Wide Stocking and Augmentation Plan*. Billings, MT. http://www.fwspubs.org/doi/suppl/10.3996/022012-JFWM-013/suppl_file/10.3996_022012-jfwm-013.s8.pdf [last accessed April 2015].

USGS 2002. *Standards for the Analysis and Processing of Surface-Water Data & Information Using Electronic Methods*, USGS, Reston, VA, Water-Resources Investigation Report 01-4044.

USGS 2011. National Water Information System: USGS data available at http://nwis.waterdata.usgs.gov/nwis [accessed 12 December 2011].

USGS 2012. Significant figures: at http://nwrc.usgs.gov/techrpt/sta18.pdf [accessed 27 June 2012].

VENTURELLI, P. & WILLIAM, T. 2006. Diet and growth of northern pike in the absence of prey fishes: initial consequences of persisting in disturbance-prone lakes. *Transactions of the American Fisheries Society*, **135**, 1512–1522.

VERY, N. M. & SHERIDAN, M. A. 2002. The role of somatostatins in the regulation of growth in fish. *Fish Physiology and Biochemistry*, **27**, 217–226.

WETHERBEE, B. M. & GRUBER, S. H. 1993. Absorption efficiency of the lemon shark *Negaprion brevirostris* at varying rates of energy intake. *Copeia*, 416–425.

WILDHABER, M. L., DELONAY, A. J. ET AL. 2007. *A Conceptual Life-History Model for Pallid & Shovelnose Sturgeon*. USGS, Reston, VA, Circular 1315.

WILDHABER, M. L., HOLAN, S. H., BRYAN, J. L., GLADISH, D. W. & ELLERSIECK, M. 2011. Assessing power of large river fish monitoring programs to detect population changes: the Missouri river sturgeon example. *Journal of Applied Ichthyology*, **27**, 282–290.

WILDHABER, M. L., WIKLE, C. K., MORAN, E. H., ANDERSON, C. J., FRANZ, K. J. & DEY, R. In prep. Hierarchical, stochastic modeling across spatiotemporal scales of large river ecosystems and somatic growth in fish populations under various climate models: Missouri River sturgeon example. *In*: RIDDICK, A. T., KESSLER, H. & GILES, J. R. A. (eds) *Integrated Environmental Modelling to Solve Real World Problems: Methods, Vision and Challenges*. Geological Society, London, Special Publications, **408**.

Hierarchical stochastic modelling of large river ecosystems and fish growth across spatio-temporal scales and climate models: the Missouri River endangered pallid sturgeon example

MARK L. WILDHABER[1]*, CHRISTOPHER K. WIKLE[2], EDWARD H. MORAN[1], CHRISTOPHER J. ANDERSON[3], KRISTIE J. FRANZ[4] & RIMA DEY[2]

[1]*United States Geological Survey, Columbia Environmental Research Center, 4200 New Haven Road, Columbia, MO 65201-8709, USA*

[2]*Department of Statistics, University of Missouri, 146 Middlebush Hall, Columbia, MO 65211-6100, USA*

[3]*Climate Science Initiative, Iowa State University, 2021 Agronomy Hall, Ames, IA 50011, USA*

[4]*Geological and Atmospheric Sciences, Iowa State University, 3023 Agronomy Hall, Ames, IA 50011, USA*

**Correspondence: mwildhaber@usgs.gov*

Abstract: We present a hierarchical series of spatially decreasing and temporally increasing models to evaluate the uncertainty in the atmosphere – ocean global climate model (AOGCM) and the regional climate model (RCM) relative to the uncertainty in the somatic growth of the endangered pallid sturgeon (*Scaphirhynchus albus*). For effects on fish populations of riverine ecosystems, climate output simulated by coarse-resolution AOGCMs and RCMs must be downscaled to basins to river hydrology to population response. One needs to transfer the information from these climate simulations down to the individual scale in a way that minimizes extrapolation and can account for spatio-temporal variability in the intervening stages. The goal is a framework to determine whether, given uncertainties in the climate models and the biological response, meaningful inference can still be made. The non-linear downscaling of climate information to the river scale requires that one realistically account for spatial and temporal variability across scale. Our downscaling procedure includes the use of fixed/calibrated hydrological flow and temperature models coupled with a stochastically parameterized sturgeon bioenergetics model. We show that, although there is a large amount of uncertainty associated with both the climate model output and the fish growth process, one can establish significant differences in fish growth distributions between models, and between future and current climates for a given model.

Recent decades have brought substantive changes in land use and climate across the Earth, prompting a need to think of population and community ecology not as a static entity, but as a dynamic process (United States Climate Change Science Program: CCSP 2003). Increasingly, there is evidence of ecological changes due to climate change (e.g. Walther *et al.* 2005; Bergengren *et al.* 2011). Although much of this evidence comes from ground-truth observations of biogeographical data, there is increasing reliance on models that relate climate variables to biological systems (CCSP 2003). Such models are used to explore potential changes to population and community-level ecological systems in response to climate scenarios as obtained from atmosphere – ocean global climate models (AOGCMs) (Nakicenvoic *et al.* 2000; CCSP 2003).

When modelling ecosystem response to climate, the resolution of AOGCMs is not typically sufficient to draw inferences at the scales of variability necessary for understanding ecological processes (e.g. Tabor & Williams 2010). Rather, the AOGCM physical variables must be 'downscaled' to local ecological/biological response scales that can be used in vulnerability and risk assessments of climate change (CCSP 2003). Traditionally, one either accomplishes the downscaling by linking the AOGCM to a smaller scale through the use of deterministic models (i.e. 'dynamical downscaling') or through the use of statistical models (e.g. Grotch & MacCracken 1991; Fowler *et al.* 2007). In the case of evaluating individual organism response to potential climate variability, one must project across multiple scales of spatial and temporal variability,

From: Riddick, A. T., Kessler, H. & Giles, J. R. A. (eds) 2017. *Integrated Environmental Modelling to Solve Real World Problems: Methods, Vision and Challenges*. Geological Society, London, Special Publications, **408**, 119–145.
First published online October 12, 2015, https://doi.org/10.1144/SP408.11

and it is not likely that the 'transfer function' that converts large-scale climate simulation results to very small spatial and temporal scales can be represented well by a single statistical downscaling model, especially given the non-linear nature of the transfer of information across scale. Rather, a more realistic transfer function would attempt to accommodate the various scales of variability, such as through a series of deterministic process models.

Consideration of the 'cascade of uncertainty' that arises in the application of multiple models is a necessary step (e.g. Henderson-Sellers 1993; Jones 2000; Wilby & Harris 2006). Several studies exploring the potential hydrological impacts from climate change have shown substantial uncertainty associated with the large-scale hydrological model forcing derived from climate models (e.g. Stone *et al.* 2003; Jasper *et al.* 2004; Wood *et al.* 2004; Salathé 2005; Chen *et al.* 2006; Wilby & Harris 2006). Uncertainty in climate models results from the initial and boundary conditions, parameter uncertainty, and the structural uncertainty in the models themselves (Knutti *et al.* 2010). There are two primary approaches used to attempt to account for this uncertainty in climate change studies, the so-called 'multi-model ensemble' (MME) and 'perturbed physics ensemble' (PPE) methods. The MME approach considers a sample of opportunity consisting of output from multiple models of similar complexity, whereas the PPE method typically assumes a model with a common core but with ensembles chosen based on multiple parameter sets, often selected based on some climate sensitivity index. For a recent comparison between uncertainties associated with MME and PPE, see Collins *et al.* (2011). In the context of hydrological downscaling MME, Fowler *et al.* (2007) found that, in general, hydrological impacts are sensitive to spatial and temporal biases in precipitation and temperature, and are, thus, sensitive to the particular GCM considered. In general, as summarized by Knutti (2010) and Knutti *et al.* (2010), there are numerous fundamental issues that arise when considering such ensembles, including lack of model independence, unresolvable structural uncertainty, calibration and choice of evaluation metrics. Yet, until such time as fully stochastically parameterized GCMs are in common use, the MME and PPE frameworks provide the most viable approach for accounting for climate model uncertainty.

In addition to the uncertainty associated with the climate models, in ecological impact studies there is typically substantial uncertainty associated with our knowledge of the ecological phenomena of interest. It is possible that ecological model uncertainty could be greater than the uncertainty associated with the climate/physical/hydrological models and/or greater than the variation that would occur under different climate scenarios, thereby limiting the ability to draw inference concerning potential impacts to the ecological system. This leads to the fundamental question that this paper seeks to address: when the uncertainty in the ecological process is taken into account through a stochastic parameter ensemble, are there still significant differences in the ecological response forced from various climate models in a MME framework? If the answer is yes, is the difference in ecological response associated with a future climate scenario and a current climate scenario also significant relative to the uncertainty associated with the ecological process? If the answer is no, then we must address in what ways the uncertainty associated with the ecological response can be reduced.

We address the questions above for the case of the Missouri River pallid sturgeon (*Scaphirhynchus albus*), which was Federally listed as endangered in 1990 and is rare in the Missouri River Basin (Dryer & Sandvol 1993). The conceptual life-history model for pallid sturgeon of Wildhaber *et al.* (2007, 2011*a*) provides the framework to illustrate how climate may interact with river management actions to affect species recovery. As Figure 1 illustrates, we add stochastic parameterization to a pallid sturgeon bioenergetics model developed by Chipps *et al.* (2008) to translate potential changes in water temperature and velocity associated with potential climate change into pallid sturgeon growth. We use data from multiple climate models to force fixed/calibrated hydrological flow and temperature models to obtain river discharge, velocity and temperature at the river reach scale. To quantify uncertainty, it is important that uncertainty associated with critical parameters be accounted for across these scales of variability.

As mentioned above, downscaling the hydrological processes from the climate scale to river reach scale could also be performed via statistical downscaling through the development of empirical relationships between basin runoff and air temperature, and mainstem flow and river temperature, respectively (Larson & Schwein 2004; Blevins 2006; Vrac *et al.* 2007). Although statistical downscaling methods have become quite sophisticated and useful in the study of climate impacts (e.g. Maraun *et al.* 2010), statistical approaches inherently imply extrapolation in the context of climate change studies and, more critically, would likely be unable to directly account for multivariate spatio-temporal variability for many important variables (e.g. evapotranspiration, snow melt, river management). Indeed, the development of multivariate non-linear spatio-temporal statistical models is beyond the current state of the art (e.g. see Cressie & Wikle 2011). Perhaps more importantly, since much of

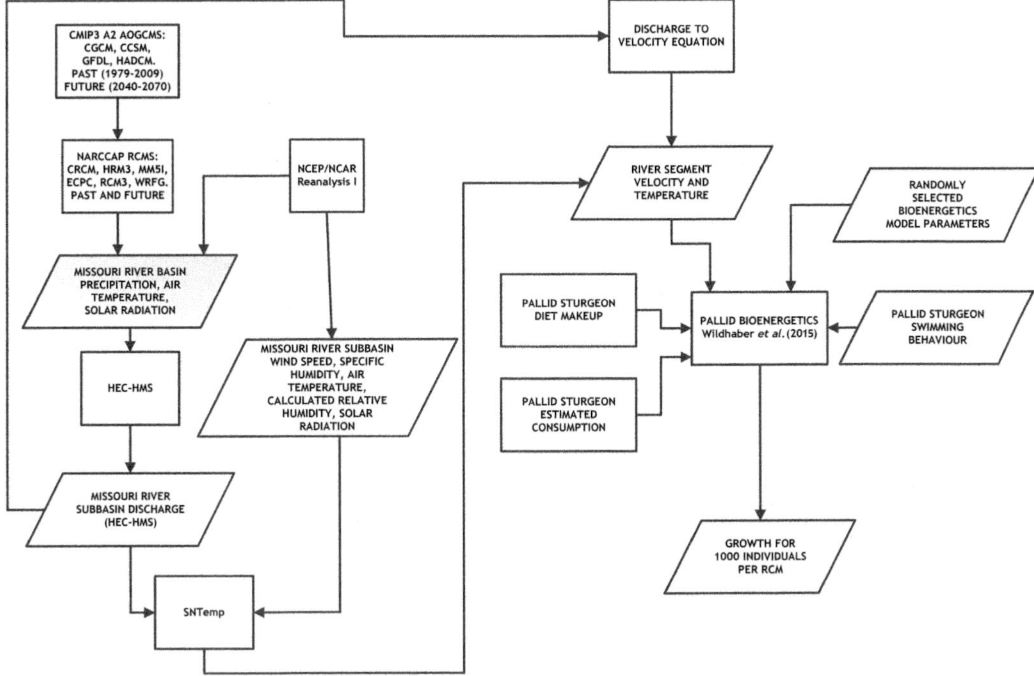

Fig. 1. Flowchart illustrating the linkages between models within and across temporal and spatial scales. NARCCAP datasets are used in the simulations. AOGCM simulations were taken from the WCRP's CMIP3 multi-model dataset (Meehl *et al.* 2007) based on the A2 emissions scenario (representing less international co-operation to reduce greenhouses gases) that projects relatively high greenhouse gas concentration increases. Model data used included four AOGCMs: CGCM (Flato 2005), CCSM, GFDL and HADCM. Model data used include six Regional Climate Models (RCM) within the AOGCMs: CRCM, HRM3, MM5I, ECPC, RCM3 and WRFG. Measured/observed data: NCEP/NCAR Reanalysis 1 (Kalnay *et al.* 1996) (NCEP/NCAR) from Earth System Research Laboratory (www.esrl.noaa.gov). River discharge simulated using a lumped-parameter precipitation runoff model HEC-HMS (USACE 2010 – see Appendix A for more details). Water temperature is simulated using the physical-process SNTemp stream-temperature model (Theurer *et al.* 1984).

our interests are in evaluating the impacts of uncertainties across a wide range of spatial and temporal scales, a hybrid deterministic/statistical approach is more useful for the present study. Although these models still require calibration to define model parameters through fitting to observed data, the underlying structure is based on known physical properties and does not solely rely on empirical associations.

We present here one of many possible approaches to address the potential for meaningful inference associated with uncertainty in AOGCMs and regional climate models (RCMs), and hydrological models v. uncertainty in the ecological response through application of a set of physical and biological models that cover a hierarchy of scales. While a complete climate prediction may be intractable at this time – for instance, the climate projections may not incorporate land use and other physical changes into the system (e.g. geomorphology) or solar fluctuations into the boundary conditions – our framework is flexible enough to adapt to advances in climate simulations.

Modelling approach

The modelling framework developed (Fig. 1) uses data and modelling results from synergistic monitoring and modelling studies. We make use of: (1) extensive and long-term biotic and abiotic data (e.g. Drobish 2008; USFWS 2008; Wildhaber *et al.* 2011*b*, 2012); (2) simulations of climate variation in high-rate rainfall and streamflow (Mauget 2004; CCSP 2008); (3) on-going analysis and modelling of spatial and temporal patterns of the benthic fish community, and their relationship to physical and chemical factors (e.g. Arab *et al.* 2008, 2012; Wildhaber *et al.* 2011*b*, 2012); (4) sturgeon life-history models developed by Wildhaber *et al.* (2007, 2011*a*) based on what is seen in the Missouri River; (5) sturgeon population models developed by

Bajer & Wildhaber (2007) as a basis for initial population viability analysis for Lower Missouri River sturgeons; and (6) on-going research identifying factors affecting Lower Missouri River sturgeon spawning physiology, behaviour, habitat choice and success as part of a comprehensive research programme designed to identify life-history bottlenecks (e.g. Wildhaber *et al.* 2007; DeLonay *et al.* 2009; Holan *et al.* 2009; McElroy *et al.* 2012). Initial models for each scale were developed using data collected by the United States Fish and Wildlife Service (USFWS), the USGS, the United States Army Corps of Engineers (USACE), the National Oceanic and Atmospheric Administration (NOAA), the United States Department of Agriculture (USDA), and numerous other federal and state natural resource agencies. The data include land use, geological and geomorphic, hydrological, water quality, atmospheric, and pallid sturgeon (*Scaphirhynchus albus*) growth and movement. Along with climatological modelling by the North American Regional Climate Change Assessment Program (NARCCAP) (Mearns *et al.* 2009; see http://www.narccap.ucar.edu/ for details) and established hydrological models (HEC-HMS: USACE 2000; SNTemp: Theurer *et al.* 1984), we use juvenile pallid sturgeon bioenergetics models (Chipps *et al.* 2008) extrapolated beyond a fork length of 700 mm in concert with observed growth of adult pallid sturgeon in the Missouri River

Global to regional climate models

For this downscaling component (Fig. 1), we focus on NARCCAP results owing to their comprehensive nature in consideration of multiple climate models and the availability of data that have already been downscaled to the regional level. The NARCCAP simulations provide physically downscaled air temperature, precipitation, solar radiation, humidity, cloud cover and wind speed on 50 km grids that are used as input to Missouri River Basin hydrological and water-temperature models. The data are generated from RCM simulations driven by AOGCMs over a domain covering most of North America. The AOGCM simulations were taken from the World Climate Research Programme's (WCRP's) Coupled Model Intercomparison Project phase 3 (CMIP3) multi-model dataset (Meehl *et al.* 2007). Simulations of AOGCM in CMIP3 spanned 1850–2100, using historical climate drivers (e.g. greenhouse gas concentration, solar radiation, volcanic eruptions) from 1850 through to 2000 and scenarios of climate drivers from 2000 through to 2100. The NARCCAP RCMs were provided data for two 30 year periods: 1971–2000 and 2041–70. Data for the mid-twenty-first century period came from the A2 emissions scenario (representing less international co-operation to reduce greenhouse gases: Nakicenvoic *et al.* 2000) that projects relatively high greenhouse gas concentration increases and is useful as an upper bound for studying realistic impacts and adaptation strategies for climate warming. The NARCCAP simulation results include various combinations of six RCMs (Canadian Regional Climate Model (CRCM); Hadley Regional Climate Model Version 3.0 (HRM3); Mesoscale Meteorological model Version 5.0 (MM5I); Experimental Climate Prediction Center Regional Spectral Model (ECPC); Regional Climate Model 3.0 (RCM3); and Weather Research and Forecast Model 3.0 (WRFG)) and four AOGCMs (Canadian Global Climate Model Version 3.1 (CGCM) (Flato 2005); Community Climate System Model Version 3.0 (CCSM); Geophysical Fluid Dynamics Laboratory Climate Model Version 2.1 (GFDL); and Hadley Centre Climate Model Version 2.0 (HADCM)) (see Mearns *et al.* 2009 and references therein for full model descriptions) (Fig. 1; Table 1).

Multi-model ensembles are critical to climate change impacts analysis because differences in model formulations can produce uncertainty in model output (Solomon *et al.* 2007). In such studies, it is also important to understand the biases of the RCMs. Model developers may have suggestions of how the biases relate to particular submodels, and that may inform their model development approach for the next generation, but it is impossible to infer from the current formulation that created these simulations what bias will occur. For this reason, each NARCCAP RCM simulated the observed climate for 1979–2004, given data from the National Center for Environmental Prediction (NCEP) – Department of Energy (DOE) Atmospheric Model Intercomparison Project (AMIP-II) Reanalysis (Kanamitsu *et al.* 2002). These data are an approximation for observations, and so the regional models are expected to replicate the climate for the 1979–2004 period associated with the NCEP data, subject to their internal variability. The deviations of retrospective RCM simulations from observations show the simulation bias.

For the purposes of this paper, we focus on three climate models (i.e. CCSM, CGCM and GFDL) with available regional models for each (i.e. CRCM and WRFG for CGCM and CCSM, HRM3 for GFDL, MM5I for GFDL, and RCM3 for CGCM and GFDL) and NCEP/NCAR Reanalysis 1 (Kalnay *et al.* 1996), which is used to develop the river hydrology and water-temperature models (described in the next subsection) – a total of 17 scenarios (Fig. 1; Table 1). The uncertainty considered here is that of the differences in model outputs which occur between AOGCM–RCM combinations.

Table 1. *NARCCAP datasets used in the simulations*

Regional model	Global model: CGCM	CCSM	GFDL	HADCM	Regional model type	NCEP/NCAR reference
CRCM	2° latitude, 2° longitude	1.5° latitude, 1.5° longitude	2° latitude, 2.5° longitude	2.5° latitude, 3.75° longitude	University of Quebec, the Ouranos Consortium (Caya & Laprise 1999)	2.5° latitude, 2.5° longitude
HRM3	2° latitude, 2° longitude	1.5° latitude, 1.5° longitude	2° latitude, 2.5° longitude	2.5° latitude, 3.75° longitude	Atmospheric only	2.5° latitude, 2.5° longitude
MM5I	2° latitude, 2° longitude	1.5° latitude, 1.5° longitude	2° latitude, 2.5° longitude	2.5° latitude, 3.75° longitude	Limited-area, non-hydrostatic, terrain-following sigma-coordinate	2.5° latitude, 2.5° longitude
ECPC	2° latitude, 2° longitude	1.5° latitude, 1.5° longitude	2° latitude, 2.5° longitude	2.5° latitude, 3.75° longitude	Limited-area atmospheric numerical	2.5° latitude, 2.5° longitude
RCM3	2° latitude, 2° longitude	1.5° latitude, 1.5° longitude	2° latitude, 2.5° longitude	2.5° latitude, 3.75° longitude	Based on MM5I, 3D, sigma-coordinate, primitive equation regional climate	2.5° latitude, 2.5° longitude
WRFG	2° latitude, 2° longitude	1.5° latitude, 1.5° longitude	2° latitude, 2.5° longitude	2.5° latitude, 3.75° longitude	Grell Convective Parameterization Scheme, atmospheric model with infrastructure (buildings, concrete, etc.), land use/landscape, biospheric components, etc.	2.5° latitude, 2.5° longitude

Model data used included four AOGCMs: Canadian Global Climate Model Version 3.1 (CGCM) (Flato 2005); Community Climate System Model Version 3.0 (CCSM); Geophysical Fluid Dynamics Laboratory Climate Model Version 2.1 (GFDL); and Hadley Centre Climate Model Version 2.0 (HADCM). Model data used include six RCMs within the AOGCMs: Canadian Regional Climate Model (CRCM); Hadley Regional Climate Model Version 3.0 (HRM3); Mesoscale Meteorological model Version 5.0 (MM5I); Experimental Climate Prediction Center Regional Spectral Model (ECPC); Regional Climate Model 3.0 (RCM3); and Weather Research and Forecast Model 3.0 (WRFG). Measured/observed data: NCEP/NCAR Reanalysis 1 (Kalnay *et al.* 1996) (NCEP/NCAR) from Earth System Research Laboratory (www.esrl.noaa.gov).

Regional- to river-level hydrological and temperature models

For this downscaling component, we focus on the Missouri River Basin (Fig. 2). This basin encompasses approximately 1370 000 km^2 (529 000 square miles) and is home to around 12 million people (USACE 2006). The basin traverses 10 states and part of Canada, and extends from the Rocky Mountains to its confluence with the Mississippi River. The Missouri River mainstem flows for about 3735 km (2321 miles). Nearly 1.9 million acres of floodplain surround the Missouri River downstream of Sioux City, Iowa. Of the 894 impoundments throughout the Missouri River Basin (National Atlas of the United States – NAUS 2006), there are six dams along the mainstem in Montana, Nebraska, North Dakota and South Dakota.

The uncertainty evaluation of AOGCM/RCM/hydrology v. fish bioenergetics uncertainty is performed at two locations on the Lower Missouri River that correspond to gauges: Station 06807000 Missouri River at Nebraska City, NE (draining about 1 062 000 km^2 at RK 906, referenced as 'Nebraska'); and Station USGS 06909000 Missouri River at Boonville, MO (draining about 1300 000 km^2 at RK 317 and at 172 m above NAVD88, referenced as 'Missouri') (USGS 2011). The Nebraska site represents the Lower Missouri River between the Gavins Point Dam and the Kansas River confluence with the Missouri River. The Missouri site represents the Lower Missouri River between the Kansas River confluence and the mouth of the Missouri River.

To translate gridded climate data from regional to river level, continuous river velocity and water temperature are simulated using a lumped-parameter precipitation runoff model HEC-HMS (USACE 2010) and the physical-process SNTemp stream-temperature model (Theurer *et al.* 1984), respectively (see Appendix A for more details).

The NCEP/NCAR Reanalysis 1 data (Kalnay *et al.* 1996) (NCEP/NCAR) (Table 1) is used as input for both models. The HEC-HMS is a surface and quasi-subsurface hydrological modelling system. Our HEC-HMS model implementation for

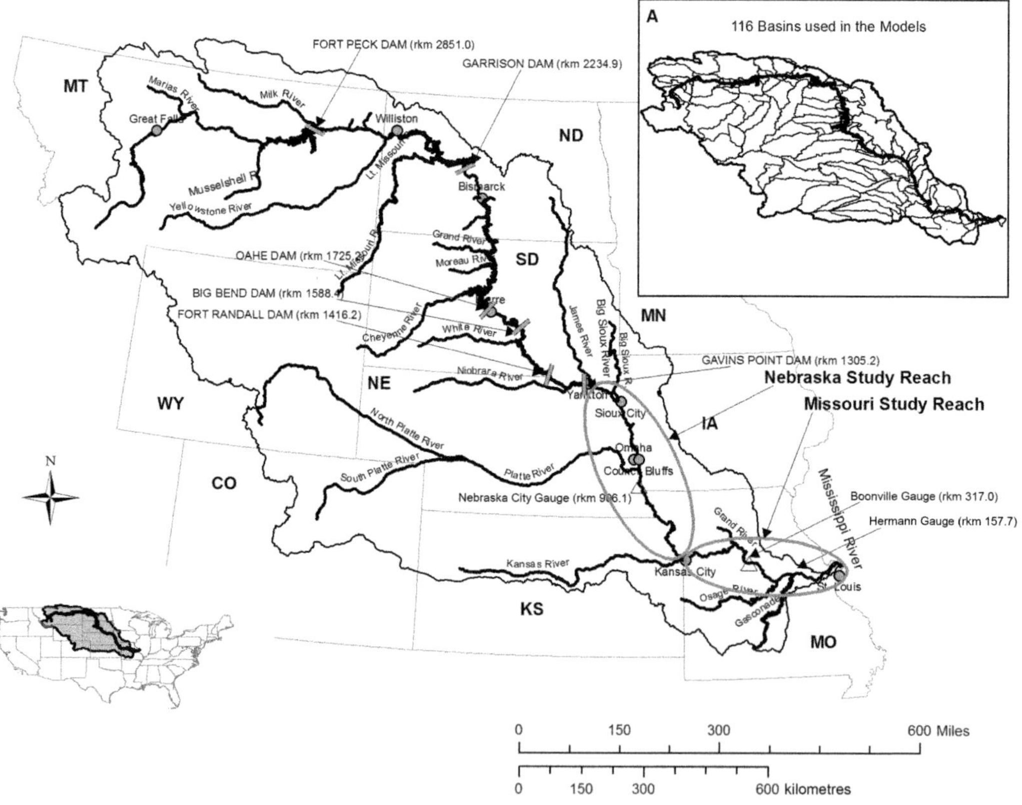

Fig. 2. Missouri River study area. The states included in the study area are: Missouri (MO), Kansas (KS), Iowa (IA), Nebraska (NE), Colorado (CO), Wyoming (WY), Minnesota (MN), South Dakota (SD), North Dakota (ND) and Montana (MT). rkm, river kilometre.

the Missouri River Basin includes 116 sub-basins and corresponding reaches. We use the Deficit and Constant Loss method for sub-basin runoff, the Priestley–Taylor potential evapotranspiration method for potential evapotranspiration, the Temperature Index method to simulate snow accumulation and melt, and the Kinematic-Wave Routing or Muskingum–Cunge methods for streamflow routing (USACE 2010) (see Appendix A for more detail). Inputs into the model are precipitation, air temperature and solar radiation data, as averaged over the defined sub-basins. The output from the HEC-HMS includes discharge at the sub-basin outlets and along the river reaches.

The SNTemp model is a physically based, steady-state one-dimensional (1D) heat transport model. The SNTemp calculates net-heat flux as the sum of heat from solar radiation, convection, conduction, evaporation, streamside shading, streambed fluid friction and the back radiation of water (Theurer *et al.* 1984). The model assumes homogeneous stream segments, where each is described by flow/discharge, length, width, slope, channel roughness or travel time. SNTemp requires continuous discharge and stream temperature values at all upstream points (upper-flow boundaries) on each modelled reach, although zero-flow headwaters are an exception and do not require actual water temperature values (Bartholow 2000). Riparian shading characteristics can be included but were not considered in this study. Model inputs are air temperature, relative humidity, wind speed, percent possible sunshine (inverse of cloud cover) and ground-level solar radiation, in addition to discharge simulations from the HEC-HMS. The model calculates within-segment streamflow accretions by mass balance, while including a component for contributing groundwater accretion flow and temperature. Simulated temperatures, however, represent the average value of all water within the dimensions of a specified channel location.

The SNTemp model has been used to investigate the effects of discharge timing, release temperature and/or release volume, riparian shading, thermal loading by power plants, and changes in channel morphology on downstream water temperatures (Bartholow 2000). When high-quality input data are used, the model simulates daily water temperatures quite accurately, typically to within less than 0.5°C of observed values, while requiring little or no calibration (Bartholow 1991). The core SNTemp model is designed to model only one meteorological station. To model the 116 sub-basins and river segments of the Missouri River Basin, each with their own set of meteorological datasets, SNTemp was programed to cascade HEC-HMS discharge and SNTemp-generated water temperatures from upper to lower sub-basins/reaches.

Both the HEC-HMS and SNTemp models were calibrated through manual parameter adjustment to find values that minimized residuals between observed and simulated discharge and water temperature, respectively (see Appendix A for details). In addition, for the HEC-HMS, sub-basin characteristics such as hydraulic conductivity and actual water capacity of soils were extracted from the National Resource Conservation Service soils property database (NRCS 2011), and elevation, land-use class and impervious area were obtained from the USGS National Land Cover Database (Fry *et al.* 2011). The calibration period was selected as 1972–95, which corresponds to a period in which mainstem reservoir operations were stable, as measured in terms of flow at the Nebraska site (USGS at Gavin's Point Dam). The model was validated for the 1996–2009 period at locations with measured discharge.

The simulated discharge estimated using HEC-HMS, velocity (calculated based on the equation, $V = aD^b$ (Leopold & Maddock 1953), where V is velocity, D is discharge, and a and b are the empirical coefficients of the non-linear equation) and water temperature were evaluated using the Nash–Sutcliffe model efficiency coefficient (NS), the Pearson product – moment correlation coefficient (r), the percentage bias (PBIAS) and the ratio of the root mean square error to the standard deviation of measured data (RSR). The possible values for NS range from 1 (perfect fit) to negative infinity, where negative values suggest that the mean value of observed data is a better predictor than the simulated value (Moriasi *et al.* 2007). Values of RSR and PBIAS closer to 0 indicate reduced root-mean squared error or residual variability and percentage bias, respectively, thus suggesting better model performance (Moriasi *et al.* 2007). Based on a monthly time step, review of previous studies found that the 'rule of thumb' for satisfactory model calibration was NS > 0.4 (Engel *et al.* 2007) to NS > 0.5, PBIAS $\pm$ 25% and RSR ≤ 0.7; however, Moriasi *et al.* (2007) did acknowledge that NS > 0.36 has also been considered satisfactory.

To establish the initial model states, a 1 year spin-up period (starting in January) was used in HEC-HMS and discarded from any analyses. The calibrated HEC-HMS produced better than satisfactory results, as set forth by Engel *et al.* (2007) and Moriasi *et al.* (2007) for discharge and velocity simulations at Nebraska and Missouri (Table 2; Fig. 3); even better results were found for seasonal patterns (i.e. days of the year averaged for a 23 year period). In addition, using the model to extrapolate (i.e. validation time period) produced near or better than satisfactory results, thus demonstrating the validity of this model for use in future climate scenario modelling.

The calibrated SNTemp model showed a correlation coefficient of 0.94, with an NS of 0.87 and an RSR of 0.13 for stream temperature (Table 2; Fig. 3). In addition, results were near perfect for seasonal patterns (i.e. days of the year averaged for a 23 year period). Moreover, using the model produced near perfect results for the verification period at Nebraska and extremely good results for Missouri, where there were no data available for calibration.

River flow and temperature to a bioenergetics model

A substantial focus of this study was the development of a stochastic bioenergetics model for fish growth. Specifically, to translate water velocity and temperature into fish growth, we implement a daily time step, individual-based, bioenergetics model for the endangered pallid sturgeon based on a basic pallid sturgeon bioenergetics model (Chipps *et al.* 2008), combined with swimming energetics described for other sturgeon species (Geist *et al.* 2005); a general description is given below (details of the model can be found in Wildhaber *et al.* 2015). The parameters of the bioenergetics model were estimated through field study of pallid and other sturgeon species, extrapolated across different size classes or kept constant throughout the population (Tetzlaff *et al.* 2011). A key source of pallid sturgeon swimming data was the on-going USGS Comprehensive Sturgeon and Research Program study of

Table 2. *Statistical results for the HEC-HMS runoff discharge model, discharge-velocity calculation and SNTemp water-temperature model*

	Discharge ($m^3 s^{-1}$)		Velocity ($m s^{-1}$)		Temperature (°C)	
Statistic	Calibration	Validation	Calibration	Validation	Calibration	Validation
Nebraska City						
Daily values						
Time period	1972–1995	1996–2009	1972–1995	1996–2009	1972–1995	2006–2009
NS	0.4	0.27	0.43	0.28	0.87	0.74
Correlation	0.63	0.53	0.66	0.56	0.94	0.92
PBIAS	1.94	−1.14	−0.21	−1.35	2.39	13.07
RSR	0.6	0.73	0.57	0.72	0.13	0.26
Count	8766	5114	8766	5114	2504	670
Seasonality						
Time period	1972–1995	1996–2009	1972–1995	1996–2009	1972–1977	
NS	0.9	0.88	0.95	0.89	0.97	
Correlation	0.95	0.97	0.96	0.97	0.98	
PBIAS	1.94	−4.44	−0.21	−2.24	1.57	
RSR	0.1	0.12	0.05	0.11	0.03	
Boonville						
Daily values						
Time period	1972–1995	1996–2009	1972–1995	1996–2009		2006–2009
NS	0.44	0.41	0.45	0.34		0.64
Correlation	0.66	0.67	0.68	0.65		0.9
PBIAS	1.46	−8.86	−1.2	−4.01		13.72
RSR	0.56	0.59	0.55	0.66		0.36
Count	8766	5114	8766	5114		896
Seasonality						
Time period	1972–1995	1996–2009	1972–1995	1996–2009		
NS	0.75	0.69	0.82	0.66		
Correlation	0.9	0.94	0.93	0.95		
PBIAS	1.45	−11.5	−1.2	−4.78		
RSR	0.25	0.31	0.18	0.34		

The results include the Nash–Sutcliffe model efficiency coefficient (NS), the Pearson product–moment correlation (*r*), the percentage bias (PPBIAS) and the root mean square error to the standard deviation (RSR) for model calibration and validation periods. Intermittent observed/measured daily temperature data were available from 1972 to 2009 for Nebraska (i.e. Nebraska City gauge) and from 2006 to 2009 for Missouri (i.e. Boonville gauge); whereas continuous daily observed/measure discharge data were available for the 1972–1995 calibration period and 1996–2009 validation period for both locations.

Daily statistics represent values for daily time steps. Seasonality (seasonal values) represent days of the years averaged over the identified period (i.e. 1972–95, 1996–2009 and 2006–09). Seasonal water temperature statistics could only be calculated for Nebraska from 1972 to 1977 owing to the lack of continuous observed/measured data.

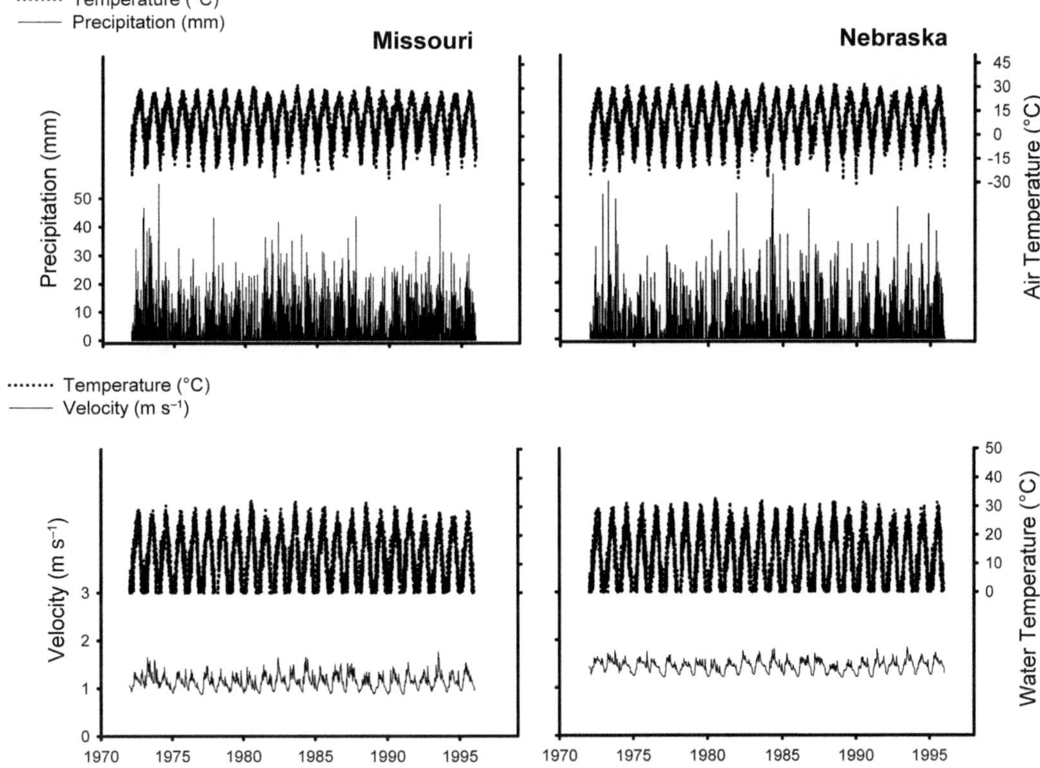

Fig. 3. Average daily precipitation, velocity, water temperature, and air temperature for Missouri (left-hand column) and Nebraska (right-hand column) for NCEP/NCAR Reanalysis I data. Ambient temperature and precipitation were daily averages. Velocity based on river discharge simulated using a lumped-parameter precipitation runoff model HEC-HMS (USACE 2010). Water temperature is simulated using the physical-process SNTemp stream-temperature model (Theurer *et al.* 1984) (see Appendix A for more details).

migration, physiology, habitat choice and spawning success of pallid sturgeon (DeLonay *et al.* 2009).

For fish growth, along with considering the uncertainty between results that occurs with AOGCM–RCM combinations, our uncertainty analysis critically depends on the incorporation of parameter stochasticity through random sampling from reported parameter distributions. Forcing uncertainties were accommodated through river temperature and flow distributions. We use output from HEC-HMS (USACE 2000) and SNTemp (Theurer *et al.* 1984) to produce Missouri River velocities (based on USGS historical channel morphology at the gauge modelled) and temperatures (see Appendix A for details). These water velocities and temperatures were used as input into the pallid sturgeon bioenergetics model to produce simulated growth trajectories. In addition, using the simulated temperature and velocity from two locations within the Lower Missouri River mainstem, we demonstrate the importance of incorporating spatial uncertainty.

Basic bioenergetics model

Our bioenergetics model (as originally described by Kitchell *et al.* 1977) (Wildhaber *et al.* 2015) is fundamentally based on the following simple relationship that conserves energy input and output:

$$C = (R + S) + (F + U) + G \tag{1}$$

where C is food consumption (g/g/day), R is standard metabolism (g/g/day), S is specific dynamic action (g/g/day), F is egestion (g/g/day), U is excretion (g/g/day) and G is gonadal or somatic growth (g/g/day). Consumption (C) is defined as a function of the maximum consumption rate ($C_{\max} = a_1 W^{b_1}$, where W is weight (g) and a_1 and b_1 are the empirical coefficients of the non-linear equation) achievable at the optimal temperature for consumption for an individual fish of a given size, as follows:

$$C = C_{\max} P\, r_c \tag{2}$$

where r_c is a temperature-dependent proportional adjustment of consumption rate, with P being the actual proportion of maximum consumption consumed. Both take values between 0 and 1.

Pallid sturgeon diet changes from macroinvertebrates to fish as the pallid sturgeon increases in size (Grohs *et al.* 2009). Gerrity *et al.* (2006) and Grohs *et al.* (2009) showed that pallid sturgeon between 350 and 500 mm in fork length consume 57% fish, and those with fork length more than 500 mm consume 90% fish. To incorporate this information into the bioenergetics model, we have pallid sturgeon shifting to a 57% diet of fish between 350 and 500 mm, and 90% fish >500 mm. These transitions from macroinvertebrates to fish were chosen randomly from a uniform distribution of size ranges (i.e. 250–350 and 500–600 mm, respectively) to incorporate the fact that the size at which the switch was observed to occur varied between individuals tested.

Respiration rate (R) is measured by oxygen consumption, which is dependent on water temperature, fish size and activity cost. Chipps *et al.* (2008) defined this as: $R = R_{max}Ar_R$, where $R_{max} = a_2W^{b_2}$, where W is weight (g) and a_2 and b_2 are the empirical coefficients of the non-linear equation, the maximum weight-specific standard respiration rate at the optimum temperature. Again, r_R is a temperature-dependent proportional adjustment of consumption rate between 0 and 1, and A is an activity parameter used to specify respiration rates.

The cost of swimming associated with varying water velocities was accounted for by replacing constant activity costs (A) in the standard model with a function that relates swimming speed to activity cost based on white sturgeon (Geist *et al.* 2005). Accuracy of swimming cost estimates was improved by including seasonality in swimming, based on observed swimming patterns of Lower Missouri River pallid sturgeon and their associated variability (DeLonay *et al.* 2009). We incorporated the cost of swimming upstream by assuming that swimming speed was equivalent to the observed upstream swimming speed plus the velocity of the water through which the fish swam. This water velocity was estimated from the water velocity of observed upstream paths travelled by pallid sturgeon in the Lower Missouri River, based on the observation by McElroy *et al.* (2012) that the chosen upstream path of pallid sturgeon was energetically less costly than the average possible based on a random sample of possible paths. The chosen path had an estimated energetic cost equivalent to an average velocity of 1.18 m s^{-1}; the average possible path velocity was 2.03 m s^{-1} (B. McElroy pers. comm.). Therefore, a constant 58% (i.e. $1.18/2.03 \times 100$) of the estimated average velocity was used as the estimate of the velocity experienced by the fish. Using a proportion is valid because when average velocity changes in a channel section, the relative distribution of high and low velocities does not (i.e. the flow is fastest in channel and slowest along edges). A proportional adjustment provides additional utility, thus making it possible to apply the model to both migrating and non-migrating pallid sturgeon if the assumption is that reproductive state does not change energetic cost choices. Here, this proportion was held constant since the actual swimming speed cost coefficient was the focus. This cost coefficient was randomly chosen on a daily basis.

Specific dynamic action (S) is a proportion of consumption (C), and is the metabolic cost of digestion, deposition and absorption of consumed energy (Kitchell *et al.* 1977). Waste losses due to egestion (F) and excretion (U) were modelled as a constant proportion of consumed energy, as given by Chipps *et al.* (2008).

We estimated P, actual proportion of maximum consumption consumed, using an empirically determined non-linear function for the relationship of fork length (FL) to P for fish >500 mm FL. To determine these functions of P to FL, we used mark – recapture data from the Pallid Sturgeon Population Assessment Program (PSPAP) (Drobish 2008) and the Pallid Sturgeon Stocking Program (PSSP) (USFWS 2008) for pallid sturgeon recaptured in the Missouri River between the Gavins Point Dam and the Kansas River confluence (i.e. Nebraska), and between the Kansas River confluence to the Missouri River mouth (i.e. Missouri) parts of the Lower Missouri River (Fig. 1). For Nebraska and Missouri separately, we estimated P using observed beginning and ending weights over corresponding time periods and location for pallid sturgeon using the fixed parameter bioenergetics model. We used those estimates to develop relationships between FL and P. The resulting functions were:

- Nebraska: $P = 1 - e^{(-FL/696.42)}$;
- Missouri: $P = 1 - e^{(-FL/764.98)}$.

For further details on the development, description and parameterization of the pallid sturgeon bioenergetics model, see Wildhaber *et al.* (2015).

Hydrological/water temperature model validation relative to bioenergetics

We tested the effectiveness and validity of our combined HEC-HMS–SNTemp modelling by using the resulting simulated temperature and velocity data in place of observed data in the previously described bioenergetics model. This is important because, as mentioned in the introduction, our downscaling here is effectively a means to transfer AOGCM–RCM multi-model uncertainty to fish growth. It is critical that simulations relative to the current climate can be downscaled to produce similar fish growth response as when the bioenergetics model is forced with observations. Thus, the simulated data were used in our stochastically parameterized bioenergetics model for known periods of time in which we had observed pallid sturgeon growth in the Lower Missouri River (Fig. 3). For validation of our combined HEC-HMS–SNTemp model-simulated temperature and velocity data, we used basin-distributed NCEP data in place of individual meteorological-station observed data (NCDC 2010) in our bioenergetics model. Using these simulated water temperatures and velocities, and the bioenergetics models developed for Missouri and Nebraska separately, we demonstrate the effectiveness and validity of our physical models using Monte Carlo simulation of 1000 individual pallid sturgeon (somewhat less than half the number of wild pallid sturgeon estimated to be in the Lower Missouri River at the time of the listing of the species) (Fig. 4). For 14 pallid sturgeon in Nebraska and 18 in Missouri, when the observed initial weight was used in the bioenergetics model, the observed final weight was found within the bounds set by the lower 2.5 percentile and the upper 97.5 percentile of the distribution of the 1000 simulated fish with the same initial weight for all but one fish in Missouri (Fig. 4).

One reason for the effectiveness of the pallid sturgeon bioenergetics model is the nature of the velocity data. A common means to obtain velocity is by using the velocity power equation relationship developed by Leopold & Maddock (1953). The nature of the power-function relationship results in a flattening out of velocity at high and extreme discharge values. Even so, our results are more remarkable than might seem at first glance when one considers the limited information known and available about actual conditions to which these fish were exposed. All that was known about these fish and the conditions they might have experienced was their initial and final weights with limited or no record of the location in which the fish were growing in the river during the time period (i.e. some for nearly 2 years). The initial and final collection locations were used to identify a presumptive restriction to either the Nebraska or Missouri segments of the Lower Missouri River for a given fish.

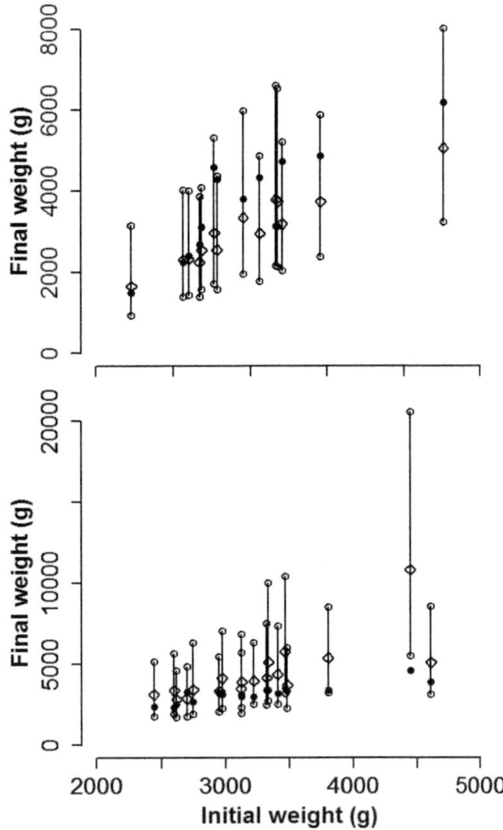

Fig. 4. Observed (filled circles) v. predicted (median, open diamond; upper and lower 95% percentiles, open circles) pallid sturgeon growth based on Missouri (bottom) and Nebraska (top) water velocity and temperature using the bioenergetics model developed by Wildhaber *et al.* (2015). Velocity is based on the river discharge simulated using a lumped-parameter precipitation runoff model HEC-HMS (USACE 2010). Water temperature is simulated using the physical-process SNTemp stream-temperature model (Theurer *et al.* 1984) (see Appendix A for more details).

Methods

Evaluation of multi-model uncertainty relative to the bioenergetics parameter uncertainty

To demonstrate whether one can draw a meaningful inference with climate simulations given multi-model uncertainty and ecological response parameter uncertainty, we present results from modelled sturgeon growth using fixed/calibrated system model outputs of water flow and temperature, and a stochastically parameterized bioenergetics model for fish growth that uses those outputs as inputs (Fig. 1). These results include recent-past and future simulated climate conditions under three climate models with available regional models for each (Table 1), with the NCEP/NCAR data considered as measured/observed data – a total of 17 datasets. The multi-model climate simulation uncertainty considered here is that of the differences in results that occurs between AOGCM–RCM combinations at two distinctively different points along the Lower Missouri River: Nebraska and Missouri (Fig. 2). We used the climate data from each regional model run as input into HEC-HMS and SNTemp in order to simulate temperature and velocity data from both points over 25 years for past (i.e. 1970–94) and future (i.e. 2039–63) climate models, along with NCEP/NCAR (i.e. 1970–94) as input into the bioenergetics model. These two 25 year time frames were chosen for consistency and for the sake of comparison based on the length of the dataset with the shortest, maximum continuous record length (i.e. 25 years).

Although the focus of this study is on the fish growth response relative to uncertainty in the multi-model climate ensemble and the bioenergetics model uncertainty, the ability of the physical downscaling system of models to represent a plausible variability in the river condition relative to the climate simulations is also important. Thus, we do consider the downscaled physical model output relative to the climate simulations in our results to help gauge the variability in these quantities relative to the climate simulation uncertainty. In doing this, bias corrections were not applied to the RCM climate data when it was downscaled to the sub-basins. We use HEC-HMS and SNTemp results based on NCEP/NCAR as the reference response for current conditions. We then compare those results with the results for each AOGCM–RCM combination to provide an assessment of the effectiveness of each AOGCM–RCM association at simulating the observed conditions (see Appendix B).

For the bioenergetics model, we demonstrate the potential effect of uncertainty associated with model parameterization. We do this by assigning distributions to each of the empirically derived parameters using their observed variation. Distributions of the parameters are based on literature, statistical analysis and pilot studies. Most of the parameters are normally (left truncated) distributed (Bevelhimer *et al.* 1985). Waste losses and specific dynamic action are assumed to follow a triangular distribution (Dowd *et al.* 2006). The swimming speed distribution was based on pallid sturgeon movement data (DeLonay *et al.* 2009). For full details of and the parameter values used in the bioenergetics model, see Wildhaber *et al.* (2015).

Simulations

We considered an initial distribution of 1000 pallid sturgeon from a truncated normal distribution with mean weight of 571.21 g (SD = 317.35 g) for the Lower Missouri River. The distribution used is based on a best fit to observed data collected by PSPAP. Once the initial population was determined, the bioenergetics model was run on a daily time step using the velocities and temperatures from a given hydrological/climate model or dataset. All bioenergetics model parameters were sampled from their respective distributions at the beginning of each day. No births or deaths were allowed. It is important to note that this was not meant to serve as a real-world population study, but rather as a comparison of the sensitivity of the bioenergetics of the fish across climate model scenarios.

Statistical analyses

To evaluate climate and hydrological variables (i.e. air and water temperature, precipitation, and velocity), we used the annual average (i.e. the sum of daily values for a year divided by the number of days in a year, either 365 or, for leap years, 366) so that seasonal variation did not mask any long-term trends. For fish growth, we used the final weights of the 1000 simulated fish. The statistical tests performed to assess differences in response between climate models were non-parametric owing to a lack of constancy of variance for climate and hydrological data, and non-normality of final fish weights. Our goal was to test differences between distributions. Therefore, we chose a combination of tests to assess differences in overall distributions and variances. To do this, the Kruskal–Wallis non-parametric one-way analysis of variance (ANOVA) and the Conover squared rank test of variances were used to test between climate scenarios at each gauge (i.e. Nebraska and Missouri). Kruskal–Wallis was used to test overall differences between distributions, while Conover was used to test differences in variances. There were three groups of tests carried out for each parameter at each of the gauges. In addition to the ANOVA tests, we considered pairwise tests within each group (note that the Kruskal–Wallis test is simply a Mann–Whitney test of two samples in this case, but requires Type I error adjustment to account for the multiplicity). In particular, the three groups of tests considered were: (1) differences between AOGCM–RCM scenarios and NCEP/NCAR for the recent past (Bonferroni-adjusted Type I errors for effective significance levels of $0.05/36 = 0.0014$); (2) differences between model scenarios for future climates (Bonferroni-adjusted Type I errors for an effective significance of $0.05/28 = 0.0018$); and (3) differences between the recent past and the future within each AOGCM/RCM scenario (Bonferroni-adjusted Type I error significance levels of $0.05/8 = 0.0063$).

Results

For all climate, hydrology and growth variables, the Kruskal–Wallis one-way ANOVA models revealed

Table 3. *Kruskal–Wallis non-parametric one-way ANOVA to test the overall differences between distributions and the Conover squared rank test of variance results*

Parameter	Gauge	Kruskal–Wallis		Conover	
		χ^2 statistic	p value	χ^2 statistic	p value
Recent past					
Ambient temperature (°C)					
	Nebraska City, NE	165.70	<0.0001	15.74	0.0463
	Boonville, MO	155.67	<0.0001	14.01	0.0815
Water temperature (°C)					
	Nebraska City, NE	174.16	<0.0001	17.00	0.0301
	Boonville, MO	159.90	<0.0001	16.01	0.0423
Velocity (m s^{-1})					
	Nebraska City, NE	194.79	<0.0001	23.93	0.0024
	Boonville, MO	179.04	<0.0001	35.50	<0.0001
Precipitation (cm)					
	Nebraska City, NE	139.92	<0.0001	43.29	<0.0001
	Boonville, MO	114.10	<0.0001	31.81	0.0001
Fish growth (g g^{-1}) in their respective models					
	Nebraska City, NE	4300.19	<0.0001	3098.56	<0.0001
	Boonville, MO	5544.67	<0.0001	6665.33	<0.0001
Fish growth (g g^{-1}) in their reverse models					
	Nebraska City, NE	4034.32	<0.0001	343.38	<0.0001
	Boonville, MO	3424.81	<0.0001	7152.55	<0.0001
Future					
Ambient temperature (°C)					
	Nebraska City, NE	139.77	<0.0001	4.89	0.6731
	Boonville, MO	137.99	<0.0001	6.66	0.4652
Water temperature (°C)					
	Nebraska City, NE	150.79	<0.0001	3.50	0.8355
	Boonville, MO	148.85	<0.0001	3.46	0.8389
Velocity (m s^{-1})					
	Nebraska City, NE	172.44	<0.0001	11.86	0.1053
	Boonville, MO	163.37	<0.0001	14.47	0.0434
Precipitation (cm)					
	Nebraska City, NE	131.74	<0.0001	24.41	0.001
	Boonville, MO	112.42	<0.0001	11.09	0.1348
Fish growth (g g^{-1}) in their respective models					
	Nebraska City, NE	3397.91	<0.0001	1097.91	<0.0001
	Boonville, MO	4254.99	<0.0001	2878.73	<0.0001
Fish growth (g g^{-1}) in their reverse models					
	Nebraska City, NE	3018.58	<0.0001	446.53	<0.0001
	Boonville, MO	4434.87	<0.0001	6201.45	<0.0001

Annual mean of climate scenario data is used as input into the hydrological model, and outputs from that model along with bioenergetics growth model outputs were individually analysed for each river gauge. The Nebraska City, NE gauge corresponds to 'Nebraska' and the Boonville, MO gauge corresponds to 'Missouri'. There were nine climate models for the recent past and eight for future tests. *p*, probability.

highly significant differences (Table 3). Air and water temperature, precipitation, water velocity, and fish growth produced a significant difference in the Conover squared rank of the variances for the recent past, but only for precipitation, velocity and fish growth for the future. For the pairwise Kruskal–Wallis test, the great majority of the Bonferroni-adjusted pairwise comparisons were significant (Table 4; Appendix B). For precipitation, >50%, around 65% and 0% of the comparisons were significant between climate models for the recent past, between climate models for the future, and between the recent past and the future within a climate model, respectively. For air temperature, >70%, >75% and 100% of the comparisons were significant with respect to the recent past, the future, and between the recent past and the future, respectively. For velocity, >75%, >75% and >37%

Table 4. *Probabilities from the pairwise climate scenario non-parametric Kruskal–Wallis tests conducted for fish growth (g) at each gauge (n = 36)*

			G	CC	CC	CC	CC	CC	CC	CG	CG	CG	CG	CG	CG	GF	GF	GF	GF
	Model		R	CR	CR	MM	MM	WR	WR	CR	CR	RC	RC	WR	WR	HR	HR	RC	RC
Site	G	R	Period	P	F	P	F	P	F	P	F	P	F	P	F	P	F	P	F
B	NC		P	+		+		−		−		+		−		+		+	
N	NC		P	NO		+		−		−		+		−		+		+	
B	CC	CR	P		+	+		−		−		+		−		+		+	
N	CC	CR	P		+	+		−		−		+		−		+		+	
B	CC	CR	F				MA		−		−		+		−		+		+
N	CC	CR	F				UN		−		−		+		−		+		+
B	CC	MM	P				+	+		−		+		+		+		+	
N	CC	MM	P				+	+		−		+		+		+		+	
B	CC	MM	F						−		−		+		−		+		+
N	CC	MM	F						−		−		+		−		+		+
B	CC	WR	P						+	+		+		NO		+		+	
N	CC	WR	P						+	UN		+		−		+		+	
B	CC	WR	F								+		+		−		+		+
N	CC	WR	F								+		+		−		+		+
B	CG	CR	P								+	+		−		+		+	
N	CG	CR	P								+	+		−		+		+	
B	CG	CR	F										+		−		+		+
N	CG	CR	F										+		−		+		+
B	CG	RC	P										+	−		−		NO	
N	CG	RC	P										+	−		−		+	
B	CG	RC	F												−		NO		−
N	CG	RC	F												−		−		MA
B	CG	WR	P												+	+		+	
N	CG	WR	P												+	+		+	
B	CG	WR	F														+		+
N	CG	WR	F														+		+
B	GF	HR	P														+	+	
N	GF	HR	P														+	+	
B	GF	HR	F																−
N	GF	HR	F																+
B	GF	RC	P																+
N	GF	RC	P																+
B	GF	RC	F																
N	GF	RC	F																

There were three groups of tests performed for each parameter at each of the gauges: (1) differences between climate models and NCEP/NCAR for the recent past (Bonferroni-adjusted alpha for significance of $0.05/36 = 0.0014$); (2) differences between climate models for the future (Bonferroni-adjusted alpha for significance of $0.05/28 = 0.0018$); and (3) differences between the recent past and the future within each climate model ($0.05/8 = 0.0063$). Missouri (i.e. Boonville gauge) designated as 'B' and Nebraska (Nebraska City gauge) as 'N'. Global climate models (G): NC, NCEP/NCAR; CC, CCSM; CG, CGCM; GF, GFDL. Regional climate models (R): CR, CRCM; MM, MM5I; RC, RCM3; WR, WRFG; HR, HRM3; P, recent past; F, future; − or +, significantly less or greater, respectively, row model than column model at Bonferroni alpha; UN, p value between Bonferroni alpha and 0.05; MA, p value between 0.05 and 0.1; NO, p value above 0.1 (p, probability).

of the pairwise comparisons were significant, respectively. For water temperature, around 78%, >75% and 100% of the tests were significant, respectively. For fish growth, around 94%, around 93% and 100% of the comparisons were significant, respectively.

Despite the variation in response observed within and between climate models for the environmental parameters (Appendix B), resulting differences in fish weight were much more consistent, with more than half of the 34 climate model combinations (past and future combined) significantly different from all other climate models, with the rest being similar to only one other climate model (Table 4). For Missouri, all non-significant comparisons for the recent past were between the same RCMs from different AOGCMs. For Missouri future and Nebraska recent past and future, two of

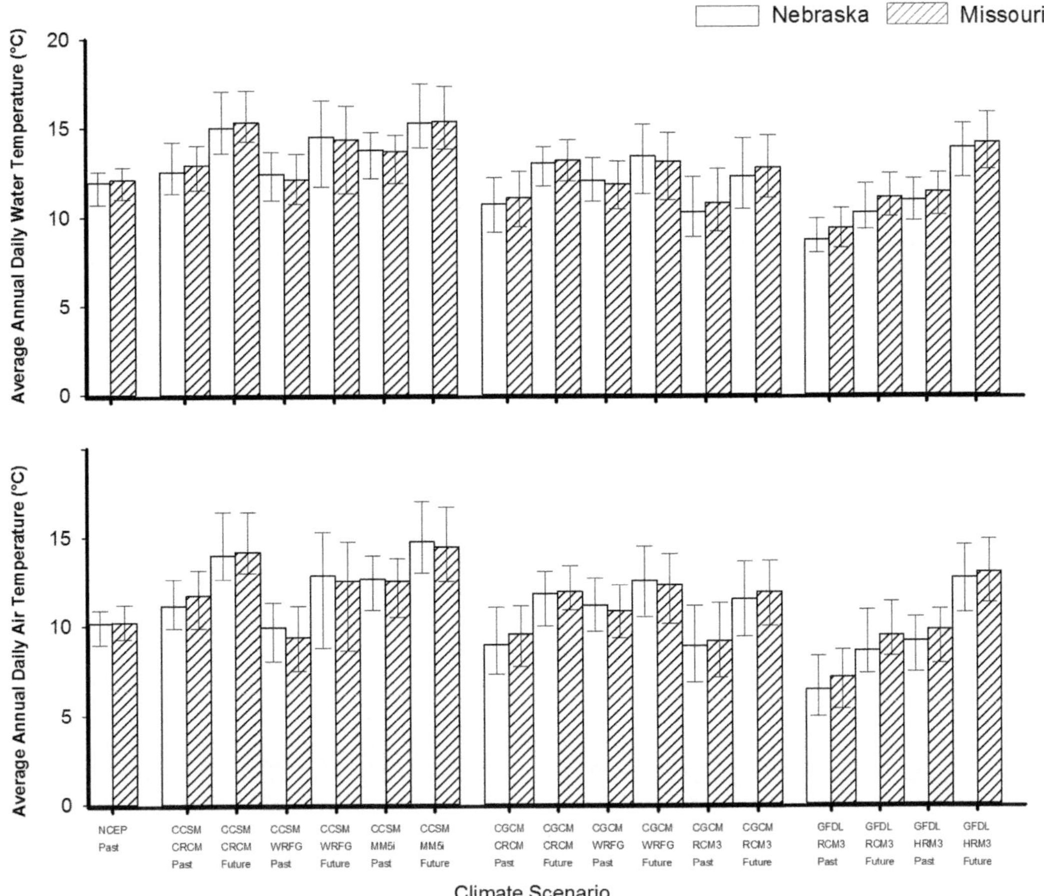

Fig. 5. Regional climate model (RCM) median, annual air temperature (lower panel) and median, annual, simulated water temperature (upper panel) for each climate scenario. Water temperature is simulated using the physical-process SNTemp stream-temperature model (Theurer *et al.* 1984) (see Appendix A for details). Bars represent minimum and maxium. North American NARCCAP datasets used in simulations. AOGCM simulations were taken from the WCRP's CMIP3 multi-model dataset (Meehl *et al.* 2007) based on the A2 emissions scenario (representing less international co-operation to reduce greenhouses gases) that projects relatively high greenhouse gas concentration increases. Model data used included four AOGCMs: CGCM (Flato 2005), CCSM, GFDL and HADCM. Model data used include six RCMs within the AOGCMs: CRCM, HRM3, MM5I, ECPC, RCM3 and WRFG. Measured/observed data: NCEP/NCAR Reanalysis 1 (Kalnay *et al.* 1996) (NCEP/NCAR) from Earth System Research Laboratory (www.esrl.noaa.gov).

the four non-significant comparisons showed the same pattern. Again, despite the inconsistency in response for the environmental variables between climate models and the fact that actual weights differed between climate models (Figs 5–7), all recent past weights were significantly greater than all future weights (Fig. 7); this was consistent with temperature differences (Fig. 5). In addition, median fish weight increased over the 25 year simulation period under all climate models except for GFDL/RCM3 and CGCM/RCM3 for the recent past and future, and for GFDL/HRM3 for the future for Nebraska (Fig. 7). The top five greatest fish weights occurred for the CGCM/WRFG recent past and future and the CGCM/CRCM recent past, the CCSM/WRFG recent past, and the NCEP/NCAR recent past at Missouri. Although gauge differences were not directly tested, in all cases the increase in weight was greater for Missouri compared to Nebraska (Fig. 7).

Discussion

We found differences in all but a few of the comparisons between climate models for the recent past and the future, and for all comparisons within climate

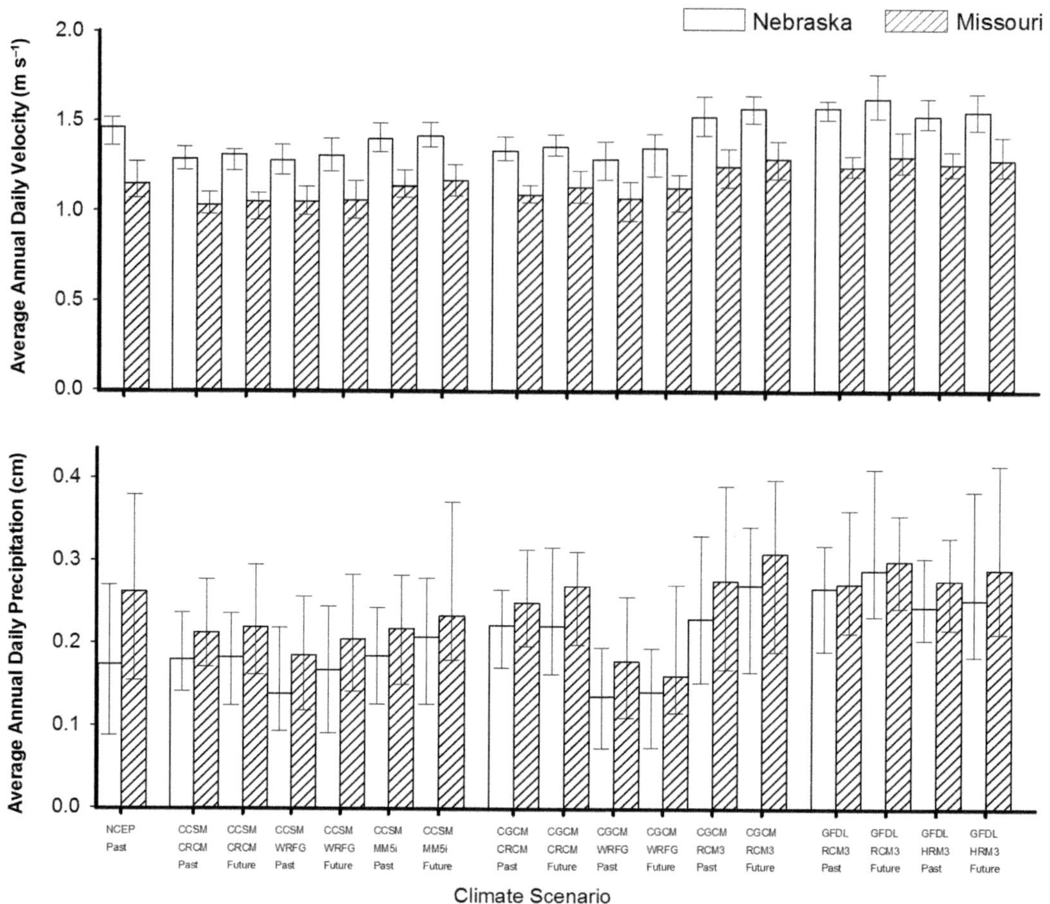

Fig. 6. Regional climate model (RCM) median, annual precipitation (lower panel) and median, annual, simulated velocity (upper panel) for each climate scenario. Velocity based on river discharge simulated using a lumped-parameter precipitation runoff model HEC-HMS (USACE 2010 – see Appendix A for more details). Bars represent minimum and maximum values. NARCCAP datasets were used in simulations. AOGCM simulations were taken from the WCRP's CMIP3 multi-model dataset (Meehl *et al.* 2007) based on the A2 emissions scenario (representing less international co-operation to reduce greenhouses gases) that projects relatively high greenhouse gas concentration increases. Model data used included four AOGCMs: CGCM (Flato 2005), CCSM, GFDL and HADCM. Model data used include six RCMs within the AOGCMs: CRCM, HRM3, MM5I, ECPC, RCM3 and WRFG. Measured/observed data: NCEP/NCAR Reanalysis 1 (Kalnay *et al.* 1996) (NCEP/NCAR) from Earth System Research Laboratory (www.esrl.noaa.gov).

models between the recent past and the future for growth rates of pallid sturgeon (Table 4). Actual predicted pallid sturgeon growth was dependent on the climate model considered (Fig. 7). However, despite the variation in responses seen between and within climate models for the various environmental parameters (Appendix B; Figs 5 & 6), the primary differences of greater growth for the recent past compared to the future, and for Missouri compared to Nebraska (Table 3; Fig. 7) held for all comparisons. Simultaneously, the model was able to capture, through predicted growth, the general pattern of GFDL compared to the CCSM and CGCM, with greater differences occurring between the AOGMs than between the recent past and future conditions within an AOGCM (Fig. 7), and the opposite pattern seen when comparing the CCSM and CGCM. For GFDL, increased velocities resulted in increased energetic costs and slowest growth rate (Fig. 7). As a result of the narrower, channelized river in the Nebraska reach compared to the Missouri reach, water velocities tended to be higher for Nebraska (Appendix B; Fig. 6), while water temperatures were similar (Appendix

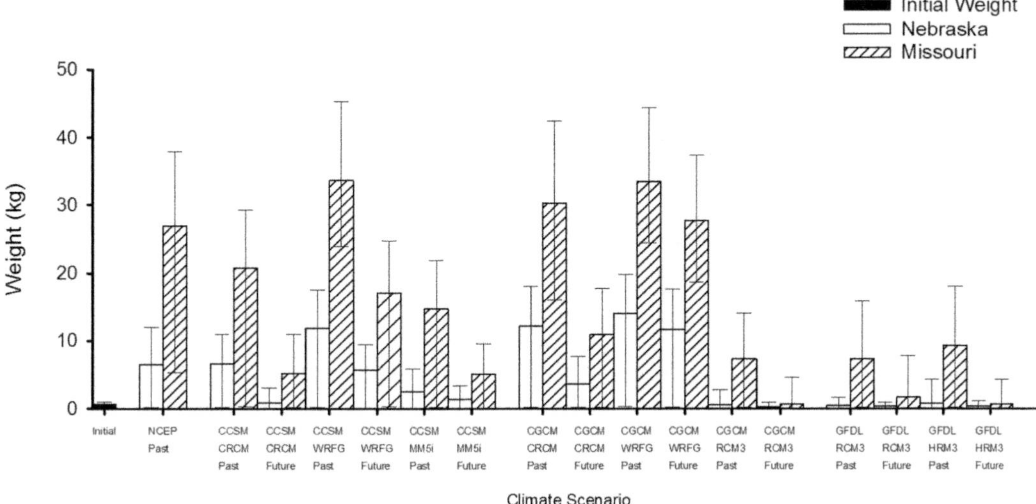

Fig. 7. Predicted, median final weight of pallid sturgeon based on Missouri and Nebraska water velocity and temperature using the bioenergetics model developed by Wildhaber *et al.* 2015). Velocity based on river discharge simulated using a lumped-parameter precipitation runoff model HEC-HMS (USACE 2010 – see Appendix A for more details). Water temperature is simulated using the physical-process SNTemp stream-temperature model (Theurer *et al.* 1984 – see Appendix A for more details). Bars represent minimum and maximum. NARCCAP datasets used in simulations. AOGCM simulations were taken from the WCRP's CMIP3 multi-model dataset (Meehl *et al.* 2007) based on the A2 emissions scenario (representing less international co-operation to reduce greenhouses gases) that projects relatively high greenhouse gas concentration increases. Model data used included four AOGCMs: CGCM (Flato 2005), CCSM, GFDL and HADCM. Model data used include six RCMs within the AOGCMs: CRCM, HRM3, MM5I, ECPC, RCM3 and WRFG. Measured/observed data: NCEP/NCAR Reanalysis 1 (Kalnay *et al.* 1996) (NCEP/NCAR) from Earth System Research Laboratory (www.esrl.noaa.gov).

B; Fig. 5). These higher velocities translated into higher energetic costs and, thus, a need for different consumption-to-length relationships between locations, not one representing pallid sturgeon growth for the entire Lower Missouri River.

In conclusion, relative to the analysis goal of considering inference on potential fish growth in the context of simulated climate data given the uncertainty associated with fish bioenergetics and the necessity for multi-model ensembles of climate simulations, it is apparent that, in the three situations considered here (i.e. recent past, future, and past v. future), uncertainty associated with the bioenergetics model does not preclude our ability to find significant differences in fish growth. Indeed, although our uncertainty in the bioenergetics model parameters is quite large, and it does lead to a fairly large 'spread' in the results, this spread is not so large as to swamp the variability in the climate model simulators. In addition, the associated differences, or lack thereof, in the physically downscaled weather and river condition variables between climate models (or within climate model when considering the past/future comparisons) suggest that the physical differences in the variables that feed into the bioenergetics model are responsible for these large differences in growth. This implies that it is possible to make inferential statements while capturing uncertainty in the climate model simulations and the ecological response. It also suggests that it is reasonable then to start considering the uncertainties associated with the intermediate scales as a next step.

Areas for improvement

As described in the methods, our bioenergetics model had to be built using a combination of results directly taken from pallid sturgeon juveniles and other sturgeon species. We had to assume that the model that described juvenile pallid sturgeon bioenergetics also described the bioenergetics of adults, which is often an incorrect assumption (Hansen *et al.* 1993). We also had to assume that the swimming energetics of other sturgeon species represented the swimming energetics of pallid sturgeon. These assumptions identified important areas of research still needed on pallid sturgeon basic and swimming energetics.

Some of the next steps for this modelling effort would be to address the existing obvious sources of error and uncertainty: land use, geological and geomorphic, hydrological, temperature, velocity and energetics models, and to add additional hydrology models to the analysis to better address the issue of error propagation across scales. For example, improved capabilities for simulation of snowmelt and base flow components, as well as more detailed accounting of anthropogenic effects, could reduce some uncertainty related to the hydrological response of the Missouri River Basin to changing climate. The framework we present works very well for the mainstem of the Missouri River where hydrological conditions are the result of aggregations of large areas and physical processes, but more detailed modelling (i.e. finer spatial discretization and more data) may be required to address locations on the Missouri River Basin other than the mainstem (e.g. headwaters and lower flow segments). Along with these additions is the need for the inclusion of more model-based analysis of the distributional results that can accommodate biases, interactions and uncertainties. As demonstrated by the place-based results we presented, there is a need to add spatial linkages and, most importantly, the need to fully develop an effective method for uncertainty propagation across scales (Wikle 2003; Clark 2007; Cressie *et al.* 2009; Cressie & Wikle 2011).

Critically, as we move into a modelling framework that seeks to evaluate probabilistic outputs, the standard simple statistical comparisons presented here will give way to more complicated assessments of distributional variability. The Bayesian paradigm is increasingly being used to address the issues associated with probabilistic impact studies (e.g. Tebaldi *et al.* 2004, 2005; Greene *et al.* 2006; Furrer *et al.* 2007; Rougier 2007; Berliner & Kim 2008; Rougier *et al.* 2009; Tebaldi & Sansó 2009; Sain *et al.* 2011). In particular, issues related to model weighting, information content and the role of expert opinion should be considered (e.g. Knutti *et al.* 2010; Knutti 2010). Such large-scale probabilistic impact studies are starting to be considered (e.g. Murphy *et al.* 2007; Rougier *et al.* 2009; Sankarasubramanian *et al.* 2009; Sokolov et al. 2009; Harris et al. 2010; Semenov & Stratonovitch 2010), but are still in their infancy when it comes to the consideration of individual scale ecological impacts.

Linking with pallid sturgeon recovery management

The life expectancy of pallid sturgeon is at least 40 years (USFWS 1993). This means individuals living under current management practices of the Pallid Sturgeon Recovery Plan will probably live to experience climate change. The most recent USFWS 5 year review of pallid sturgeon recovery efforts urges an update of the Pallid Sturgeon Recovery Plan to include the most recent information regarding genetics, distribution, life history, abundance and trends, threats, and conservation measures (USFWS 2007). Climate change should be added to this list, and it should be recognized that understanding the sensitivity of pallid sturgeon to climate change means continuing to learn a great deal about the sensitivity of its habitat to climate variability and about conservation measures.

The capability to directly link projections of climate change with current and hypothetical management practices for pallid sturgeon is a critically important aspect of a flexible hierarchical approach to quantifying uncertainty. Although we illustrate integrating multi-model climate uncertainty and parameter-uncertainty species models using a bioenergetics growth model, the mathematical framework could also be applied to models for habitat (Jacobson *et al.* 2009), migratory constraints (Jacobson *et al.* 2009) and prey availability (Spindler *et al.* 2012). In addition, the hierarchical structure can use, as either parameter distributions or structural models, the knowledge gained from sampling programmes that have speculated on preferred habitats for spawning (Koch *et al.* 2012), migratory channels (DeLonay *et al.* 2009; Jacobson *et al.* 2009; McElroy *et al.* 2012) and habitats altered by sediment transport (DeLonay *et al.* 2009).

Since the Pallid Sturgeon Recovery Plan identifies as an immediate threat the present or threatened destruction, modification or curtailment of its habitat or range, a logical next step, then, should be to link climate projections not only to bioenergetics models but also to habitat models. However, the results of this preliminary study indicate that a primary change to the pallid sturgeon environment expected of climate warming is a tendency towards increased water temperature and velocity. Temperature change could alter the timing of spawning cues, and this suggests the possibility of a decoupling between the seasonality of flow pulses and water temperature appropriate for embryo development. Simultaneously, increased velocities could decrease growth, and, therefore, survival and reproductive potential.

Ultimately, any modelling approach used to help guide pallid sturgeon recovery efforts needs to be ecosystem based, such as that of the marine ecosystem modelling approach of Ecopath with Ecosim developed at the University of British Columbia (http://www.ecopath.org/). Without key ecosystem components (e.g. prey) incorporated into the modelling approach, important non-linear feedbacks are ignored. Therefore, these additional components need to be included as submodels that have a feedback loop with the primary model. For fish species fed on by adult pallid sturgeon, a separate set of bioenergetics models is needed for each species to estimate their mass at any given time point. That mass then becomes the estimate of available food for adult pallid sturgeon. As a feedback, this prey mass is then affected by the predation rate of pallid sturgeon, and so on.

Conclusions

In climate impact studies, there is uncertainty in our models and observations at each scale from the global to the individual. Attempting to account for these uncertainties is critical in trying to evaluate potential effects of climate change since lakes and streams have been identified as sentinels of environmental change (Williamson *et al.* 2008). Through this work, we have advanced efforts to understand and accommodate model uncertainty based on stochastic individual model parameterization and multi-model climate simulation. In particular, we allow information from the multi-model climate simulators to propagate to the individual scale in a deterministic manner that accounts for multiple variables with realistic spatial and temporal variability. Ultimately, this allows us to examine whether there is a meaningful variation evident across climate model and scenario relative to the uncertainty associated with our knowledge of fish bioenergetics relative to the Missouri River sturgeon population dynamics. In support of the need for such an approach, we show how we have combined available data from numerous agencies (e.g. the USGS, the NOAA, the USACE, the USDA and the USFWS) and models (e.g. the North American Climate Change Assessment Program (NARCCAP), and sturgeon population, movement and bioenergetics models) across a variety of scales to try to develop initial simulations of potential pallid sturgeon population change.

Although not considered in this study, the ultimate goal is a framework for describing the potential consequences of regional and global climate changes on large riverine ecosystems and the related influence on fish growth using the uncertainty associated with various climate model scenario data sources. This work is the basis for such an approach that provides a tool for understanding the potential consequences of regional to global change and the resulting changes for large river ecological systems, and the options for sustaining and improving ecological systems and related goods and services, given projected regional to global changes – both questions identified in the USGS Strategic Science Plan (USGS 2007). The expectation is a framework for identifying potential consequences of climate change on large riverine ecosystems (key ecosystems), such as the Missouri River, to alert decision-makers to the most likely consequences to such ecosystems. In addition, it should provide the basis for the development of a new suite of indicators of large riverine ecosystem change and health.

The Missouri River is a highly regulated river as a result of a series of mainstem reservoirs put in place for flood control and hydropower that influence water flow and temperature (Wright *et al.* 1999; Galat & Lipkin 2000). The result is natural flow regimes that are influenced by this reservoir network. Therefore, how releases/abstractions are managed today and in the future is/will be important. The framework presented here provides the basis for development of a model that allows comparisons of the effectiveness of such management practices and others (e.g. habitat changes, propagation), especially in the presence of climate change. Any future work should be designed to specifically address these management questions.

The Missouri Department of Conservation (V. Travnichek pers. comm.), the USFWS (C. Scott pers. comm.) and researchers (Quist *et al.* 2004; Bergman *et al.* 2008; Doyle *et al.* 2011) have indicated the need for this work for the very reasons just described. This might be achieved through the development of a predictive model to evaluate management options (e.g. reservoir management) for coping with global change effects and by providing a powerful tool for assessing the associated uncertainty. This model could be used to help assess sensitivity and adaptability of the Missouri River, document methods that could be applied to other large river ecosystems, and assess the potential effectiveness of adaptation and mitigation methodologies at minimizing the effects of directional and non-linear climate change on the Missouri River ecosystem and resident fishes. The ultimate model would refine, apply and interpret basin and ecosystem process models to assist natural resource managers.

Funding for this work was provided by the USGS through the National Climate Change Wildlife Science Center. We would like to thank the PSPAP and the USACE, the USFWS and the USGS for providing access to PSPAP, the Pallid Sturgeon Propagation Program and the Comprehensive Sturgeon Research Program data, respectively, and Dr S. R. Chipps of the USGS for use of the pallid sturgeon basic bioenergetics model parameterization to help parameterize the initial bioenergetics model. We would also like thank the following University of Missouri undergraduate students (S. Childers, L. Crawford, Q. Hall, J. Kastman, T. Schrautemeier, R. Scott and A. Wettlaufer) and the USGS staff (J. Albers, E. Beahan and L. Johnson) for their assistance in compiling data and running model simulations. Thanks to R. Jacobson, J. Peterson and S. Markstrum of the USGS for reviewing earlier versions of this manuscript. Any use of trade, product or firm names is for descriptive purposes only and does not imply endorsement by the United States Government.

Appendix A

HEC-HMS and SNTemp model parameterization

To model the hydrological components (i.e. river velocity and water temperature) needed for the bioenergetics model, the Missouri River Basin was divided into 116 sub-basins and corresponding reaches delineated using the ArcGIS Geo-HEC program from the HYDRO1K hydrologically corrected dataset (last accessed 8 December 2011, http://eros.usgs.gov/#Find_Data/Products_and_Data_Available/gtopo30/hydro/namerica). The 116 sub-basins were based on 84 USGS stream gauges, with 32 additional self-generated based on topography and determined using ArcGIS Geo-HMS. Sub-basin size ranged from 82.9 to 60 780 km^2 (32 to 23 469 square miles), with stream length varying from 6.4 to 493.7 km (3.98–306.75 miles). Sub-basin boundaries were used to determine stream segmentation (Fig. 1) using a 10 000 km^2 threshold. Approximately 13 820 km (8584 miles) of stream channels were simulated, including 3460 km (2150 miles) of the Missouri River mainstem. All meteorological data were distributed or downscaled across each basin using the Thiessen Polygon technique (USACE 2000), giving 116 individual meteorological data points – a source of uncertainty not accounted for here but would be if other sources of uncertainty were considered.

For hydrological modelling using HEC-HMS, to account for initial soil moisture and soil-water loss rate (hydraulic conductivity), we used the Deficit and Constant Loss precipitation-runoff method with constant hydraulic conductivity available in the model. This method adequately allows for long-term simulation for use with potential evapotranspiration processes (USACE 2010) and requires the least number of unknown parameters when unsaturated/saturated aquifer interactions are not quantified. While the model might not predict losses well within a specific storm event due to the assumption of a constant loss rate, total loss is well simulated, and the model 'has been implemented successfully in many research projects conducted throughout the United States' (USACE 2001). Evapotranspiration was simulated through the Priestley–Taylor method, which is the only method included in HEC-HMS at this time (USACE 2010). Snow accumulation and melt were simulated using the Temperature Index method (USACE 2010), a simpler method when assuming that solar radiation dominates the snowmelt process (Debele *et al.* 2010). Sub-basin outflows were computed using the Clark unit hydrograph (UH) (Clark 1945), a mathematical approach that relates direct runoff to one unit of uniformly distributed excess precipitation occurring over a specific duration. The Clark UH requires two parameters calculated from land use/class and soil class: the time of concentration of channel water from the sub-basin and a sub-basin storage coefficient, which accounts for the temporary storage of precipitation excess. Stream baseflow was computed using the HEC-HMS Recession method combined with the Ratio to Peak Threshold option (a constant determining when baseflow is reset on the falling limb of the hydrograph). This method best represents baseflow quantity when ground and surface water interactions are not quantified; HEC-HMS does not simulate groundwater conditions. However, the Monthly Baseflow HEC-HMS option was modified by using averaged daily baseflow from 1972 to 1995 (calibration period) for all headwater basins, as baseflow in HEC-HMS is computed only from simulated flow and does not automatically add additional channel water contributed directly from groundwater input. Baseflow volume was computed using baseflow separation techniques provided by the Web-based Hydrograph Analysis Tool (WHAT) (Lim *et al.* 2005).

In HEC-HMS, one of two channel routing methods was used, which represents the lag and attenuation of a flood wave as it moves downstream. Our preferred method was the Kinematic-Wave Routing method, a mathematical method relating streamflow dynamics to the shape, channel roughness (channel Manning's *n*) (Arcement & Schneider 1989) and slope of the channel (USACE 1993). If, however, the channel flow computation did not converge while using the Kinematic-Wave method, the Muskingum–Cunge Routing method was applied. The Muskingum–Cunge method is based on the conservation of mass equation and the diffusive form of the momentum equation (USACE 2001) and uses a transformed Kinematic-Wave diffusion equation that numerically attenuates an imperfectly centred finite-difference solution (USACE 1991). These methods allow for the use of more physical channel characteristics, such as channel shape, than other methods provided in HEC-HMS. All Missouri River mainstem reservoirs were simulated using HEC-HMS Reservoir Elements while applying Outflow Curve and Storage Discharge methods. The Storage Discharge function in HEC-HMS is a mathematical expression used to represent reservoir management. Following the HEC-HMS manual, the parameters that define this function were estimated using observed discharge and reservoir storage values, and calibrated on 1972–95 data.

Velocity was calculated from observed and HEC-HMS simulated discharge based on the equation, $V = aD^b$ (Leopold & Maddock 1953), where V is velocity, D is discharge, and a and b are the empirical coefficients of the non-linear equation, which are held constant. The estimated constants a and b for Nebraska were 0.173 and 0.302, and for Missouri 0.117 and 0.303, respectively. These relationships vary with changes in channel form and do not account for the 15% measurement error known to be present in the observed discharge data (Moriasi *et al.* 2007). Therefore, we used the longest available record of field measured data (i.e. 1950–2009) to produce long-term average estimates; additional sources of uncertainty that could be incorporated if known. Using these fixed, estimated parameter values, we used the discharge to velocity equation to convert daily discharge to daily velocity data. Computed velocity data were rounded to 0.5 dm s^{-1} to account for reporting precision (USGS 2002, 2012).

In SNTemp, required groundwater (baseflow) and ground temperatures are generally assumed to be at average annual ambient temperatures (Freeze & Cherry 1979; Bartholow 2000). However, we do acknowledge that groundwater temperatures do vary and, therefore, this study used a 365 day-centred moving average of ambient air temperatures for daily groundwater temperatures for each sub-basin/reach. This option is allowable in SNTemp and greatly influences stream water temperatures. Ground temperature (restricted by SNTemp to one value for each sub-basin for the entire simulation period) was estimated using average air temperatures for the entire period. Accounting for varying groundwater temperatures using the 365 day moving average method, however, accounts for much of the influence that changing soils temperatures have on surface water temperatures.

SNTemp was calibrated to water temperature data collected at the Nebraska USGS gauging station location, which is an upper point on the Lower Missouri River mainstem. The Nebraska location had the longest observed stream temperature record of all USGS gauging stations used in this study. To maintain model simplicity, the only SNTemp parameter calibrated in this study was the Bowen ratio, as this was considered the most uncertain value and results were most sensitive to this parameter. Bowen ratios range between 0.1 for open water (specifically, ocean surfaces), 0.2 for mixed forests and wetlands, 2.0 for deserts and up to 10 for arid – shrubland regions (USEPA 2008). Bowen ratios can be negative, but such values are not allowed in SNTemp. A Bowen ratio of 0.05, appropriate for land-based surface-water bodies, was used for all reaches. This value was determined through trial and error calibration for river reaches in Nebraska where the only usable long-term temperature was available for the Lower Missouri River.

Appendix B

Precipitation, air temperature, water velocity and water temperature

Except for GFDL/RCM3 recent past and future, and CCSM/MM5I recent past air temperature, all climate models (recent past and future combined) were not significantly different from that of one to four other climate models for Nebraska and Missouri, and one–three climate models for water temperature; for all climate models, however, air and water temperatures were significantly different between the recent past and the future (Table B1). Except for NCEP/NCAR recent past, CCSM MM5I recent past and future, and CGCM/CRCM recent past and GFDL/RCM future for Nebraska, velocity for all climate models was not significantly different to that from one to three other climate models (Table B2). Velocity was only different between the recent past and the future for CGCM/CRCM, CGCM/WRFG and GFDL/RCM for Nebraska and Missouri, and CCSM/MM5I for Nebraska. For precipitation, in comparisons between climate models for the recent past and the future, all climate models were not significantly different from at least one other climate model, while no comparison between the recent past and future within a climate model was significant. Although gauge differences were not directly tested, air and water temperature tended to be similar between Missouri and Nebraska, while precipitation tended to be less for Nebraska than Missouri, and velocity tended to be less for Missouri than Nebraska (Table 4).

Table B1. *Probabilities from the pairwise climate scenario non-parametric Kruskal-Wallis tests conducted for air temperature (above diagonal) and water temperature (below diagonal) at each gauge (n = 36)*

	Model		G	NC	CC	CC	CC	CC	CC	CC	CG	CG	CG	CG	CG	CG	GF	GF	GF	GF
			R		CR	CR	MM	MM	WR	WR	CR	CR	RC	RC	WR	WR	HR	HR	RC	RC
Site	G	R	Period	P	P	F	P	F	P	F	P	F	P	F	P	F	P	F	P	F
B	NC		P		−		−		+		+		+		UN		UN		+	
N	NC		P		−		−		NO		+		+		−		+		+	
B	CC	CR	P	+		−	UN		+		+		+		UN		+		+	
N	CC	CR	P	−		−	−		+		+		+		NO		+		+	
B	CC	CR	F		+			NO		+		+		+		+		+		+
N	CC	CR	F		+			MA		+		+		+		+		+		+
B	CC	MM	P	+	UN	−		−	+		+		+		−		+		+	
N	CC	MM	P	+	+	−		−	+		+		+		−		+		+	
B	CC	MM	F			NO	+			+		+		+		+		+		+
N	CC	MM	F			NO	+			+		+		+		+		+		+
B	CC	WR	P	NO	−		−			−	NO		NO		−		NO		+	
N	CC	WR	P	UN	−		−			−	UN		UN		−		UN		+	
B	CC	WR	F			−		−	+			NO		MA		−		UN		+
N	CC	WR	F		+	UN		−	+			UN		+		+		NO		+
B	CG	CR	P	−	−		−		−	−		−	NO		−		NO		+	
N	CG	CR	P	−	−		−		−	−		−	NO		−		NO		+	
B	CG	CR	F			−		−		−	+			NO		NO		−		+
N	CG	CR	F			−		−		−	+			NO		−		−		+
B	CG	RC	P	−	−		−		−		NO			−	−		NO		+	
N	CG	RC	P	−	−		−		−		UN			−	−		NO		+	
B	CG	RC	F			−		−		−		MA	+			NO		−		+
N	CG	RC	F			−		−		−		MA	+			UN		−		+
B	CG	WR	P	NO	−		−		NO		−		+			−	+		+	
N	CG	WR	P	NO	UN		−		NO		−		+			−	+		+	
B	CG	WR	F			−		−		−		NO		NO	+			−		+
N	CG	WR	F			−		−		−		UN		UN	+			NO		+
B	GF	HR	P	−	−		−		−		NO		UN		UN			−	+	
N	GF	HR	P	−	−		−		−		NO		UN		−			−	+	
B	GF	HR	F			−		−		NO		+		+		+	+			+
N	GF	HR	F			−		−		UN		+		+		UN	+			+
B	GF	RC	P	−	−		−		−		−	−	−	−	−	−	−	−		−
N	GF	RC	P	−	−		−		−		−	−	−	−	−	−	−	−		−
B	GF	RC	F			−		−		−		−		−		−		−	+	
N	GF	RC	F			−		−		−		−		−		−		−	+	

There were three groups of tests performed for each parameter at each of the gauges: (1) differences between climate models and NCEP/NCAR for recent past (Bonferroni-adjusted alpha for significance of 0.05/36 = 0.0014); (2) differences between climate models for future (Bonferroni-adjusted alpha for significance of 0.05/28 = 0.0018); and (3) differences between the recent past and the future within each climate model (alpha: 0.05/8 = 0.0063). Missouri (i.e. Boonville gauge) designated as 'B' and Nebraska (Nebraska City gauge) as 'N'. Ambient temperature is the daily averages. Water temperature is the daily value simulated using SNTemp. Global climate models (G): NC, NCEP/NCAR; CC, CCSM; CG, CGCM; GF, GFDL. Regional climate models (R); CR, CRCM; MM, MM5I; RC, RCM3; WR, WRFG; HR, HRM3; P, recent past; F, future; − or +, significantly less or greater, respectively, row model than column model at the Bonferroni alpha; UN, p value between the Bonferroni alpha and 0.05; MA, p value between 0.05 and 0.1; NO, p value above 0.1 (p, probability).

Table B2. *Probabilities from the pairwise climate scenario non-parametric Kruskal–Wallis tests conducted for precipitation (above diagonal) and velocity (below diagonal) at each gauge* (n = 36)

			G	NC	CC	CC	CC	CC	CC	CC	CG	CG	CG	CG	CG	CG	GF	GF	GF	GF
	Model		R		CR	CR	MM	MM	WR	WR	CR	CR	RC	RC	WR	WR	HR	HR	RC	RC
Site	G	R	Period	P	P	F	P	F	P	F	P	F	P	F	P	F	P	F	P	F
B	NC		P		UN		UN		+		NO		NO		+		NO		NO	
N	NC		P		NO		NO		UN		UN		−		UN		−		−	
B	CC	CR	P	−		NO	NO		+		−		−		+		−		−	
N	CC	CR	P	−		NO	NO		+		−		−		+		−		−	
B	CC	CR	F		NO			MA		+		−		−		NO		−		−
N	CC	CR	F		NO			MA		+		−		−		MA		−		−
B	CC	MM	P	NO	+			NO	+		UN		−		+		−		−	
N	CC	MM	P	−	+			MA	+		−		−		+		−		−	
B	CC	MM	F			+	UN			+		UN		−		UN		−		−
N	CC	MM	F			+	+			+		NO		−		UN		−		−
B	CC	WR	P	−	+		+			NO	−		−		NO		−		−	
N	CC	WR	P	−	−		+			NO	−		−		NO		−		−	
B	CC	WR	F			NO		−	NO		−	−		−		NO		−		−
N	CC	WR	F			NO		−	MA		−	−		−		NO		−		−
B	CG	CR	P	−	+		−		UN			NO	MA		+		UN		UN	
N	CG	CR	P	−	+		−		+			NO	MA		+		−		−	
B	CG	CR	F			+		UN		+	+			UN		+		UN		−
N	CG	CR	F			+		−		+	+			UN		+		−		−
B	CG	RC	P	+	+		+		+		+			NO	+		NO		NO	
N	CG	RC	P	+	+		+		+		+			NO	+		NO		NO	
B	CG	RC	F			+		+		+		+	UN		+	+		NO		NO
N	CG	RC	F			+		+		+		+	UN		+	+		NO		MA
B	CG	WR	P	−	NO		−		NO		UN		−			UN	−		−	
N	CG	WR	P	−	NO		−		NO		−		−			MA	−		−	
B	CG	WR	F			+		−		NO		NO		−	+			−		−
N	CG	WR	F			UN		−		NO		NO		−	+			−		−
B	GF	HR	P	+	+		+		+		+		NO		+			NO	NO	
N	GF	HR	P	+	+		+		+		+		NO		+			NO	NO	
B	GF	HR	F			+		+		+		+		NO		+	UN			NO
N	GF	HR	F			+		+		+		+		NO		+	MA			UN
B	GF	RC	P	+	+		+		+		+		NO		+		NO			UN
N	GF	RC	P	+	+		+		+		+		UN		+		+			UN
B	GF	RC	F			+		+		+		+		NO		+		MA	+	
N	GF	RC	F			+		+		+		+		*		+		+	+	

There were three groups of tests done for each parameter at each of the gauges: (1) differences between climate models and NCEP/NCAR for the recent past (Bonferroni-adjusted alpha for significance of 0.05/36 = 0.0014); (2) differences between climate models for the future (Bonferroni-adjusted alpha for significance of 0.05/28 = 0.0018); and (3) differences between the recent past and the future within each climate model (0.05/8 = 0.0063). Missouri (i.e. Boonville gauge) designated as 'B' and Nebraska (Nebraska City gauge) as 'N'. Ambient temperature is the daily averages. Water temperature is the daily value simulated using SNTemp. Global climate models (G): NC, NCEP/NCAR; CC, CCSM; CG, CGCM; GF, GFDL. Regional climate models (R): CR, CRCM; MM, MM5I; RC, RCM3; WR, WRFG; HR, HRM3; P, recent past; F, future; − or +, significantly less or greater, respectively, row model than column model at the Bonferroni alpha; UN, p value between the Bonferroni alpha and 0.05; MA, p value between 0.05 and 0.1; NO, p value above 0.1 (p, probability).

Correction notice: The original version was incorrect. This was due to an error in the caption for Figure 1.

References

Arab, A., Wildhaber, M. L., Wikle, C. K. & Gentry, C. N. 2008. Zero-inflated modeling of fish catch per unit area resulting from multiple gears – application to channel catfish and shovelnose sturgeon in the Missouri River. *North American Journal of Fisheries Management*, **28**, 1044–1058.

Arab, A., Holan, S. H., Wikle, C. K. & Wildhaber, M. L. 2012. Semiparametric bivariate zero-inflated Poisson models with application to studies of abundance for multiple species. *Environmetrics*, **23**, 183–196.

Arcement, G. J. Jr. & Schneider, V. R. 1989. *Guide for Selecting Manning's Roughness Coefficients for Natural Channels and Flood Plains.* U.S. Geological Survey Water Supply Paper 2339.

Bajer, P. G. & Wildhaber, M. L. 2007. Population viability analysis of Lower Missouri River shovelnose sturgeon with initial application to the pallid sturgeon. *Journal of Applied Ichthyology*, **23**, 457–464.

Bartholow, J. M. 1991. A modeling assessment of the thermal regime for an urban sport fishery. *Environmental Management*, **15**, 833–845.

Bartholow, J. M. 2000. *The Stream Segment and Stream Network Temperature Models: A Self-Study Course (Version 2.0).* United States Geological Survey, Open-File Report, **99-112**.

Bergengren, J. C., Waliser, D. E. & Yung, Y. L. 2011. Ecological sensitivity: a biospheric view of climate change. *Climatic Change*, **107**, 433, https://doi.org/10.1007/s10584-011-0065-1

Bergman, H. L., Boelter, A. M. et al. 2008. *Research Needs and Management Strategies for Pallid Sturgeon Recovery. Research Needs and Management Strategies for Pallid Sturgeon Recovery.* William D. Ruckelshaus Institute of Environment and Natural Resources. University of Wyoming, Laramie, WY.

Berliner, L. M. & Kim, Y. 2008. Bayesian design and analysis of superensemble-based climate forecasting. *Journal of Climate*, **21**, 1891–1910.

Bevelhimer, M. S., Stein, R. A. & Carline, R. F. 1985. Assessing significance of physiological differences among three esocids with a bioenergetics model. *Journal of Fisheries and Aquatic Sciences*, **42**, 57–69.

Blevins, D. W. 2006. *The Response of Suspended Sediment, Turbidity, and Velocity to Historical Alterations of the Missouri River.* United States Geological Survey, Circular, **1301/8**.

Caya, D. & Laprise, R. 1999. A semi-implicit semi-lagrangian regional climate model. The Canadian RCM. *Montana Weather Review*, **127**, 341–362, https://doi.org/10.1175/1520-0493(1999)127<0341:ASISLR>2.0.CO;2

CCSP. 2003. *Vision for the Program and Highlights of the Scientific Strategic Plan. A Report by the Climate Change Science Program and the Subcommittee on Global Change Research.* Climate Change Science Program, Washington, DC, 14, 19, 20, 22, 26.

CCSP. 2008. *Weather and Climate Extremes in a Changing Climate. Regions of Focus: North America, Hawaii, Caribbean, and U.S. Pacific Islands. A Report by the U.S. Climate Change Science Program and the Subcommittee on Global Change Research.* Department of Commerce, NOAA National Climatic Data Center, Washington, DC.

Chen, D., Achberger, C., Räisänen, J. & Hellstrom, C. 2006. Using statistical downscaling to quantify the GCM-related uncertainty in regional climate change scenarios: a case study of Swedish precipitation. *Advances in Atmospheric Sciences*, **23**, 5460.

Chipps, S. R., Klumb, R. A. & Wright, E. B. 2008. *Development and Application of a Juvenile Pallid Sturgeon Bioenergetics Model. Final Report.* Federal Aid Project Number T-24-R, Study 2424. South Dakota Department of Game, Fish and Parks, Pierre, SD.

Clark, C. O. 1945. Storage and the unit hydrograph. *Transactions of the American Society of Civil Engineers*, **110**, 1419–1447.

Clark, J. S. 2007. *Models for Ecological Data: An Introduction.* Princeton University Press, Princeton, NJ.

Collins, M., Booth, B. B. B., Bhaskaran, B., Harris, G. R., Murphy, J. M., Sexton, D. M. H. & Webb, M. J. 2011. Climate model errors, feedbacks and forcings: a comparison of perturbed physics and multi-model ensembles. *Climate Dynamics*, **36**, 1737–1766.

Cressie, N. & Wikle, C. K. 2011. *Statistics for Spatio-Temporal Data.* Wiley, Hoboken, NJ.

Cressie, N., Calder, C. A., Clark, J. S., VerHoef, J. M. & Wikle, C. K. 2009. Accounting for uncertainty in ecological analysis: the strengths and limitations of hierarchical statistical modeling. *Ecological Applications*, **19**, 553–570.

Debele, B., Srinivasan, R., & Gosain, A. K. 2010. Comparison of process-based and temperature-index snowmelt modeling in SWAT. *Water Resources Management*, **24**, 1065–1068.

DeLonay, A. J., Papoulias, D. M. et al. 2009. *Ecological Requirements of Pallid Sturgeon Reproduction and Recruitment in the Lower Missouri River – A Research Synthesis 2005–08.* United States Geological Survey, Scientific Investigation Report, **2009-5201**.

Dowd, W. W., Brill, R. W., Bushnell, P. G. & Musick, J. A. 2006. Estimating consumption rates of juvenile sandbar sharks, (*Carcharhinus Plumbeus*), in Chesapeake Bay, Virginia, using a bioenergetics model. *Fisheries Bulletin*, **104**, 332–342.

Doyle, M., Murphy, D., Bartell, S., Farmer, A., Guy, C. S. & Palmer, M. 2011. *Final Report on Spring Pulses and Adaptive Management. Missouri River Recovery Program Independent Science Advisory Panel.* United States Institute for Environmental Conflict Resolution, Washington, DC.

Drobish, M. R. 2008. *Pallid Sturgeon Population Assessment Project, Volume 1.3.* United States Army Corps of Engineers, Yankton, SD.

Dryer, M. P. & Sandvol, A. J. 1993. *Recovery Plan for the Pallid Sturgeon (Scaphirhynchus albus).* United States Fish and Wildlife Service, Bismarck, ND.

Engel, B., Storm, D., White, M. & Arnold, J. G. 2007. A hydrologic/water quality model application protocol. *Journal of the American Water Resources Association*, **43**, 1223–1236.

FLATO, G. M. 2005. *The Third Generation Coupled Global Climate Model (CGCM3)*, http://www.ec.gc.ca/ccmac-cccma/default.asp?lang=En&n=1299529F-1

FOWLER, H. J., BLENKINSOP, S. & TEBALDI, C. 2007. Linking climate change modelling to impacts studies: recent advances in downscaling techniques for hydrological modelling. *International Journal of Climatology*, **27**, 1547–1578.

FREEZE, R. A. & CHERRY, J. A. 1979. *Groundwater*. Prentice-Hall, Englewood Cliffs, NJ.

FRY, J., XIAN, G. *ET AL.* 2011. Completion of the 2006 national land cover database for the conterminous United States. *Photogrammetric Engineering & Remote Sensing*, **77**, 858–864.

FURRER, R., SAIN, S. R., NYCHKA, D. & MEEHL, G. A. 2007. Multivariate Bayesian analysis of atmosphere-ocean general circulation models. *Environmental and Ecological Statistics*, **14**, 249–266.

GALAT, D. L. & LIPKIN, R. 2000. Restoring the ecological integrity of great rivers: historical hydrographs aid in defining reference conditions for the Missouri River. *Hydrobiologia*, **422/423**, 29–48.

GEIST, D. R., BROWN, R. S., CULLIMAM, V., BRINK, S. R., LEPLA, K., BATES, P. & CHANDLER, J. A. 2005. Movement, swimming speed, and oxygen consumption of juvenile white sturgeon in response to changing flow, water temperature, and light level in the Snake River, Idaho. *Transactions of the American Fisheries Society*, **134**, 803–816.

GERRITY, P. C., GUY, C. S. & GARDNER, W. M. 2006. Juvenile pallid sturgeon arepiscivorous: a call for conserving native cyprinids. *Transactions of the American Fisheries Society*, **135**, 604–609.

GREENE, A. M., GODDARD, L. & LALL, U. 2006. Probabilistic multi-model regional temperature change projections. *Journal of Climate*, **19**, 4326–4343.

GROHS, K. L., KLUMB, R. A., CHIPPS, S. R. & WANNER, G. A. 2009. Ontogenetic patterns in prey use by pallid sturgeon in the Missouri River, South Dakota and Nebraska. *Journal of Applied Ichthyology*, **25**, 48–53.

GROTCH, S. L. & MACCRACKEN, M. C. 1991. The use of general circulation models to predict regional climatic change. *Journal of Climate*, **4**, 286–303.

HANSEN, M. J., BOISCLAIR, D., BRANDT, S. B., HEWETT, S. W., KITCHELL, J. F., LUCAS, M. C. & NEY, J. J. 1993. Applications of bioenergetics models to fish ecology an management: where do we go from here? *Transactions of the American Fisheries Society*, **122**, 1019–1030.

HARRIS, G. R., COLLINS, M., SEXTON, D. M. H., MURPHY, J. M. & BOOTH, B. B. B. 2010. Probabilistic projections for 21st century European climate. *Natural Hazards and Earth System Sciences*, **10**, https://doi.org/10.5194/nhess-10-2009-2010

HENDERSON-SELLERS, A. 1993. An antipodean climate of uncertainty? *Climatic Change*, **25**, 203–224.

HOLAN, S. H., DAVIS, G. M., WILDHABER, M. L., DELONAY, A. J. & PAPOULIAS, D. M. 2009. Hierarchical Bayesian Markov switching models with application to predicting spawning success of shovelnose sturgeon. *Journal of the Royal Statistical Society – Series C (Applied Statistics)*, **58**, 47–64.

JACOBSON, R. B., JOHNSON, H. E., III & DIETSCH, B. J. 2009. *Hydrodynamic Simulations of Physical Aquatic Habitat Availability for Pallid Sturgeon in the Lower Missouri River, at Yankton, South Dakota, Kenslers Bend, Nebraska, Little Sioux, Iowa, and Miami, Missouri, 2006–07*. United States Geological Survey, Scientific Investigation Report, **2009-5058**.

JASPER, K., CALANCA, P., GYALISTRAS, D. & FUHRER, J. 2004. Differential impacts of climate change on the hydrology of two alpine river basins. *Climate Research*, **26**, 113–129.

JONES, R. N. 2000. Analysing the risk of climate change using an irrigation demand model. *Climate Research*, **14**, 89–100.

KALNAY, E., KANAMITSU, M. *ET AL.* 1996. The NCEP/NCAR 40-Year Reanalysis Project. *Bulletin of the American Meteorological Society*, **77**, 437–471.

KANAMITSU, M., EBISUZAKI, W., WOOLLEN, J., YANG, S., HNILO, J. J., FIORINO, M. & POTTER, G. L. 2002. NCEP–DOE AMIP-II Reanalysis (R-2). *Bulletin of the American Meteorological Society*, **83**, 1631–1643, https://doi.org/10.1175/BAMS-83-11-1631

KITCHELL, J. F., STEWART, D. J. & WEININGER, D. 1977. Applications of a bioenergetics model to yellow perch *(Perca Flavescens)* and walleye (*Stizostedion vitreum vitreum*). *Journal of the Fisheries Research Board of Canada*, **34**, 1922–1935.

KOCH, B., BROOKS, R. C. *ET AL.* 2012. Habitat selection and movement of naturally occurring pallid sturgeon in the Mississippi River. *Transactions of the American Fisheries Society*, **141**, 112–120.

KNUTTI, R. 2010. The end of model democracy? An editorial comment. *Climatic Change*, **102**, 395–404.

KNUTTI, R., FURRER, R., TEBALDI, C., CERMAK, J. & MEEHL, G. A. 2010. Challenges in combining projections from multiple climate models. *Journal of Climate*, **23**, 2739–2758.

LARSON, L. W. & SCHWEIN, N. O. 2004. Temperature, precipitation, and streamflow trends in the Missouri Basin, 1895 to 2001. *In*: *Combined Preprints: 84th American Meteorological Society (AMS) Annual Meeting*, 11–15 January, 2004, Seattle, WA, United States. American Meteorological Society, Boston, MA.

LEOPOLD, L. B. & MADDOCK, T., JR. 1953. *The Hydraulic Geometry of Stream Channels And Some Physiographic Implication*. United States Geological Survey, Professional Paper, **252**.

LIM, K. J., ENGEL, B. A., TANG, Z., CHOI, J., KIM, K.-S., MUTHUKRISHNAN, S. & TRIPATHY, D. 2005. Automated web GIS based hydrograph analysis tool. *JAWRA – Journal of the American Water Resources Association*, **41**, 1407–1416, https://10.1111/j.1752-1688.2005.tb03808.x

MARAUN, D., WETTERHALL, F. *ET AL.* 2010. Precipitation downscaling under climate change: recent developments to bridge the gap between dynamical models and the end user. *Reviews of Geophysics*, **48**, RG3003, https://doi.org/10.1029/2009RG000314

MAUGET, S. A. 2004. Low frequency streamflow regimes over the central United States. *Climatic Change*, **63**, 121–144.

MCELROY, B., DELONAY, A. & JACOBSON, R. 2012. Optimum swimming pathways of fish spawning migrations in rivers. *Ecology*, **93**, 29–34.

MEARNS, L. O., GUTOWSKI, W. J., JONES, R., LEUNG, L.-Y., MCGINNIS, S., NUNES, A. M. B. & QIAN, Y.

2009. A regional climate change assessment program for North America. *Eos, Transactions of the American Geophysical Union*, **90**, 311–312.

Meehl, G. A., Covey, C. *et al.* 2007. The WCRP CMIP3 multi-model dataset: a new era in climate change research. *Bulletin of the American Meteorological Society*, **88**, 1383–1394.

Moriasi, D. N., Arnold, J. G., Van Liew, M. L., Bingner, R. L., Harmel, R. D. & Veith, T. L. 2007. Model evaluation guidelines for systematic quantification of accuracy in watershed simulations. *Transactions of the ASAB*, **50**, 885–900.

Murphy, J. M., Booth, B. B. B., Collins, M., Harris, G. R., Sexton, D. M. H. & Webb, M. J. 2007. A methodology for probabilistic predictions of regional climate change from perturbed physics ensembles. *Philosophical Transactions of the Royal Society*, **365**, 1993–2028.

Nakicenvoic, N. & Swart, R. (eds) 2000. *Special Report on Emissions Scenarios. A Special Report of Working Group III of the Intergovernmental Panel on Climate Change.* Cambridge University Press, Cambridge.

NAUS 2006. *Major Dams of the United States. National Atlas of the United States.* https://catalog.data.gov/dataset/usgs-small-scale-dataset-major-dams-of-the-united-states-200603-shapefile

NCDC 2010. *Federal Climate Complex Global Surface Summary Of Day Data, Version 7.* National Climate Data Center, NOAA, Asheville, NC, ftp://ftp.ncdc.noaa.gov/pub/data/gsod/readme.txt.

NRCS 2011. *Web Soil Survey.* Natural Resources Conservation Service, United States Department of Agriculture, Washington, DC, http://websoilsurvey.nrcs.usda.gov/app/ [accessed 8 August 2011].

Quist, M. C., Boelter, A. M. *et al.* 2004. *Research and Assessment Needs for Pallid sturgeon recovery in the Missouri River – Final Report to the U.S. Geological Survey, U.S. Army Corps of Engineers, U.S. Fish and Wildlife Service, and U.S. Environmental Protection Agency.* William D. Ruckelshaus Institute of Environment and Natural Resources. University of Wyoming, Laramie, WY.

Rougier, J. C. 2007. Probabilistic inference for future climate using an ensemble of climate model evaluations. *Climatic Change*, **81**, 247–264.

Rougier, J. C., Sexton, D. M. H., Murphy, J. M. & Stainforth, D. A. 2009. Analysing the climate sensitivity of the HadSM3 climate model using ensembles from different but related experiments. *Journal of Climate*, **22**, 3540–3557.

Sain, S. R., Furrer, R. & Cressie, N. 2011. A spatial analysis of multivariate output regional climate models. *The Annals of Applied Statistics*, **5**, 150–175.

Salathé, E. P. 2005. Downscaling simulations of future global climate with application to hydrologic modeling, International. *Journal of Climatology*, **25**, 419–436.

Sankarasubramanian, A., Lall, U., Filho, F. A. S. & Sharma, A. 2009. Improved water allocation utilizing probabilistic climate forecasts: short-term water contracts in a risk management framework. *Water Resources Research*, **45**, W11409, https://doi.org/10.1029/2009WR007821

Semenov, M. & Stratonovitch, P. 2010. Use of multi-model ensembles from global climate models for assessment of climate change impacts. *Climate Research*, **41**, https://doi.org/10.3354/cr00836

Sokolov, A. P., Stone, P. H. *et al.* 2009. Probabilistic forecast for twenty-first-century climate based on uncertainties in emissions (without policy) and climate parameters. *Journal of Climate*, **22**, 5175–5204.

Solomon, S., Qin, D. *et al.* 2007. Technical summary. *In*: Solomon, S., Qin, D. *et al.* (eds) *Climate Change 2007: The Physical Science Basis.* Contribution of Working Group I to the Fourth Assessment Report of the Intergovernmental Panel on Climate Change. Cambridge University Press, Cambridge.

Spindler, B. D., Chipps, S. R., Klumb, R. A., Graeb, B. D. S. & Wimberly, M. C. 2012. Habitat and prey availability attributes associated with juvenile and early adult pallid sturgeon occurrence in the Missouri River, USA. *Endangered Species Research*, **16**, 225–234.

Stone, M. C., Hotchkiss, R. C. & Mearnes, L. O. 2003. Water yield responses to high and low spatial resolution climate change scenarios in the Missouri River Basin. *Geophysical Research Letters*, **30**, 1186.

Tabor, K. & Williams, J. W. 2010. Globally downscaled climate projections for assessing the conservation impacts of climate change. *Ecological Applications*, **20**, 554–565.

Tebaldi, C. & Sansó, B. 2009. Joint projections of temperature and precipitation change from multiple climate models: a hierarchical Bayesian approach. *Journal of the Royal Statistical Society*, **172A**, 83–106.

Tebaldi, C., Mearns, L. O., Nychka, D. & Smith, R. L. 2004. Regional probabilities of precipitation change: a Bayesian analysis of multimodel simulations. *Geophysical Research Letters*, **31**, L24213, https://doi.org/10.1029/2004GL021276

Tebaldi, C., Smith, R. L., Nychka, D. & Mearns, L. O. 2005. Quantifying uncertainty in projections of regional climate change: a Bayesian approach to the analysis of multimodel ensembles. *Journal of Climate*, **18**, 1524–1540.

Tetzlaff, D., Soulsby, C., Hrachowitz, M. & Speed, M. 2011. Relative influence of upland and lowland headwaters on the isotope hydrology and transit times of larger catchments. *Journal of Hydrology*, **400**, 438–447.

Theurer, F. D., Voos, K. A. & Miller, W. J. 1984. *Instream Water Temperature Model. Instream Flow.* Instream Flow Information Paper 16, FWS/OBS-84/15. United States Fish and Wildlife Service, Fort Collins, CO.

USACE 1991. *A Muskingum-Cunge Channel Flow Routing Method for Drainage Networks.* Technical Paper 135.

USACE 1993. *Introduction and Application of Kinematic Wave Routing Techniques Using HEC-1.* Technical Document 10.

USACE 2000. *Hydrologic Modeling System HEC-HMS. Technical Reference Manual.* United States Army Corps of Engineers, Washington, DC.

USACE 2001. *Ala Wai Flood Study, Island of O'ahu, Honolulu, HI, Planning Assistance to States Study Report.* Fort Shafter, HI.

USACE 2006. *Missouri River Mainstem Reservoir System. Master Water Control Manual.* United States Army

Corps of Engineers, Portland OR, http://digitalcommons.unl.edu/usarmyceomaha/71/

USACE 2010. *Hydrologic Modeling System HEC-HMS, User's Manual, Technical Reference Manual*. United States Army Corps of Engineers, Washington, DC, http://www.hec.usace.army.mil/software/hec-hms/documentation/HEC-HMS_Users_Manual_3.5.pdf

USEPA 2008. *AERSURFACE User's Guide*. United States Environmental Protection Agency, Research Triangle Park, NC, http://www.epa.gov/scram001/7thconf/aermod/aersurface_userguide.pdf [last accessed December 2011].

USFWS 1993. *Recovery Plan for the Pallid Sturgeon (Scaphirhynchusalbus)*. United States Fish and Wildlife Service, Denver, CO.

USFWS 2007. *Recovery Plan for the Pallid Sturgeon (Scaphirhynchusalbus)*. United States Fish and Wildlife Service, Denver, CO.

USFWS 2008. *Pallid Sturgeon (Scaphirhynchus Albus) Range-Wide Stocking and Augmentation Plan*. United States Fish and Wildlife Service, Billings, MT.

USGS 2002. *Standards for the Analysis and Processing of Surface-Water Data & Information Using Electronic Methods*. USGS, Reston, VA, Water-Resources Investigation Report 01-4044.

USGS 2007. *Facing Tomorrow's Challenges – U.S. Geological Survey Science in the Decade 2007–2017*. United States Geological Survey, Circular, **1309**.

USGS 2011. *National Water Information System*. http://nwis.waterdata.usgs.gov/nwis

USGS 2012. *Significant Figures*. http://nwrc.usgs.gov/techrpt/sta18.pdf [accessed 27 June 2012].

Vrac, M., Stein, M. & Hayhoe, K. 2007. Statistical downscaling of precipitation through nonhomogeneous stochastic weather typing. *Climate Research*, **34**, 169–184.

Walther, G.-R., Berger, S. & Sykes, M. T. 2005. An ecological 'footprint' of climate change. *Proceedings of the Royal Society, Biological Sciences*, **272**, 1427–1432.

Wikle, C. K. 2003. Hierarchical models in environmental science. *International Statistical Review*, **71**, 181–199.

Wilby, R. L. & Harris, I. 2006. A framework for assessing uncertainties in climate change impacts: low-flow scenarios for the River Thames, UK. *Water Resources Research*, **42**, W02419.

Wildhaber, M. L., DeLonay, A. J. *et al.* 2007. *A Conceptual Life-History Model for Pallid and Shovelnose Sturgeon*. United States Geological Survey, Circular, **1315**.

Wildhaber, M. L., DeLonay, A. J. *et al.* 2011*a*. Identifying structural elements needed for development of a predictive life-history model for pallid and shovelnose sturgeons. *Journal of Applied Ichthyology*, **27**, 462–469.

Wildhaber, M. L., Holan, S., Bryan, J. L., Gladish, D. W. & Ellerseick, M. R. 2011*b*. Assessing power of large river fish monitoring programs to detect population changes: the Missouri River sturgeon example. *Journal of Applied Ichthyology*, **27**, 282–290.

Wildhaber, M. L., Gladish, D. W. & Arab, A. 2012. Distribution and habitat use of the Missouri River and Lower Yellowstone River benthic fishes from 1996 to 1998: a baseline for fish community recovery. *River Research and Applications*, **28**, 1780–1803.

Wildhaber, M. L., Dey, R., Wikle, C. K., Moran, E. H., Anderson, A. J. & Franze, K. J. 2015. A stochastic bioenergetics model-based approach to translating large river flow and temperature into fish population responses: the pallid sturgeon example. *In*: Riddick, A. T., Kessler, H. & Giles, J. R. A. (eds) *Integrated Environmental Modelling to Solve Real World Problems: Methods, Vision and Challenges*. Geological Society, London, Special Publications, **408**. First published online May 28, 2015, https://doi.org/10.1144/SP408.10

Williamson, C. E., Dodds, W., Kratz, T. K. & Palmer, M. A. 2008. Lakes and streams as sentinels of environmental change in terrestrial and atmospheric processes. *Frontiers in Ecology and the Environment*, **6**, 247–254.

Wood, A. W., Leung, L. R., Sridhar, V. & Lettenmaier, D. P. 2004. Hydrologic implications of dynamical and statistical approaches to downscaling climate model outputs. *Climatic Change*, **62**, 189216.

Wright, S. A., Holly, F. M., Jr, Bradley, A. A. & Krajewski, W. 1999. Long-term simulation of thermal regime of Missouri River. *Journal of Hydraulic Engineering*, **125**, 242–252.

Geological map fusion: OneGeology-Europe and INSPIRE

J. L. LAXTON

British Geological Survey, Murchison House, West Mains Road, Edinburgh EH9 3LA, UK
john.laxton@lineone.net

Abstract: Geological maps can be seen as a type of model and can be implemented in digital systems as geological spatial databases. In this context, geological map fusion can be implemented at different levels: harmonization of the conceptual data model describing the map objects; the use of shared concepts to describe properties in the model to give semantic harmonization; and ensuring geometric consistency. GeoSciML has been developed as an interchange language for geosciences information, derived from a common conceptual data model, along with common vocabularies of concepts to populate the object properties. GeoSciML and the vocabularies were used in the OneGeology-Europe project where a 1:1 million scale geological map of Europe was delivered using disseminated web services from 20 different data providers. The lessons learnt from the OneGeology-Europe project informed the development of the INSPIRE Geology Data Specification. The INSPIRE data specification is used to define what information must be made available through web services under the INSPIRE legislation, so has to be kept simple. The INSPIRE data model can be extended with GeoSciML and will provide a basis for geological map fusion.

A model is a simplified conceptual representation of some real-world phenomena, designed to meet some specific objective. Modelling has grown in importance in geoscience with the need to handle increasing volumes of digital information accompanied by greatly enhanced computing power and improved visualization technologies. Process modelling, usually aimed at making predictions, and 3D spatial modelling, designed to increase understanding of the disposition of geological objects through improved visualization, are two common categories of model. 3D spatial models commonly categorize rock bodies using some classification system, such as lithostratigraphy, and then use interpolation algorithms from data points, such as boreholes, to generate the bounding surfaces of the rock bodies. As the interpolation and 3D visualization are intensively computational processes, the results are seen as 3D geological spatial 'models' in contrast with the earlier 2D geological spatial 'maps'. Although geological maps are constructed using different technologies, they too are 'a simplified conceptual representation of some real world phenomena', and thus a type of model. This paper will look at what model fusion means in the context of geological maps with respect to OneGeology-Europe and INSPIRE in particular.

OneGeology-Europe background

OneGeology-Europe (OneGeology-Europe 2010) was an eContentPlus project running from 2008 to 2010 with the objective of bringing together a web-accessible, interoperable geological (map) spatial dataset for the whole of Europe at the 1:1 million scale. The aim was not to create a single dataset, but rather to build a portal through which web services provided by the partner organizations, predominantly geological surveys, could be viewed together. As well as delivering a 1: 1 million scale map, OneGeology-Europe also looked at the need for delivery of higher-resolution information based on the identified requirements of the user community.

The project had 29 partners, 20 of whom were data providers, and developed from the OneGeology-Global initiative originally set up as a geological survey contribution to the International Year of Planet Earth. At the time of the OneGeology-Europe project, OneGeology-Global simply delivered a map image, with very limited attribution, using the OGC (Open Geospatial Consortium) WMS (Web Mapping Service) standard (ISO/TC211 2005). In OneGeology-Europe, the objective was to extend this by delivering richer information about the geological objects shown on the map using the OGC WFS (Web Feature Service) standard (ISO/TC211 2010).

One of the project's key objectives was to act as a testbed for the INSPIRE technical specifications, which were in the process of being developed, and to investigate the requirements for successful geological map integration prior to the development of the INSPIRE geological data specifications.

INSPIRE background

INSPIRE (Infrastructure for Spatial Information in the European Community) (EC 2016) is an initiative of the European Community to enable the sharing of

From: Riddick, A. T., Kessler, H. & Giles, J. R. A. (eds) 2017. *Integrated Environmental Modelling to Solve Real World Problems: Methods, Vision and Challenges*. Geological Society, London, Special Publications, **408**, 147–160.
First published online September 28, 2016, https://doi.org/10.1144/SP408.16

environmental spatial information among public sector organizations and better facilitate public access to spatial information across Europe. INSPIRE was set up through a directive of the European Parliament (European Parliament 2007) and defines 34 spatial data themes, one of which is geology. It provides both data specifications for the themes and rules on the technical implementation, the latter largely based on the ISO series of spatial data standards. INSPIRE can be seen as the specification for an environmental Spatial Data Infrastructure (SDI) for Europe.

The implementation of INSPIRE is specified in a series of legally binding Implementing Rules (IR), which include a timescale for implementation. All public sector organizations that hold digital data within the scope of the INSPIRE data specifications will be required to make them available through INSPIRE-compliant web services. There is no requirement to create new data or make non-digital data available.

The INSPIRE directive defines the geology theme as 'Geology characterized according to composition and structure. Includes bedrock, aquifers and geomorphology'. The task of expanding this definition into a full data specification was the responsibility of a Thematic Working Group (TWG) set up for each of the INSPIRE themes. The TWG responsible for the Geology specification comprised 13 members drawn mainly, but not exclusively, from European geological surveys. The principal deliverable from the TWG are 'Guidelines for the Data Specification on Geology' (INSPIRE 2013), which describe the data specification in detail, including Universal Modeling Language (UML) diagrams and feature catalogues. The essential components of the data specification are incorporated into the legally binding Implementing Rule.

What does geological map fusion mean?

Traditional paper geological maps were generally produced in series that embodied two of the requirements for map fusion: that the maps should display the same objects; and that they should be at the same scale. These two factors are related, so that certain types of object might only be shown on maps of a particular scale: for example, fold axes might be shown on large-scale maps but not on small-scale maps. In addition, map scale determines the size of object that can be shown on a paper map based on factors such as line width, although the selection of objects for inclusion was commonly left to cartographic judgement rather than being based on a firm set of rules.

In paper maps, ensuring that maps within a map series followed consistent standards, and ideally that neighbouring maps would join up, was as far as map fusion could go. In practice, it was common for this level of map fusion not to be achieved, even within a single map series. This was due to the long revision cycle that led to different map sheets being produced decades apart, during which geological mapping techniques and scientific understanding, and thus classification systems, had developed and changed. This long revision cycle was one of the factors that drove the digitization of geological maps, first to aid map production, but then to enable the transformation of geological maps into spatial databases that could serve a wide range of objectives (Laxton & Becken 1996). The digitization of geological maps also enables more comprehensive levels of map fusion.

The main objects displayed on most geological maps are bodies of rock classified according to some system, such as lithostratigraphy or chronostratigraphy. The Earth's crust in any particular area is made up of rock material with spatially distributed physical, chemical and temporal (age and genesis) properties. The categorization of the rock body into individual rock units is based on a detailed study of these properties, along with the spatial relationship between rock units. Commonly, the categorization systems used will be hierarchical, and the boundary of any unit will depend both on the categorization system and the hierarchical level being used within that system. If a different system or level is used, then the overall body of rock will be subdivided into different individual rock units, with different spatial extents.

The fusion of geological maps therefore depends on a wide range of factors and different levels of fusion, or harmonization, can be considered. The level of map fusion that is required will depend on individual use cases. Each level of map fusion gives increased benefits, but also generally requires increased effort to achieve, so it is important to consider what the map fusion is for. It is also important to consider what is meant by a 'map' in a digital context. A narrow view would be that it is simply a visualization of some part of the spatial database, so map fusion would then only require the fusion of those elements that can be visualized. A more complete interpretation, on the basis that the map is a type of model, would be to see map fusion as fusion of the underlying spatial database, both that which is visualized and that which is not.

Map fusion implies the fusion of individual geological objects, so the first level of harmonization requires the maps to contain the same geological objects. As these objects are defined on the basis of their properties, it follows that objects need to be described using the same properties. In a digital system, this means that all maps to be fused must share a common conceptual data model with an agreed set of objects (features) and agreed properties (attributes) of those objects. Fusion of the data model facilitates data transfer and encourages software development against a single, widely used, data model. In practice, many geological surveys will have already established their own internal geological map data models which it will be difficult to change, so a common data model must be developed in parallel with this specifically for the purpose of data sharing. Data can then be mapped from the internal model to the shared model, a process that can generally be automated. However, the reverse, transferring data from the shared model to the internal one, will not necessarily be possible: for example, if two attribute fields in the internal model have to be mapped to a single field in the shared model; or if an attribute field in the shared model has no equivalent in the internal model. There is generally some loss of information in the mapping process, so over time organizations may choose to adopt the shared model, possibly with extensions, internally.

A common conceptual data model will not, of itself, lead to map fusion, however. As well as an agreed set of attributes describing an object, each data provider will generally have an agreed set of valid values for each attribute, and these will be held in a vocabulary. Rock unit objects from different data providers might both have a 'lithology' property, but if the lithology concepts used to populate the property in the two maps are drawn from different vocabularies then fusion is not possible. It is important here to distinguish between concepts and terms. Two different vocabularies might both contain concepts termed 'granite', but unless the definition for granite in the two vocabularies is identical then they are different concepts. Using common vocabularies leads to semantic harmonization and the ability to merge objects, leading to a higher level of map fusion. Semantic harmonization requires both agreement on the common vocabularies to be used and then translation to these. Translation from concepts used in internal databases to those in common vocabularies generally requires more effort than mapping to a common database structure, in that it usually requires domain experts to assess the concept definitions in the two vocabularies and decide on individual translations. Where there is not an exact correspondence between concepts there should be agreed rules as to how the translation should be carried out: for example, by matching to the concept deemed closest or by going up the hierarchical level until a broader concept that includes the concept to be mapped is reached. This process can lead to a significant loss of information, and if not done consistently by different data suppliers will result in maps that are not as harmonized as they may appear.

Where maps are described using concepts in hierarchical vocabularies it is also important they use concepts drawn from the same level in the vocabulary where possible. For example, two maps might populate a geological age property with concepts from a common vocabulary of chronostratigraphic units, but if one map is using units at the series level and the other at the stage level then only incomplete fusion will result. Full harmonization would require all maps to use the lowest hierarchical level in the vocabulary that was common to all but, depending on the use case, it might be preferable for all maps to use information at the highest semantic level available, even if this resulted in inconsistencies at map boundaries and a lower level of map fusion. In a digital system, it would be possible to link the map object to a hierarchical vocabulary: thus, providing both the highest semantic level available, and allowing dynamic generalization and fusion as required.

The final requirement for map fusion is geometrical consistency. Maps to be fused need to have a similar spatial resolution, which is generally the case for maps of the same scale. In addition, polygon boundaries of adjacent maps need to join up. Although, in principle, this is also a requirement for a paper map series, in practice paper maps of different ages or produced by different geologists commonly did not join. Minor geometrical mismatches at map borders can be adjusted cartographically, but larger mismatches may indicate a difference in interpretation of the mapped units, or geological mapping errors, and may need correction in the field. Achieving full spatial consistency over a large area can be a time-consuming and expensive task, so the extent to which this is required needs to be assessed in light of the use cases.

Most digital geological maps have been created from paper originals where many of the issues were not addressed, and incorporating the requirements for fusion retrospectively can be time-consuming and expensive, and potentially lead to the loss of information. Many of the issues arising in the fusion of geological maps will also arise when fusing geological spatial models, although where geological models are created within a digital environment, rather than by extending a pre-existing geological map, the requirements for fusion can be addressed at the outset.

The role of GeoSciML

The increasing requirement to interchange harmonized geological map information led to the creation of the Interoperability Working Group (IWG) of the IUGS Commission for the Management and Application of Geoscience Information (CGI) in 2004.

The ultimate objective of the IWG is to enable seamless web integration of select geosciences information hosted by different data providers. The more specific objectives are:

- To develop a conceptual model of geoscientific information drawing on existing data models.
- To implement an agreed subset of this model in an agreed schema language.
- To implement an XML (eXtended Markup Language)/GML (Geography Markup Language) encoding of the model subset.
- To develop a testbed to illustrate the potential of the data model for interchange.
- To identify areas that require standardized classifications in order to enable interchange.

This work has led to the development of GeoSciML as an interchange language for geosciences information, underpinned by a UML conceptual data model (CGI 2012*a*) from which it has been generated automatically. The development of GeoSciML has been a global initiative with active participants drawn from Europe, North America, Asia and Australia.

The scope of the content of GeoSciML covers those geoscience objects that form the main components of a geological map (geological units, faults, contacts and geomorphology), as well as boreholes and the laboratory analysis of specimens. GeoSciML v3.0 is described in this paper and was the basis for the INSPIRE data specification. The current version of GeoSciML, v4.0, has similar scope but is modelled a little differently. In modelling the geological map, the concept of the 'Geologic Feature', which can be either a 'Geologic Unit', a 'Geologic Structure' or a 'Geomorphologic Feature' is central. The model separates the concept of the 'Geologic Feature' from its individual representations on a geological map, which are termed 'Mapped Features'.

A typical geological map is made up of 'Mapped Features', which are in turn specified by 'Geologic Features'. A single 'Mapped Feature' can be specified by only one 'Geologic Feature', but a 'Geologic Feature' can have several 'Mapped Feature' occurrences, on maps of different scales or in a 3D spatial model, for example. The main function of a 'Mapped Feature' therefore is to hold geometry, which it inherits from standard types defined in GML. Figure 1 shows the UML diagram for the 'Geologic Feature' package, which illustrates the relationship between 'Mapped Features' and 'Geologic Features'.

The GeoSciML data model allows for the detailed description of 'Geologic Units' (Fig. 2). The composition can be described in detail using the related 'Earth Material' package, which allows the definition of 'Rock Materials' in terms of their constituent parts, or more simply using the lithology property constrained by a vocabulary of rock types. The age of a 'Geologic Feature' is defined in terms of a 'Geologic History', which relates one or more 'Geologic Events' to a 'GeologicFeature'.

'Geologic Units' can be classified using 'Controlled Concepts' drawn from vocabularies such as stratigraphic lexicons (Fig. 3). The 'Controlled Concepts' can, in turn, be defined using prototypes, which in the case of a stratigraphic lexicon would be a normative 'Geologic Unit' definition.

GeoSciML provides a conceptual data model and interchange language for geological maps, and therefore enables the first level of geological map fusion. Apart from the International Commission on Stratigraphy (ICS) international chronostratigraphic chart (ICS 2015), there are relatively few internationally agreed geosciences vocabularies. In order to enable semantic harmonization, a CGI GeoScience Terminology Working Group has been set up with the specific purpose of developing vocabularies for interoperability. The aim is to develop a vocabulary for each property in the GeoSciML data model for which vocabulary constraint is appropriate (CGI 2012*b*). It should be noted that these vocabularies are not seen as definitive for their particular domains, but rather are designed to provide a set of concepts suitable for interoperability and to which data-providing organizations should be able to map from concepts in their own vocabularies.

The OneGeology-Europe approach to map fusion

Three of the objectives of the OneGeology-Europe project (Laxton *et al.* 2010) were:

- To bring together a web-accessible, interoperable geological spatial dataset for the whole of Europe at the 1:1 million scale.
- To develop a harmonized specification for basic geological map data and make significant progress towards harmonizing the dataset.
- To accelerate the development and deployment of a nascent international interchange standard for geological data – GeoSciML.

The project also investigated what extra properties would be required for 'Geologic Units' and 'Geologic Structures' for higher resolution maps. The first two objectives indicate that the level of geological map fusion that was being aimed for required an agreed conceptual data model with the use of common vocabularies, and that progress should be made towards geometrical consistency. The third objective indicated that GeoSciML should be used for the data model, and therefore the GeoSciML vocabularies were also strong candidates for use. Use of GeoSciML had been made a specific objective of

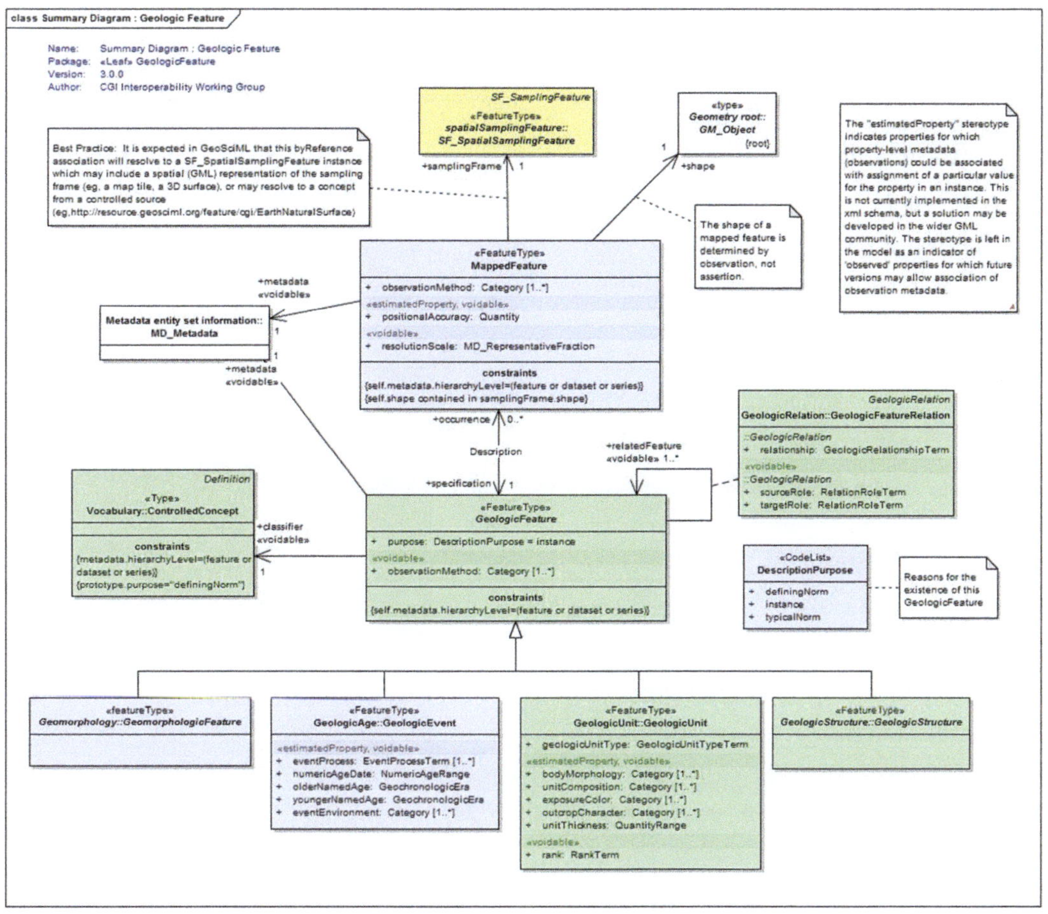

Fig. 1. Summary UML diagram for the GeoSciML 'Geologic Feature' package.

the project so that it could be tested by a wide range of different data providers using different internal data models. It was hoped that GeoSciML could be used as the basis for the INSPIRE geology data specification and use in OneGeology-Europe would indicate how practical that would be. The version of GeoSciML used in the OneGeology-Europe project was v2.1, which, although generally similar, has some distinct differences from version v3.0 described in the preceding section and used as the basis for the INSPIRE geology data specification. It was in part the experience of using GeoSciML v2.1 in OneGeology-Europe that led to the changes incorporated into GeoSciML v3.0.

OneGeology-Europe used a subset of the GeoSciML data model determined by the geological objects that it was considered practical to show on a 1:1 million scale map, along with those extra properties identified as being required for high-resolution maps. As the main purpose of OneGeology-Europe was to deliver geological maps, the web services delivered 'Mapped Features' that could be specified by either 'Geologic Units' or 'Geologic Structures'. Many geological maps display lithostratigraphy, as this is commonly the basis for geological mapping, but lithostratigraphic units are rarely correlated over large areas so were not considered suitable for a map of Europe. It was decided therefore that 'Geologic Units' should be displayed in terms of their age and composition, as separate visualizations. Other properties available in GeoSciML would also be provided, but would not be used for visualization. The principal 'Geologic Structures' delivered were faults, and these were displayed according to their type.

Fusion using chronostratigraphy

In GeoSciML, age is described in terms of the time at which a particular 'Geologic Event' took place,

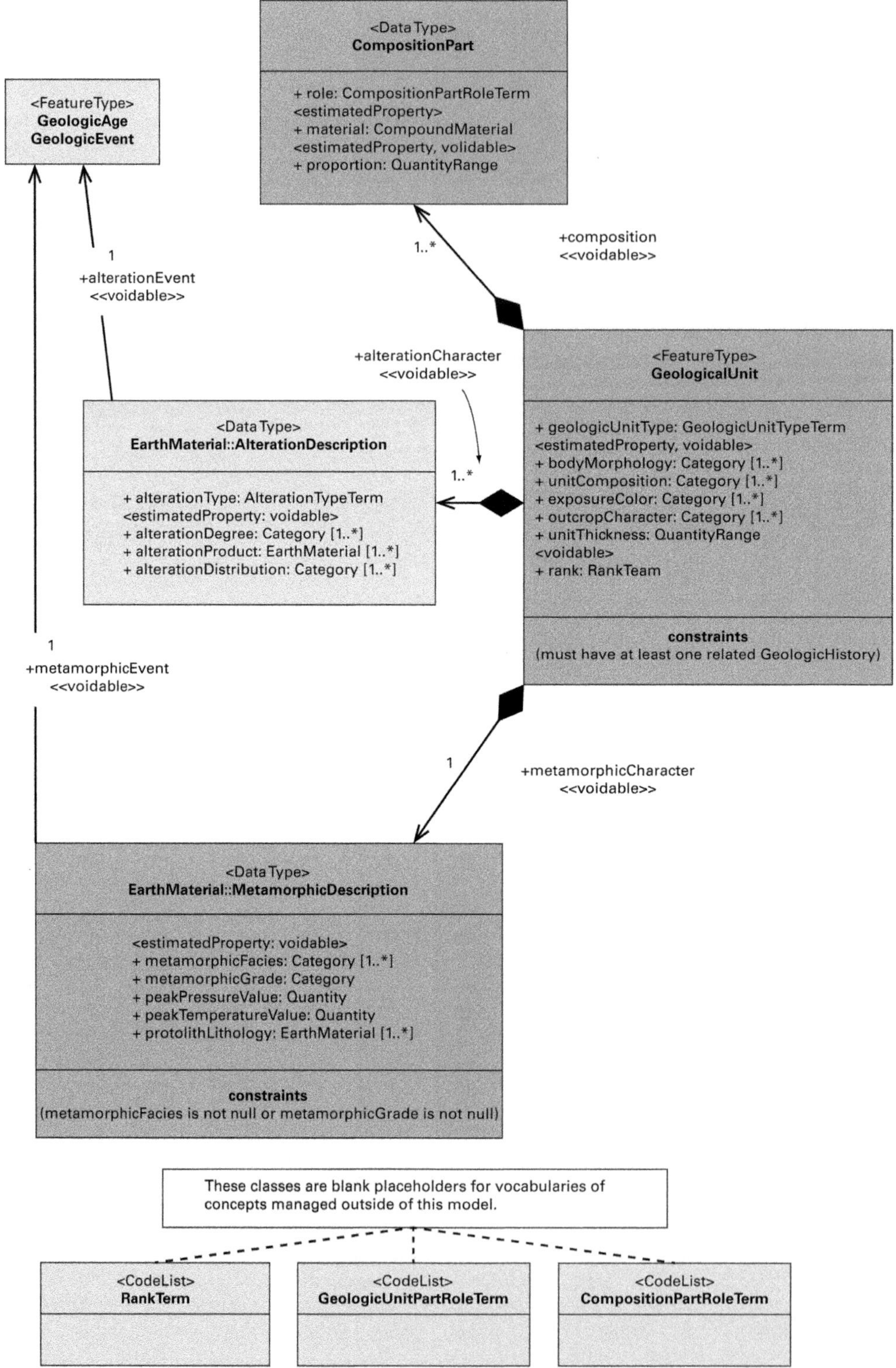

Fig. 2. Summary UML diagram for the GeoSciML 'Geologic Unit' package.

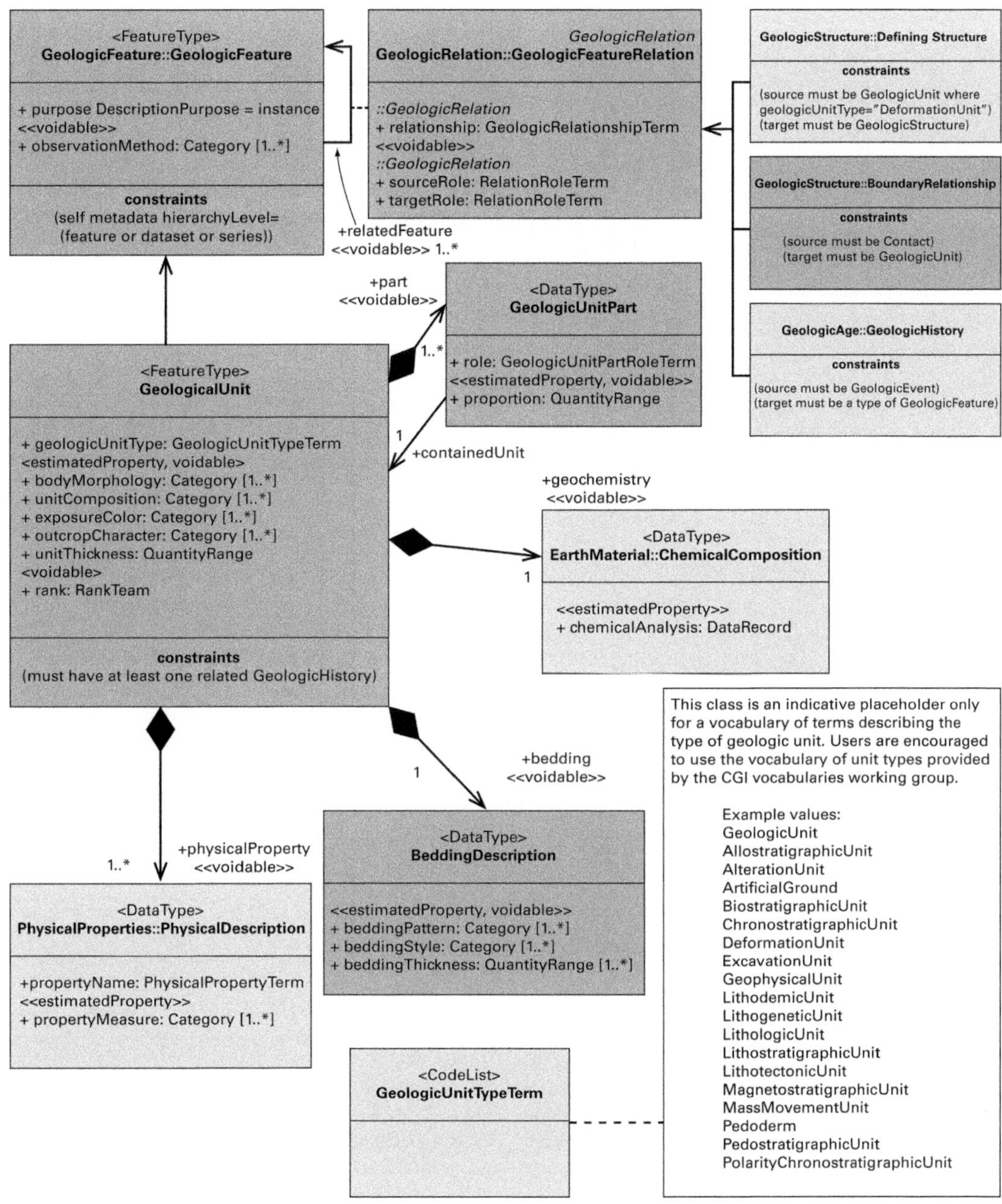

Fig. 2. *Continued.*

with properties for the event process and, optionally, event environment, as well as the age (Fig. 1). It was decided to describe the age of units as a time period with an older and younger chronostratigraphic unit. As well as providing a single age range for the 'Geologic Unit', OneGeology-Europe data providers could optionally also provide a geological history as a series of 'Geologic Events'.

It was decided to standardize on the use of the chronostratigraphic units defined in the ICS international chronostratigraphic chart, as these were widely used by the data providers. However, several data providers depart from this standard on their geological maps for parts of the geological column, where it was deemed not best suited to the local geology. For the purposes of map fusion, it was

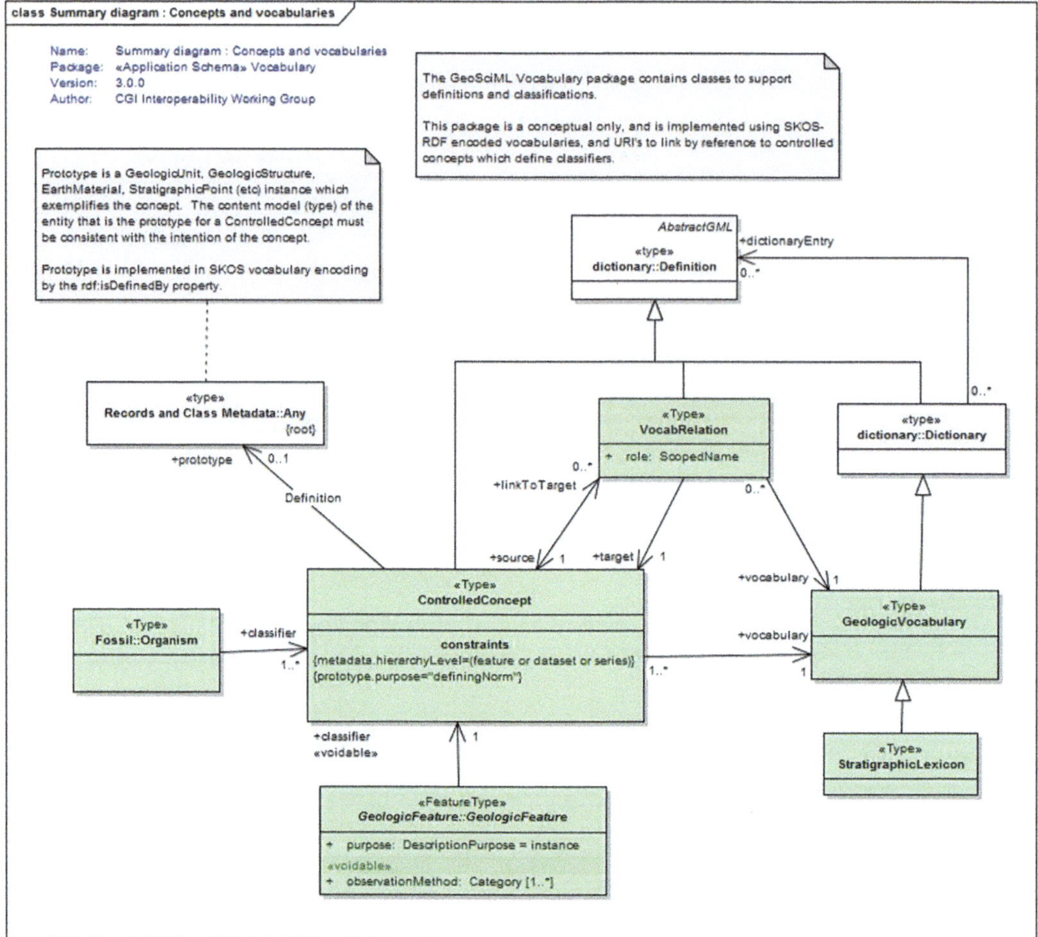

Fig. 3. Summary UML diagram for the GeoSciML vocabulary package.

necessary to map from these local chronostratigraphic units to the standard ICS ones, and some temporal resolution was lost as a result of this. For the Pre-Cambrian rocks of Scandinavia, the ICS units were used down to system level, but new stages were proposed below this, as the ICS units provided insufficient temporal resolution. The new Pre-Cambrian units were evolved from the classifications used on the Fennoscandian Shield map (Koistinen *et al.* 2001), and a vocabulary combining these with the ICS units has been made available (CGI 2012*b*).

It was necessary to have a single age for 'Geologic Units' for the purpose of visualization, but this raised several issues. First, what 'Geologic Event' marks the age of the 'Geologic Unit'? For sedimentary rocks, this was generally taken as being the age of deposition but for metamorphic units was more problematic, relying on individual interpretation. Displaying an age range also raised issues, with options including selecting either the older or younger age for display, or displaying the hierarchically lowest chronstratigraphic unit that included both the older and younger units. The latter option is arguably the most 'correct' but could result in a significant loss of temporal resolution, so it was decided to arbitrarily display the older endpoint of the age range, although the full information was delivered by the web service.

Fusion using lithology

In GeoSciML, composition is described using the 'CompositionPart' data type that allows description using either concepts drawn from a lithology vocabulary or a full description using the 'EarthMaterial' package (Fig. 2). For OneGeology-Europe, the latter was deemed too complex and too difficult to harmonize, so it was decided to use concepts from a lithology vocabulary. The concepts used were drawn from the CGI simple lithology vocabulary (CGI 2012*b*) but a subset was selected so as to give only a single hierarchy of concepts. This is because in the full CGI simple lithology vocabulary there are multiple hierarchies allowing for the different lithological classification systems in use. Simplifying the vocabulary increased harmonization, forcing all to use the same classification system, and made portrayal easier. The disadvantage was that it made translation from the internal vocabularies of some data providers more difficult and potentially reduced semantic resolution.

A 'GeologicUnit' can have multiple 'CompositionParts', such as interbedded layers, with the role and proportion of each described by specific properties. Each 'CompositionPart' can also have multiple lithologies, such as a laminated mudstone and siltstone. Where a unit has approximately equal amounts of two lithologies it may be described as, for example, 'mudstone and sandstone' or 'sandstone and mudstone'. This doesn't matter in terms of the data delivered where the relationship can be described, but for visualization one lithology had to be chosen for display. At the boundaries where maps from different data providers were fused, portrayal discontinuities could arise depending on the way composite lithologies had been described. This could have been addressed with portrayal categories for composite lithologies but would have led to a very complex portrayal scheme.

Metamorphic rocks were described using the 'MetamorphicDescription' data type from GeoSciML (Fig. 2) using common vocabularies for metamorphic grade, metamorphic facies and protolith lithology (CGI 2012*b*). This provided some level of harmonization of complex descriptions but there was no attempt to provide portrayal for these properties. Some complex rock types, such as flysch and ophiolite, are defined by a combination of lithology, genesis and other concepts, and, in order to achieve harmonization, the way these rock types should be encoded was precisely specified.

Fusion of geological structures

Faults were the only significant structure delivered in OneGeology-Europe and they were displayed according to their type, with types defined by a common vocabulary. There proved little difficulty in mapping to these types as concepts in the common vocabulary related closely to those used by data providers. A more important harmonization issue arose from different data providers delivering faults at differing levels of 'significance'. There was no definition given on which faults should be delivered and which should not, so there was little consistency between data providers. Defining fault 'significance' would be quite difficult but is something that needs to be further considered in map fusion.

Lessons learned

One of the main lessons learnt from OneGeology-Europe was that, in general, it is much easier for data providers to map from their internal data models to a common one for interoperability, than it is to map concepts defined in internal vocabularies to different ones in common vocabularies. For example, there are nearly 50 different classification schemes for sandstone (Okada 1971) and converting between them is not always straightforward. The latter requires domain experts to decide on the translation of individual concepts and, ideally, for the translation to be done according to agreed rules to ensure consistency. In OneGeology-Europe, examples and advice were provided, but the translation still relied on the individual judgement of domain experts, so is unlikely to have been entirely consistent.

Another issue arising in OneGeology-Europe was the need to populate properties that were mandatory in the GeoSciML model, but for which there was no equivalent property value in the database of the data provider, or for which the same value applied to all objects of a particular class on the map. In the former case, encoding guidelines were provided, for example, for simplicity, the 'role' of a 'CompositionPart' was set to either the 'only part', where there was only one 'CompositionPart', or 'unspecified', where there were multiple 'CompositionParts'. Where the same value applied to all instances this was encoded into the software carrying out the mapping to the common data model, thus ensuring consistency and aiding map fusion.

The data being used in OneGeology-Europe were the map series closest to the 1:1 million scale for each of the participating countries. Generally, geometrical consistency within these map series had already been achieved, so the principal issues of geometrical map fusion arose at the national borders where the maps from different countries abutted. Minor geometrical mismatches can be addressed by arbitrarily snapping lines within an agreed distance, but is more difficult to achieve when the maps to be joined are being provided by different web services, as was the case in OneGeology-Europe. The web service technology did not allow web editing, so individual databases would have had to be edited, probably in conjunction with the data from the neighbouring country. This would have been a very time-consuming activity and was only attempted in limited areas.

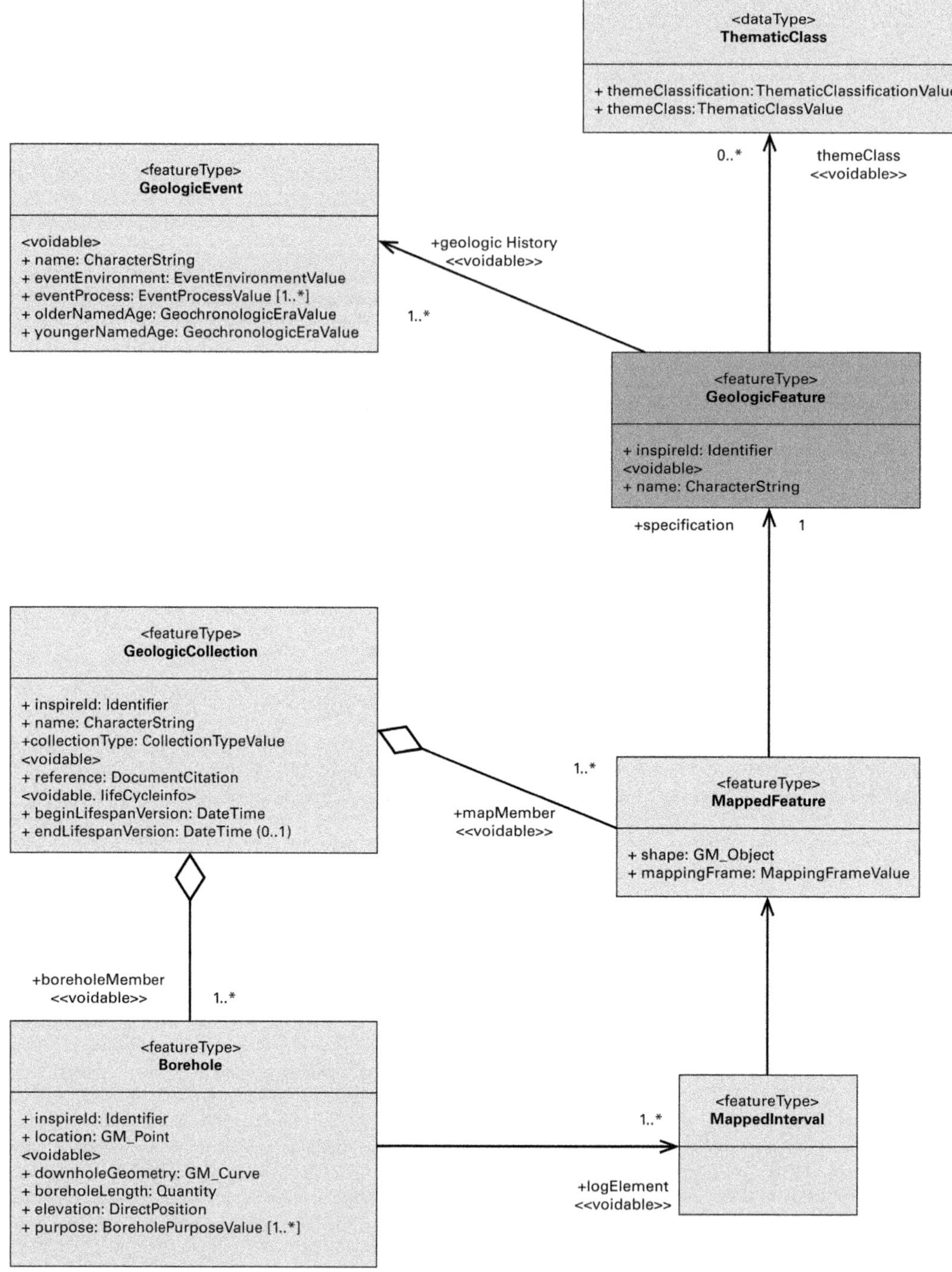

Fig. 4. UML class diagram for the INSPIRE core geological model.

Where a geometrical mismatch is large there is a bigger problem as arbitrary snapping of lines could introduce significant errors. Here, the only approach is to revise the geological mapping, probably requiring fieldwork. This was beyond the scope of the OneGeology-Europe project.

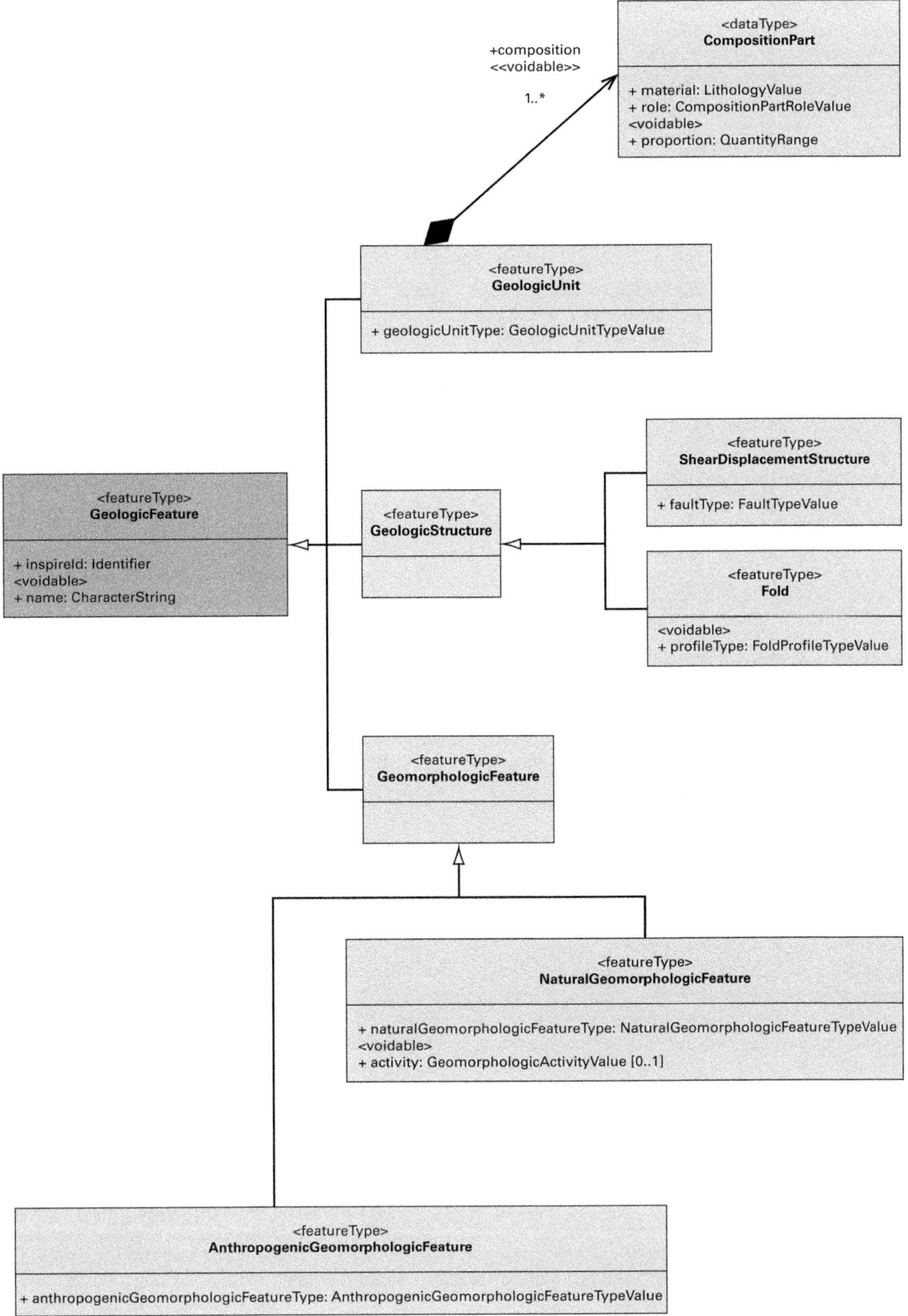

Fig. 4. *Continued.*

The results of the OneGeology-Europe project have been taken forward into the OneGeology-Global project and can be best visualized using the OneGeology portal (OneGeology 2016).

The INSPIRE approach to map fusion

The purpose of the INSPIRE geology data specification (INSPIRE 2013) is to define the geological data that should be made available as web services under the INSPIRE legislation. Analysis of use cases relevant to INSPIRE indicated that the information shown on geological maps is one of the most important classes of geological information, and therefore forms a key part of the data specification, but the overall scope of INSPIRE geology is broader than geological maps. OneGeology-Europe had, in part, been designed as a trial for INSPIRE, and its success in providing integrated geological web map services suggested that the GeoSciML data model would be a good basis for the INSPIRE geology data model. It is also a stated objective of INSPIRE to use internationally accepted domain standards, such as GeoSciML, where these exist. A Thematic Working Group (TWG) was set up for each INSPIRE theme for the purpose of defining the data specification. The combined TWG for the Geology and Mineral Resources themes contained several people who had been involved in developing GeoSciML, and who had worked on the OneGeology-Europe project.

There are, however, important differences between OneGeology-Europe and INSPIRE. There is no requirement under INSPIRE to create new data or alter existing data, other than that required to make it available in conformance with the INSPIRE data model. This means that there is specifically no requirement to achieve geometrical consistency between datasets. The INSPIRE directive (European Parliament 2007) states that the objective is to achieve 'interoperability and, where practicable, harmonization of spatial data sets and services'. Achieving interoperability is taken to mean having a common data model and harmonization means semantic harmonization, which is therefore desirable, but optional, in INSPIRE.

The INSPIRE data specifications are used to create an Implementing Rule (IR) specifying the objects and their properties that must be delivered by data providers, along with any mandatory codelists (vocabularies) that must be used. The IR is legally binding on all EU member states, and any object and property specified in it must be delivered by any public sector organization that is the primary holder of that data. Some properties may be defined as 'voidable' in the data specification, but in the context of the IR this means that the object can be delivered without data for the voidable property where such data do not exist. It does not mean that the data provider can choose whether or not to provide data for properties defined as voidable.

In developing an interoperability specification, such as GeoSciML, the aim is to include all those objects and properties that may be required for a wide range of possible use cases. A data provider can choose which part of the specification to use to deliver data in any particular circumstance and ignore those that are not relevant. There is a clear distinction therefore between the objectives of GeoSciML and the INSPIRE geology data specification, in that inclusion of an object or property in the former means it may be delivered, whereas inclusion in the latter means it must be delivered. This had significant implications for the design of the INSPIRE geology data specification.

The TWG considered that the information shown on geological maps was one of the most important geological datasets and should form a key component of the INSPIRE geology data specification. It was agreed that the GeoSciML data model should be the starting point for this part of the INSPIRE data model, but that the INSPIRE model should be simpler both in terms of the number of features and in the number of properties of those features. The TWG also had to consider requirements from the user community derived from an INSPIRE consultation, which identified requirements beyond those of OneGeology-Europe. The INSPIRE data model needed to be a compromise between a broad range of user requirements and the need to ensure that the burden on data providers was not excessive.

Figure 4 provides an overview of that part of the INSPIRE geological data model that includes geological map information. By comparison with Figures 1 and 2, it can be seen that the INSPIRE model follows the overall pattern of GeoSciML, with 'MappedFeatures' specified by different types of 'GeologicFeature', age described as an aggregation of 'GeologicEvents' and composition described with one or more 'CompositionParts'. The main difference with the GeoSciML model is that the model has been simplified and non-essential properties removed. For example, in light of experience with Onegeology-Europe, lithology can only be described with a vocabulary of rock types and the option of a full 'EarthMaterial' description has been removed.

The simplified INSPIRE data model will not meet all use cases, but, because it follows the pattern of GeoSciML, it should be easy for data providers to extend the model with extra objects and features from GeoSciML as required. To facilitate this, it is recommended that INSPIRE data be encoded using GeosciML, although this is not mandatory.

Although it was not necessary to incorporate the use of mandatory vocabularies into the INSPIRE data specification, it was felt that the harmonization benefits that would result were worth the extra burden this would place on data providers. In order to maximize harmonization internationally, and because of their availability, the CGI vocabularies were used where possible (CGI 2012*b*).

The use of a common data model for geological maps in INSPIRE will ensure that maps delivered by INSPIRE services will achieve the first level of map fusion – they will contain the same type of geological object described with the same properties. The use of common vocabularies will go a long way in achieving the second level of map fusion – semantic harmonization. However, there will be no encoding guidelines telling data providers exactly how to map from their internal vocabularies to the INSPIRE ones or requiring, for example, the use of a particular hierarchical level in the vocabulary. This should make it easier for data providers to conform with the INSPIRE specifications, but will reduce the level of map fusion that is achieved. There are no INSPIRE requirements for geometrical consistency, so that level of map fusion is unlikely to be achieved. Similarly, there are no rules for how the INSPIRE model is extended with GeoSciML, so such data services will achieve the first level of map fusion but not more than that unless further encoding rules are agreed. INSPIRE geological map data providers may wish to agree among themselves, outside of the INSPIRE legislation, encoding guidelines similar to those used in OneGeology-Europe to achieve higher levels of map fusion.

Conclusions

Geological maps are a type of geological model and have evolved from being implemented as visualizations on paper, where only limited levels of map fusion were possible, to geological spatial databases from which traditional geological maps, along with a wide range of other products, can be produced. The implementation of a geological map as a spatial database means that map fusion is no longer concerned solely with ensuring the harmonization of the map visualization, but also with harmonizing the underlying conceptual data model and the vocabularies of concepts used to populate properties in the model. This requires the development of agreed standard data models and common vocabularies of shared geological concepts. Full map fusion also requires geometrical harmonization, with the polygon boundaries of all spatial objects fully joining up.

The increase in level of map fusion, from model to vocabulary concepts to geometry, requires an increased level of effort to achieve. Not all use cases require full map fusion, so the specific user requirements need to be carefully considered before implementing a particular level of map fusion.

GeoSciML, underpinned by a conceptual data model, has been developed as an interchange language for geological map information and was trialled in the OneGeology-Europe project. This showed that GeoSciML could be used successfully by a wide range of different data providers to provide integrated geological map web services. The project also trialled the use of a limited set of common vocabularies to achieve semantic harmonization and, thus, a higher level of map fusion. While this was successful, it involved significantly more work for data providers than simply mapping to the common GeoSciML data model. It was also clear that, as well as specifying common vocabularies, there needed to be some rules as to their use – for example, specifying a particular level in a hierarchical vocabulary – if semantic harmonization is to be fully achieved.

The INSPIRE geology data specification was developed in light of the experience gained on the OneGeology-Europe project. The objective of INSPIRE is to specify a legally mandated SDI for Europe and this led to a model based on GeoSciML, but simpler. The INSPIRE geology data specification mandated the use of a set of common vocabularies, although there are no specific rules on their use. INSPIRE services should, therefore, partially achieve the second level of harmonization. There is no requirement for geometrical harmonization in INSPIRE. As the INSPIRE data model is closely based on GeoSciML, and it is recommended INSPIRE use GeoSciML for encoding, it will be possible to extend INSPIRE services to meet additional use cases in conformance with the GeoSciML data model.

Developments in GeoSciML, OneGeology-Europe and INSPIRE have developed a framework enabling geological map fusion, at a range of levels for different purposes, and the delivery and visualization of the fused maps in web services.

I would like to thank the CGI Interoperability Working Group, responsible for the development of GeoSciML, the OneGeology-Europe project team, and the INSPIRE Geology and Mineral Resources Thematic Working Group, responsible for the development of the INSPIRE geology data specification. I would also like to thank the reviewers, whose comments have led to improvements in the paper. This paper is published with the permission of the Executive Director, British Geological Survey (NERC).

References

CGI 2012*a*. *GeoSciML Resource Repository*. Commission for the Management and Application of

Geoscience Information (CGI) World, http://www.geosciml.org/

CGI 2012*b*. *Resources for Geosciences XML*. Commission for the Management and Application of Geoscience Information (CGI), http://resource.geosciml.org/

EC 2016. INSPIRE. Infrastructure for Spatial Information in the European Community, http://inspire.ec.europa.eu/

European Parliament 2007. *Directive 2007/2/EC of the European Parliament and of the Council of 14 March 2007 establishing an Infrastructure for Spatial Information in the European Community (INSPIRE)*. European Parliament, Council of the European Union.

ICS 2015. *International Chronostratigraphic Chart*. International Commission on Stratigraphy, http://www.stratigraphy.org/index.php/ics-chart-timescale

INSPIRE 2013. *INSPIRE Data Specification for the Spatial Data Theme Geology*. Infrastructure for Spatial Information in the European Community, http://inspire.ec.europa.eu/documents/Data_Specifications/INSPIRE_DataSpecification_GE_v3.0.pdf

ISO/TC211 2005. *Geographic Information – Web Map Server Interface: ISO 19128*. International Organization for Standardization (ISO)/Technical Committee 211 (TC211).

ISO/TC211 2010. *Geographic information – Web Feature Service: ISO 19142*. International Organization for Standardization (ISO)/Technical Committee 211 (TC211).

Koistinen, T., Stephens, M.B., Bogatchev, V., Nordgulen, Ø., Wennerström, M. & Korhonen, J. 2001. *Geological Map of the Fennoscandian Shield, Scale 1:2,000,000*. Geological Surveys of Finland, Norway and Sweden and the North-West Department of Natural Resources of Russia.

Laxton, J.L. & Becken, K. 1996. The design and implementation of a spatial database for the production of geological maps. *Computers & Geosciences*, **22**, 723–733.

Laxton, J.L., Serrano, J.-J. & Tellez-Arenas, A. 2010. Geological applications using geospatial standards – an example from OneGeology-Europe and GeoSciML. *International Journal of Digital Earth*, **3**, 31–49, https://doi.org/10.1080/17538941003636909

Okada, H. 1971. Classification of Sandstone: analysis and Proposal. *Journal of Geology*, **79**, 509–525, https://doi.org/10.1086/627673

OneGeology 2016. OneGeology portal, http://portal.onegeology.org/OnegeologyGlobal/

OneGeology-Europe 2010. OneGeology-Europe, http://www.onegeology-europe.org/

Integrated Environmental Modelling: human decisions, human challenges

PIERRE D. GLYNN

US Geological Survey, 432 National Center, Reston, VA 20191, USA
pglynn@usgs.gov

Abstract: Integrated Environmental Modelling (IEM) is an invaluable tool for understanding the complex, dynamic ecosystems that house our natural resources and control our environments. Human behaviour affects the ways in which the science of IEM is assembled and used for meaningful societal applications. In particular, human biases and heuristics reflect adaptation and experiential learning to issues with frequent, sharply distinguished, feedbacks. Unfortunately, human behaviour is not adapted to the more diffusely experienced problems that IEM typically seeks to address. Twelve biases are identified that affect IEM (and science in general). These biases are supported by personal observations and by the findings of behavioural scientists. A process for critical analysis is proposed that addresses some human challenges of IEM and solicits explicit description of (1) represented processes and information, (2) unrepresented processes and information, and (3) accounting for, and cognizance of, potential human biases. Several other suggestions are also made that generally complement maintaining attitudes of watchful humility, open-mindedness, honesty and transparent accountability. These suggestions include (1) creating a new area of study in the behavioural biogeosciences, (2) using structured processes for engaging the modelling and stakeholder communities in IEM, and (3) using 'red teams' to increase resilience of IEM constructs and use.

The experiences and education that scientists receive invariably affect their perspectives and create bias. Sarewitz (2004, p. 392) states this reality well in his provocative article on 'How science makes environmental controversies worse':

> Even the most apparently apolitical, disinterested scientist may, by virtue of disciplinary orientation, view the world in a way that is more amenable to some value systems than others. That is, disciplinary perspective itself can be viewed as a sort of conflict of interest that can never be evaded.

Indeed, Sarewitz (2004) argues that the very act of making choices, and of being sentient human beings, force humans to acquire bias. Scientists and numerical modellers cannot escape this reality. At best, they can try to acknowledge and examine their sources of bias.

After an introduction to the author's experiential biases and professional background, this paper will discuss the needs and use of Integrated Environmental Modelling (IEM) for the improved management of society's natural resources and environments. [Our definition of natural resources includes all resources provided by nature, regardless of their biologic, geologic, hydrologic, or atmospheric origins or characteristics.] Following sections will consider the balance between: (1) the inherent complexity of the integrated transdisciplinary numerical models and tools of IEM, and (2) the simplifications that are required for effective human construction and use of IEM, and that often reflect, or are influenced by, human limitations, biases and heuristics. Several of these human biases and heuristics will be individually recognized and examined. A reference frame, 'the eye of reality', will also be introduced that may be useful in thinking about and classifying our human pursuit of knowledge, while keeping in mind our human biases and our related creative intuitions. Lastly, the paper will suggest some ideas and approaches that may help address IEM's 'human challenges', that is, those distinctly human challenges that we need to recognize and overcome to effectively use IEM. Human biases and heuristics are a large part of these challenges.

Some personal experiences and biases

Many of my own biases were formed through my management experiences gained while directing a hydrology research group within the US Geological Survey, and proposing new science directions, for example, in the areas of groundwater studies (Glynn & Plummer 2005; Konikow & Glynn

From: Riddick, A. T., Kessler, H. & Giles, J. R. A. (eds) 2017. *Integrated Environmental Modelling to Solve Real World Problems: Methods, Vision and Challenges*. Geological Society, London, Special Publications, **408**, 161–182.
First published online May 21, 2015, https://doi.org/10.1144/SP408.9

2013; Plummer & Glynn 2013; Plummer *et al.* 2013), small watershed-basin research and monitoring (Glynn *et al.* 2009), 3D and 4D modelling and visualization (Glynn *et al.* 2011; Jacobsen *et al.* 2011; Pantea *et al.* 2013), and most recently in the domain of the behavioural biogeosciences (Glynn 2014). My perspectives were also formed through some of my research experiences:

(1) modelling, and attempting to predict, the reactive transport of acidic heavy-metal contamination in groundwaters of the Pinal Creek Basin in Arizona;
(2) providing geochemical understanding, and scenario and process modelling of oxygen and radionuclide reactive transport in support of performance assessments for high-level nuclear waste disposal in the Fennoscandian Shield in Sweden.

The experiences discussed in this section show the development of my appreciation for the need to more fully consider nature's complexities, as well as the inevitable surprises that nature invariably provides that diminish our hubris as modellers. The surprises and experiences generated a personal set of experiential biases, a set that overlies more innate biases, some of which will be described in later sections.

Combined inverse and forward modelling to assess and reduce knowledge gaps

Glynn & Brown (1996, 2012) used inverse geochemical modelling to deduce the possible sets of reactions that were affecting the chemical evolution of contaminated groundwater at the Pinal Creek site. [Inverse modelling uses observations and data to infer, past or current, process and system information. In contrast, forward modelling assumes process information and a set of initial system conditions and system parameters to predict a future state, see Glynn & Brown (1996, 2012).] The authors also conducted forward reactive transport modelling, using the possible sets of reaction processes obtained through inverse modelling, to examine the resulting migration speed and sequencing of contaminant fronts at the site. They compared these results to the migration of fronts observed at the site, which helped further constrain the sets of potential reaction processes and associated geochemical conditions that applied to the site. Glynn & Brown's (1996, 2012) integration of inverse and forward geochemical modelling helped determine the information that was critically needed to further improve understanding of contaminant transport at the site. The 1996 study provided a basis for further field investigations and numerical simulations of contaminant transport (Brown *et al.* 1998, 2000). These additional studies added further understanding on the hydrodynamics, reaction mechanisms, and kinetics controlling contaminant transport at the site. They also led the authors to design some *in situ* field experiments to further test their knowledge (Brown & Glynn 2003). *'Lessons learned' were: (1) modelling could be used to determine knowledge gaps and to guide field data collection and experiments, and (2) different modelling approaches were highly informative when used synergistically.*

An early surprise

Nature provided a 'Black Swan' surprise (i.e. a high-impact low-probability event, according to Taleb 2007) at the Pinal Creek site, before the completion of the Glynn & Brown (1996) study. An early assumption that the site had relatively steady groundwater flow dynamics was revised in the winter of 1993. Massive flooding over the course of a few months during that winter resulted in water table rises of up to 16 m and a complete reorganization of the usually dry Pinal Creek channel bed with up to 60 m of lateral bank erosion. Critical wells that had been emplaced on the banks were lost, and there was a sudden 'cleanup' or flushing of about a third of the contaminated groundwater system that completely dwarfed the pumping and remediation efforts that had proceeded to date. Site study designs, field investigations and modelling plans were changed by the 1993 event, as were contaminant remediation plans. *Key lesson: catastrophic geomorphic changes (and external forces) can cause abrupt change to groundwater systems and waylay the best-laid plans and the overly narrow, overly static, perspectives of a groundwater scientist, in this case myself.*

A wrong prediction

Nature confounded one of the predictions made by the Glynn & Brown (1996) study. Glynn & Brown (1996) had predicted that pyrolusite (MnO_2) that had been carefully weighed and suspended in research wells emplaced in the contaminant plume would undergo reductive dissolution and loss of material. Instead, the samples acquired mass. A new reactive mechanism discovered through careful laboratory experiments (Villinski *et al.* 2001) was found to best explain the observed gain in mass (Brown & Glynn 2003). *Key lesson: don't get too attached to your 'predictions'.*

Glynn & Brown (2012) provide a 15-year retrospective on Glynn & Brown (1996) and later studies conducted at the Pinal Creek site. Some of their key conclusions, pertinent to IEM, are provided below:

> Constructing, analyzing and interpreting numerical models, regardless of the type of model (hydrologic vs. geochemical; inverse vs. forward), forces the modeler(s), and hopefully the user(s) of the models, to reexamine and revise their conceptual model and perceptions of the available information. The modelling process forces the modelers and users to assemble, structure, transform, and assess a wide variety of information ... The studies conducted at the Pinal Creek site illustrate the fact that nature always keeps surprises in reserve for its observers and interpreters. Humility, and frequent testing of assumptions, are needed in modelling nature's systems ... Given our often limited knowledge of natural systems, it behooves us to model these systems by considering *general* system behavior before interpreting, matching, and predicting *specific* system behavior.

Long-term climate scenarios and performance assessments for nuclear waste disposal

I gained experience relevant to IEM through my work for a small interdisciplinary team of investigators tasked by the Swedish Nuclear Power Inspectorate (SKI) to review extensive investigations conducted by the Swedish Nuclear Fuel and Waste Management Company (SKB) for the Äspö Hard-Rock Laboratory (HRL). The Äspö HRL was studied to assess the potential performance of high-level nuclear waste disposal at 500 m depth in the Fennoscandian Shield. The SKB investigations involved a large group of contractors (i.e. a few hundred) who constructed climate scenarios and examined the many factors that could affect the performance of the waste disposal site over the next 120 000 years. The analyses and performance assessments conducted by SKB were impressive and sophisticated. Milankovich astronomical cycles were used to construct a climate scenario that included the occurrence of three glacial cycles over the next 120 000 years. During two of the cycles, a 2–3 km-high ice sheet was expected to be present on the landscape above the Äspö HRL site. SKB's performance assessments, at least initially, assumed that geochemical conditions were going to remain close to chemical steady state at repository depth. The redox regime was assumed to remain relatively constant: not sufficiently reducing, or sufficiently oxidizing, to cause either sulphidic or oxidative corrosion of the copper canisters used for waste disposal. In particular, a continuous absence of dissolved oxygen in the groundwaters outside the repository was assumed. Dissolved oxygen would have been a problem because of (1) its potential corrosion of the copper canisters and (2) its potential to mobilize radionuclides such as those of U, Tc, Pu and Np, should the nuclear waste become exposed to the groundwaters. *Key point: SKB performance assessments were IEM constructs that considered both external forcings and internal processes and were detailed in their simulations of complexity.*

Questioning a conceptual model: a small independent team effort

SKI conducted its own performance assessments for a deep repository for high-level waste disposal, the SITE-94 project (Swedish Nuclear Power Inspectorate 1997). The SKI effort was based on the development of risk scenarios constructed after an exhaustive identification of 'features, events, and processes' that could potentially affect the integrity of the disposal site and its ability to keep the high-level nuclear waste products isolated from the human-living environment. Relatively complex and visually impressive 3D hydrodynamic modelling was conducted by both SKB and SKI for their performance assessments and scenario building. Nonetheless, the SITE-94 project showed (Glynn & Voss 1999, Glynn *et al.* 1999) that a scenario that assumed the presence of 2-3 km-high warm-based ice sheets over the Äspö HRL could potentially entail the relatively rapid transport of highly oxygenated glacial meltwaters to 500 m depth, because of the large head gradient and because of low fracture porosity. Observations from the base of the Greenland ice sheet suggested that dissolved oxygen concentrations in the glacial meltwaters, under the base of the ice sheet, could be as high as four to five times the concentrations that would normally be expected under equilibrium with the atmosphere. This possibility disrupted the initial SKB (and SKI) concept scenario of a stable redox regime at repository depth. SKB mounted an extensive research effort to evaluate this possibility. *Lesson learned: despite their sophistication, the performance assessments initially left key assumptions unexamined; an independent team helped point out potential problems.*

Assuming constancy: a recurring problem

Additionally, I conducted numerical simulations that investigated the conditions under which the transport of radionuclides, such as those of Pu and Np, might be reasonably modelled by assuming constant partitioning of the radionuclides between aqueous phases and solid surfaces. The results, applicable to nuclear waste disposal in Sweden (Glynn 2003) and also to radioactive waste at the Idaho National Laboratory in the USA (Nimmo *et al.* 2004; Rousseau *et al.* 2005), indicated that constant partitioning was generally not a reasonable assumption in simulating actinide transport in geological media. *Key point: humans (including scientists) are wired, often unreasonably, to seek constancy and simplicity.*

A summary of personal lessons

Several lessons resulted from my experiences assessing contaminant transport and nuclear waste disposal issues in Arizona, in Sweden, and later at the Idaho National Laboratory. First, independent analysis by small teams or individuals can be critical in avoiding 'groupthink' (Janis 1972) in the development of conceptual models or numerical models. Second, there is no better way of using a model to develop greater understanding of a system than to obtain additional observations and information (but this is not always possible). Third, it is important to keep an open mind for the 'Black Swan' surprises and (or) invalidation of assumptions that will invariably occur. *In summary, we have a natural tendency to assume constancy, to simplify and to seek confirmation of our mental models.* Those tendencies can easily lead us into error when trying to model complex, dynamic, systems. Indeed, we may construct highly sophisticated, publicly impressive, numerical models that can nonetheless incorporate problematic simplifying assumptions or preconceptions. When properly utilized, however, structured, interdisciplinary, integrated modelling frameworks may help reduce failures of our human imagination. They can help us organize our knowledge as we gain information and understanding. They can help us uncover process interplays or model sensitivities not previously considered.

What is IEM?

According to Laniak *et al.* (2013, p. 4):

> Integrated Environmental Modelling (IEM) is a discipline [that] provides a science-based structure to develop and organize multidisciplinary knowledge. It provides a means to apply this knowledge to explain, explore, and predict environmental-system response to natural and human-induced stressors. By its very nature, it breaks down research silos and brings scientists from multiple disciplines together with decision makers and other stakeholders to solve problems for which the social, economic, and environmental considerations are highly interdependent.

Moore *et al.* (2013) further state: 'At its most basic level, integrated modelling (IM) is about linking computer models that simulate different processes to help understand and predict how those processes will interact in particular situations.' The authors add that IEM applies IM to the analysis of environmental problems. The present paper takes a broader view: IEM is needed not only for studies of environmental systems, but also for studies of the natural resources and of the human activities that are linked to the state of natural and built environments. Despite our innate tendency to do otherwise (Glynn 2014), use and management of natural resources should be integrated, or at least considered, in the simulation of environmental stresses.

Simulation of human activities, and therefore simulation of Coupled Human and Natural Systems (CHANS), requires an understanding of human behaviour, its drivers, commonalities and range of variability in a diversity of social settings (e.g. individual, family, communities, nations) and for a wide range of spatial and temporal scales. My educational and professional background creates a bias towards consideration of biophysical processes above the simulation of human activities and behaviours as might be done in CHANS modelling. *However, my claim here (also in Glynn 2014) is that scientists, including behavioural scientists, often do not consider how human biases and heuristics affect human interactions with, and human study of, natural resources and environments.* This paper focuses on human biases and heuristics that affect the study of natural resources and environments, and therefore the construction and use of IEM simulations – whether or not those simulations also include human and social processes.

Processing integrated information: are computers required?

Computers are not inherently required to construct and use IEM. As individuals, we gather, process, integrate, and act on information and beliefs, often unconsciously. We construct and use a personal form of IEM that is based on a diversity of cognitive inputs, memories and reactions acquired from our past experiences, ingrained social rituals, and innate responses acquired from our evolutionary past. When confronted by unusual events or situations that are not in our experience base or in our genetic code, we often 'infer' our responses or actions through logical deduction, induction, or through 'fuzzy' analogies to other situations.

Computer models, and other structured and distributable information frameworks, however, can help us share information and knowledge with other people, and can potentially provide greater structure, traceability and accountability for the sources of our knowledge, and ultimately for our actions. In the past, communities and individuals used maps as information frameworks and aids that could help them quickly assess:

(1) the boundaries, locations, types and quantities of resources and communities (e.g. the oldest 'modern' world atlas, the *Theatrum Orbis Terrarum* by Abraham Ortelius, 1570);

(2) the temporal trends in those resources and (or) communities (e.g. Charles Minard's 1869 flow map of Napoleon's march through Europe).

Maps have been, and still are, highly successful information frameworks because of their portability, their ability to convey a diversity of information in a highly accessible manner, and because of their ability to segregate information into different levels: the user of a map does not necessarily need to take in all the information presented in the map at once. Instead, the user can choose to access only some elements, while ignoring others, or leaving other elements for a later, more detailed, assessment. Usable simplicity and scalable visualization is a feature of well-constructed maps.

We are now at a stage where the 2D structure of maps, and often their lack of a temporal or dynamic representation of information, are too restrictive. New multi-dimensional, computer-based or web-based, IEM tools are required to help us assess, share and cooperatively use the vast amounts of information that are often available. Proper use of these tools requires consideration of IEM goals and needs, and cognizance of the human challenges and limitations that affect IEM (and much of modern science).

Why is IEM needed? How can it be used?

IEM is needed to simulate complex, dynamic systems with multiple processes at multiple scale

Natural resources and both natural and built environments are affected and linked by a complex diversity of processes made dynamic through natural variability, climatic change, population expansion, human behaviour and land-use change. Improved management of resources and environments requires improved understanding of these complex, dynamic, systems. Tools and structured processes are needed that can: (1) help forecast, predict or explore potential system changes, (2) inform policy actions and support decision making, and (3) track impacts of policy actions (or of their absence). IEM provides some new ways to investigate connections, couplings and feedbacks that generally would not be explored in traditional discipline-focused numerical simulations.

Reductionist science and overly simplified models do not suffice

Assuming that resources and environments are not linked, are not complex, and are not subject to dynamic changes is not a suitable approach to manage the longer term, larger scale, well-being of society (Sterman 2001, 2002). Normal reductionist scientific approaches are insufficient in the face of the complexity and uncertainties associated with natural resources and environments; instead, a 'post-normal' integrative science is needed that acknowledges complexity and helps deal with uncertainty (Funtowicz & Ravetz 1993). Later sections in this paper will expand on these issues, including the balancing of complexity and simplicity.

How can IEM be used?

IEM can help assemble the information that we possess, and the knowledge that we believe to have, in logical, structured constructs (numerical models and databases). IEM can provide a dynamic, adaptive, integrated information framework for the improved management of natural resources and environments. Specifically, IEM can be used to:

- assemble and organize large sets of disparate information, both quantitative and qualitative;
- transform information (e.g. convert, interpolate, extrapolate, integrate, differentiate) to calculate stocks, flows or other system properties;
- design monitoring networks to effectively observe and quantify the stocks, flows, properties or qualities needed for the assessment of natural resources and environments at different scales;
- assess correlations and patterns in observations (i.e. through statistical modelling tools);
- test causality of correlations, suggest testable hypotheses, or help design or interpret field experiments or natural experiments through deterministic modelling approaches;
- explore effects of including or excluding given processes, the equations by which they are represented, the parameters that control them, or the spatial and temporal scales to which they are applied;
- examine sensitivities, thresholds, tipping points, and non-linear behaviours of system processes, representative parameters, boundaries, or system components;
- predict, forecast, or test results of system changes, or explore different scenarios of change.

Generally, IEM can help devise and implement better-considered, more useful, policies to help manage landscapes and natural resources. IEM has the potential to help communities mitigate and adapt to increasingly complex environmental stresses. IEM's complexity arises because of the need to:

- consider, analyse, compile and synthesize multiple types and sources of information;

- interpolate and extrapolate available information across geographic landscapes;
- extrapolate information through time to make forecasts (or hindcasts), or to fill in time gaps;
- transform information into more useful types of information that lend themselves to policy decision making;
- consider and assess assumptions, biases and uncertainties that are inherent in constructed models or information frameworks, and
- assess the potential impact of knowledge gaps and low-probability high-impact events (i.e. 'Black Swans') in a given information framework. Policy decisions and management actions that seemed reasonable at the time of their being taken have often proved problematic later on, usually because longer term impacts, external processes or cascading impacts were not sufficiently considered.

Most of the complexity of IEM comes from the manipulation, analysis, assessment, synthesis and focusing of needed information for IEM construction and use. Additionally, complexities can arise in the processes needed to assess whether IEM-derived policy actions are useful, harmful or need to be adapted to better meet societal needs. For example, adaptive management (Ladson & Argent 2001; Argent 2009; Williams *et al.* 2009; Williams & Brown 2012) and structured decision making have started to be applied in the management of natural resources, replacing 'implement and forget' policy actions, and including stakeholders throughout the policy study and decision process. Adaptive staging, a form of adaptive management, has also been proposed as a potential implementation strategy for nuclear waste disposal (McCombie *et al.* 2003).

Why does simplification remain critical to IEM progress and implementation?

Good management of resources and environments requires (1) getting sufficient information of useful quality and consistency, (2) assembling, transforming and filtering the information to understand it and to help make decisions, (3) getting feedback on the simulated information and on the impact of the implemented decisions, and, most importantly, (4) getting community support for the entire process. Simplification of system complexities and dynamics is needed for many reasons, including the following:

- monitoring and observation systems are restricted by funding and by other practical constraints – not everything can be observed or measured everywhere at any time;
- technology and practical considerations may also limit the acquisition and transmittal of measurements as well as the computer-based processing, archiving and retrieval of information;
- humans have limits (and biases) in their cognitive capabilities, in their abilities to sense, perceive, retrieve and store information, in their abilities to process and transform information into knowledge and in their abilities to act on their knowledge.

Additionally, simplification and shortcuts are essential to human behaviour. 'Fast and frugal' heuristics are evolutionary and experience-based features that, most of the time, provide essential highly efficient guides for human behaviour and decision making (Gigerenzer & Brighton 2009; Marewski *et al.* 2010; Kruglanski & Gigerenzer 2011). Simplifications and heuristics help us avoid the 'paradox of choice' (Schwartz 2004): too much complexity or too many choices can lead to paralysis in decision making.

Lastly, but perhaps most importantly, a commonality of understanding and support is needed at all phases of IEM studies, and especially for IEM-derived decision making and implementation of management actions. A commonality of understanding and support invariably implies that the greater and differing understanding of many individuals gets subsumed to a 'minimum common denominator' of broadly accepted and explainable knowledge. Analogous conclusions, pointing to the benefits of 'small' system dynamics models (i.e. relatively simple and easy to understand), were reached by Ghaffarzadegan *et al.* (2011) in their study on the use of models to address social policy questions.

What are some downfalls or biases related to simplification?

'Simplification' often represents evolutionary adaptation or learned or acquired behaviours that may be expressed as human biases or heuristics. These simplifying biases and heuristics allow human management of complex processes, often with surprising accuracy (Gigerenzer & Goldstein 1996; Gigerenzer & Brighton 2009; Marewski *et al.* 2010). Nonetheless, simplification is not reality, and may be especially poorly suited when confronting modern issues that may have not been experienced or were infrequently experienced in our evolutionary or experiential past (Glynn 2014). Simplification can lead to significant errors of man or of machine, to wrong or misleading simulation results or interpretations, to poor decision making. Human over-reliance on intuitive thoughts and reactions can lead to highly biased and ineffective decision making (Tversky & Kahneman 1974; Kahneman 2003*a*, 2003*b*, 2011). Similarly, lack of consideration of low-probability high-impact events, i.e. 'Black Swans', can also lead to poor decision making (Taleb 2007).

Poor decision making, at least in the context of a longer term or larger scale perspective, may occur when there is a lack of immediate, sharply distinguished, feedbacks at the level of the individual or of a local (i.e. tightly knit) community; for example, when human decisions or reactions have subtle, large scale or delayed impacts on resources and environments (Sterman 1994). It can also occur when available feedbacks are overprinted by more pressing needs (i.e. more immediate, more local), or by other often irrational considerations (Gilbert 2011). Over-exploitation of common resources (e.g. overfishing, groundwater depletion) is a typical problem that can occur (Hardin 1968; Ostrom *et al.* 2002). Poor choices or judgments may also simply result from poor human cognition of important aspects or processes in complex ecosystems.

Deficient cognition, or lack of cognition, will occur not only because of human memory limitations or limitations of experience. It may also occur because some important ecosystem components are relatively hidden from us (e.g. groundwater, microbes), or because we have natural preferences to track biota or animate entities rather than relatively inanimate entities. I suspect that human cognition preferentially tracks, in decreasing order of importance: (1) oneself, (2) other humans, (3) other biota, (4) physical objects and landscapes. On a parallel track, human cognition is also probably adapted to recognize, in decreasing order of importance: (1) immediate local threats to human security (e.g. aggressive humans or large animals, extreme weather), (2) basic day-to-day resource needs and opportunities (food and water), and (3) the potential of social relations that enable our reproductive success, and help us get respect or esteem. These needs are essential components of Maslow's theory of human motivation or 'hierarchy of needs' (Maslow 1943; Koltko-Rivera 2006).

What are some general human biases that may occur in the application of IEM to ecosystem management?

Our cognitive limitations and adaptive heuristics are responsible for a multitude of human biases that affect the functioning of our minds, our judgments and actions. Here, however, I consider some general human biases that may affect the construction and application of IEM when seeking to improve management of ecosystem resources and environments. Because I am not trained in the behavioural sciences, some of the biases listed below are speculative, and reflect personal, non-quantitative, observations.

The 'temporal insensitivity' bias

Humans are better at representing, understanding and utilizing spatially distributed information than time-distributed information. Spatial distributions seem to be more frequently used, now more than ever with the advent of the internet and our increasingly connected world. Retrieving historical or even older information about past conditions often requires greater effort, or may be impossible. Additionally, the uncertainties (and surprises) associated with forecasts or future scenarios, and the lack of feedback and/or our personal 'lack of skin' in any long-term predictions (beyond one or two generations, i.e. 20–40 years) tend to limit actionable human interest in the distant future. (By 'lack of skin', I mean the lack of a near–immediate, bodily experienced, personal stake.) The longer the timescale of the available (or modelled) information, the lower the degree to which scientists and society are able to easily appreciate, understand and use the information to manage the environment and natural resources. The fields of system dynamics and industrial dynamics have demonstrated the difficulties that human societies have in dealing with systems that have multiple, complex, non-linear feedbacks even on the relatively short management timescales of companies and organizations (Forrester 1968, 1971, 1994). Longer delayed feedbacks make appropriate societal responses even more difficult. Society has generally not understood, or applied, the fundamental reality that our environmental systems and resources (e.g. watersheds, forests, groundwater systems) have a diversity of lagged or delayed responses that range from days to months, years, decades, centuries, millennia and more. The widely varying timescales of ecosystem processes do not generally harmonize with political cycles, or with timescales of societal decision making and feedback. This does not mean that considering long-term ecosystem dynamics is not important, especially when making major societal investments. Following local weather predictions is important to us on a daily basis because it helps us dress appropriately, or tells us of extreme weather events that may be coming towards us. Considering hydrological and climatic variability, or the risk of extreme events, on the timescale of decades to centuries or millennia may also be important if we want to make smart investments in infrastructure (e.g. installing or removing dams and reservoirs, installing tsunami protection barriers of appropriate height) or in relatively rigid legal compacts between communities or countries (e.g. the Colorado River Compact 1922).

The 'steady-state' bias

This bias is related to the previous one. As we observe the world around us, I believe that we are programmed to seek, and see, stability and simplicity, and to extrapolate current knowledge of active processes and their rates into the future. This makes it easier for us to make decisions in response to short-term imperatives. The historical prevalence of statist-, equilibrium-, or steady-state-oriented perceptions and interpretations of ecosystem processes in the scientific literature (and therefore in management and policy applications) may also partly result from discomfort or avoidance in thinking about our ultimate demise. Improved management of our ecosystems increasingly requires dynamic models, in which steady states may occur and persist for some length of time, and may sometimes recur over longer timescales, but are generally intrinsically unstable because of the many different sources of disturbances or perturbations that can occur, of either natural or human origin (Botkin & Sobel 1975; Botkin 2012). 'Non-stationarity' (Milly *et al.* 2008; Hirsch 2011) and the increasing realization that ecosystems are highly dynamic in all their characteristics and can easily exceed previously considered 'historical ranges of variation', have important implications for the management of our environment and natural resources (Betancourt 2012).

The 'man v. nature' bias

Our sense of exceptionalism often leads us to consider ourselves, and our actions, as removed from the rest of the natural world. 'Natural' systems have been studied and modelled as if they were isolated from human systems and the built environment (Botkin 2000). The concept of the 'independent observer' and the development of the scientific method inherently assume that we are removed from nature. Under many conditions, for example, for small-scale simple systems observed over short time frames, this is not a problem. It is a problem, however, when modelling the larger scale, longer term, integrated processes of complex dynamic ecosystems. Conceptual and numerical models of these systems, and generally of the environment and its resources, have rarely considered the complexity of human behaviour and human decisions, and their full impacts on the environment and its resources. Such models often do not include humans and their behaviour. This artificial separation has negatively impacted our ability to model and manage the environment and its resources (Force & Machlis 1997; Machlis *et al.* 1997; Machlis & McNutt 2010).

The 'anthropomorphic' bias

Human nature relates best to itself, and commonly seeks to anthropomorphize entities that it does not understand well, including computers and their associated technology (Nass & Moon 2000). Computer 'personalities' that relate to human personalities can be created relatively easily and humans respond socially to technologies (Nass *et al.* 1995). I suspect that integrated models, as complex multi-dimensional dynamic information frameworks, will not escape our tendency for personification, especially if they become accessible to the average person and acquire 'black box' or 'artificial intelligence' characteristics. Their predictions or output may be treated like the prophecies of the Oracle of Delphi: that is, generally held in great respect by many believers, subject to obscure pronouncements that frequently need translation from acolytes and high priests, occasionally amenable to providing additional prophetic details (or more obscure pronouncements) when consulted again with suitable accompanying 'gifts'. It is likely that integrated models will be both shaped and related to as if they were human entities. Hopefully, they will provide a better record of transparency and more widely understood meaning than the Oracle of Delphi. (There is no doubt though that the Oracle was considered by society as a useful source of wisdom and prophecy: she is believed to have been in place for 12 centuries, from 800 BCE to 395 CE; Wikipedia 2015*a*).

The 'single species' bias

This bias is related to our need for simplicity. Management actions, policies and regulations have often focused on single species, disregarding their interactions or dependencies on other species. Ecosystem management has often ignored the complexities of food webs and focused on individual species, or on a short list of species of interest: threatened and endangered species, 'keystone' species or 'indicator' species. Charismatic species have often received greater study than species that did not appeal as much to the public, or were not as visible (or were not as scary). Less charismatic or less visible species, however, often have great functional relevance in ecosystem processes. As illustrated in the management of sea otters and many other threatened species, ignoring species–species interactions and the need for monitoring and modelling multiple populations has often led species management actions astray (Botkin 2012). Despite many studies on large apex predators, the trophic cascades and ecosystem processes that they often control or influence remain areas of much needed study (Ripple *et al.* 2014).

Cognitive perceptions and the 'visible is credible' bias

Vision is such an active sense that it may overwhelm our other cognitive inputs, and possibly also diminish our ability for conscious logical thought (Glynn 2014). Our senses include four other 'classic' receptor senses (hearing, smell, taste, touch or skin sensation), as well as many others, including equilibrioception (balance, acceleration, gravity), proprioception (kinesthetic sense), thermoception (heat flux), chronoception (time), nociception (pain) and other internal 'interoception' and chemoception senses. Vision is a privileged sense that allows near–immediate human response to impactful events. It allows quick assessments of situations. People near us are able to immediately share our visual perceptions. In contrast, other senses (1) involve internal perceptions that are not as easily shared, and/or (2) are associated with greater transmittal delays between emission and reception of the sensory signal (e.g. hearing), and/or (3) require greater time for reception and human processing before action can be taken (e.g. smell). It is no coincidence that vision is a most important sense, especially when it comes to the shaping of human beliefs. 'To see is to believe' is a common human expression. Conversely, the invisible requires a greater effort of belief, or of education and knowledge, and consequently often ends up unrepresented in our mental models, in our conceptual models, and therefore in our numerical models.

Lack of accounting for groundwater processes, and the regulation and management of groundwater and surface water resources as if they were separate resources, are pervasive problems (Winter *et al.* 1999) that, according to Glennon (2002), have contributed to poor management of groundwater resources in several regions of the USA. As another example, integrated environmental watershed models of water, sediment and nutrient inputs to the Chesapeake Bay have, generally, inadequately represented groundwater processes. The models have not accounted for the decadal timescales of groundwater processes (Sanford & Pope 2013), or for the decadal to millennial timescales of sediment processes (Pizzuto *et al.* 2014). As a result, the IEM watershed models for the Chesapeake Bay watershed have, for the most part, ignored the response times required for best management practices and other efforts to limit sediment and nutrient transport to the Bay.

The 'creeping normality' bias

This is a bias described by Jared Diamond in his study on the collapse of human societies (Diamond 2004). It is also sometimes referred to as the 'boiling frog syndrome' (Wikipedia 2015*b*). Humans are conditioned to respond quickly to immediate, clear and specific risks to themselves and their present communities. Individuals are not conditioned to make decisions that impact them, their descendants, or society in general (present and especially future) in response to diffuse risks. By diffuse risks, I mean risks that are either not perceived or at best perceived as diffuse by individuals because the risks (1) increase slowly or are spread over a longer timeframe, (2) occur at a large spatial scale without clearly noticeable local feedbacks or (3) are buried by uncertainty or variability or diluted by too many other factors affecting human perception. Integrated modelling (IM), in many ways, seeks to upset this conditioned, evolutionary, reality of human perception and response. IM by its very nature seeks to broaden human perception, while still aiming for consequent action. IM often seeks to simulate a greater number or diversity of processes than may be otherwise considered. In many instances, it will provide information that does not relate to immediate and local impacts, or that simulates gradual changes or changes that may be otherwise 'buried away' from human, or at least societal, perception, and therefore from consequent action.

The 'disciplinary' biases

Biases relating to our disciplinary expertise, or to the social communities or peer groups that we associate with, are plentiful. These associations are generally beneficial in that they help us develop expertise and knowledge in specific areas, and they also provide professional or personal security that we would not have as isolated individuals. However, these associations also have the potential to create biases that can skew our perspectives, our mental models, and therefore the way that numerical models are assembled, interpreted and applied or used. IEM does not escape from these biases, but it does have the potential advantage of bringing in a diversity of perspectives. IM seeks and needs to provide broader and more integrated perspectives for management or policy actions; users of IM must be mindful of using the best possible disciplinary knowledge and expertise, while being very open to different or alternative perspectives. Ecosystems should not be studied and modelled only by biologists with little training in the physical sciences, or by physical scientists with little training in biology. Definition and quantification of ecosystem functions and services, or assessments of ecosystem health, clearly require the engagement of a broad community. Economic valuation of ecosystem services, or the construction of trading frameworks for various ecosystem crediting plans (e.g. wetland trading, nutrient trading, carbon trading), offer examples of the breadth and depth of interdisciplinary collaboration required for useful applications of IEM. Ultimately, however, broad and diverse collaborations require a core of common understanding, helped by the use of common ontologies and semantics, but also by useful and appropriate simplifications.

The 'dominant stature' bias

This bias reflects the common occurrence where more aggressive individuals, justifiably or not, seek to assert their leadership or dominance, while others, justifiably or not, follow their lead. Relatively similar behaviour can be observed in the leadership/power relationships of wolf packs and ungulate herds. In human gatherings, a widely acknowledged expert or leader may persuade, or subdue, less aggressive participants into accepting an opinion or course of action, sometimes regardless of appropriate justification, i.e. without the requisite level of expertise, knowledge and logical thinking needed from both the leader(s) and the followers. There are excellent reasons for this behaviour in many situations. However, the breadth of reasoned participation required by IEM will generally argue for minimizing, or at least controlling, such behaviour. The 'dominant stature' bias is related to our tendency to follow people who exhibit confidence, even when unwarranted: Chabris & Simons (2010) call this the 'illusion of confidence'.

The 'managed expectations' bias

The 'managed expectations' bias relates to the common need that scientists, policy makers and other professionals have to present their results and conclusions in ways that are more likely to be accepted by their colleagues and/or in a form that avoids jeopardy to their employment (i.e. their security and access to food and other resources, the bottom levels of Maslow's 'hierarchy of needs'). There are probably many psychological factors at play in this general bias, both at the level of the individual and his/her relations with a peer group. To take one well-known example, the 'loss aversion' heuristic (Kahneman 2011) affects our ability to have the most objective judgments. Our high sensitivity to avoidance of potential losses is what causes stockholders to sell their stocks more often than not at the bottom of a market swing. Similarly, weather newscasters often announce with great confidence that a major snowstorm is most likely the next day, when the greater probability is that it will not occur. The newscaster (and perhaps also the forecaster) is managing societal expectations, and minimizing personal risk to continued employment, by over-weighting the likelihood of a negative event. Scientists are increasingly expected, beyond their duties, to seek objective knowledge, to be careful communicators of risk and to be sensitive to the management of public expectations. This reality is illustrated by the recent trial and conviction of seven Italian scientists to six-year prison terms because of their insufficient attention to public sensitivities (Cartlidge 2012; Marzocchi 2012; Boschi 2013). Managing expectations and the public's perception to risks and its need to find culprits to blame (or scapegoats: a basic social need according to Girardian anthropology, e.g. Girard 1987) is likely to affect how IEM results are used and presented to a broader public. It also has the potential to affect what science is conducted and presented by scientists.

The 'confirmation' bias

This type of bias, also referred to as 'myside' bias, is one of our most important biases. It allows us to quickly, and often efficiently, pursue or act on our beliefs. It also leads us to use, and filter, observations that seek to confirm our pre-existing mental models or conceptual models, rather than to try to discredit those models (Bacon 1620; Nickerson 1998). Confirmation bias minimizes the 'cognitive dissonance' between our existing beliefs and our behaviours and cognitive inputs: we tend to align our behaviour, and the information that we consider, with our pre-existing beliefs (Festinger 1957). Confirmation bias is likely to mislead us in the design and use of IEM in cases where behavioural responses have not been sufficiently conditioned from feedbacks and experiential learning, gained either earlier in our lives or in the human evolutionary past. We need to be cognizant and vigilant of a natural tendency to indulge in confirmation bias. A conscious effort to disprove or rigorously test our conceptual models, and our constructed IEM frameworks, is needed in our pursuit of improved ecosystem management actions for broad societal benefits.

A polymorphous complexity of human biases and limitations

There is a large literature of knowledge that relates to memory biases, human heuristics, social biases and the limits of our attention and logical thinking. Many of these biases and human limitations have been discovered through the examination of human rationality and the boundaries and conditions that affect its application (Simon 1990). Discussions of the many memory biases, cognitive biases and heuristic strategies that commonly affect human thinking and decision making can be found in textbooks such as Baron (2006); reference books such as Stanovich (2010); popular books such as Kahneman (2011), Ariely (2010) and Chabris & Simons (2010); scientific reviews aimed at applications such as medical diagnosis (Anderson 2012) or forecasting (Stewart 2001), and on the internet (e.g. Wikipedia 2015*c*, *d*). While biases and heuristics relate to the behaviour of individuals, social forces strongly affect them and end up also controlling the behaviour of social groups. Ariely (2010) gives the example of a small group that goes to dinner together, where each person sequentially orders their preference. He argues that the only person that undoubtedly orders what he or she desires is the first one to place their order. All others are generally affected by wanting to either support ordering decisions already made by others, or differentiate themselves by ordering something different. 'Framing' biases affect our judgments, not only through the company we keep, or the environments and habitats that house us, but also through the way information is presented to us. Even as highly educated individuals, we are more likely to decide in favour of a course of action when told that it has a 70% chance of success, than when told that it has a 30% chance of failure (Kahneman 2011).

Framing ourselves as scientists 'conducting objective science' ignores the frequent subjectivity of our judgments

Such judgments are often called into play, assuming that we are not complete robots and that we serve some purpose beyond that of inanimate computers. Instead of deluding ourselves through an 'objectivity frame', we would be better served acknowledging that we often make subjective judgments in the pursuit of science. *We should strive to discern, examine and understand our biases and subjectivity, and take appropriate countermeasures if needed.* As Stanovich & West (2003, p. 171) state:

> People assess probabilities incorrectly, they display confirmation bias, they test hypotheses inefficiently, they violate the axioms of utility theory, they do not properly calibrate degrees of belief, they overproject their own opinions onto others, they display illogical framing effects, they uneconomically honor sunk costs, they allow prior knowledge to become implicated in deductive reasoning, and they display numerous other information processing biases.

Francis Bacon (1620) similarly realized that human minds distort reality when he introduced his classification of the 'idols of the mind': (1) the 'idols of the tribe' that innately affect all human beings; (2) the 'idols of the cave' that distinctively mould individuals (e.g. through their experiences or education); (3) the 'idols of the market place' that reflect the distortions of human communications; and (4) the 'idols of the theatre' that impose falsely constraining philosophies or mythologies. Scientists are not immune from these 'idols of the mind', especially when they remain unperceived and unacknowledged.

The 'eye of reality': a frame for knowledge simplicity, complexity and uncertainty

Human biases and limitations are most likely to affect, by being harder to counteract, our evaluations of uncertain, complex, dynamic systems, rather than our evaluations of simpler systems containing mostly factual, static, information. Finding the right level of simplification or of representation for different systems under different circumstances means estimating the level of unrepresented (or unknown) complexity. In terms of visual artistry, it means not just understanding the 'positive space' occupied by simulation model(s) and associated data, it also means having a reasonable level of understanding of the 'negative space' that is unoccupied by a modelling or information construct, i.e. the assumptions and other simplifications that have been made, the reality that is not represented or simulated.

As human beings, we may distinguish ourselves from other animal species through our propensity to derive abstract simplifications and symbolisms of reality. Although these abstract 'simple' constructions can sometimes lead us astray (Stanovich 2013), our mental models and conceptual frames provide essential guides, often unconsciously used, for our thinking and behaviour. Figure 1 illustrates what I call the 'eye of reality', a reference frame that may be useful in thinking about and classifying our human pursuit of knowledge. My frame depicts information that is:

(1) known to be known, or that can at least be easily, logically or factually determined; or
(2) known to be unknown – information that is indistinctly sensed or perceived or that is not easily available or determined; or
(3) part of the 'unknown unknowns' – unknown information that we may really need but that we do not even know that we need.

The terminology of knowns and unknowns referred to above was presented by Donald Rumsfeld, US Secretary of Defense, at a news briefing

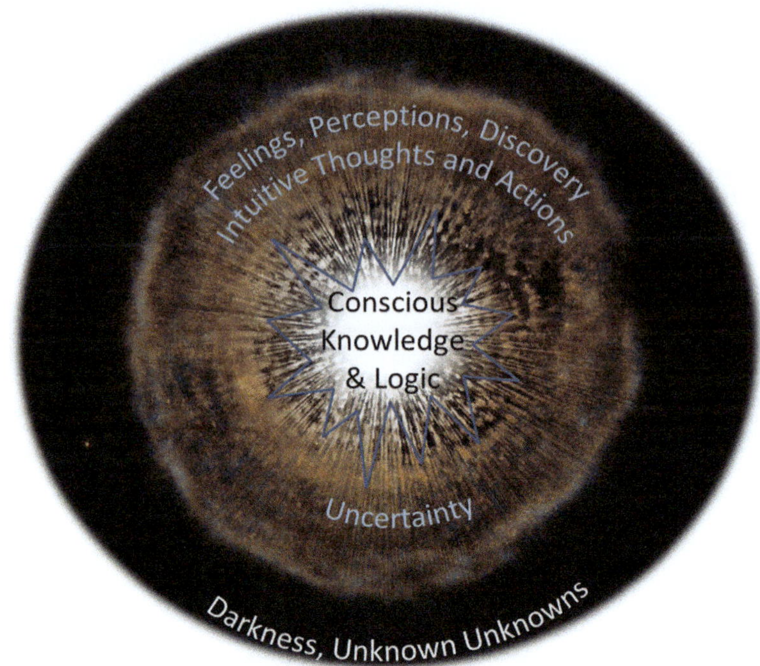

Fig. 1. The 'eye of reality': a representation of knowledge, perceptions, uncertainties and unknowns based on a photograph taken from the Hubble Space Telescope by the National Aeronautics and Space Administration (NASA) of an exploding Red Giant star, a dying unstable star that periodically 'blows a bubble', a nearly spherical shell of gas (http://www.nasa.gov/multimedia/imagegallery/image_feature_2302.html). Image used by permission from NASA (http://www.nasa.gov/audience/formedia/features/MP_Photo_Guidelines.html).

in February 2002 (Wikipedia 2015*e*). However, there are many antecedents to Rumsfeld's pronouncement going back all the way to the saying 'I know that I know nothing', i.e. the Socratic Paradox, attributed to Plato's accounts of the Greek philosopher (Wikipedia 2015*f*), which may in turn have originated from the Oracle of Delphi (Wikipedia 2015*g*).

The white core at the centre of the image represents the most objective and factual knowledge (or information) that we either have or can obtain relatively easily, for example, by quantitative monitoring of our resources and environments, or by applying the scientific method in its strictest form (Popper 1959, 1972), i.e. by seeking to refute a single hypothesis at a time through experiments or logical deductions. 'Normal' (as opposed to 'post-normal') science is part of this white core.

The irregular, star-shaped, blue line around the white core in Figure 1 represents the fact that we all make choices (often unconsciously) as to what hypotheses to test, what properties or entities to notice, observe or quantify, what expertise or fields of study to pursue, what minds to prod. Often our decisions, conscious or unconscious, are made on the basis of vague perceptions, feelings or intuitions about what might be profitable pursuits in our seeking to explore the 'known unknowns', the speckled and striped brownish area in our diagram. This area of gaseous and increasingly unsubstantive materials represents our decreasing base of knowledge and perceptions as we move away from our white, most factual, core of information. Last but not least, the outer black area in the diagram represents the 'unknown unknowns', the dark matter that we can perhaps reduce through experience but, by definition, that we can never uncover (at least in the immediate).

How can our 'eye of reality' frame be used to address the human challenges of IEM? There are no easy, universal solutions. I would argue that explicitly recognizing and better defining the partitions between the three areas of knowledge are essential, as are attempts to recognize and analyse our human biases, limitations and heuristics that influence our judgments and actions, and invariably also affect our intuitions and creative thoughts. These intuitions and creative thoughts serve as whispered introductions to the 'known unknowns', and possibly eventually to the 'unknown unknowns'.

Effectively using these 'whispers' requires cognizance and understanding of our human biases and limitations, their evolutionary, individual and/or societal origins, and a comparison with the issues and systems at hand. This understanding is critical. Are the 'whispers' helpful and appropriate in evaluating or thinking about a particular issue or system? Or do they need to be counteracted?

Addressing the human challenges of IEM

Devising methods and processes to address the human challenges of IEM, and the complexities and uncertainties of the studied systems and issues, is a major area of study. I can only provide initial suggestions that may be helpful in addressing these challenges, beyond the first steps, which are to take greater, explicit, cognizance of human biases and behaviours and to use appropriate knowledge frames, such as 'the eye of reality' discussed above.

Appropriate simplicity, adaptive compensations

The efficiency of simplicity is compelling. Simplicity breeds clear understanding by a large community, efficiency of study and minimization of short-term costs. Management and policy actions are generally not taken if a simple enough understanding cannot be achieved by a non-specialist community. Integrated modelling will necessarily breed complexity. The success of IEM is dependent on its ability to model complex and multiple processes; but it is also strongly dependent on being able to reduce that complexity into simple enough descriptions and processes that can be clearly understood by a large community, and that will therefore lead to implementation of reasonable management and policy actions.

The downside of simplicity is that it includes and/or engenders human (or technical) biases, oversights or errors that may need to be compensated for. Adaptive management, sometimes simply called 'learning by doing', theoretically provides an iterative way to evaluate information, outline expectations and take actions in the face of uncertainty and complexity, while allowing for later modifications of the actions taken or policies developed as more information and knowledge accrue. As pointed out earlier, adaptive management supported by IEM can improve the management of our natural resources and environments. Adaptive management is not a panacea that will be suitable for all types of situations (Norton & Reckhow 2008; Craig & Ruhl 2014). It will not be a suitable strategy for situations where follow-through monitoring, evaluations and adaptive actions are unlikely. Ethical or legal reasons can also prevent its use. Adaptive management (and the use of IEM) may also fail in managing systems that have lagged or highly non-linear responses, that have too high a complexity or unobservable causative drivers and feedbacks, or that have threshold responses from which there may be no recovery.

Structured processes to address complexity

The intrinsic complexity of integrated modelling is unavoidable. It requires adequate understanding and simulation of a wide diversity of processes studied (or monitored) by a wide diversity of communities. It involves simulation of a web of interdependent cascading processes that sometimes have threshold behaviours or other non-linear behaviours, and where the importance, or lack of importance, of any given process in the simulation may be something that varies depending on system conditions, spatial and temporal variations, and the consideration (or lack thereof) of other processes. The key to addressing and managing this complexity is to structure, layer, compartmentalize and abstract it in a scalable manner (i.e. simplify it). There are multiple reasons for doing this, which include not only the technical traceability, accountability and use of IEM, but also our fundamental human needs for appropriate and scalable simplicity in visualizing, understanding and sharing the information and knowledge provided by IEM amongst a user community. Additionally, structured complexity and scalable simplicity/abstraction may help quicken understanding and response to unanticipated impacts from decisions or actions taken as a result of an IEM process.

Another way of keeping complexity manageable and understandable is to build it gradually over time into integrated models, perhaps by building on a series of simple models. Alternatively, a more problematic approach is to reduce complexity, that is, to develop general, highly inclusive, complex models, and to reduce their complexity from the top down. While this approach may lead to significant errors and problems related to a lack of detailed understanding of the models, it has the merit of allowing for the simulation of processes that might not have been accounted for in models built from the bottom up. Modelling results from this top-down approach must be carefully considered to avoid their misuse.

Participatory processes

Education of the scientific community and of the wider public to enable them to achieve a greater level of comfort and understanding of the capabilities, and limits, of IEM is also needed. Learning may be enhanced through the advent and propagation of new visualization tools, gaming methods and other learning technologies. Perhaps most importantly, education of a wider community can be achieved by encouraging their greater participation in IEM (Voinov & Gaddis 2008; Voinov & Bousquet 2010) in the assembly and use of models to explore and foster understanding of complex systems, or to build scenarios and modelling forecasts (Alcamo 2008; Alcamo & Henrichs 2008; Pahl-wostl 2008). Mediated modelling (Van den Belt 2004), also known as cooperative modelling or participatory modelling, can often provide a useful, structured approach to engage stakeholders and the interested public, together with scientists and other professionals, in helping manage resources and environments. Cockerill *et al.* (2006) provide an excellent account of the strengths and weaknesses of a cooperative modelling project that was used to inform and help select water management options for the Middle Rio Grande (MRG) basin in New Mexico. The cooperative modelling simulated a wide diversity of hydrological and ecosystem processes, including human infrastructure and the impacts of human activities on the landscape. The model and water management scenarios developed helped gain public and stakeholder understanding of the complex system dynamics in the basin and of the trade-offs involved in different management options. The modelling effort ultimately informed the water management plan developed for the MRG basin in 2004 (MRG Water Assembly 2004).

Lessons from a painting: structuring community interactions, using new modes of representation, finding the missing, seeking the 'unknowns'

Jakeman *et al.* (2008, p. 4) state:

> Modelling should be about the systematic organization of data, assumptions and knowledge for a specific purpose. In the environmental domain the main reasons for modelling are for knowledge generation and sharing in order to inform a decision that could be for operational management or strategic policy development and implementation.

Clearly, human behaviour and judgment enter the process of assembling and using models. Consequently, just as it is important to build structured, scalable simplicity and abstraction in IEM tools and representations, it is also important to structure and systematically trace and account for human and social interactions during the assembly and use of IEM. These human and social interactions are inherently complex. They include: (1) stakeholder engagement and learning processes; (2) the accounting and management of multiple perspectives or human information sources; (3) the creative exploration of individual intuitions or perceptions; (4) the recognition, understanding and possible counteraction of human biases and limitations; and (5) decision trees or other methods to trace the construction, evolution and use of IEM.

We should consider using a greater diversity of media and forms of communication and representation to creatively explore our conceptual models, frames of reference and our simplifications in the pursuit of the 'known unknowns' and the 'unknown unknowns' that ideally should be considered in IEM. Here, I use a painting by the American neo-impressionist painter Maurice Prendergast to illustrate some points about the structuring of community and social interactions and the building and use of integrated models (cf. Fig. 2). *A painting represents an integrated expression of mind, experiences and vision, sometimes conscious, sometimes not; in other words, it has many of the characteristics of an integrated model.* The painting in Figure 2 depicts a colourful community enjoying what looks like a leisurely weekend day in the wonderfully structured city of Venice. The painting shows at least three bridges and a multitude of buildings, located alongside the Venetian lagoon, that share similar architectural designs and height and width constraints while avoiding blandness of uniformity. An analogy can be drawn here with the modular structures and connection standards of IEM. There is a sense of purpose in the movements of the crowd, as most (but not all) people seem to be moving away from the painter, towards some common objective or attractor. The painting, through its depiction of Venice and its ordered community, celebrates the spirit of human enterprise and organization. Can IEM achieve a similar spirit of purpose, enterprise and organization?

Now, what is the missing from the information frame of Figure 2? What is the 'negative space' unrepresented but perhaps defined by the painting? The incomplete outline of a person in the lower right is a symbol of missing information and knowledge. There seems to be a lack of diversity in the social classes, professions, origins and other characteristics of the people represented. Does this mean that the collective knowledge base of the community has an impoverished diversity of perspectives? Does the relative uniformity and beauty of their constructs, buildings and bridges point to an attractive form of groupthink? Would a greater diversity of people and constructions provide feedback mechanisms that could help the community and the city avoid a future tragedy, such as one caused by climate change, sea-level rise, and an overconcentration of people and infrastructure in a vulnerable area?

Fig. 2. Maurice Prendergast's (1858–1924) painting of the Ponte della Paglia in Venice. The American post-impressionist painter started the painting during his visit to Venice in 1898–1899 but extensively repainted and completed it two decades later in 1922. Photograph provided, with permission to reproduce it in this article, by the Phillips Collection, Washington DC (http://www.phillipscollection.org/research/american_art/artwork/Prendergast-Ponte_Paglia.htm).

Planning for the future requires understanding past conditions that might have given rise to currently observed realities. Venice developed in marshlands near the sea. The community initially took advantage of the fishing opportunities provided by the lagoon and the marshes, and also periodically sought refuge from Germanic and Hun invasions. As Venice grew in power and infrastructure, so did its commerce and trade. Venice was positioned in an ecotone, an ecologically rich transition area between two biomes. Venice shows man's taming and use of this rich natural ecotone. Because it creates bridges across spatial and temporal scales and across knowledge domains, IEM also has a form of 'ecological richness'. IEM can also serve as a monument to human enterprise and organization. Both Venice and IEM, however, are susceptible to possible failure, possibly made worse or more catastrophic by initial achievements and by groupthink (Janis 1972). Can/should we provide resilience to our IEM constructs? Providing structured processes, transparent assumptions, shared knowledge, structured communities and inclusive participation is essential, but insufficient in testing for resilience of IEM constructs.

Red teams storming

As exemplified – so far – by the history of Venice and its frequent flooding, cities and communities have the potential to grow more resilient and/or to adapt to changes when they are tested by disturbances or perturbations of suitable magnitude. A well-known hypothesis in ecology, the 'Intermediate Disturbance Hypothesis' (e.g. Connell 1978), similarly suggests that smaller disturbances of appropriate intermediate frequency can sometimes, but not always, help avert much bigger catastrophes. This hypothesis is still the subject of significant debate (e.g. Fox 2013; Sheil & Burslem 2013). By analogy, IEM frameworks, and the associated understanding and prediction generated through IEM, may also develop greater resiliency and confidence if they are tested through the efforts of small teams that question or seek to invalidate aspects or assumptions related to a particular IEM effort and its application. Groupthink is the enemy of IEM resilience, honesty and transparency, and of the effort to improve management of resource and environment systems. There are several ways to avoid groupthink in IEM. One way is to set up small groups that compete to achieve some stated IEM objectives. A problem with this approach is that it probably requires metrics of success or of achievement to be made explicitly ahead of the competition, when the end results needed may still be relatively undefined. Another way is to conduct 'in-process reviews' by independent panels that seek to assess and constructively address IEM limitations. In my view, the 'constructionist' expectations mean that these panels may be insufficiently autonomous and insufficiently vigorous in their identification, testing and review of IEM vulnerabilities. A better approach may be to create independent, innovative, highly focused 'red teams' that actively fight groupthink and confirmation bias, and try to 'storm' the IEM construct during its assembly, during interpretation of its outputs or during discussions of its use or applications. 'Red Teaming' is a structured testing approach commonly used to test security processes in military and intelligence operations, or to provide 'alternative analyses' of a given situation (United Kingdom Ministry of Defence 2013).

Behavioural biogeosciences: a new area of study supported by IEM

We are not well adapted to address resource and environment issues that differ from those experienced in our human evolutionary past, and/or that have not provided frequent, sharply experienced feedbacks at the level of the individual, or of a local community. The behavioural sciences can help us understand the extent to which our biases are the result of our evolutionary adaptation to threats and opportunities in our ecosystems. They can tell us when those adaptations may not provide the best solutions to managing our ecosystems (and ourselves), for example, because the temporal and spatial scales of reference, or the dynamics of change, are outside of our natural adaptive capabilities. The behavioural sciences can help us take cognizance of, and when appropriate compensate for, human limitations in our organized pursuit of knowledge and its applications, including our construction and use of IEM. These limitations extend beyond the biases and heuristics that are discussed in this paper, and beyond the natural, but sometimes inappropriate, prioritizations of our human cognition and attention, as also mentioned earlier. The behavioural sciences can help us understand the useful, but often biased, wellsprings of human intuition, creativity and abstract thought. These are characteristics of our species that help us explore new frontiers of knowledge.

Understanding the full complexity and dynamics of ecosystem processes, correctly assessing their relative importance under different conditions, and therefore their appropriate prioritizations and simplifications during model construction and use, should not be an area of study where only biologists and ecologists participate while primarily focused on biota and on the easily visible. Understanding the dynamic complexity of the physical processes affecting our habitats, our security and our access to water, energy and minerals is equally important. We need to quantitatively monitor those physical processes, and also strive to quantitatively assess or monitor the dynamics of food webs and biotic populations. IEM can provide tools that help organize and maximize our knowledge of the biogeosciences, while also taking into explicit account our human needs, responses and biases. Study of the 'behavioural biogeosciences' needs to become an integral part of IEM.

Expanding our knowledge beyond our current human limitations: some additional thoughts

The suggestions provided above are complementary to what we already know is important in the proper construction and use of numerical models. The construction, interpretation and application of IEM should follow the standards of good modelling practice and scenario building (examples of some excellent reviews are: Alcamo & Henrichs 2008; Crout *et al.* 2008; Schmolke *et al.* 2010; Saltelli & Funtowicz 2014). *Most importantly, models should be considered as tools for gaining system understanding, rather than revered as providing near-absolute truth(s) once they happen to have been calibrated and tested (i.e. considered 'validated' in some engineering terminology).* Models must also strike the right balance between simplicity and complexity, a balance that will probably vary depending on modelling objectives and available knowledge.

These principles have been well articulated in several reviews and commentaries on the topic (e.g. Konikow & Bredehoeft 1992; Bredehoeft & Konikow 1993, 2012; Konikow 2011; Voss 2011*a*, 2011*b*; Nordstrom 2012). Models should also reflect a balance between (1) the description of process theory or knowledge and (2) available observations and quantitative measurements. As Kirchner (2006, p. 1) states:

> . . . scientific progress will mostly be achieved through the collision of theory and data, rather than through increasingly elaborate and parameter-rich models that may succeed as mathematical marionettes, dancing to match the calibration data even if their underlying premises are unrealistic.

For IEM in particular, this means that we must put greater emphasis on the most objective part of the scientific enterprise – observations and monitoring – even though we should still use models to help organize available information and decide how, when and where to possibly collect more. Human biases and limitations affect our conceptual models that, together with practical realities, commonly drive our observations and monitoring programmes. Statistical analyses, visualization and other technology tools, when honestly and efficiently used (Tufte 1983, 1990, 1997), may provide useful techniques to help us take cognizance of our biases and subjectivity, whether (1) organizing available information or (2) transforming, agglomerating/reducing or extending information through our models. But fundamentally, improved management of our natural resources and environments through the use of IEM will depend on accessibility to well-characterized, multi-scale, long-term, observations and monitoring (Lovett *et al.* 2007; Keeling 2008; Lins *et al.* 2010). Factual observations and monitoring are critical parts of the white core of our 'eye of reality'.

Summary comments

Integrated Environmental Modelling (IEM) is needed to help communities better manage the complex and dynamic ecosystems that provide natural resources and form their environments. IEM can (1) help organize and transform basic information (observations, quantitative measurements), (2) complement (or test) existing knowledge, and (3) sometimes provide new knowledge or insights that can help society manage its resources and environments. To be understandable, and therefore usable, by a broad community, IEM will always involve simplifications. Those simplifications will sometimes be consciously made, and sometimes will be unconsciously decided. This paper has provided examples of a diversity of human biases and heuristics that may also affect how IEMs are assembled, interpreted and applied.

Essentially, there are three steps that are needed at every stage of the IEM construction, interpretation and application process.

- First, all available information and knowledge needs critical examination. In artistic terms, this corresponds to examining the 'positive space' occupied by our knowledge base.
- Second, conscious, critical examination is required, to the extent possible, of what is not included in an IEM construct or application. This can be thought of as examining the 'negative space' of the IEM construct and application.
- Third, IEM developers, interpreters and users need to take active cognizance, to the extent possible, of the inherent human biases and heuristics that may have (1) affected their definitions of positive and negative space, or (2) influenced their information and knowledge base and, consequently, any modelling constructs and uses.

There will always be 'unknown unknowns' that surprise or confound IEM developers and users. The three-step process suggested here seeks to decrease the number of surprises, while maintaining an attitude of watchful humility. Several other specific suggestions may help address the human challenges of IEM. These include:

(1) testing/auditing model predictions and management policies through an adaptive management iterative process, when feasible;
(2) using structured processes – such as progressive complexity and/or progressive model reduction – to test for appropriate simplicity, and to maximize understanding and transparency of IEM constructs and use;
(3) participatory modelling to engage a diversity of perspectives, and grow stakeholder and expert understanding and use of IEM;
(4) developing structured processes to systematically account for human behaviour (including human biases) and social interactions, and to maximize effective use of IEM for larger scale, longer term issues, i.e. for problems that humans are not naturally adapted to address;
(5) using creative forms of communication and representation to explore intuitions, conceptual models, simplifications and transient frames of reference, in the pursuit of the 'known unknowns' and the 'unknown unknowns';
(6) soliciting 'red team' raids at all stages of IEM construction and use to elicit greater critical thinking, and avoid (or at least control) groupthink and confirmation bias;

(7) forming a new area of study in the 'behavioural biogeosciences' that melds the knowledge of behavioural scientists with that of biologists and ecologists, and especially with the expertise of physical scientists engaged in the quantitative description and monitoring of ecosystem processes;
(8) maximizing smart, efficient and honest practices, not only in modelling activities, but also in information gathering (i.e. observations and monitoring) and visualization.

Formal journal reviews by Nina Burkardt (USGS) and Gerry Laniak (Environmental Protection Agency (EPA)), editorial review by Andrew Riddick (British Geological Survey (BGS)), and additional reviews by Lenny Konikow, Mary Jo Baedecker and Kevin Breen of USGS, greatly helped clarify and improve this manuscript. The author is also grateful for the important comments, insights and references provided by Ed Cokely (Michigan Technological University), Olivier Delahaye (Central University of Venezuela), Fred Glynn, Gary Shenk (EPA), Ferris Webster (University of Delaware), and USGS colleagues Ken Eng, Karl Haase, Dianna Hogan, Greg Noe and Ward Sanford. The author retains full responsibility for any errors of fact, expression or judgment. Any use of trade, product or firm names in this publication is for descriptive purposes only and does not imply endorsement by the US Government.

References

ALCAMO, J. 2008. Introduction: the case for scenarios of the environment. *In*: ALCAMO, J. (ed.) *Developments in Integrated Environmental Assessment. Environmental Futures: The Practice of Environmental Scenario Analysis*. Elsevier BV, Amsterdam, **2**, 1–11.

ALCAMO, J. & HENRICHS, T. 2008. Towards guidelines for environmental scenario analysis. *In*: ALCAMO, J. (ed.) *Developments in Integrated Environmental Assessment. Environmental Futures: The Practice of Environmental Scenario Analysis*. Elsevier BV, **2**, 13–35.

ANDERSON, C. 2012. On the nature of thought processes and their relationship to the accumulation of knowledge, Part XVI – the process of making a diagnosis. *Dermatology Practical & Conceptual*, **2**, 47–62, https://doi.org/10.5826/dpc.0204a12

ARGENT, R. 2009. Components of adaptive management. *In*: ALLAN, C. & STANKEY, G. H. (eds) *Adaptive Environmental Management: A Practitioner's Guide*. Springer, Berlin, 11–38, https://doi.org/10.1007/978-1-4020-9632-7_2

ARIELY, D. 2010. *Predictably Irrational: The Hidden Forces That Shape Our Decisions*. Harper Perennial, New York.

BACON, F. 1620. *The New Organon (or True Directions Concerning the Interpretation of Nature)*. Retrieved from http://eBooks@Adelaide.edu.com (from Taggard and Thompson 1863 edn).

BARON, J. 2006. *Thinking and Deciding*. Cambridge University Press, New York.

BETANCOURT, J. L. 2012. Reflections on the relevance of history in a nonstationary world. *In*: WIENS, J. A., HAYWARD, G. D., SAFFORD, H. D. & GIFFEN, C. (eds) *Historical Environmental Variation in Conservation and Natural Resource Management*. Wiley-Blackwell, Hoboken, New Jersey, 307–318, https://doi.org/10.1002/9781118329726

BOSCHI, E. 2013. L'Aquila 's aftershocks shake scientists. *Science*, **341**, 1451.

BOTKIN, D. B. 2000. *No Man's Garden*. 1st edn. Island Press, Washington DC.

BOTKIN, D. B. 2012. *The Moon in the Nautilus Shell: Discordant Harmonies Reconsidered*. Oxford University Press, New York.

BOTKIN, D. B. & SOBEL, M. J. 1975. Stability in time-varying ecosystems. *The American Naturalist*, **109**, 625–646.

BREDEHOEFT, J. D. & KONIKOW, L. F. 1993. Ground-water models: validate or invalidate. *Groundwater*, **31**, 178–179.

BREDEHOEFT, J. D. & KONIKOW, L. F. 2012. Ground-water models: validate or invalidate. *Groundwater*, **50**, 493–495, https://doi.org/10.1111/j.1745-6584.2012.00951.x

BROWN, J. G. & GLYNN, P. D. 2003. Kinetic dissolution of carbonates and Mn oxides in acidic water: measurement of in situ field rates and reactive transport modeling. *Applied Geochemistry*, **18**, 1225–1239, https://doi.org/10.1016/S0883-2927(03)00010-6

BROWN, J. G., BASSETT, R. L. & GLYNN, P. D. 1998. Analysis and simulation of reactive transport of metal contaminants in ground water in Pinal Creek Basin, Arizona. *Journal of Hydrology*, **209**, 225–250, https://doi.org/10.1016/S0022-1694(98)00091-2

BROWN, J. G., BASSETT, R. L. & GLYNN, P. D. 2000. Reactive transport of metal contaminants in alluvium – model comparison and column simulation. *Applied Geochemistry*, **15**, 35–49, https://doi.org/10.1016/S0883-2927(99)00004-9

CARTLIDGE, E. 2012. Prison terms for L'Aquila experts shock scientists. *Science*, **338**, 451–452.

CHABRIS, C. & SIMONS, D. 2010. *The Invisible Gorilla*. Harmony, New York.

COCKERILL, K., PASSELL, H. & TIDWELL, V. 2006. Cooperative modeling: building bridges between science and the public. *Journal of the American Water Resources Association*, **42**, 457–471.

COLORADO RIVER COMPACT 1922. *The Law of the River*. US Department of the Interior. http://www.usbr.gov/lc/region/g1000/lawofrvr.html [last accessed May 2015].

CONNELL, J. H. 1978. Diversity in tropical rain forests and coral reefs. *Science*, **199**, 1302–1310.

CRAIG, R. K. & RUHL, J. B. 2014. Designing Administrative Law for Adaptive Management. *Vanderbilt Law Review*, **67**, 1(48).

CROUT, N., KOKKONEN, T. *ET AL*. 2008. Good modelling practice. *In*: JAKEMAN, A. J., VOINOV, A. A., RIZZOLI, A. E. & CHEN, S. H. (eds) *Developments in Integrated Environmental Assessment. Environmental Modelling, Software and Decision Support*. Elsevier BV, Amsterdam, **3**, 15–31.

DIAMOND, J. 2004. *Collapse: How Societies Choose to Fail or Succeed*. 1st edn. Viking Adult, New York.

FESTINGER, L. 1957. *A Theory of Cognitive Dissonance*. Stanford University Press, Stanford, CA.

FORCE, J. E. & MACHLIS, G. E. 1997. The human ecosystem Part II: social indicators in ecosystem management. *Society & Natural Resources*, **10**, 369– 382, https://doi.org/10.1080/08941929709381035

FORRESTER, J. 1968. Industrial dynamics – after the first decade. *Management Science*, **14**, 398–415.

FORRESTER, J. 1971. Counterintuitive behavior of social systems. *MIT Technology Review*, **73**, 52–69, https://doi.org/10.1016/S0040-1625(71)80001-X

FORRESTER, J. 1994. System dynamics, systems thinking, and soft OR. *System Dynamics Review*, **10**, 1–14.

FOX, J. W. 2013. The intermediate disturbance hypothesis should be abandoned. *Trends in Ecology & Evolution*, **28**, 86–92, https://doi.org/10.1016/j.tree.2012.08.014

FUNTOWICZ, S. & RAVETZ, J. 1993. Science for the post-normal age. *Futures*, 739–755.

GHAFFARZADEGAN, N., LYNEIS, J. & RICHARDSON, G. P. 2011. How small system dynamics models can help the public policy process. *System Dynamics Review*, **27**, 22–44, https://doi.org/10.1002/sdr.442

GIGERENZER, G. & BRIGHTON, H. 2009. Homo heuristicus: why biased minds make better inferences. *Topics in Cognitive Science*, **1**, 107–143, https://doi.org/10.1111/j.1756-8765.2008.01006.x

GIGERENZER, G. & GOLDSTEIN, D. G. 1996. Reasoning the fast and frugal way: models of bounded rationality. *Psychological Review*, **103**, 650–669.

GILBERT, D. T. 2011. Buried by bad decisions. *Nature*, **474**, 275–277, https://doi.org/10.1038/474275a

GIRARD, R. 1987. *Things Hidden Since the Foundation of the World*. Stanford University Press, Stanford, CA (originating publisher of English edn: The Athlone Press, London).

GLENNON, R. 2002. *Water Follies: Groundwater Pumping and the Fate of America's Fresh Water*. 1st edn. Island Press, Washington DC.

GLYNN, P. D. & BROWN, J. 2012. Integrating field observations and inverse and forward modeling: application at a site with acidic , heavy-metal-contaminated groundwater. *In*: BUNDSCHUH, J. & ZILBERBRAND, M. (eds) *Geochemical Modeling of Groundwater, Vadose and Geothermal Systems*. CRC Press, Boca Raton, Florida, 181–234.

GLYNN, P. D., VOSS, C. & PROVOST, A. 1999. Deep penetration of oxygenated meltwaters from warm based ice sheets into the Fennoscandian shield. *In*: *Use of Hydrological Information in Testing Groundwater Flow Models: Technical Summary and Proceedings of a Workshop*. Borgholm, Sweden, September 1–3, 1997. Nuclear Energy Agency, Issy-les-Moulineaux, 201–241.

GLYNN, P. D. 2003. Modeling Np and Pu transport with a surface complexation model and spatially variant sorption capacities: implications for reactive transport modeling and performance assessments of nuclear waste disposal sites. *Computers & Geosciences*, **29**, 331–349, https://doi.org/10.1016/S0098-3004(03)00009-8

GLYNN, P. D. 2014. W(h)ither the oracle? Cognitive biases and other human challenges of integrated environmental modeling. *In*: AMES, D. P., QUINN, N. W. T. & RIZZOLI, A. E. (eds) *7th International. Congress on Environmental Modelling and Software*. International Environmental Modelling and Software Society, San Diego, **8**, https://doi.org/10.13140/2.1.3919.8089

GLYNN, P. D. & BROWN, J. G. 1996. Reactive transport modeling of acidic metal-contaminated ground water at a site with sparse spatial information. *Reviews in Mineralogy*, **34**, 377–438.

GLYNN, P. D. & PLUMMER, L. N. 2005. Geochemistry and the understanding of ground-water systems. *Hydrogeology Journal*, **13**, 263–287, https://doi.org/10.1007/s10040-004-0429-y

GLYNN, P. D. & VOSS, C. I. 1999. *Geochemical Characterization of Simpevarp Ground Waters near the Äspö Hard Rock Laboratory*. SKI, Stockholm, Report **96**, 29.

GLYNN, P. D., LARSEN, M. C. ET AL. 2009. Selected achievements, science directions, and new opportunities for the WEBB small watershed research program. *In*: WEBB, R. M. & SEMMENS, D. J. (eds) *Third Interagency Conference on Research in the Watersheds: Planning for an Uncertain Future: Monitoring, Integration, and Adaptation*. USGS Scientific Investigations, Estes Park, CO, Report **2009–5049**, 39–52.

GLYNN, P. D., JACOBSEN, L. J. ET AL. 2011. 3D/4D modeling, visualization and information frameworks: current US Geological Survey practice and needs. *In*: RUSSELL, H. A. J., BERG, R. C. & THORLEIFSON, L. H. (eds) *Three-Dimensional Geological Mapping; Workshop Extended Abstracts*. Geological Survey of Canada, Minneapolis, MN. Open File 6998, https://doi.org/10.4095/289609

HARDIN, G. 1968. The tragedy of the commons. *Science*, **162**, 1243–1248.

HIRSCH, R. M. 2011. A perspective on nonstationarity and water management. *Journal of the American Water Resources Association*, **47**, 436–446, https://doi.org/10.1111/j.1752-1688.2011.00539.x

JACOBSEN, L. J., GLYNN, P. D., PHELPS, G. A., ORNDORFF, R. C., BAWDEN, G. W. & GRAUCH, V. J. S. 2011. US Geological Survey: a synopsis of three-dimensional modeling (chapter 13). *In*: BERG, R. C., MATHERS, S. J., KESSLER, H. & KEEFER, D. A. (eds) *Synopsis of Current Three-Dimensional Modeling in Geological Survey Organizations*. Illinois State Geological Survey Circular 578. Illinois State Geological Survey, Champaign, Il, 69–79.

JAKEMAN, A. J., CHEN, S. H., RIZZOLI, A. E. & VOINOV, A. A. 2008. Modelling and software as instruments for advancing sustainability. *In*: JAKEMAN, A. J., VOINOV, A. A., RIZZOLI, A. E. & CHEN, S. H. (eds) *Developments in Integrated Environmental Assessment. Environmental Modelling, Software and Decision Support*. Elsevier BV, Amsterdam, **3**, 1–13.

JANIS, I. L. 1972. *Victims of Groupthink: A Psychological Study of Foreign-Policy Decisions and Fiascoes*. Houghton Mifflin, Oxford.

KAHNEMAN, D. 2003*a*. A perspective on judgment and choice: mapping bounded rationality. *The American Psychologist*, **58**, 697–720, https://doi.org/10.1037/0003-066X.58.9.697

KAHNEMAN, D. 2003*b*. A psychological perspective on economics. *The American Economic Review*, **93**, 162–168.

KAHNEMAN, D. 2011. *Thinking, Fast and Slow*. Farrar, Straus and Giroux, New York.

Keeling, R. F. 2008. Atmospheric science. Recording Earth's vital signs. *Science*, **319**, 1771–1772, https://doi.org/10.1126/science.1156761

Kirchner, J. W. 2006. Getting the right answers for the right reasons: linking measurements, analyses, and models to advance the science of hydrology. *Water Resources Research*, **42**, W03S04, https://doi.org/10.1029/2005WR004362

Koltko-Rivera, M. E. 2006. Rediscovering the later version of Maslow's hierarchy of needs: self-transcendence and opportunities for theory, research, and unification. *Review of General Psychology*, **10**, 302–317, https://doi.org/10.1037/1089-2680.10.4.302

Konikow, L. F. 2011. The secret to successful solute-transport modeling. *Ground Water*, **49**, 144–159, https://doi.org/10.1111/j.1745-6584.2010.00764.x

Konikow, L. F. & Bredehoeft, J. D. 1992. Ground-water models cannot be validated. *Advances in Water Resources*, **15**, 75–83, https://doi.org/10.1016/0309-1708(92)90033-X

Konikow, L. F. & Glynn, P. D. 2013. Modeling groundwater flow and quality. *In*: Selinus, O. (ed.) *Essentials of Medical Geology*. Springer, Dordrecht, Heidelberg, New York, London, 727–753.

Kruglanski, A. W. & Gigerenzer, G. 2011. Intuitive and deliberate judgments are based on common principles. *Psychological Review*, **118**, 97–109, https://doi.org/10.1037/a0020762

Ladson, A. & Argent, R. 2001. Adaptive management of environmental flows: lessons for the Murray-Darling Basin from three large North American rivers. *Australian Journal of Water Resources*, **5**, 89–101.

Laniak, G. F., Olchin, G., *et al.* 2013. Integrated environmental modeling: a vision and roadmap for the future. *Environmental Modelling & Software*, **39**, 3–23, https://doi.org/10.1016/j.envsoft.2012.09.006

Lins, H., Hirsch, R. & Kiang, J. 2010. *Water – the Nation's Fundamental Climate Issue: a White Paper on the US Geological Survey Role and Capabilities*. USGS Circular 1347, https://pubs/usgs.gov/circ/1347

Lovett, G. M., Burns, D. A. *et al.* 2007. Who needs environmental monitoring? *Frontiers in Ecology and the Environment*, **5**, 253–260.

Machlis, G. E. & McNutt, M. K. 2010. Scenario-building for the Deepwater Horizon oil spill. *Science*, **329**, 1018–1019, https://doi.org/10.1126/science.1195382

Machlis, G. E., Force, J. E. & Burch, W. R. 1997. The human ecosystem Part I: the human ecosystem as an organizing concept in ecosystem management. *Society & Natural Resources*, **10**, 347–367, https://doi.org/10.1080/08941929709381034

Marewski, J. N., Gaissmaier, W. & Gigerenzer, G. 2010. Good judgments do not require complex cognition. *Cognitive Processing*, **11**, 103–121, https://doi.org/10.1007/s10339-009-0337-0

Marzocchi, W. 2012. Putting science on trial. *Physics World*, **25**, 17–18.

Maslow, A. H. 1943. A theory of human motivation. *Psychological Review*, **50**, 370–396.

McCombie, C., Daniel, D. E. *et al.* 2003. *One Step at a Time: The Staged Development of Geologic Repositories for High-Level Radioactive Waste*. Committee on Principles and Operational Strategies for Staged Repository Systems, Board on Radioactive Waste Management, National Research Council, The National Academies Press, Washington DC, http://www.nap.edu/catalog/10611.html

Milly, P. C. D., Betancourt, J. *et al.* 2008. Stationarity is dead: whither water management? *Science*, **319**, 573–574, https://doi.org/10.1126/science.1151915

Moore, R., Hughes, A. *et al.* 2013. *International Summit on Integrated Environmental Modeling* U.S. Environmental Protection Agency, Washington, DC, EPA/600/R/12/728. http://cfpub.epa.gov/si/si_public_file_download.cfm?p_download_id=509013

MRG Water Assembly 2004. *MRG Water Plan*. http://www.waterassembly.org/waterplan.htm [last accessed May 2015].

Nass, C. & Moon, Y. 2000. Machines and mindlessness: social responses to computers. *Journal of Social Issues*, **56**, 81–103, https://doi.org/10.1111/0022-4537.00153

Nass, C., Moon, Y. & Fogg, B. 1995. Can computer personalities be human personalities? *International Journal of Human Computer Studies*, **43**, 228–229.

Nickerson, R. S. 1998. Confirmation bias: a ubiquitous phenomenon in many guises. *Review of General Psychology*, **2**, 175–220, https://doi.org/10.1037/1089-2680.2.2.175

Nimmo, J. R., Rousseau, J. P., Perkins, K. S., Stollenwerk, K. G., Glynn, P. D., Bartholomay, R. C. & Knobel, L. L. 2004. Hydraulic and geochemical framework of the Idaho National Engineering and Environmental Laboratory Vadose Zone. *Vadose Zone Journal*, **3**, 6–34, https://doi.org/10.2136/vzj2004.6000

Nordstrom, D. K. 2012. Models, validation, and applied geochemistry: issues in science, communication, and philosophy. *Applied Geochemistry*, **27**, 1899–1919, https://doi.org/10.1016/j.apgeochem.2012.07.007

Norton, J. P. & Reckhow, K. H. 2008. Modelling and monitoring environmental outcomes in adaptive management. *In*: Jakeman, A. J., Voinov, A. A., Rizzoli, A. E. & Chen, S. H. (eds) *Developments in Integrated Environmental Assessment. Environmental Modelling, Software and Decision Support*. Elsevier BV, Amsterdam, 181–204.

Ostrom, E., Dietz, T., Dolsak, N., Stern, P. C., Stonich, S. & Weber, E. U. (eds) 2002. *The Drama of the Commons*. National Academy Press, Washington, DC.

Pahl-wostl, C. 2008. Participation in building environmental scenarios. *In*: Alcamo, J. (ed.) *Developments in Integrated Environmental Assessment. Environmental Futures: The Practice of Environmental Scenario Analysis*. Elsevier BV, Amsterdam, **2**, 105–122.

Pantea, M., Glynn, P. D., Phelps, G., Bawden, G. & Soller, D. 2013. Three-dimensional visualization and modeling at the US Geological Survey: examples and issues. *In*: Thorleifson, L. H., Berg, R. C. & Russell, H. A. J. (conveners) *Three-Dimensional Geological Mapping; Workshop Extended Abstracts*. Geological Society of America Annual Meeting, Denver, CO, 26 October 2013. Minnesota Geological Survey Open File Report OFR 13-2, 69–73. http://conservancy.umn.edu/bitstream/handle/11299/159772/3dmapping_workshop_ofr_13_02.pdf?sequence=1

PIZZUTO, J., SCHENK, E. R. *ET AL*. 2014. Characteristic length scales and time-averaged transport velocities of suspended sediment in the mid-Atlantic region, USA. *Water Resources Research*, **50**, 790–805, https://doi.org/10.1002/2013WR014485

PLUMMER, L. N. & GLYNN, P. D. 2013. Radiocarbon dating in groundwater systems. *In*: SUCKOW, A., AGGARWAL, P. K. & ARAGUAS-ARAGUAS, L. (eds) *Isotope Methods for Dating Old Groundwaters*. International Atomic Energy Agency, Vienna, 33–90.

PLUMMER, L. N., SANFORD, W. E. & GLYNN, P. D. 2013. Characterization and conceptualization of groundwater-flow systems. *In*: SUCKOW, A., AGGARWAL, P. K. & ARAGUAS-ARAGUAS, L. (eds) *Isotope Methods for Dating Old Groundwaters*. International Atomic Energy Agency, Vienna, 5–20.

POPPER, K. R. 1959. *The Logic of Scientific Discovery*. Harper & Row, New York (originally published in 1934 as *Logik der Forschung*).

POPPER, K. R. 1972. *Objective Knowledge: An Evolutionary Approach*. Oxford University Press, Oxford.

RIPPLE, W. J., ESTES, J. A. *ET AL*. 2014. Status and ecological effects of the world's largest carnivores. *Science*, **343**, 1241484, https://doi.org/10.1126/science.1241484

ROUSSEAU, J., LANDA, E. *ET AL*. 2005. *The Radioactive Waste Management Complex, Waste Area Group 7 Operable Unit 7-13/14, Idaho National Engineering and Environmental Laboratory, Idaho*. USGS Scientific Investigations Report 2005-5026.USGS, Idaho Falls, ID.

SALTELLI, A. & FUNTOWICZ, S. 2014. *When all Models are Wrong*. Issues in Science and Technology, Winter 2014, 79–85.

SANFORD, W. E. & POPE, J. P. 2013. Quantifying groundwater's role in delaying improvements to Chesapeake Bay water quality. *Environmental Science & Technology*, **47**, 13330–13338, https://doi.org/10.1021/es401334k

SAREWITZ, D. 2004. How science makes environmental controversies worse. *Environmental Science & Policy*, **7**, 385–403, https://doi.org/10.1016/j.envsci.2004.06.001

SCHMOLKE, A., THORBEK, P., DEANGELIS, D. L. & GRIMM, V. 2010. Ecological models supporting environmental decision making: a strategy for the future. *Trends in Ecology & Evolution*, **25**, 479–486, https://doi.org/10.1016/j.tree.2010.05.001

SCHWARTZ, B. 2004. *The Paradox of Choice*. Harper Collins, New York.

SHEIL, D. & BURSLEM, D. F. R. P. 2013. Defining and defending Connell's intermediate disturbance hypothesis: a response to Fox. *Trends in Ecology & Evolution*, **28**, 571–572, https://doi.org/10.1016/j.tree.2013.07.006

SIMON, H. A. 1990. Invariants of human behavior. *Annual Review of Psychology*, **41**, 1–19.

STANOVICH, K. E. 2010. *Rationality and the Reflective Mind*. 1st edn. Oxford University Press, New York.

STANOVICH, K. E. 2013. Why humans are (sometimes) less rational than other animals: cognitive complexity and the axioms of rational choice. *Thinking & Reasoning*, **19**, 1–26, https://doi.org/10.1080/13546783.2012.713178

STANOVICH, K. E. & WEST, R. F. 2003. Evolutionary versus instrumental goals: How evolutionary psychology misconceives human rationality. *In*: OVER, D. E. (ed.) *Evolution and the Psychology of Thinking: The Debate*. Gilhooly, K. J. (series ed.) *Current Issues in Thinking and Reasoning*, Psychological Press, Hove (UK) and New York, 171–223.

STERMAN, J. D. 1994. Learning in and about complex systems. *System Dynamics Review*, **10**, 291–330.

STERMAN, J. D. 2001. System dynamics modeling: tools for learning in a complex world. *California Management Review*, **43**, 8–25, https://doi.org/10.2307/41166098

STERMAN, J. D. 2002. All models are wrong: reflections on becoming a systems scientist. *System Dynamics Review*, **18**, 501–531, https://doi.org/10.1002/sdr.261

STEWART, T. 2001. Improving reliability of judgmental forecasts. *In*: ARMSTRONG, J. S. (ed.) *Principles of Forecasting: A Handbook for Researchers and Practitioners*. International Series in Operations Research & Management Science **30**, Springer, New York, 81–106.

SWEDISH NUCLEAR POWER INSPECTORATE (SKI) 1997. *SKI Site-94: Deep Repository Performance Assessment Project*. SKI Report 96:36, SKI, Stockholm

TALEB, N. N. 2007. *The Black Swan: The Impact of the Highly Improbable*. Random House, New York.

TUFTE, E. R. 1983. *The Visual Display of Quantitative Information*. 2nd edn. Graphics Press, Cheshire, CT.

TUFTE, E. R. 1990. *Envisoning Information*. Graphics Press, Cheshire, CT.

TUFTE, E. R. 1997. *Visual Explanations: Images and Quantities, Evidence and Narrative*. Graphics Press, Cheshire, CT.

TVERSKY, A. & KAHNEMAN, D. 1974. Judgment under uncertainty: heuristics and biases. *Science*, **185**, 1124–1131, https://doi.org/10.1126/science.185.4157.1124

UNITED KINGDOM MINISTRY OF DEFENCE 2013. *Red Teaming Guide*. 2nd edn. United Kingdom Ministry of Defence, Bicester.

VAN DEN BELT, M. 2004. *Mediated Modeling: A System Dynamics Approach to Environmental Consensus Building*. Island Press, Washington, DC.

VILLINSKI, J. E., O'DAY, P. A., CORLEY, T. L. & CONKLIN, M. H. 2001. In situ spectroscopic and solution analyses of the reductive dissolution of $Mn0_2$ by Fe(II). *Environmental Science and Technology*, **35**, 1157–1163.

VOINOV, A. & BOUSQUET, F. 2010. Modelling with stakeholders. *Environmental Modelling & Software*, **25**, 1268–1281, https://doi.org/10.1016/j.envsoft.2010.03.007

VOINOV, A. & GADDIS, E. J. B. 2008. Lessons for successful participatory watershed modeling: a perspective from modeling practitioners. *Ecological Modelling*, **216**, 197–207, https://doi.org/10.1016/j.ecolmodel.2008.03.010

VOSS, C. I. 2011*a*. Editor's message: groundwater modeling fantasies – part 1, adrift in the details. *Hydrogeology Journal*, **19**, 1281–1284, https://doi.org/10.1007/s10040-011-0789-z

VOSS, C. I. 2011*b*. Editor's message: groundwater modeling fantasies – part 2, down to earth. *Hydrogeology Journal*, **19**, 1455–1458, https://doi.org/10.1007/s10040-011-0790-6

Wikipedia 2015*a*. *The Oracle of Delphi*. http://en.wikipedia.org/wiki/Pythia [last accessed May 2015].

Wikipedia 2015*b*. *Boiling Frog*. http://en.wikipedia.org/wiki/Boiling_frog [last accessed May 2015].

Wikipedia 2015*c*. *List of cognitive biases*. http://en.wikipedia.org/wiki/List_of_cognitive_biases [last accessed May 2015].

Wikipedia 2015*d*. *Heuristic*. http://en.wikipedia.org/wiki/Heuristic [last accessed May 2015].

Wikipedia 2015*e*. *Known unknowns*. http://en.wikipedia.org/wiki/There_are_known_knowns [last accessed May 2015].

Wikipedia 2015*f*. *Socratic Paradox*. http://en.wikipedia.org/wiki/I_know_that_I_know_nothing [last accessed May 2015].

Wikipedia 2015*g*. *Oracular statements from Delphi*. http://en.wikipedia.org/wiki/List_of_oracular_statements_from_Delphi [last accessed May 2015].

Williams, B. K. & Brown, E. D. 2012. *Adaptive Management: The US Department of the Interior Applications Guide. Adaptive Management Working Group, US Department of the Interior*. Washington DC.

Williams, B., Szaro, R. C. & Shapiro, C. D. 2009. *Adaptive Management: The US Department of the Interior Technical Guide. Adaptive Management Working Group, US Department of the Interior*. Washington DC.

Winter, T., Harvey, J. W., Franke, O. L. & Alley, W. M. 1999. *Ground Water and Surface Water: A Single Resource*. USGS Circular 1139. USGS, Denver, CO.

Socio-hydrology modelling for an uncertain future, with examples from the USA and Canada

PATRICIA GOBER[1,2]*, DAVE D. WHITE[3], RAY QUAY[3], DAVID A. SAMPSON[3] & CRAIG W. KIRKWOOD[4]

[1]*School of Geographical Sciences and Urban Planning, Arizona State University, Tempe, AZ 85287-5302, USA*

[2]*Johnson-Shoyama Graduate School of Public Policy, University of Saskatchewan, 101 Diefenbaker Place, Saskatoon, SK S7N 5B8, Canada*

[3]*Decision Centre for a Desert City, Arizona State University, Tempe, AZ 85287-8209, USA*

[4]*W. P. Carey School of Business, Arizona State University, Tempe, AZ 85287-4706, USA*

**Correspondence: gober@asu.edu*

Abstract: Socio-hydrology brings an interest in human values, markets, social organizations and public policy to the traditional emphasis of water science on climate, hydrology, toxicology and ecology. It also conveys a decision focus in the form of decision support tools, stakeholder engagement and new knowledge about the science–policy interface. This paper demonstrates how policy decisions and human behaviour can be better integrated into climate and hydrological models to improve their usefulness for decision support. Examples from SW USA and western Canada highlight uncertainties, vulnerabilities and critical tradeoffs facing water decision makers in the face of rapidly changing environmental and societal conditions. Irreducible uncertainties in downscaled climate and hydrological models limit the usefulness of climate-driven, predict-and-plan methods of water resource planning and management. Thus, it is argued that such methods should be replaced by approaches that use exploratory modelling, scenario planning and risk assessment in which the emphasis is on managing uncertainty rather than on reducing it. Model fusion supports all of these processes in integrating human and biophysical aspects of water systems, allowing policy impacts to be quantified and clarified, and fostering public engagement with water resource modelling.

The limited impact of scientific information on policy making and climate adaptation in North America has raised awareness of the need for new modelling strategies and knowledge transfer processes (National Research Council 2010, 2012). The field of socio-hydrology responds to this need in the water sector by adding two dimensions: modelling strategies that treat humans, their activities, and policy decisions as an endogenous part of the water system; and research about how decision makers use scientific knowledge for policy making. This paper provides examples of a socio-hydrological perspective in water resources modelling from the Colorado River Basin (CRB) in the USA and the Saskatchewan River Basin (SRB) in Canada. After a brief discussion of the field of socio-hydrology, we provide an overview of societal and environmental change in the two basins, arguing that new modelling techniques and outreach efforts are needed for water resources management. Specifically, there is a critical need to integrate or fuse models that represent climatic, hydrological and ecological processes with models of social systems representing demographics, economics and policy processes. Such fused models are necessary to understand dynamic interactions and feedbacks between natural and social systems. The paper includes examples of new tools for, and approaches to, decision making under uncertainty (DMUU) for long-term water planning.

From hydro-sociology to socio-hydrology

Some 30 years ago the Swedish hydrologist Malin Falkenmark drew attention to the human dimensions of water science, noting the complex and dynamic interactions between human activities and the natural environment (Falkenmark 1979). Human effects on the water environment include withdrawals for irrigation, industrial and residential use, construction of water infrastructure to facilitate abstraction, hydropower production, regulation of river flow and land use modification. In turn, humans are affected by: water quantity and quality; floods, drought, erosion and landslides;

From: Riddick, A. T. Kessler, H. & Giles, J. R. A. (eds) 2017. *Integrated Environmental Modelling to Solve Real World Problems: Methods, Vision and Challenges*. Geological Society, London, Special Publications, **408**, 183–199.
First published online June 23, 2014, https://doi.org/10.1144/SP408.2

and ecosystem function. Falkenmark argued for social scientists to be more deeply involved in water planning, and christened this new field 'hydro-sociology'. She defined hydro-sociology to include studies of the way cultural, religious and social differences between regions introduce difficulties in the transfer of water knowledge and technology from one region to another.

In a recent commentary in *Hydrological Processes*, Sivapalan *et al.* (2012) underscored the importance of interactions between human and natural forces for water science and planning. Using the twentieth century development of the Murrumbidgee River Basin (a tributary of the Murray-Darling) in New South Wales, Australia, as an example, they drew attention to early upstream abstractions for irrigated agriculture, subsequent competition between human and environmental needs, and eventually the deterioration of river health and recognition that existing management strategies were unsustainable. Finally in 2007, there was government action to buy back high-value agricultural water rights upstream to protect downstream environmental assets. The authors argued that it would not have been possible to predict water cycle dynamics over decadal or longer time periods without considering the interactions and feedback among natural and human components of the water system. This feedback moved the system beyond critical tipping points into new, previously unobserved states. Sivapalan *et al.* used the term 'socio-hydrology' to describe the study of the 'co-evolution of human-natural coupled systems'.

The forgotten piece of Falkenmark's early call for collaboration between hydrologists and social scientists is the knowledge transfer process–the fact that evidence-based water planning requires not only knowledge about human–natural coupled systems but also about the processes that transfer this knowledge through social systems embedded in place-based contexts. There is growing evidence to suggest that this knowledge transfer process is not functioning well. A 2007 report by the National Research Council (of the USA National Academy of Sciences) declared that 'inadequate progress has been made in synthesizing research results assessing impacts on human systems, or providing knowledge to support decision making and risk analysis' (National Research Council 2007*b*, p. 34). For example, recent advances in forecasting El Niño Southern Oscillation (ENSO) events provide information that is potentially usable for climate-sensitive sectors such as agriculture, water and disaster response, but advances have not been utilized to their full potential (Miles *et al.* 2006, p. 19616). Reasons presented for this include inflexible decision rules, informal arrangements that prefer established and tested practices over innovative prediction tools, organizational culture and reward structures, risk averse and vulnerable cultural contexts, lack of meaningful interaction between scientists and decision makers, hard to interpret presentation of scientific information, cascading uncertainties when climate model results are used for regional and local prediction, and user difficulty in translating probability information into action. Dilling & Lemos (2011) emphasized the importance of two-way, iterative engagement between producers and users of scientific information to build trust and better understand the needs of policy and what scientists can provide to assist policy making. Cash *et al.* (2003) found that credibility (whether the information meets rigorous scientific standards), salience (whether it is usable for decision making), and legitimacy (whether it is developed and disseminated in a fair and transparent way) are strong determinants of whether information is used by policy makers. Clark & Clarke (2011) highlighted the importance of active negotiation processes to support the creation of usable scientific knowledge as well as social networks between researchers and decision makers. Crona & Parker (2011) found that policy makers who have greater social interactions with researchers were more likely to utilize scientific information to govern water resources.

Pahl-Wostl & Borowski (2007) noted similar problems in implementing the requirement for participatory water management as required by the European Union Water Framework Directive. Despite claims of usability and problem solving by scientists and the European Commission, many of the tools designed for decision support did not meet user needs (Borowski & Hare 2007). The reasons involved fundamental misunderstandings between the science and user communities about a range of modelling and decision support issues. Recommendations emphasized the processes of social learning, which involves two-way exchange of knowledge and co-production of water models.

Environmental and societal change in the Colorado and Saskatchewan River Basins

Climate change presents formidable challenges to water managers in southwestern USA and western Canada. Arid and semi-arid lands in these regions experience inherently highly variable interannual steamflows. Thus, their water management practices have traditionally focused on managing seasonal and interannual climatic variability, which can be estimated based on historical data. The effects of potential long-term climate change, where uncertainties are high and the historical record is less useful, are only beginning to be integrated into management and planning processes. In addition, there is increasing pressure on water systems from population growth, urbanization, land intensification and economic development. The uncertainties associated with societal management of these processes are equally difficult to integrate into traditional modelling and management practice.

Spanning a length of 2 330 km from the Central Rocky Mountains to the Gulf of California (Sea of Cortez), the Colorado River drains an expansive, arid watershed of 640 000 km^2 (Fig. 1). It is a lifeline providing water for large-scale urban growth (Los Angeles, Phoenix and Las Vegas), irrigated agricultural production, recreation and hydropower

Fig. 1. Colorado River Basin. Source: Decision Center for a Desert City, Arizona State University.

generation. A 2007 National Research Council Report, *Colorado River Basin Water Management: Evaluating and Adjusting to Hydroclimatic Variability,* reported significant warming over the past century. Climate model projections indicate that temperatures in the CRB will continue to rise in the foreseeable future (National Research Council 2007*a*). Higher regional temperatures are shifting snowmelt, and thus runoff, to earlier in the spring and increasing water demand, especially during the summer. A 2010 issue of the *Proceedings of the National Academy of Sciences of the United States of America* devoted to the future of the CRB pointed to increasing aridity, more intense and frequent droughts, and increased forest and woodland mortality due to fires and pathogenic outbreaks (MacDonald 2010).

Coupled with environmental change, the CRB has experienced rapid population growth and economic development. Between 1960 and 2009, population in the seven-state CRB region (Wyoming, Utah, Colorado, New Mexico, Arizona, Nevada and California) grew by more than 166.4% compared with 77.2% for the USA as a whole. Arizona grew by a factor of four and Nevada by a factor of eight, while California added more than 21 million new residents during that period (US Census Bureau 2011). The growth imperative in the western USA has traditionally assumed that water supplies would always be available to support an expanding population. The reality, however, of an over-allocated basin is now apparent. The river was apportioned in the 1922 Colorado River Compact, using early twentieth century flows that have since been shown to be the product of unusually wet years. Over-allocation is manifest in increasing competition for water supplies, urban and agricultural conservation efforts, emerging water markets, and the transfer of water rights from agricultural to urban users (National Research Council 2007*b*). A 2012 report by the USA's Bureau of Reclamation projected a long-term imbalance in supply and demand of up to 3.2 million acre feet (3.96 km^3) in the Colorado River Basin by mid-century without additional future management actions (Bureau of Reclamation 2012*a*).

These climate–population dynamics are similar in the SRB, albeit on a somewhat smaller scale and in a different cultural and policy context. The SRB covers 405 864 km^2, encompassing a large portion of western Canada (Fig. 2). It is responsible for 75% of Canada's irrigated agriculture and is a major global exporter of food (Agriculture and Agri-Food Canada 2011). Headwaters in the Canadian Rocky Mountains in Alberta provide some 80% of the SRB's total runoff (Pomeroy *et al.* 2005). The Saskatchewan River's two major tributaries flow east from the continental divide. The South Saskatchewan River passes through the Canadian Prairies, a region of high natural climate variability. The North Saskatchewan River passes through Prairie landscapes and Boreal forest (Partners FOR the Saskatchewan River Basin 2009; Toth *et al.* 2009). Like the Colorado, the Saskatchewan River is experiencing rapid environmental change. A warming climate is causing mountain glaciers to retreat, changing the rain/snow balance and the processes of snow accumulation and melt, and hence influencing the magnitude and timing of river flows (Comeau *et al.* 2009; DeBeer & Sharp 2009; Moore *et al.* 2009). Changing climate also has caused a mountain pine beetle infestation resulting in widespread forest loss in British Columbia moving eastward into the SRB (Natural Resources Canada 2012).

The population of the SRB is only around three million, the majority concentrated in the Province of Alberta, with 2.5 million people living in the Edmonton to Calgary Corridor (Partners FOR the Saskatchewan River Basin 2009). Water is fully allocated in three of Alberta's most densely populated sub-basins (Bow, Oldman and South Saskatchewan). Alberta Environment ceased granting new licences for water abstractions in these three sub-basins in 2003. Rapid growth in Alberta and downstream in Saskatchewan has stressed the river's capacity to support additional population growth, irrigated acreage and natural resource development, and compromised water quality in many places (Schindler & Donahue 2006). Alberta has introduced water markets as a mechanism to reconcile competing demands; Saskatchewan is in the process of developing a 25-year water plan to address inevitable conflicts about future use among irrigators, hydropower suppliers, mining companies, municipal water providers, First Nations communities and environmentalists (Alberta Environment 2006; Saskatchewan Water Security Agency 2013).

Consequences for management and decision making

Rapid societal and environmental changes present significant challenges for water management, especially if Sivapalan *et al.* (2012) are right and the interactions between biophysical and social forces cause systems to shift to previously unobserved states. Also challenging from a decision making perspective are the uncertainties associated with these biophysical and social forces (Camacho 2009). Milly *et al.* (2008) drew attention to the distinction between uncertainty and variability in their discussion of the stationarity assumption in water resources management. Stationarity is the idea that natural systems function over time within a known

Fig. 2. Saskatchewan River Basin. Source: Global Institute for Water Security, University of Saskatchewan.

envelope of variability. Water resource management based on stationarity presumes that hydrological variables such as annual streamflow or peak flow have probability density functions that adhere to the observed historical record. See, for example, annual streamflows on the Colorado (Fig. 3a) and Saskatchewan Rivers (Fig. 3b). Applying these probability functions traditionally has served as the basis for training and practice in water engineering and management. Water decision making in the two basins has evolved to manage high seasonal and interannual variability but within the historical envelope of variability. Milly *et al.* assert, however, that 'stationarity is dead' because climate change is inexorably altering the means and extremes of precipitation, evapotranspiration and rates of river discharge. We have therefore entered a new era when the probability density functions based on historical records are no longer adequate to manage risk and plan for the long term.

Trenberth (2010) warned that this period of deep uncertainty is unlikely to end in the near future. As knowledge of the climate system has increased, so also has the scientific understanding of factors that were previously unaccounted for in global models, such as the release of greenhouse gases from melting permafrost, the fertilizing effect of atmospheric carbon dioxide on vegetation, and the effects of aerosols on clouds. As climate models begin to incorporate these processes, they will inevitably disagree about the nature, extent and geographic patterns of climate change impacts. Wilby & Dessai (2010) have described a 'cascade of uncertainty' with each stage in the climate impact assessment process–from imagining what a future society will look like, reconciling the different results of the climate models, downscaling climate results to regional and local levels, to assessing impacts and making adaptation decisions.

There is sociological evidence that the water sector is not adequately prepared to deal with this new cascade of uncertainties. Lach *et al.* (2005) identified cultural forces within water organizations that impede change and prevent innovation. In 120 semi-structured interviews with representatives of the Columbia River system of the Pacific Northwest, the Metropolitan Water District of southern California and the Potomac River Basin/Chesapeake

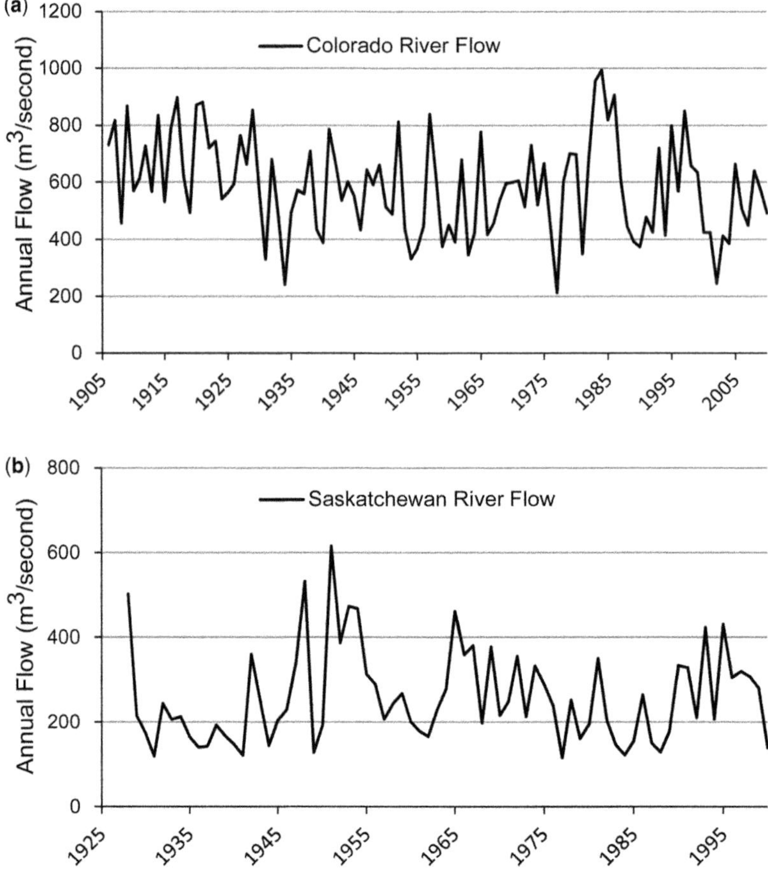

Fig. 3. (**a**) Annual flows on the Colorado River, 1906–2010; (**b**) Annual flows on the Saskatchewan River at the Alberta–Saskatchewan border, 1928–2000.

Bay in the Washington DC area, they found agencies geared toward conservative decision making and slow to accept new practices. The work culture and organizational practices have evolved to make an inherently unreliable resource more reliable through infrastructure and redundancy. Management within this paradigm relies heavily on 'craft' skills that are acquired through long, place-based experience. New employees are mentored in the organization until they are fully imbued with its outlook and practices. Emphasis is on the long-term agency mission – to provide a reliable resource to customers. Water organizations have advanced without serious public engagement because they have associated public attention with unreliability and system failure. The climate adaptation community argues, however, that just such attention and intervention is crucial to ensuring that public values are included as communities make difficult choices about how water is to be redistributed in a fully allocated system, what environmental standards are to be met, what adaptation strategies are best, and how much risk can be absorbed in a non-stationary environment (Brunner 2010; Bierbaum *et al.* 2013).

Also relevant to the decision making capacities of water organizations are uncertainties about the human dimensions of water systems including: lifestyles, such as preferences for lawns and pools; economic and population growth; institutional capacity; and environmental policy. In a study of water managers in Central Arizona, White *et al.* (2008) found that respondents thought that climate change was worthy of their consideration, but lamented the lack of usable scientific information compatible with decision processes and timelines. They also were concerned about the status of Native American water claims, endangered species designations, environmental permitting processes, and the nature and pace of economic growth (see, for example,

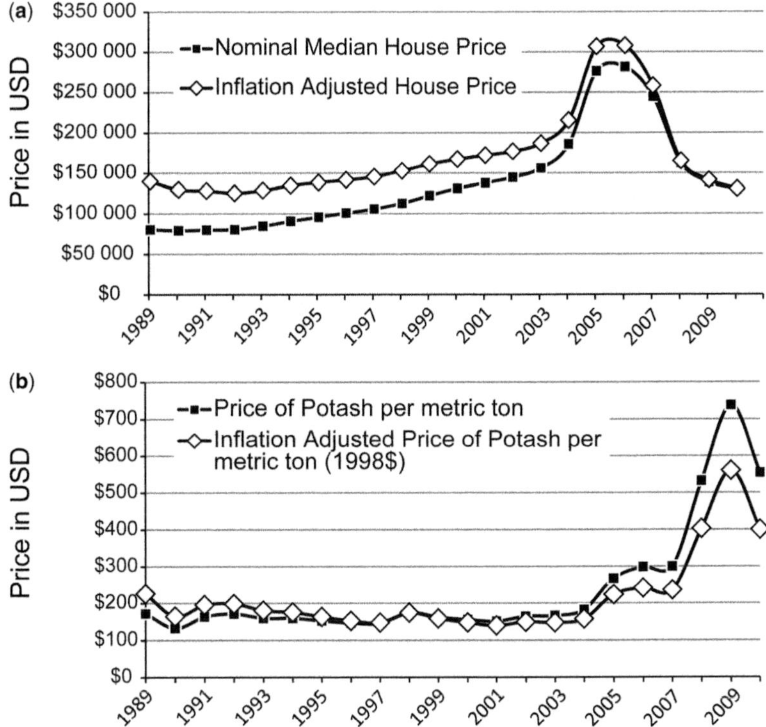

Fig. 4. (**a**) Phoenix Metropolitan Area house prices for October, 1989–2010; (**b**) Potash price per unit value, 1989–2010.

the variability in Phoenix area home prices in Fig. 4a). Similarly in Saskatchewan, where royalties from potash sales average 20% of provincial tax revenues, uncertainties about global resource markets, energy prices and trade policy are highly relevant to water decision makers (Fig. 4b).

Both the CRB and SRB face the similar challenge of coordinated basin-wide climate adaptation with highly fragmented and decentralized systems of water governance. Central Arizona is not governed by a single regional water authority but by 40 large water providers and 80 small ones, each maintaining its own supply portfolio, making its own water conservation plans, and responding to customers in its own service area (Gober 2007). There are fewer suppliers in the SRB, but regulation is dispersed over many layers of government from the international (a small portion of the basin extends beyond the USA/Canadian border into Glacier National Park in Montana) and national, down to provincial, regional and local levels, including First Nations communities. For example, current water quality problems occur on small First Nations reserves where water quality is the responsibility of the national government but supervision of source lands (e.g. fertilizer use on agricultural lands) is in the hands of provincial and local authorities (Patrick 2011).

Modelling experiments

The need to integrate the human and biophysical aspects of water systems, address the inherent uncertainties in these systems, and provide effective decision support tools requires new frameworks for integrated water resource modelling (Gober *et al.* 2010; Wilby & Dessai 2010; Gober 2013). These frameworks need to be less focused on predicting and optimizing, and more oriented toward exploring the future. Exploratory modelling, scenario analysis and stakeholder engagement are proven tools for DMUU (Lempert *et al.* 2003). These tools facilitate stakeholder discussion of difficult choices that lie ahead and enable vulnerability assessments, sensitivity analyses and robust decision making, leading to choices that have acceptable outcomes across a range of future climate and societal conditions. Exploratory modelling does not aim to perfectly replicate the water system or predict in the face of non-stationarity. Rather, it aims to characterize essential features, identify critical weaknesses and support policy discussion.

With these goals in mind, this section reports on modelling efforts to explore system vulnerabilities of, and policy options for, socio-hydrological systems in Central Arizona and western Canada. Efforts draw attention to key vulnerabilities and highlight policy options that can reduce them.

Vulnerability at the urban fringe of Phoenix

Groundwater is heavily regulated in Arizona's urban regions because of a history of severe overdrafts. Arizona's 1980 Groundwater Management Act granted groundwater rights to cities, towns and private providers with the expectation that they would reduce overdrafts and achieve 'safe yield' (i.e. withdrawals equal natural plus artificial recharge) by 2025 (Arizona Department of Water Resources 2012). According to policy, new sub-divisions are not allowed to proceed unless developers demonstrate that they have an assured 100-year water supply or are located in a provider service area that has an assured supply. The 1980 Groundwater Management Act assigned providers an initial allocation of groundwater based on their use in 1980; any future overdraft use was supposed to be phased down to zero in 30 years. The 100-year assured supply can be achieved through renewable surface water and groundwater, and effluent credits. Additional groundwater credits are earned from recharging excess surface water into an underground storage facility. Incidental recharge, primarily from outdoor water use, enters the vadose layer and also adds to the providers' annual groundwater credits. Treated effluent can also be stored as groundwater and used later to augment water supplies.

The core cities of metropolitan Phoenix are not expected to grow as fast as outlying suburban communities during the next 20 years. If present trends continue, the majority of regional growth will occur in urban-fringe communities where land is available for low- and mid-density property development (Maricopa Association of Governments 2007). Official projections assume that communities have adequate water supplies to meet projected growth. Arizona's 100-year assured water supply (AWS) rule concerning sub-division platting is assessed every 10 years and all but a few communities have been able to meet the AWS rule. It is not clear, however, that situation will pertain in the future if growth unfolds as projected. We used WaterSim 5.0 to explore whether each community has a 100-year assured water supply under historical climate conditions, in other words, will they meet the AWS rule 20, 40, 60, 80 and 100 years from now.

WaterSim 5.0 represents a modification and broader implementation of an earlier version developed and modified over time by Decision Centre for a Desert City at Arizona State University (Gober *et al.* 2011; Sampson *et al.* 2011) (Fig. 5). It consists of several sub-models that work together to characterize water supply and demand in the Phoenix region. WaterSim 5.0 is written as modules in FORTRAN controlled by a user interface written in Microsoft C#. It uses a mass balance approach to track regional-, local- and provider-level estimates of surface water, groundwater, and water demand and use. Difference equations are used to represent states and rates of water supply and water demand on an annual timestep, although some driver variables operate at finer temporal resolutions. The current model focuses on urban water demand (indoor and outdoor) using provider-specific representations of the water supply. It includes water treatment facilities, and residential, commercial and industrial water use from supply source to waste treatment and water reuse. The model uses estimates based on historical data to model reservoir operations for the three surface water sources (Salt, Verde, and Colorado Rivers). Water rights for 10 local providers are used within the model to allocate water from the Salt and Verde systems immediately upstream from Phoenix. Groundwater estimates from the Arizona Department of Water Resources (ADWR) are used to initialize available provider-level credits. Annual provider pumping credits are determined from groundwater rights.

Using WaterSim 5.0, 100-year simulations of normal surface water conditions (replication of actual past climate conditions projected into the future) revealed four patterns of groundwater management exemplified by the communities of Buckeye, Paradise Valley, Scottsdale and Phoenix (Fig. 6).

(1) Buckeye is an urban-fringe community slated for rapid growth and heavily reliant on groundwater. Buckeye's 2010 population was 74 906; official projections in 2010 show Buckeye growing to 218 591 in 2020 and 419 146 in 2030. Model results suggest that it has sufficient groundwater credits to support growth only until 2040 when it may no longer be able to meet the AWS rules. Thus, the community enters a period of high uncertainty about climate, growth and regulation with little flexibility to deal with unanticipated conditions.

(2) The Town of Paradise Valley is served by three providers, the municipalities of Phoenix and Scottsdale along with Arizona American, a private provider that manages a number of small systems throughout the region. The Town of Paradise Valley has little land

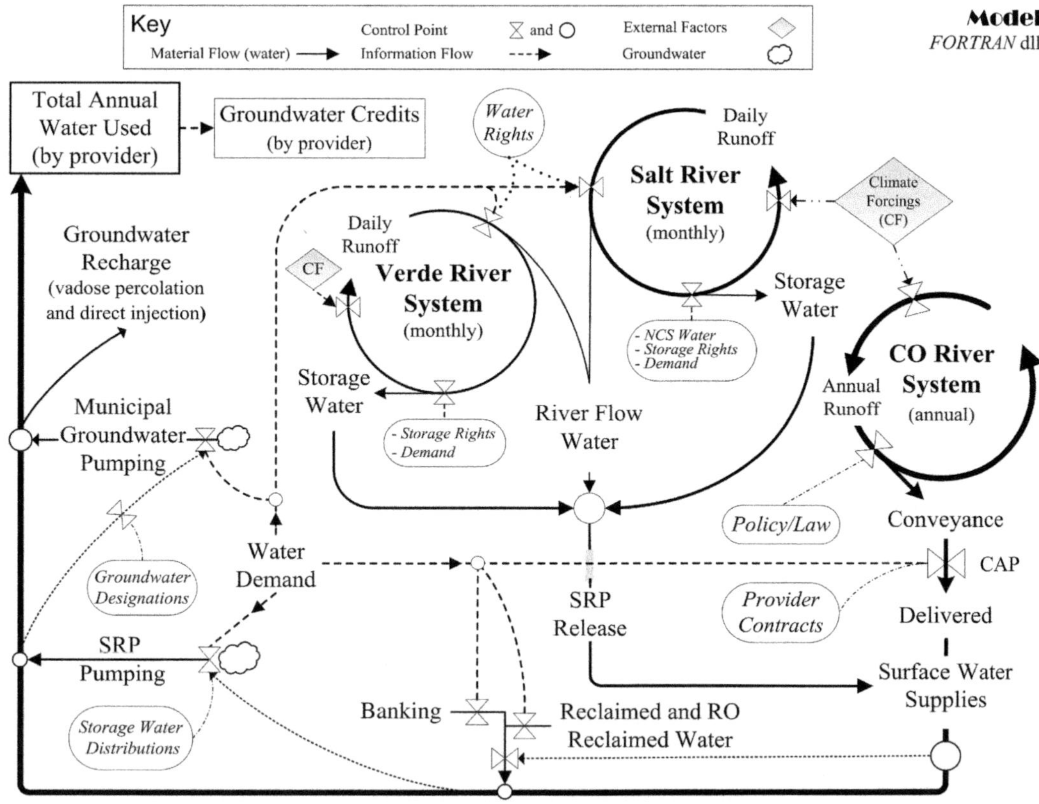

Fig. 5. WaterSim 5.0. CAP, Central Arizona Project; NCS, New Conservation Space; RO, reverse osmosis; SRP, Salt River Project.

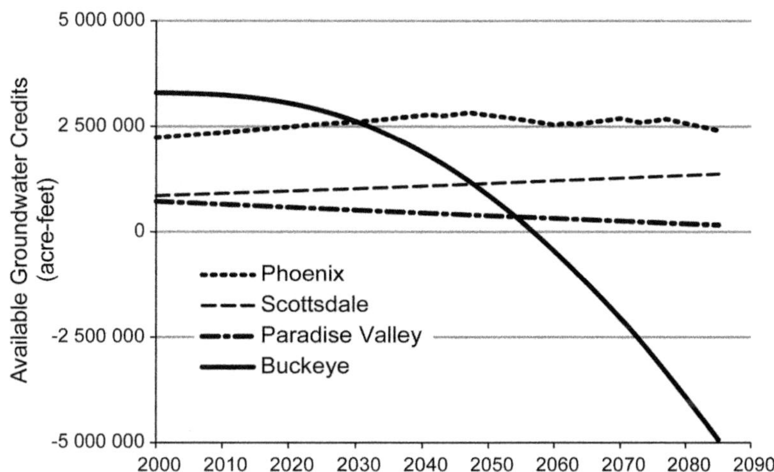

Fig. 6. Available assured water supply groundwater credits for four Phoenix metropolitan cities, assuming no climate change.

available for new growth, and given its one-house-per-acre zoning code, is not likely to experience high-density development in the foreseeable future. Though the Arizona American system currently meets the official mandate for a 100-year assured supply for existing development, WaterSim 5.0 projects that the Town of Paradise Valley will experience a slow decline in groundwater credits indicating that its groundwater supply may not be sustainable in the long term. Thus, toward the end of the 100-year period, barring no changes in technology or prevailing lifestyles, the town's population would not be able to absorb growth and still meet AWS rules.

(3) Scottsdale's groundwater credits are moving in a positive direction, replenishing groundwater by recharging excess surface water from the Colorado River and recycling effluents to the aquifer. Recognizing the limits of its surface water supplies, Scottsdale constructed its own water reclamation plant, known as the Water Campus, in 1998 and uses treated effluent to irrigate turf lawns on its 23 local golf courses and for groundwater recharge.

(4) Phoenix, the region's largest city with population estimate of 1.5 million residents in 2011 (US Census Bureau 2012) has one of the most robust surface water supplies in the region and thus exhibits a stable pattern of groundwater use. Using current population projections to 2.1 million residents in 2030, historical climate conditions and the city's supply portfolio, WaterSim 5.0 projects that the city's groundwater credits will be stable across the next 100 years, fluctuating up and down as groundwater is used as a buffer to supplement surface supplies in dry years.

Following this analysis that assumes the climate remains stationary for the next 100 years, a subsequent study explored the effects of reduced surface water allocation, resulting from climate change. Specifically, we focused on the geographical consequences of enforcing the groundwater code each year to disallow development when a community lacks a 100-year sustainable supply of groundwater for the current population. The implications of redistributing future population growth from communities with insecure to secure supplies were simulated. The analysis assumed that regional population growth will occur at projected levels, but would be spatially redistributed from deficit to surplus communities to maintain safe yield for all water providers.

We tested this policy under three different climate change assumptions. There are three aspects of surface water flow related to precipitation changes resulting from potential climate change: (1) the changes in the average river flow over time (modelled for 75 years); (2) the pattern of variation (wet to dry) in interannual river flow change; and (3) the differences (or similarities) of change over time between the Colorado River and the Salt/Verde sub-basins immediately north of Phoenix. We created scenarios of future climate change for (1) and (3) by creating combinations of river flow change for the Colorado River (from 40% decline to 20% increase by 2085, four scenarios) and from Salt and Verde Rivers, (80% decline to 20% increase by 2085, six scenarios), producing 24 scenarios. These ranges reflect the extent of change represented by the various Global Climate Models that have been downscaled to surface flow hydrology for the region (Ellis *et al.* 2008). We then addressed (2) by using three different traces from the historic records that showed low (dry), normal and high flow (wet) conditions which, when combined with the other scenarios, produced a total of 72 scenarios. We ran WaterSim 5.0 using each scenario and recorded the net shift in population for each city (i.e. how much potential population growth was lost and how much additional potential population growth was gained by each city as a result of implementing the AWS policy). Then, the net change across the 72 scenarios was averaged for each city to summarize the simulation results.

From a market development perspective, these simulations explored how growth may be distributed in favour of communities with adequate water supplies. From a policy perspective, it explored how future population growth and land development patterns could be manipulated to reduce the risk of future shortage in any one or set of providers. Although such an approach is politically unlikely, it illustrates the vulnerabilities of current growth patterns that entail unsustainable groundwater use. Simulations revealed that a sizable portion of metropolitan population growth would be redistributed away from rapidly growing westside communities in favour of the urban core and eastside communities (Fig. 7).

The simulations showed that the current decentralized system of water governance and management, if coupled with lax enforcement of the groundwater code, would allow some cities to benefit from short-term land development while transferring the long-term risks of future water shortage to neighbours through unsustainable groundwater practices. They also demonstrated that stricter state-level enforcement of existing policy would enable continued growth (albeit in a different form), reduce the overall vulnerabilities of water shortage, reward communities for sustainable water practices, and bring land development and water policies into conformance.

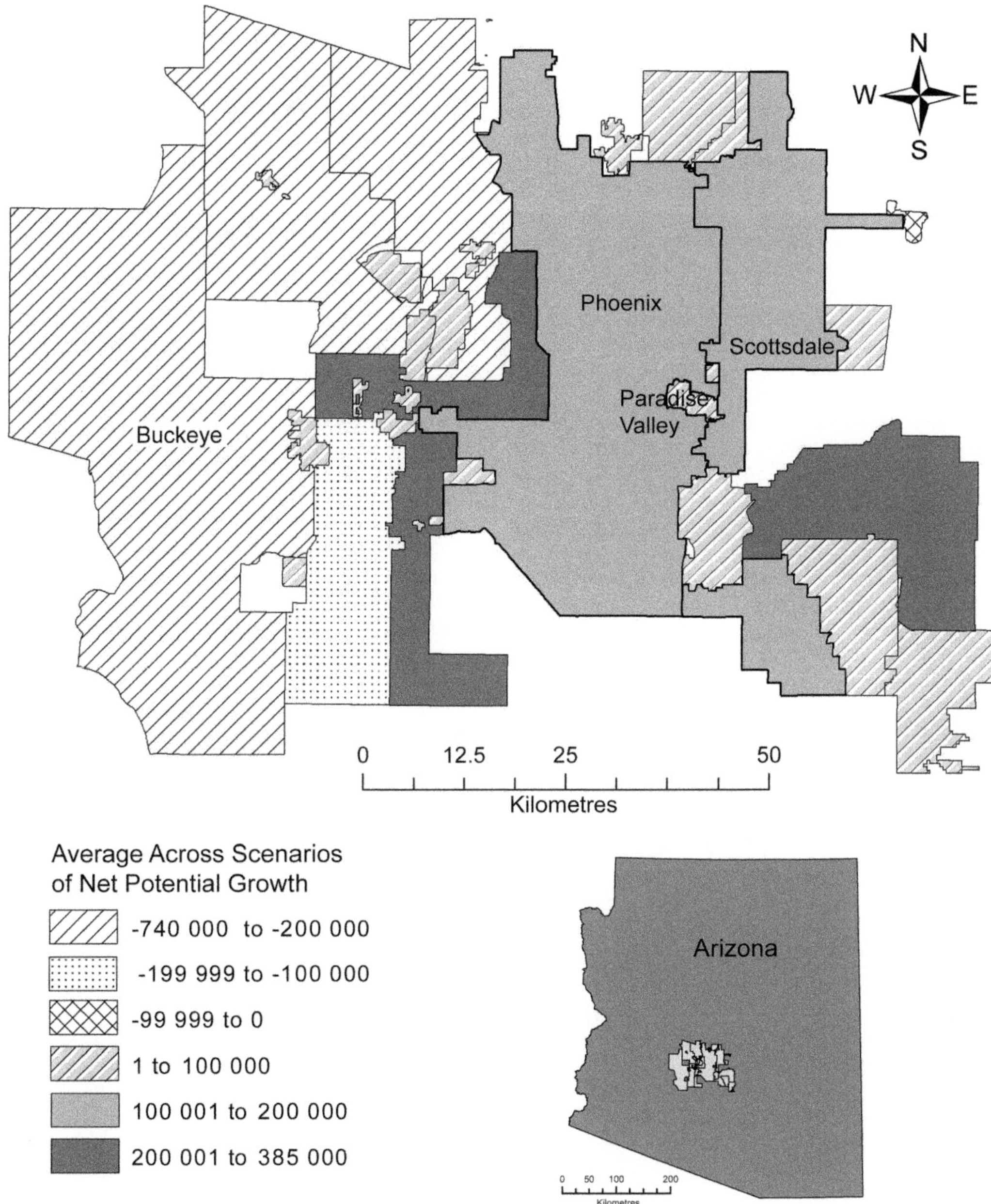

Fig. 7. Changing urban growth pattern (average net potential growth across all climate scenarios) in the Phoenix Metropolitan Area if water-short cities lost their 100-year assured water supply designation.

Lake Mead stress test

A second modelling experiment used WaterSim 5.0 to investigate the impact of uncertain hydroclimatic conditions on the ability of the Bureau of Reclamation to honour historical water compacts. The Colorado River is managed by a series of agreements, compacts, federal laws, decrees and regulatory guidelines known jointly as the 'Law of the River' (Bureau of Reclamation 2012*c*). In 1922, the seven states in the Basin agreed to separate it into an Upper (Wyoming, Colorado, Utah, and

New Mexico) and Lower (California, Arizona, and Nevada) Basin, with each group of states assigned 7.5 million acre feet (9.25 km^3) of annual withdrawal. Upper Basin states agreed not to deplete the flow below 75 million acre feet (92.5 km^3) across 10 consecutive years. Apportionment of the Lower Basin states' share occurred with the Boulder Canyon Project Act of 1928 (Bureau of Reclamation 2012*c*). Assignments were 4.4 million acre feet (5.43 km^3) to California, 2.8 million acre feet (3.45 km^3) to Arizona and 0.3 million acre feet (0.370 km^3) to Nevada. Lake Powell is operated to help the Upper Basin meet its obligation to the Lower Basin states. Lake Mead, created by the construction of the Hoover Dam, holds water to supply deliveries to the Lower Basin states (Fig. 1).

Early twenty-first century drought conditions raised awareness that it may not be possible for the river to provide the assumed water flows under sustained drought conditions. In 2007 the three Lower Basin states agreed to a Shortage Sharing Agreement under which the Secretary of Interior and Bureau of Reclamation would manage allocations if levels in Lake Mead fell below preset standards (Bureau of Reclamation 2012*b*). Guidelines were described as interim because they extend through 2026, and are intended to allow the system operators to gain experience with low reservoir conditions. The agreement specifies three levels of shortage:

(1) Light shortage. When the surface elevation at Lake Mead is below 1 075 feet (328 m) relative to mean sea level but above 1 050 feet (320 m), the Lower Basin states will be reduced from 7.5 million acre feet to 7.167 million acre feet (8.84 km^3) per year: 4.4 million acre feet (5.43 km^3) to California, 2.48 million acre feet (3.06 km^3) to Arizona and 287 000 acre feet (0.354 km^3) to Nevada.

(2) Heavy shortage. When the surface elevation of Lake Mead is below 1 050 feet (320 m) but above 1 025 feet (312 m), only 7.083 million acre feet (8.74 km^3) per year will be delivered to the Lower Basin states: 4.4 million acre feet (5.43 km^3) to California, 2.4 million acre feet (2.96 km^3) to Arizona and 283 000 acre feet (0.349 km^3) to Nevada.

(3) Extreme shortage. The most severe shortage considered in the interim guidelines is when the level of Lake Mead drops below 1 025 feet (312 m), in which event 7.0 million acre feet (8.63 km^3) per year will be delivered to the Lower Basin states: 4.4 million acre feet (5.43 km^3) to California, 2.32 million acre feet (2.86 km^3) to Arizona and 280 000 acre feet (0.345 km^3) to Nevada.

As these guidelines show, the largest burden of reduction falls to Arizona which draws Colorado River water through its Central Arizona Project (CAP). The CAP is a 336-mile (541 km) diversion canal that conveys water from the Colorado River along the western boundary of the state to municipal and agricultural users in Central Arizona (Fig. 1). In 1968, Arizona agreed to make its water rights junior to those of California in order to secure support from California's powerful congressional delegation for a federal loan to construct the CAP. The agreement was signed before climate change was under serious discussion or flows on the Colorado River were forecast in the scientific literature. Therefore, the potential for long-term shortfalls seemed small. However, Lake Mead in June 2010 was at 39% of capacity, and on 30 November 2010, it reached 1 081.94 feet (330 m) in elevation, setting a new record low (Fig. 8). Although Lake Mead has rebounded since, the low levels in 2010 triggered serious discussion about the effects of extended drought on the Colorado River and what the 2007 Shortage Sharing Agreement would mean for Central Arizona.

The second simulation experiment investigated how Lake Mead water levels would vary with

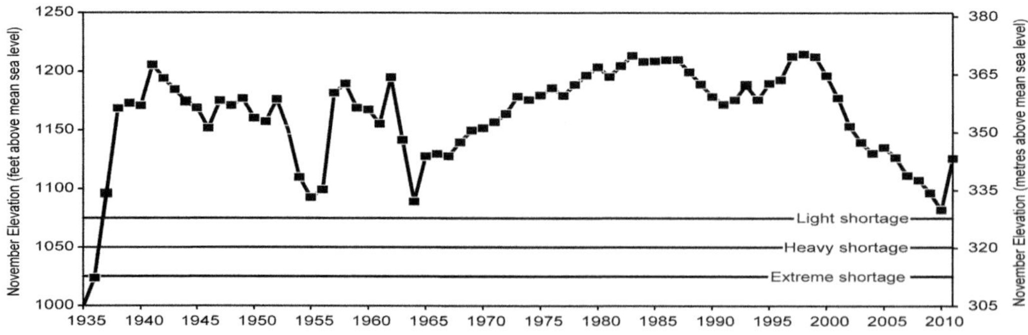

Fig. 8. Lake Mead elevations in November, 1935–2011.

flows reconstructed from prehistoric climate conditions in the Colorado River Basin that include periods of sustained drought. Meko *et al.* (2007) reconstructed Colorado River flows from AD 762 to 2005 using tree ring data. For our simulation experiments, four 30-year traces (historical sequences) from the paleo-reconstruction were used to 'stress test' Lake Mead and address potential cutbacks in CAP water supplies under a wider range of possible future climate conditions than have been seen since streamflow record-keeping began in 1906. The four traces represent a variety of different climate conditions: (1) starting in AD 1148, the first represents persistent drought–low flow and low variance conditions (this trace is often referred to in the SW as the 'Medieval drought' period); (2) starting in AD 1406, the second contains persistently wet years and is characterized by low variance; (3) starting in AD 1471, the third represents dry conditions with high variance; and (4) beginning in AD 1837, the fourth represents wet years with high variance.

Under Medieval drought conditions (AD 1148 trace), assuming similar interannual patterns and variability as found in the tree-ring record, Lake Mead would be largely unusable after 2040 (Fig. 9). The persistently wet trace (AD 1406 AD) keeps reservoir levels above the threshold that would trigger reduced allocations to CAP, but only marginally. The wet trace with high variance (AD 1837 trace) produces larger fluctuations and flirts with the 1 075-foot (328 m) mark but never triggers a reduction in Arizona's 1928 apportionment of 2.8 million acre feet (3.45 km^3). The dry trace with high variability (AD 1471 trace) provides very large swings in lake levels, plunging levels below the official guidelines for light, heavy and extreme shortage conditions for extended periods.

These results demonstrate that Central Arizona's water supply from Lake Mead and the Colorado River is highly sensitive to climate conditions. Even without climate change, the region would lose its share of Colorado River supplies during periods of persistently dry conditions that have been experienced in the past. Even intermittently dry conditions that have been experienced in the past would cause periodic shortages which would prove problematic for a highly urbanized water system with demand that would be hard to reduce. These results demonstrate that the 1968 agreement to assume junior water rights behind California and the Shortage Sharing Agreement could involve considerably more risk for Central Arizona's water supplies than previously thought. The results highlight the importance of considering scenarios in which Colorado River supplies are significantly reduced and developing regional-wide coordination and cooperation that could mitigate risk and reduce shortage under these conditions. As noted earlier, each provider has a unique portfolio of supply and thus is differentially exposed to Basin-wide climate conditions.

Decision tradeoffs in Alberta

Alberta Environment in Canada conducted an exploratory, scenario-based modelling experiment to evaluate the potential consequences of two types of policy decisions: increasing irrigated acreage; and restoring environmental flows to levels that protect biodiversity and restore healthy aquatic ecosystems (Alberta Environment 2003).

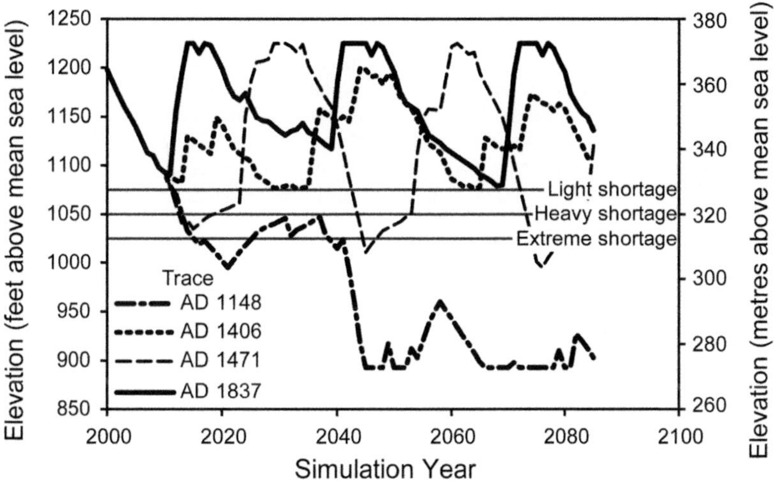

Fig. 9. Simulation results of Lake Mead water levels using four 30-year traces (historical sequences).

The importance of these policy issues came to light during a period of sustained drought in 2001–2002 which has been characterized as western Canada's worst natural disaster since the 1930s (Wheaton *et al.* 2008). Crop production in Alberta plummeted, the scarcity and cost of feed forced livestock producers to cull their herds, and agricultural communities suffered declining revenues. Low river flows exposed a host of environmental problems stemming from agricultural intensification, resource exploitation, industrial development, forestry, dam and reservoir construction, hydroelectric development, urbanization and population growth. Drought conditions stimulated an expanded public discussion of whether there is adequate water available to support continued agricultural development and what would be the consequences of restoring environmental flows to levels that protect the environment during drought conditions.

Although Alberta sits upstream from Saskatchewan and Manitoba and is the source for 80% of the Saskatchewan River's flows, its water use is constrained by an allocation agreement similar to the Colorado River Compact that is called the Prairie Province Master Agreement on Apportionment. Approved in 1969, the Agreement requires Alberta to pass 50% of the annual natural flow (i.e. flow that would occur in the absence of anthropocentric interventions in the flow regime) to Saskatchewan, which is in turn is required to pass 50% of that flow to Manitoba. Alberta, like the CRB, adheres to the principle of 'first-in-time-first-in-right' for water allocation. This principle assigns priority to use on the basis of where the water was first put to beneficial use. The 2001–2002 droughts led naturally to questioning of whether there is enough water under full allocation in this environment of high variability to meet the needs of all users in all years and how the restoration of environmental flows would impact water available to senior and junior licensees.

Alberta investigated these questions using its Water Resources Management Model (WRMM). WRMM is a steady-state, water resource model that optimally allocates the water to competing demands based on the system's physical boundaries and state, water supply and operational policies. The WRMM has been extensively used in western Canada to characterize the response of water resource systems to different natural flow conditions and/or operational alternatives (e.g. Alberta Environment 2010; Nazemi *et al.* 2013). Inputs to the model are the historical or projected streamflow series, demands, reservoir capacities, licence priorities, operational policies, instream flow needs and the apportionment requirements.

In the WRMM, the water resource system is represented as an interconnected network of nodes and links. Nodes represent storage and junctions in the system, and links represent paths and flow directions to achieve ideal levels and satisfy all water demands under normal flow conditions. The policies are based on attempting to achieve ideal release and storage to meet water demand at all locations. Water allocation priorities are modelled in the WRMM by penalty (cost) values, which show the cost of not meeting each water demand, with a higher penalty for not meeting a higher priority water allocation.

Under ideal supply conditions, where there is sufficient supply to meet all water demands, the total penalty calculated by the WRMM for the system is zero. During water shortage periods, deviations from the ideal levels will occur and the calculated penalties measure the extent to which the demands are not met. The water allocation determined by the WRMM during such periods is based on a linear programming approach that tries to minimize the overall water deficit within the system at any given time, as measured by the total system penalty. Flow scenarios investigated for the SRB using WRMM addressed the physical availability of water to meet 'instream objectives' (IOs) that are written into licence agreements to protect fish habitat and 'instream flow needs' (IFNs), which are quantitative estimates of the flows required to protect aquatic environments (Alberta Environment 2003). Licences were categorized into junior and senior, depending upon the timing of development. Different scenarios explored how well the system would work if IOs and IFNs were imposed onto existing licences, whether it would be physically feasible to increase licenced acreage with existing IOs and IFNs, and whether licence agreements could still be satisfied while establishing environmental flows at higher levels than had been used in the past.

The baseline scenario evaluated system performance (licensed consumption and IOs), assuming perpetuation of existing IOs. The second scenario considered the effects of meeting legal licences that have not yet been used, and the third looked at expanding irrigated agriculture beyond currently authorized limits. Other scenarios explored system performance if policies were changed by switching from IOs to IFNs, conserving 20% of consumptive use from all licensees, or issuing new licences to enable future land development. Model results showed that the Province could not continue its current path of economic development and achieve its environmental goals. There would be frequent and substantial deficits for junior licences in three of the sub-basins when dams and reservoirs were operated at existing IOs. The system would be seriously stressed in drought years that are a naturally occurring feature in this semi-arid environment. Additional development would put even more pressure on current allocations and result in deficits for junior licensees and compromised IOs in some years. The model showed that additional growth beyond current allocations (specifically a 10% increase in development in the Oldman and 20% in the Bow) would result in deficits during 13 of the 68 years. None of the sub-basins would meet minimum required environmental flows. Under these conditions, IFNs cannot be achieved in most years without creating deficits in most years for junior licensees and in some years for senior licensees. Even conservation efforts in the form of a 20% reduction for all users would not support IFNs or meet the commitments of all users in all years.

Based on these results, Alberta's officials concluded that the existing regulatory framework was not sustainable in either economic or environmental terms. In 2006 it issued a Water Management Plan for the South Saskatchewan River Basin (Alberta Environment 2006, p. v) calling for 'a balance between protecting the aquatic environment and water allocation of rivers in the South Saskatchewan River Basin'. *Alberta's Water for Life Strategy* called for strict limits to the number of new licences issued and required new licences issued after 2005 in parts of the basin that were not fully allocated to meet water conservation objectives of 45% of the natural rate of flow and improve the efficiency of water use. Voluntary water transfers from lower to higher value uses were allowed to occur, but there was a requirement for a holdback of up to 10% of the water to remain in the natural body or held in a water conservation licence.

Concluding comments

The CRB and SRB modelling efforts presented above illustrate that it is important for models to consider the interactions among physical, technical and socioeconomic factors if these models are to adequately support water decision making in the non-stationary environment brought on by climate change and human development. Both CRB and SRB modelling efforts showed that enforcement of existing or proposed restrictions plays a critical role in determining the future availability of adequate water supplies. Although fusing physical water models with strategies for human management has been addressed in more informal ways, formalizing these linkages enables assessment of alternative water futures and policies that work best under a range of climate conditions.

Water decision making increasingly employs exploratory modelling, including optimization models for exploratory purposes, as Alberta Environment did in the scenario exercises described above. These new modelling frameworks enable policy makers to examine the consequences of particular policy decisions (e.g. to enforce the groundwater code, restore environmental flows or change irrigated agricultural acreage) on water supply and demand under a variety of possible future conditions, including some that are outside the envelope of the historical climate record. They also support the process of 'backcast' planning where policy makers decide what kind of future they want and explore what combination of policies allows them to get there under a wide range of future scenarios. Inherent in this approach is the notion of DMUU (Lempert *et al.* 2003).

DMUU acknowledges the need for public debate about water futures, and thus it is increasingly common for modellers to incorporate stakeholders' perspectives on model development and visualization (Cliburn *et al.* 2002; Serrat-Capdevila *et al.* 2009; Burch *et al.* 2010) and the social processes occurring during the design and implementation of decision support systems (White 2013). This trend reflects a growing awareness of how individual actors and social groups frame issues and how their perspectives play out through social processes in participatory, integrated water resources management (Pahl-Wostl 2006; Isendahl *et al.* 2009, 2010). Model fusion has the potential to link land use, hydrology, climate and ecology with human values, policy opportunities and market processes, and to enliven social learning processes, interdisciplinary and synthetic scientific inquiry, and public discussion about the future of water systems.

This material is based on work supported by the National Science Foundation under Grant SES-0951366, Decision Centre for a Desert City II: Urban Climate Adaptation. Any opinions, findings, and conclusions or recommendations expressed in this material are those of the authors and do not necessarily reflect the views of the National Science Foundation. Special thanks go to S. Wittlinger at DCDC and H. Wilson at the Global Institute for Water Security at the University of Saskatchewan for their editorial and graphic support.

References

AGRICULTURE AND AGRI-FOOD CANADA 2011. An overview of the Canadian agriculture and agri-food system 2011. http://www4.agr.gc.ca/AAFC-AAC/display-afficher.do?id=1295963199087\&lang=eng [accessed 3 June 2014]

ALBERTA ENVIRONMENT 2003. *South Saskatchewan River Basin Water Management Plan Phase 2, Scenario Modelling Results Part 1*. Alberta Environment Publication No. 710. Alberta Environment, Edmonton, AB. http://ssrb.environment.alberta.ca/pubs/SSRB_Scenario_Modelling_Results.pdf [accessed 3 June 2014]

ALBERTA ENVIRONMENT 2006. *Approved Water Management Plan for the South Saskatchewan River Basin (Alberta).* Alberta Environment Publication No. I/011. Alberta Environment, Edmonton, AB. http://environment.alberta.ca/documents/SSrb_Plan_Phase2.pdf [accessed 3 June 2014]

ALBERTA ENVIRONMENT 2010. *SSRP Water Quantity and Quality Modelling Result.* Alberta Environment and Sustainable Resource development Publication No. ENV-065-OP. Alberta Environment, Edmonton, AB. http://www.environment.gov.ab.ca/info/posting.asp?assetid=8283&categoryid=5 [accessed 3 June 2014]

ARIZONA DEPARTMENT OF WATER RESOURCES 2012. *Overview of Arizona Groundwater Management Code.* Arizona Department of Water Resources, Phoenix, AZ. http://www.azwater.gov/AzDWR/WaterManagement/documents/Groundwater_Code.pdf [accessed 3 June 2014]

BIERBAUM, R., SMITH, J. B. *ET AL.* 2013. A comprehensive review of climate adaptation in the United States: more than before, but less than needed. *Mitigation and Adaptation Strategies for Global Change*, **18**, 361–406.

BOROWSKI, I. & HARE, M. 2007. Exploring the gap between water managers and researchers: difficulties of model-based tools to support practical water management. *Water Resources Management*, **21**, 1049–1074.

BRUNNER, R. D. 2010. Adaptive governance as a reform strategy. *Policy Sciences*, **43**, 301–341.

BURCH, S., SHEPPARD, S. R. J., SHAW, A. & FLANDERS, D. 2010. Planning for climate change in a flood-prone community: municipal barriers to policy action and the use of visualizations as decision-support tools. *Journal of Flood Risk Management*, **3**, 126–139.

BUREAU OF RECLAMATION 2012*a*. *Colorado River Basin Water Supply and Demand Study: Study Report.* U.S. Department of Interior, Bureau of Reclamation, Boulder City, NV.

BUREAU OF RECLAMATION 2012*b*. Colorado River interim guidelines for Lower Basin shortages and coordinated operations for Lake Powell and Lake Mead. http://www.usbr.gov/lc/region/programs/strategies/about.html [accessed 3 June 2014]

BUREAU OF RECLAMATION 2012*c*. The Law of the River. http://www.usbr.gov/lc/region/g1000/lawofrvr.html [accessed 3 June 2014]

CAMACHO, A. E. 2009. Adapting governance to climate change: managing uncertainty through a learning infrastructure. *Emory Law Journal*, **59**, 1–77.

CASH, D. W., CLARK, W. C. *ET AL.* 2003. Knowledge systems for sustainable development. *Proceedings of the National Academy of Sciences of the United States of America*, **100**, 8086–8091.

CLARK, J. R. A. & CLARKE, R. 2011. Local sustainability initiatives in English national parks: what role for adaptive governance? *Land Use Policy*, **28**, 314–324.

CLIBURN, D. C., FEDDEMA, J. J., MILLER, J. R. & SLOCUM, T. A. 2002. Design and evaluation of a decision support system in a water balance application. *Computers & Graphics*, **26**, 931–949, https://doi.org/10.1016/s0097-8493(02)00181-4

COMEAU, L. E. L., PIETRONIRO, A. & DEMUTH, M. N. 2009. Glacier contribution to the North and South Saskatchewan Rivers. *Hydrological Processes*, **23**, 2640–2653.

CRONA, B. I. & PARKER, J. N. 2011. Network determinants of knowledge utilization: preliminary lessons from a boundary organization. *Science Communication*, **33**, 448–471.

DEBEER, C. M. & SHARP, M. J. 2009. Topographic influences on recent changes of very small glaciers in the Monashee Mountains, British Columbia, Canada. *Journal of Glaciology*, **55**, 691–700.

DILLING, L. & LEMOS, M. C. 2011. Creating usable science: opportunities and constraints for climate knowledge use and their implications for science policy. *Global Environmental Change*, **21**, 680–689.

ELLIS, A. W., HAWKINS, T. W., BALLING, R. C., JR. & GOBER, P. 2008. Estimating future runoff levels for a semi-arid fluvial system in central Arizona, USA. *Climate Research*, **35**, 227–239.

FALKENMARK, M. 1979. Main problems of water use and transfer of technology. *GeoJournal*, **3**, 435–443.

GOBER, P. 2007. Water, climate, and the future of Phoenix. *Geographische Rundschau*, **3**, 20–25.

GOBER, P. 2013. Getting outside the water box: the need for new approaches to water planning and policy. *Water Resources Management*, **27**, 955–957, https://doi.org/10.1007/s11269-012-0222-y

GOBER, P., KIRKWOOD, C. W., BALLING, R. C., JR., ELLIS, A. W. & DEITRICK, S. 2010. Water planning under climatic uncertainty in Phoenix: why we need a new paradigm. *Annals of the Association of American Geographers*, **100**, 357–372.

GOBER, P., WENTZ, E. A., LANT, T., TSCHUDI, M. K. & KIRKWOOD, C. W. 2011. WaterSim: a simulation model for urban water planning in Phoenix, Arizona, USA. *Environment and Planning B*, **38**, 197–215, https://doi.org/10.1068/b36075

ISENDAHL, N., DEWULF, A., BRUGNACH, M., FRANCOIS, G., MOELLENKAMP, S. & PAHL-WOSTL, C. 2009. Assessing framing of uncertainties in water management practice. *Water Resources Management*, **23**, 3191–3205, https://doi.org/10.1007/s11269-009-9429-y

ISENDAHL, N., DEWULF, A. & PAHL-WOSTL, C. 2010. Making framing of uncertainty in water management practice explicit by using a participant-structured approach. *Journal of Environmental Management*, **91**, 844–851, https://doi.org/10.1016/j.jenvman.2009.10.016

LACH, D., INGRAM, H. & RAYNER, S. 2005. Maintaining the status quo: how institutional norms and practices create conservative water organizations. *Texas Law Review*, **83**, 2027–2053.

LEMPERT, R. J., POPPER, S. W. & BANKES, S. C. 2003. *Shaping the Next One Hundred Years: New Methods for Quantitative, Long-term Policy Analysis.* RAND Corporation, Santa Monica, CA.

MACDONALD, G. M. 2010. Water, climate change, and sustainability in the southwest. *Proceedings of the National Academy of Sciences of the United States of America*, **107**, 21256–21262.

MARICOPA ASSOCIATION OF GOVERNMENTS 2007. *Socioeconomic Projections of Population, Housing and Employment by Municipal Planning Area and Regional Analysis Zone.* Maricopa

Association of Governments, Phoenix, AZ. http://www.workforce.az.gov/pubs/demography/2006MAGprojectionsJURI.pdf [accessed 3 June 2014]

Meko, D. M., Woodhouse, C. A., Baisan, C. A., Knight, T., Lukas, J. J., Hughes, M. K. & Salzer, M. W. 2007. Medieval drought in the Upper Colorado River Basin. *Geophysical Research Letters*, **34**, L10705.

Miles, E. L., Snover, A. K., Whitely Binder, L. C., Sarachik, E. S., Mote, P. W. & Mantua, N. 2006. An approach to designing a national climate service. *Proceedings of the National Academy of Sciences of the United States of America*, **103**, 19616–19623.

Milly, P. C. D., Betancourt, J., Falkenmark, M., Hirsch, R. M., Kundzewicz, Z. W., Lettenmaier, D. P. & Stouffer, R. J. 2008. Stationarity is dead: whither water management? *Science*, **319**, 573–574.

Moore, R. D., Fleming, S. W. *et al.* 2009. Glacier change in western North America: influences on hydrology, geomorphic hazards and water quality. *Hydrological Processes*, **23**, 42–61.

NATIONAL RESEARCH COUNCIL 2007*a*. *Colorado River Basin Water Management: Evaluating and Adjusting to Hydroclimatic Variability*. The National Academies Press, Washington, DC.

NATIONAL RESEARCH COUNCIL 2007*b*. *Evaluating Progress of the U.S. Climate Change Science Program: Methods and Preliminary Results*. The National Academies Press, Washington, DC.

NATIONAL RESEARCH COUNCIL 2010. *Informing an Effective Response to Climate Change*. The National Academies Press, Washington, DC.

NATIONAL RESEARCH COUNCIL 2012. *The National Weather Service Modernization and Associated Restructuring: A Retrospective Assessment*. The National Academies Press, Washington, DC.

NATURAL RESOURCES CANADA 2012. The threat of mountain pine beetle to Canada's boreal forest. http://www.nrcan.gc.ca/forests/insects-diseases/13381 [accessed 3 June 2014]

Nazemi, A., Wheater, H. S., Chun, K. P. & Elshorbagy, A. 2013. A stochastic reconstruction framework for analysis of water resource system vulnerability to climate-induced changes in river flow regime. *Water Resources Research*, **49**, 291–305. https://doi.org/10.1029/2012WR012755

Pahl-Wostl, C. 2006. The importance of social learning in restoring the multifunctionality of rivers and floodplains. *Ecology and Society*, **11**, 10, http://www.ecologyandsociety.org/vol11/iss1/art10/

Pahl-Wostl, C. & Borowski, I. 2007. Special issue: methods for participatory water resources Management – Preface. *Water Resources Management*, **21**, 1047–1048.

PARTNERS FOR THE SASKATCHEWAN RIVER BASIN 2009. *From the Mountains to the Sea: Summary of the State of the Saskatchewan River Basin*. Partners FOR the Saskatchewan River Basin, Saskatoon, SK.

Patrick, R. J. 2011. Uneven access to safe drinking water for First Nations in Canada: connecting health and place through source water protection. *Health & Place*, **17**, 386–389.

Pomeroy, J. W., de Boer, D. & Martz, L. M. 2005. *Hydrology and Water Resources of Saskatchewan, Centre for Hydrology Report #1*. Centre for Hydrology, Saskatoon, SK.

Sampson, D. A., Escobar, V., Tschudi, M. K., Lant, T. & Gober, P. 2011. A provider-based water planning and management model – WaterSim 4.0 – for the Phoenix Metropolitan Area. *Journal of Environmental Management*, **92**, 2596–2610, https://doi.org/10.1016/j.jenvman.2011.05.032

SASKATCHEWAN WATER SECURITY AGENCY 2013. 25-Year Saskatchewan Water Security Plan. https://www.wsask.ca/Global/About%20WSA/25%20Year%20Water%20Security%20Plan/WSA_25YearReportweb.pdf [accessed 3 June 2014]

Schindler, D. & Donahue, W. 2006. An impending water crisis in Canada's western prairie Provinces. *Proceedings of the National Academy of Sciences of the United States of America*, **103**, 7210–7216.

Serrat-Capdevila, A., Browning-Aiken, A., Lansey, K., Finan, T. & Valdés, J. 2009. Increasing social–ecological resilience by placing science at the decision table: the role of the San Pedro Basin (Arizona) decision-support system model. *Ecology and Society*, **14**, 37, http://www.ecologyandsociety.org/vol14/iss1/art37/

Sivapalan, M., Savenije, H. H. G. & Blöschl, G. 2012. Socio-hydrology: a new science of people and water. *Hydrological Processes*, **26**, 1270–1276.

Toth, B., Corkal, D. R., Sauchyn, D., van der Kamp, G. & Pietroniro, E. 2009. The natural characteristics of the South Saskatchewan River Basin: climate, geography and hydrology. *In*: Marchildon, G. P. (ed.) *A Dry Oasis: Institutional Adaptation to Climate on the Canadian Plains*. Canadian Plains Research Center Press, Regina, SK, 95–127.

Trenberth, K. 2010. More knowledge, less certainty. *Nature Reports*, **4**, 20–21.

US CENSUS BUREAU 2011. *Statistical Abstract of the United States: 2012*, 131st edn. US Census Bureau, Washington, DC. http://www.census.gov/compendia/statab/

US CENSUS BUREAU 2012. State and county quick facts. http://quickfacts.census.gov/qfd/states/04/0455000.html

Wheaton, E., Kulshreshtha, S., Wittrock, V. & Koshida, G. 2008. Dry times: hard lessons from the Canadian drought of 2001 and 2002. *Canadian Geographer*, **52**, 241–262.

White, D. D. 2013. Framing water sustainability in an environmental decision support system. *Society & Natural Resources*, **26**, 1365–1373. https://doi.org/10.1080/08941920.2013.788401

White, D. D., Corley, E. A. & White, M. S. 2008. Water managers' perceptions of the science-policy interface in Phoenix, Arizona USA: implications for an emerging boundary organization. *Society and Natural Resources*, **21**, 230–243.

Wilby, R. L. & Dessai, S. 2010. Robust adaptation to climate change. *Weather*, **65**, 180–185.

Thinking platforms for smarter urban water systems: fusing technical and socio-economic models and tools

CHRISTOS MAKROPOULOS

School of Civil Engineering, National Technical University of Athens, Iroon Polytechniou 5, Athens, Greece

cmakro@mail.ntua.gr

Abstract: Engineering is currently expanding its conceptual boundaries by accepting the challenge of interdisciplinarity, while often adopting social and biological concepts in developing tools (e.g. evolutionary optimization or interactive autonomous agents) or even world views (e.g. co-evolution, resilience, adaptation). The emerging socio-technical knowledge domain is still very much restricted by partial knowledge associated with the lack of long-term transdisciplinary research effort and the unavailability of robust, integrated tools able to cover both the technical and the socio-economic domains and to act as 'thinking platforms' for long-term scenario planning and strategic decision making under (high-order) uncertainties. Here we present an example of a toolkit that attempts to bridge this gap focusing on urban water (UW) systems and their management. The toolkit consists of three tools: the UW Optioneering Tool (UWOT); the UW Agent Based Modelling Platform (UWABM); and the UW System Dynamic Environment (UWSDE). The tools are briefly presented and discussed, focusing on interactions and data flows between them and their typical results are illustrated through a case study example. A further tool (a Cellular Automata Based Urban Growth Model) is currently under development and an early coupling with the other tools is also discussed. It is argued that this type of extended model fusion, beyond what has traditionally been thought of as 'integrated modelling' in the engineering domain is a new frontier in the understanding of environmental systems and presents a promising, emerging field in modelling interactions between our societies and cities, and our environment.

The city is, arguably, a major interface between humans and the environment, becoming more important every year after decades of continuing increase in the proportion of human populations in urban centres (UN 2012). The urban interface is a prime example of, and test bed for, the need for a systems approach to the analysis (and modelling) of drivers, pressures, impacts and responses (EEA 1999) stemming from the co-existence and co-evolution (Jeffrey & McIntosh 2006) of multiple systems: social, economic, technical and environmental. Research effort, in this context, is focusing on understanding, analysing and managing complexity, often looking at the city (and region) as a 'living organism' and attempting to capture interactions between the various urban subsystems (e.g. water, energy, communications, urbanization) in order to identify key (leverage) points for change. This approach, sometimes termed 'metabolism modelling' is gaining increased attention in the research community (González *et al.* 2013). The term, coined within the urban/industrial ecology discipline in the 1960s (see Brunner 2007; Kennedy *et al.* 2007) is meant to highlight the continuously changing, dynamic nature of multiple interactions of 'flows' within the urban environment (such as flows of water, energy, carbon, people, information, knowledge, etc.). Arguably, this is only one of the possible mental models of what has been more widely called the 'socio-technical system' (Abbott 2001) and whose transition to more sustainable equilibriums (or at least pathways) though adaptive approaches (Pahl-Wostl 2006) is the real focus of most of the sustainability targeted policy (European Commission 2012) and research work today (see, for example, the European Innovation Partnership (EIP) on Water or the Climate Knowledge and Innovation Community (KIC)).

For the urban water sector in particular, such a requirement for transition to more sustainable pathways is an indispensable condition due to the limited nature of the resource and the uncertainty of its provision (Kossida *et al.* 2012), the deteriorating state of related infrastructure (ASCE 2012) and the challenges in both asset investment and governance (Bos & Brown 2012). It has been argued (Domènech & Saurí 2010; Makropoulos & Butler 2010) that moving to more decentralized urban water management solutions (including, for example, low water using appliances, rainwater harvesting (RWH), greywater recycling (GWRC), sustainable urban drainage systems (SUDS), blue–green infrastructure, etc.) and indeed a more water-aware (or else water-sensitive) approach to city planning (Wong 2006; Rozos & Makropoulos 2013) is a rational way forward to address, in a truly metabolism-minded fashion, the water challenges to be faced by the cities of tomorrow.

From: Riddick, A. T. Kessler, H. & Giles, J. R. A. (eds) 2017. *Integrated Environmental Modelling to Solve Real World Problems: Methods, Vision and Challenges*. Geological Society, London, Special Publications, **408**, 201–219.
First published online July 17, 2014, https://doi.org/10.1144/SP408.4

However, this approach presents the difficulty that the more distributed the solutions, the greater the need for experimentation (and modelling) of the complete socio-technical system; this is because, as the interventions get closer to the users, they become more dependent on their adoption and acceptance (Domènech & Saurí 2010; Bos & Brown 2012). This is the context for the conceptualization and development of the urban water (UW) socio-technical toolkit, presented in this chapter, which currently contains three main tools: the UW Optioneering Tool (UWOT); the UW Agent Based Modelling Platform (UWABM); and the UW System Dynamic Environment (UWSDE). A further tool (a Cellular Automata Based Urban Growth Model) is currently under development and will be added to the toolbox in due course.

In the following section, we present an overview of the toolkit, followed by a brief description of each tool. Next, a sample of indicative modelling experiments is presented, around the case study of the city of Athens in Greece, whose combination of climatic conditions, water scarcity, transient populations, aging infrastructure and limited investment potential make it an ideal 'laboratory' for such investigations. The chapter continues with a discussion on the types of questions the toolkit can answer, alone, or in combination with existing models and tools (through, for example, open standards, such as OpenMI) and concludes with an outlook of research directions and possible synergies with other fields.

The urban water socio-technical toolbox

Within the context discussed above, the toolbox was developed with a number of specific aims in mind:

- to **design** (optimal) **demand management strategies** (under climatic and behavioural uncertainty) with an emphasis on distributed urban water management solutions and infrastructure;
- to **design** (optimal) **supply management strategies** (under climatic uncertainty) including alternative resources and their management;
- **while exploring interactions** between:
    - technical and social systems that affect the choices above
    - centralized infrastructure and decentralized interventions
    - new built and retrofit solutions
    - end-users and their behaviour
- **within an environment** of policy choices (including pricing).

A schematic of the toolbox can be seen in Figure 1. In this schematic the System Dynamics (SD) platform (left-hand side) is used to represent the broader socio-economic, technical and natural environment in an aggregate (top–down) manner (Baki *et al.* 2012). This is where effects of, for example, new environmental legislation, or new pricing policies can be investigated. The Agent Based model (ABM) (top right-hand side of Fig. 1) performs social simulation by modelling the actions, preferences and decisions of individual households (water users) whose environmental behaviour is influenced by both their social characteristics and context (Koutiva & Makropoulos 2012) and the broader (socio-economic and natural) environment in which they operate, which is modelled by the SD. The emerging water use behaviour of the end-users calculated by the ABM then affects the broader environment by feeding back into the SD model's domestic demand. The actual demand is not calculated by the ABM, but by the Urban Water Optioneering Tool (UWOT) (bottom right-hand side of Fig. 1) (Makropoulos *et al.* 2008*a*), which simulates the complete urban water cycle at multiple scales (household, neighbourhood, city) from the source (reservoir or water body) to the tap (Rozos & Makropoulos 2013). The individual modules are briefly described below.

The Urban Water Optioneering Tool (UWOT)

UWOT is a decision support tool, of the metabolism modelling type, which simulates the complete urban water cycle by modelling individual water uses and technologies for managing them and assessing their combined effects at multiple scales across a number of indications (or performance metrics). UWOT simulates both 'standard' urban water flows (potable water, wastewater and runoff) as well as more integrated interventions (including, for example, greywater reuse and rainwater harvesting). UWOT simulates water flows based on the generation, aggregation and transmission of a demand signal, starting from the household (and indeed building from the individual water appliance level) and moving towards the source (Rozos *et al.* 2010; Rozos & Makropoulos 2012). This demand-oriented conceptualization enables UWOT to simulate the whole urban water system from source to tap (Rozos & Makropoulos 2013). Figure 2 presents the complete urban water cycle: the left-hand side is a conceptual schematic while the right-hand side is the equivalent UWOT representation. The upper part (inside the ellipse) includes abstractions, transmission and treatment of raw water (the external water system) whereas the lower part displays the generation of the demand and the disposal of wastewater and stormwater to water bodies (the internal water system). The two main parts of the UWOT model (the internal and the external water

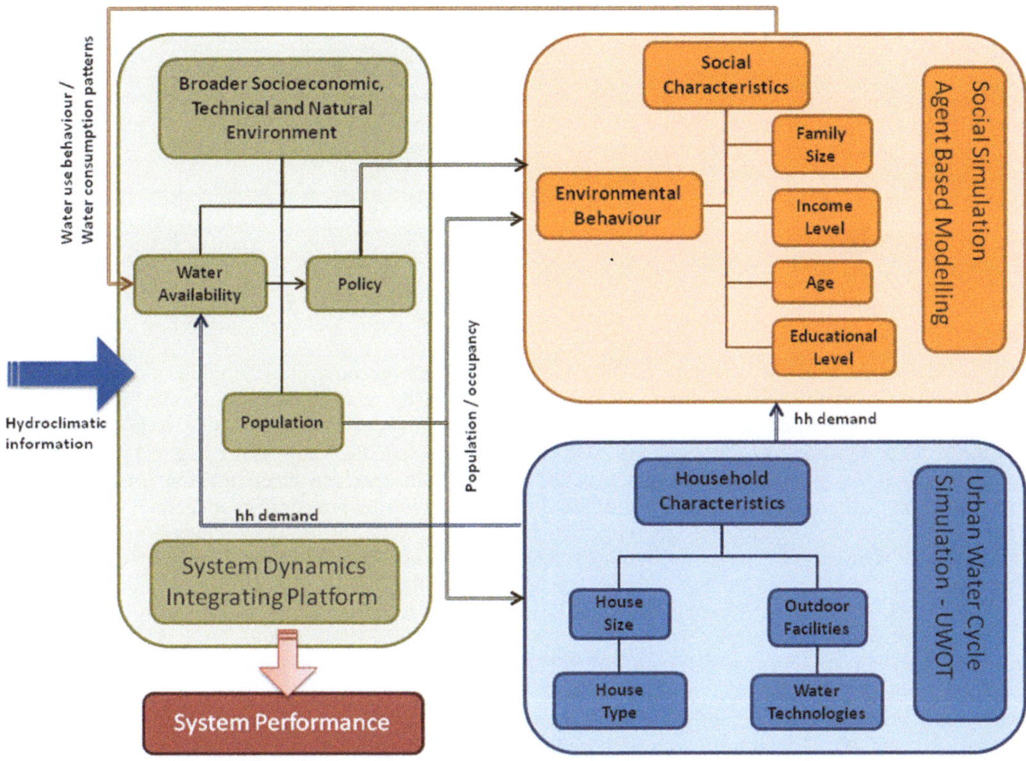

Fig. 1. The urban water socio-technical toolbox (Baki *et al.* 2012). hh, household.

systems) are discussed next (based on Rozos & Makropoulos 2013):

Modelling the internal water system. UWOT's demand-oriented modelling concept is explained with the help of a simple hypothetical urban water network, modelled in UWOT, as displayed in Figure 3. In this simplified example, the household only has two water appliances: a shower and a toilet. The demand of both appliances, in this example, is assumed to be covered from rainwater stored in a local tank. The tank is filled by runoff from impervious areas in the household (e.g. from the roof) with rainfall time series being provided as inputs. (Stochastic time series are generated through additional tools, in the periphery of the toolkit, but

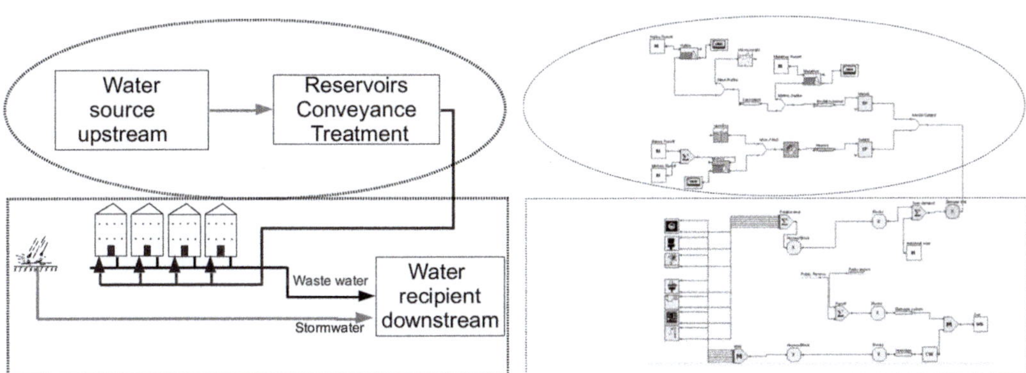

Fig. 2. Source-to-tap modelling of the complete urban water cycle.

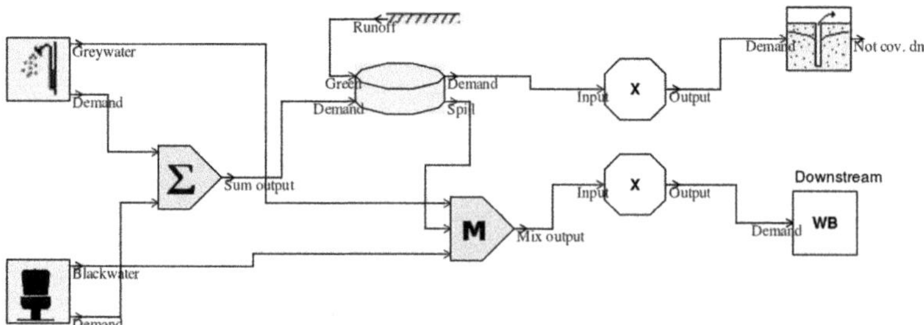

Fig. 3. A simplified example of a water supply network representation in UWOT. WW, waste water.

are not discussed here.) If the mass balance at the tank level indicates that the tank overflows, the tank spills excess water, which along with the water used in the toilet and the shower, are disposed of downstream. When the water inside the tank reaches a minimum level, the tank emits a demand signal for potable water from the centralized network which, in this case, abstracts water from a borehole. Any configuration of appliances can be set up and evaluated within UWOT, using a simple drag-and-drop interface (see discussion below).

Although not seen in this example, UWOT is also capable of providing an estimation of the water quality of the development's outflows (wastewater and runoff). The information required by UWOT to simulate the operation of water components (water consumption, required energy, biochemical oxygen demand (BOD) values, capital and operational cost, etc.) is stored in a database which is called a 'technology library'. The technology library is populated with information obtained from surveys, practitioner manuals, scientific publications, laboratory and pilot tests, etc. This database allows combinations of specific technological options to be selected for evaluation by the user (Makropoulos *et al.* 2008*a*). The technology library is implemented as a relational database, whose schema can be seen in Figure 4.

The user can decide to include a specific technology in the household (e.g. a toilet flush), select a specific type of toilet flush (e.g. a dual valve flush) with a specific water usage (e.g. 6 l/use) and estimate its performance, making the process of setting up and testing new technological options both generic (at the design stage) and detailed (at the simulation stage). This approach offers itself to evaluation of predefined technological configurations and optimization for identification of the best options (Makropoulos *et al.* 2008*a*). An example of an evaluation run for a specific configuration of a specific case study can be seen in Figure 5.

A technology setup graphical user interface (GUI) (Fig. 6) allows the user to define the topology using usual drag-and-drop and point-and-click functions. When the design is completed, the GUI automatically generates XML files. In addition to these matrices, UWOT also needs:

- the specifications of the installed components, that is, appliances/technologies (these are extracted from the technology library);
- time series of rainfall (to provide the input to reservoirs, rainwater harvesting schemes as well as urban runoff);
- time series of occupancy (to account for seasonal variations of population);

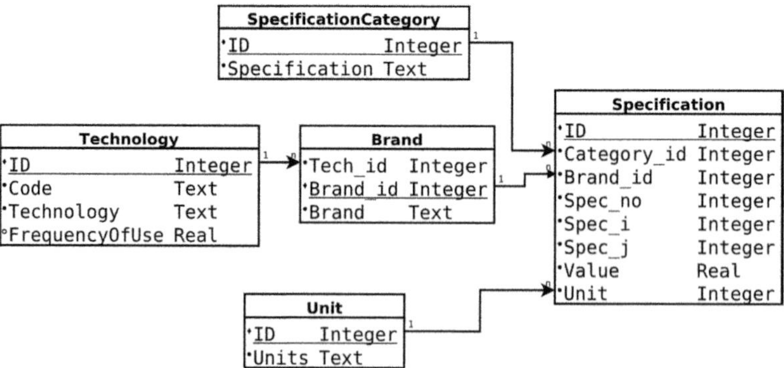

Fig. 4. Schematic of the technology library database.

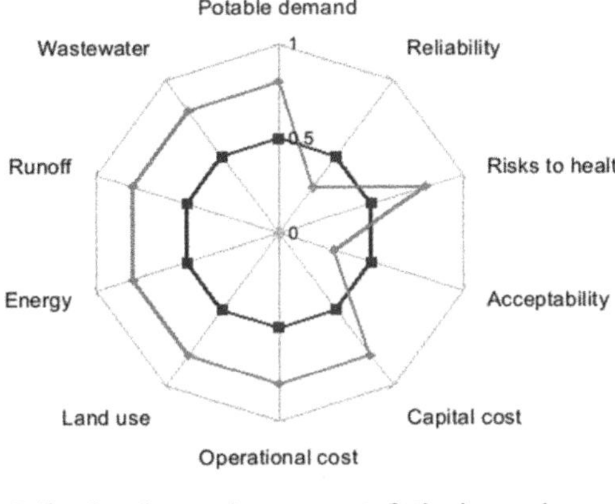

Fig. 5. An example UWOT evaluation.

- time series of numbers of households (to account for urbanization–also provided by the Cellular Automata (CA) model; see discussion below);
- frequency of use of the household appliances: although a constant value of the frequency of use of each appliance is included in the technology library, the user is also allowed to define a time series of frequency of use to enable simulation of changes in behaviour (also provided by the ABM model; see discussion in following sections).

An image of UWOT's geographical information system (GIS) GUI can be seen in Figure 7. In this figure the high-level view (city scale) can be observed, where only 'hyper-components' are visible to assist the user in designing the whole system; a hyper-component can, for example, be a district metered area (DMA). An additional window of the technology setup GUI (see Fig. 6) is also open, where a more detailed view (within each hyper-component) can be seen and where the specifics of the subsystems' configuration can be defined.

Modelling the external water system. UWOT also addresses the supply management problem, often referred to in the literature as 'optimal control of reservoir systems'. It consists of deciding at each time step how to cover multiple demands using multiple water sources. The most common method to study this kind of problem is stochastic dynamic programming (e.g. Su & Deininger 1972). This method has the disadvantage that it requires a large number of control variables, typically the sequences of releases from all reservoirs for all time steps of the control period. Motivated by this disadvantage, Nalbantis & Koutsoyiannis (1997) recommended an alternative method, the parametrization–simulation–optimization (PSO) approach. This method uses a handful of control variables, which are parameters of a simple rule. The parameterization of the rule is linked to a simulation model (here a water reservoir system), which enables the

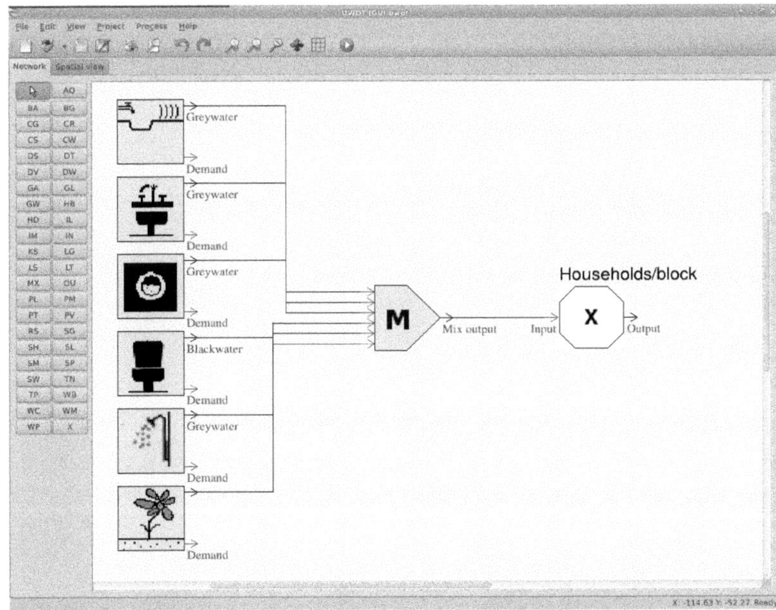

Fig. 6. UWOT technology setup GUI.

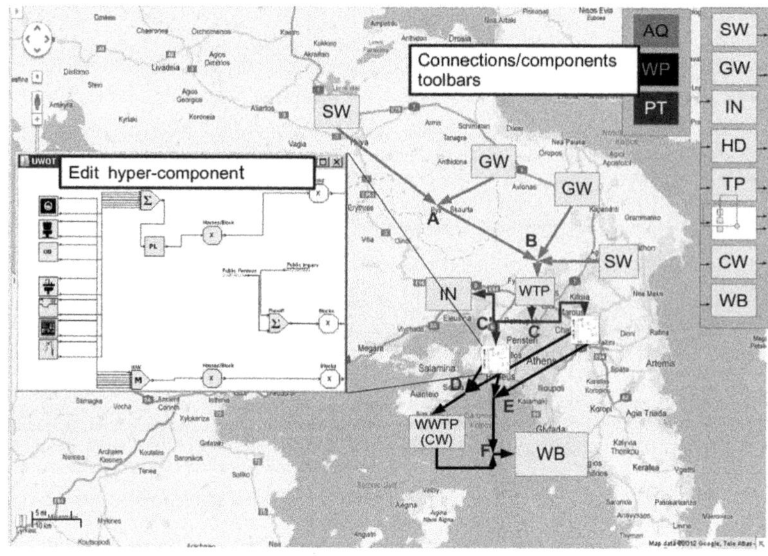

Fig. 7. UWOT GIS GUI.

calculation of a performance measure of the system for given parameter values and nonlinear optimization, which enables determination of the optimal parameter values. The exact form of the rule depends on the problem at hand (see also Koutsoyiannis & Economou 2003). This rule, referred hereafter as the 'parametric rule', is the type of rule calculated by UWOT to optimize the operation of an external water supply system. For example, Figure 8 displays a representation in UWOT of the Athens external water supply system, which is especially complex and long.

To test the performance of UWOT in simulating the external supply system and optimizing its performance, the tool was compared (Rozos & Makropoulos 2013) with the tool currently used operationally by Hydronomeas, the water company of Athens, to manage its own multi-reservoir external water supply system. The results were very positive. It should be noted that the tool currently used

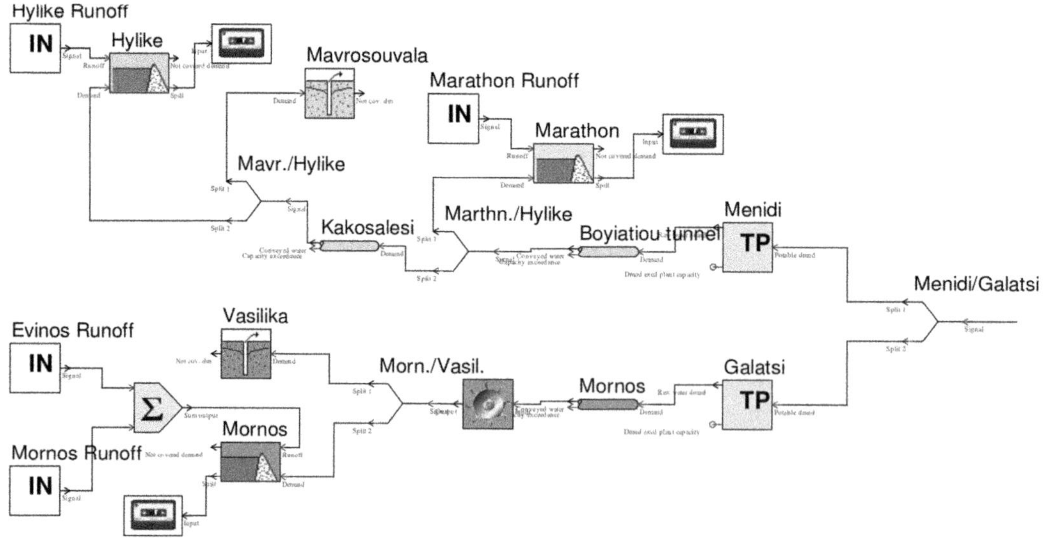

Fig. 8. Representation of Athens external water supply system in UWOT.

by Hydronomeas can only simulate the external network; the advantage of UWOT is that it can simulate and optimize the complete urban water cycle from source to tap and back again, from the water supply source (and any storage reservoirs), to the conveyance system, through treatment, to the households, all the way down to the disposal of stormwater and wastewater to water bodies, with a performance at least as good as more specialized sub-system models. This potential makes it an ideal tool for integrated demand and supply management studies, directly supporting both centralized and distributed water management solutions. In other words, the tool can now answer questions of potential trade-off between a new supply intervention (such as a new desalination plant or water reservoir) and a demand management strategy– something that is so far lacking in the portfolios of water utilities.

Although UWOT models the urban water cycle, it accepts technology adoption or implementation as an input to be evaluated. However, as suggested earlier, in the process of technology adoption, the household plays a crucial role. To get the user into the modelling framework, the UW Agent Based Modelling Platform was developed. This is presented in the next section.

The UW Agent Based Modelling Platform (UWABM)

Agent Based Modelling (ABM) is a computational intelligence application based on agents which are 'computer systems situated in some environment, capable of autonomous action in this environment in order to meet their design objectives' (Wooldridge 1999). ABM is used to address problems where the focus is on the simulation of emergence of group behaviour arising from interactions between a system's individual components and their environment (Railsback & Grimm 2009). The purpose of UWABM is to capture the effects of policies and environmental pressures on the water demand behaviour of urban households focusing on water conservation attitudes and the diffusion of domestic water saving technologies. The modelling focuses on the urban population that resides or moves into a specific area. The house types are set up to match those of the specific area under investigation, but the model does not include geographic information per se. It includes social ties within the area by representing them as a social network, as well as external social ties (relationships outside the neighbourhood) by representing them using an external social network variable. The model has a monthly time step, with the mean household water demand being either user defined or an input from UWOT. The mean demand can be a time series to take into effects of changes in weather conditions.

The 'agents' in this case are the households: Household agents make monthly decisions about their water demand and their domestic water technologies configurations. A household agent is characterized by those parameters that shape its water demand behaviour (see, for example, Fig. 9) (Koutiva & Makropoulos 2012; Koutiva *et al.* 2012).

The indicative data needed to set up the agent based model are as follows:

- number of households, distribution of household sizes, household population change;
- social network structure (open or closed scenario, type of social structure model);
- socio-demographics: income or class, education, age and ethnicity (from statistical office or normally distributed);
- level of environmentalism in society;
- water saving attitudes (percentage of water savers);
- demand elasticity (based on water demand level);

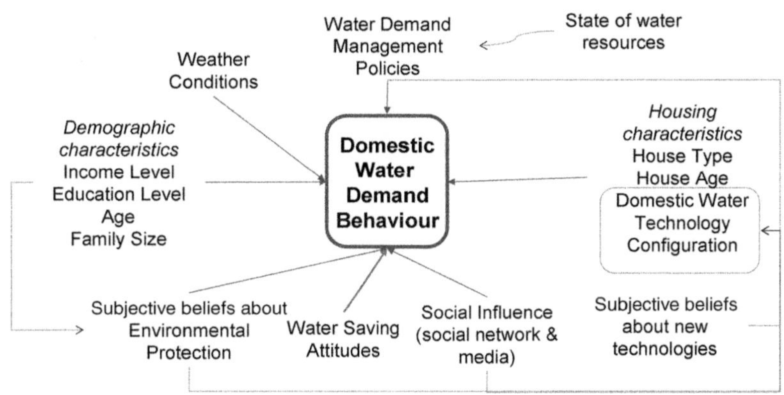

Fig. 9. Parameters shaping domestic water demand.

- social influence level in society (e.g. percentage of opinion leaders);
- innovation acceptance level in society (percentage of total innovators, percentage of total laggards, i.e. individuals who are slow to adopt an innovation);
- selected external social influence and media influence;
- selected percentage of total in favour of water conservation;
- selected percentage of total in favour of alternative water technologies.

Agents have access to external information that is either user-defined or obtained from other modelling platforms:

- water demand and investment costs for different water technologies (link with UWOT or user-defined)
- policies, including awareness campaigns, pricing policies, water restrictions and incentives (link with SD platform or user-defined)

Based on the selected 'society' scenario, the socio-demographic characteristics and the number of new and existing households, a social network is created representing the ties between existing and new residents. Two households connect based on their distance on these networks, which is computed by the weighted difference between their socio-demographic characteristic. An external social influence parameter is added to include the effects of the social network of each household that is not included in the model's social network. The social network (Fig. 10) is constructed using alternative network structures, such as the free-scale network (Albert *et al.* 1999), the random network (Erdös & Rényi 1960) and the small world network (Watts & Strogatz 1998).

A household's behaviour is based on its opinion on water conservation (positive or negative), which is affected by the opinion and influence of its social network (both local and external). After creating the social network, the setup process runs a social influence procedure based on Latane (1981), Sobkowicz (2003) and Wragg (2006), and re-assigns the opinions on water conservation taking into account social pressure. In order to include a degree of randomness in the deterministic force of pressure we use Bahr & Passerini (1998) opinion formation mechanics derived from Glauber Dynamics (Glauber 1963).

Households with a positive water conservation opinion follow a water conservation behaviour procedure to decide whether to conserve water (Fig. 11). Probabilities are assigned to water demand behaviour which are connected to the household's environmental awareness level, water saving attitude, socio-demographic characteristics and the effect of water policies (i.e. an awareness campaign raises the probability of conserving water, increase of water price raises the probability of conserving water, etc.). The probability of a household behaving in such a way is calculated by rules which follow the Bahr & Passerini (1998) opinion formation mechanics derived from Glauber Dynamics (Glauber 1963).

The household's acceptance of the technologies is calculated as a function of the household's innovation acceptance and its environmentalist type. The diffusion of technologies is assumed to follow the principles of social contagion (Young 2009) and is regarded as a two-stage process, separating the effect of causal variables on awareness and evaluation (van den Bulte & Lilien 1999).

Initially, households decide, based on (a) their innovation acceptance type, (b) the diffusion of the technologies in their social network and (c) the external marketing of the alternative technologies, whether they are open to consider a new configuration of domestic water technology. Following this initial choice, households decide whether to consider the use of another (alternative) water

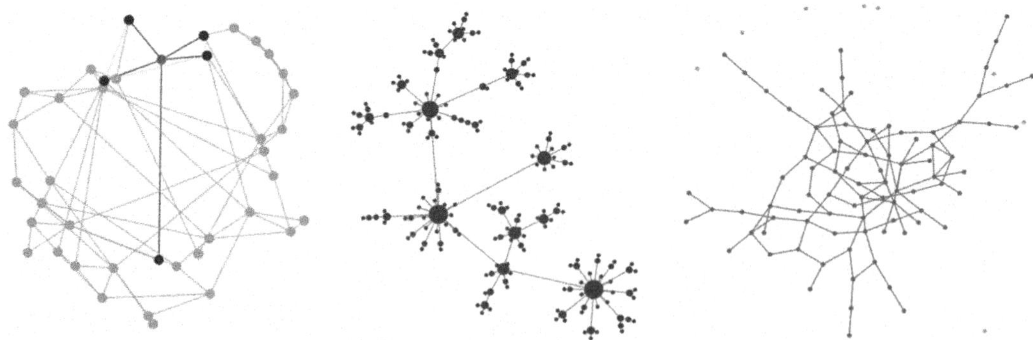

Fig. 10. Example of a small world, a free-scale and a random graph network structure.

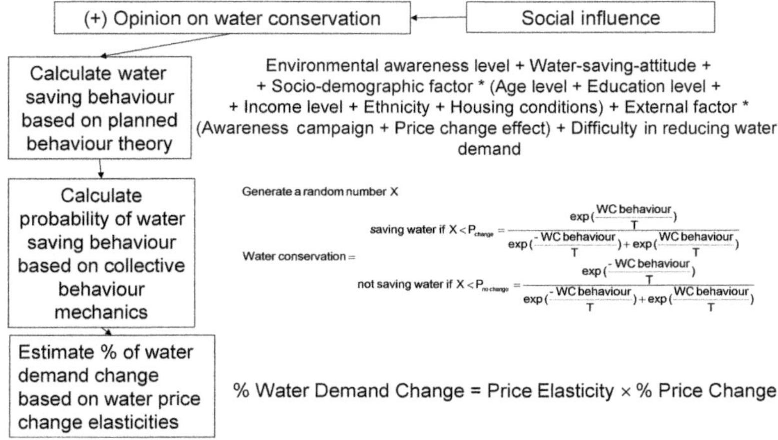

Fig. 11. Water conservation behaviour procedure.

technology. If a household is indeed open to consider new technologies, it then decides which alternative technology it prefers. This decision is the result of individual preferences. Households use their subjective preferences to calculate the utility of each alternative domestic water technology configuration. Subsidizing policies are added at this stage, reducing the investment cost needed for alternative configurations (possibly derived from the SD simulation, see discussion below). Households calculate utilities for all the alternative domestic water technology configurations and select the one that has the maximum utility. Examples of the application of UWABM in combination with UWOT to investigate the adoption of alternative domestic water technologies in an urban population on the basis of a particular decision process assumption and under different technology subsidizing policies are shown in Figures 12 and 13, respectively. In Figure 13 the population's cumulative percentage of adoption of rainwater harvesting schemes for a series of simulations for different subsidizing policies can be observed. Note that the typical S-curve of new technology adoption (reproduced in the figure) is not predefined in the model, but is rather an emerging behaviour of the agents and their individual decision making processes. This result is very encouraging with regard to the validity of the model itself, particularly since these models are notoriously difficult to verify.

Higher order parameters such as decision to change pricing policy, institutional changes or subsidies are considered an input in the model, but as suggested in the preceding section, these are key

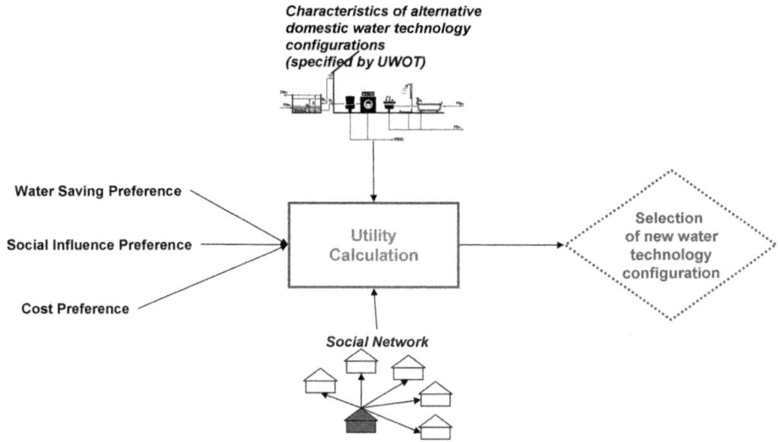

Fig. 12. Household decision process of alternative domestic water technology configuration.

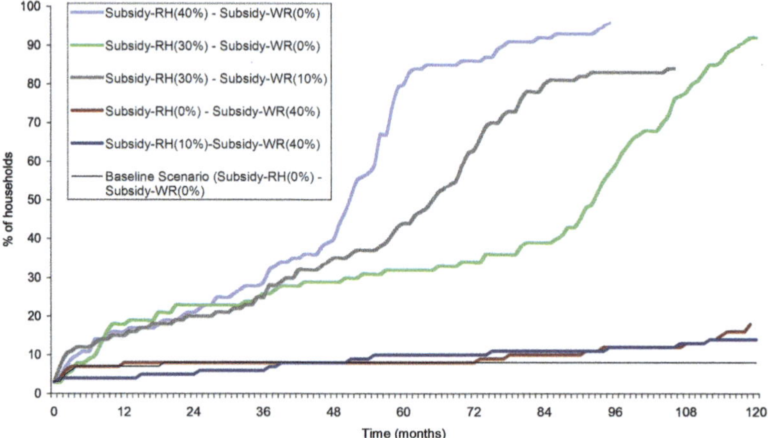

Fig. 13. Cumulative adoption of rainwater harvesting (RH) technology for different subsidizing policies. WR, Water Recycling technologies.

drivers to which the agents respond. To include these processes and interaction within the same modelling framework a System Dynamics (SD) approach was adopted. This is discussed in the following section.

The UW System Dynamic Environment (UWSDE)

Since this third module is not a standalone tool, but rather a platform which is used to connect the elements of the toolbox and facilitate both data exchange between them and high-level variables modelling (as suggested earlier), it is described here with the help of a specific model setup (described in more detail in Baki *et al.* 2012). The setup examines how early a specific water company (in this case in Athens) needs to trigger an awareness raising campaign, in the face of forecasted drought, to make sure the effects of the campaign are felt before the reservoirs reach a critical point.

An example of a typical causal loop diagram used for the development of the system dynamics model along with the indirect causal links through the agent based model and UWOT can be seen in Figure 14. This specific causal loop diagram has been developed for testing environmental awareness campaigns as a policy for demand management. The simplified representation of urban population change takes into account a net growth rate, while population increase is limited by a negative density dependant feedback as it approaches a specified population capacity (but could also be derived from the CA model; see discussion in next section). The model variable 'availability index' has been introduced as an indicator of how much water is available in the hydro-system (at the catchment scale) by calculating how many additional monthly 'consumptions' could be covered by the available water supply after having satisfied the various demands and losses. The indicator 'risk index' counts the times within a year when the availability indicator falls below a certain threshold and, when its value exceeds a set target, an environmental awareness campaign is triggered. The agent based model receives from the SD model the information on water availability and the existence or not of an environmental awareness campaign.

In turn, the ABM investigates the decision of the urban population to decrease water demand based on subjective attitudes triggered by their individual environmental behaviour (see discussion in the ABM section). The final outcome of the ABM is the total number of households per month that have decided to decrease water demand. This output feeds back into the SD, which then calculates total domestic demand based on values of conventional and reduced household water demand provided by UWOT (the reduced household water demand in this example is a result of reduced use for showers and garden watering as calculated by UWOT). Indicative examples of this experiment can be seen in Figure 15

It can be observed that, as was to be expected, as the threshold that triggers the launch of the policy increases (in other words, as risk aversion rises) and awareness campaigns are launched sooner rather than later, their effectiveness rises. Demand reductions resulting from these campaigns are incurred early enough to significantly reduce the number of system failures. Although several parameters in this analysis need to be calibrated carefully to provide added value, it is evident that such

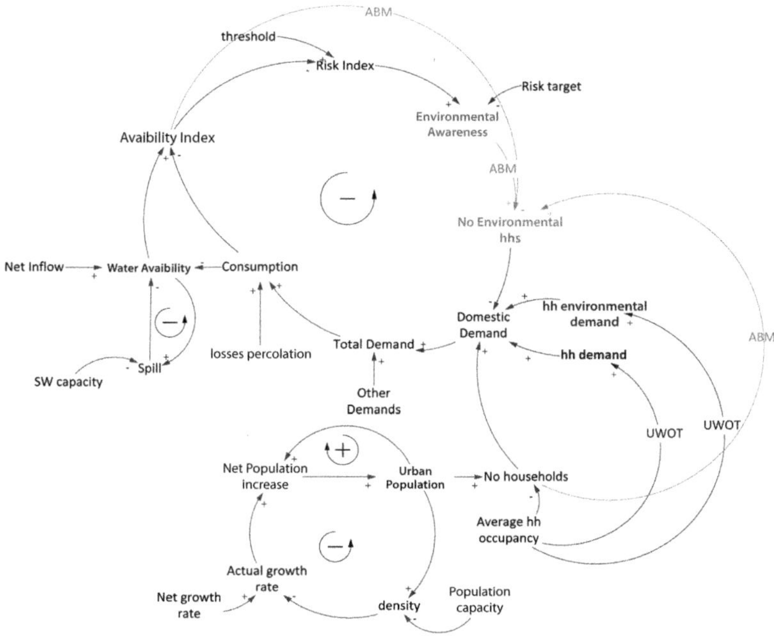

Fig. 14. Hybrid model causal loop diagram. ABM, Agent Based Modelling; hh, household; SW, stormwater; UWOT, Urban Water Optioneering Tool.

an approach can assist in setting case-specific, thresholds for water scarcity and drought management that are technically and sociologically sound, corresponding to, for example, acceptable levels of risk. The same analysis framework can also identify critical leverage points in the process of launching customized campaigns to improve their effectiveness as part of a broader intervention strategy. As discussed, however, the population itself (its location, migration and number) is included in the tools above as an external input. To account for this within the toolbox, a custom (simple) CA model for urban growth is being developed.

The Cellular Automata Urban Growth Model

The final module, currently under development, is a simple urban growth model, based on cellular automata (CA). Cellular automata are mathematical models for complex systems containing large numbers of simple identical components with local

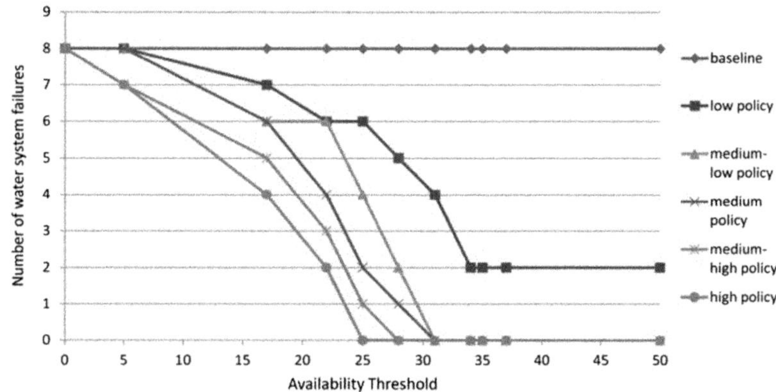

Fig. 15. Water system failures vs. water availability threshold under varying policy (awareness raising campaign) effects.

interactions (Wolfram 1984). The purpose of the CA module presented here (see also Rozos *et al.* 2011) is the explicit inclusion within the same socio-technical modelling framework of the dynamic nature of a city's development. Land-use models based on cellular automata are widely considered as effective tools, able to simulate complex urban growth dynamics and to predict future urban growth.

In general, the principle behind cellular automata is simple: the area is divided into a grid of cells, with each cell A being defined by a set of cell states $\mathbf{S} = \{S1, S2, \ldots, Sn\}$, that identify the cell properties, as well as a set of transition rules T that define how each cell changes:

$$A \leftarrow (\mathbf{S}, T).$$

Transition rules define an automaton state S_{t+1} at time step $t+1$ depending on its state at time step t, S_t (S_t, S_{t+1} elements of $\mathbf{S}$) and input information $\boldsymbol{I}_t$:

$$T: (S_t, \boldsymbol{I}_t) \rightarrow S_{t+1}.$$

The input is defined mainly by the states of 'neighbouring' cells. This definition of neighbourhood is essentially a predetermined set of cells that influences the state of the cell under consideration. Each cell A follows the connection:

$$A \leftarrow (\mathbf{S}, T, \boldsymbol{R})$$

where $\boldsymbol{R}$ denotes automata neighbouring A, and establishes the main boundary (excluding distance effects and similar interactions that may appear in some other CA models) for drawing input information $\boldsymbol{I}$ which is necessary for the application of transition rules T.

It should be noted that probabilistic or even stochastic parameters can be infused into transition rules (White & Engelen 1997), leading to single variable models whose complexity relies not on the interplay between different variables but the sophistication built in to the transition rule(s). In the case of urban growth, cellular automata models have been widely used to predict spatial development of urbanized areas, both regarding urban fabric and, in more general cases, land-use (Liu 2009). Often, however, in these models the set of cell states is simplified to follow a binary logic: the cells are characterized by two states, namely 'urban' and 'non-urban'. While this offers a simple modelling approach, it does now allow for more elaborate properties within the cells, as is required in the case in urban water management, where data like urban density, occupancy, socio-economic properties of the household and so on are also needed (see discussion on the ABM model above). This means that a CA model used for urban water management should be able to provide not only general urban growth patterns but also different cell conditions meeting specific cell properties directly related to water demands and hydrology, such as occupants, pervious and impervious areas, and so on. Therefore, a more sophisticated tool was developed and implemented that outputs a multiple-state view of urban growth. The tool is a fuzzy constrained cellular automata model (Liu 2009), based on the work of Mantelas *et al.* (2012). The key characteristics of the tool are the combined use of fuzzy logic with CA techniques, as well as the multiple-state (i.e. non-binary) nature of the CA model.

The fuzzy logic framework is deployed by forming a number of independent, parallel fuzzy inference systems (FIS), each focusing on one specific, distinct set of urban growth factors. Multiple-input-single-output and single-input-single-output FIS were employed, with the use of linguistic terms derived from human reasoning (Makropoulos *et al.* 2003). The FIS inputs that can be used depend on available data, with physical restrictions (slope, land-use, water bodies) and accessibility (transportation network) being of primary importance. Spatial variability of risk perceptions is also taken into account, with the use of the spatial ordered weighted averaging (SOWA) method (Makropoulos & Butler 2006). The final result is the raster map of overall suitability.

The urban growth model assumes three types of urban growth, which represent varying degrees of urban density and can then be divided into different housing types and consequently distinct cell states. More specifically, three urban density states are assumed (termed 2, 3 and 4), with state '2' having mild density and '4' having the heaviest one. The value '1' corresponds to a non-urban cell, while the value '0' corresponds to a cell that cannot be occupied, and denotes a physical boundary–such as the sea. The inputs of the CA model are the suitability map (which is the output of the FIS analysis suggested above), as well as the initial urban fabric image, giving the $t = 0$ state of the urban environment. This initial state needs to be extracted from available data, such as satellite images. Other input data include historic population time series which are used to calibrate the model and to validate its outputs in terms of population growth rate and influx.

The mechanics behind multiple-state urban growth follow a minimalistic pattern, which attempts to clarify the interaction that drives cellular-based urban growth without adding unneeded complexity or heuristic rules. More specifically, the following stages are followed in each simulation time step:

(1) An Urban Growth Algorithm, similar to the one presented and successfully tested by Mantelas *et al.* (2012) and Rozos *et al.* (2011), decides which non-urban cells are to be urbanized in each time step. Two rules of urban expansion and one rule of (distance-based) spontaneous growth (in areas without neighbouring urban cores) are applied (Mantelas *et al.* 2012). This rule is based on an (intermediate) binary urban image of each time step (including distraction between urban and non-urban cell types only), so in step 1, different cell states are ignored.
(2) The State Allocation Algorithm assigns different cell states to all cells which were urbanized with the previous rule. This rule applies only to cells that were turned from non-urban to urban at the specific time step.
(3) An Intensification Mechanism assigns denser urban states to existing, urban cells. This allows cells that are already urban to transform into urban cell with greater urban density, that is, from cell state 2 to cell state 3 or 4.

The basic stages of the CA model are shown in Figure 16. First the suitability factor (SF) of the area is calculated with the use of FIS, which accept as input various spatial parameters such as, inter alia, distance from the road network, elevation and land cover. Then the initial urban fabric image is produced, with the aid of available input such as land-cover/land-use data and satellite images. Finally the SF matrix and the initial urban fabric image, as well as statistics about the population of the area, are used as inputs for the CA model in order to estimate future urban growth patterns ('final urban fabric image' in Fig. 16).

Examples of the CA tools application, in combination with UWOT, to investigate distributed water demand management (WDM) measures and to quantify their effect in a growing peri-urban area in Athens can be seen in Figures 17 and 18.

Specifically, Figure 17 presents the UWOT configuration for households which employ in-house greywater reuse. In this configuration it can be observed, for example, that two in-house appliances (hand basins and showers) contribute to the production of greywater while three (toilets, washing machines and gardens) make use of it. Figure 18a is a snapshot of the CA simulation of growth in the eastern suburbs of Athens at given point within the 20-year simulation, while Figure 18b is a calculation (by UWOT) of the total water demand, both with conventional and innovative appliances (in this case the use of greywater recycling and rainwater harvesting). This specific study was developed to assess whether expansions of water provision by the water company of Athens towards the eastern suburbs was feasible, given the available water resources (at the supply side) and under what (demand management) conditions.

Component integration

The models/components of the socio-technical toolkit need to be linked together to achieve the synergies presented above. In the work undertaken so far, this has been accomplished in an ad hoc manner, in which appropriate 'gluing routines' have been developed for the particular problems at hand. For example, the ABM and SD models presented here, were both written in NETLOGO and were able to be seamlessly integrated, while UWOT is primarily written in C. Calling UWOT from NETLOGO required the development of an additional executable. In some cases, such as the integration between the CA model and UWOT, this integration was not dynamic, in that UWOT was initially run for the number of alternative housing configurations that the CA was able to simulate and the results were kept in a database, which the CA used to calculate demands at each time step. Additional interaction between the models and advanced multi-objective optimization algorithms also required linking of the models with mathematical environments such as MATLAB, which also required additional ad hoc coding. This process of model coupling has the disadvantage of requiring significant customization, dependent on both the tools and the problem at hand and is certainly an area where future development of the toolkit is looking to invest. In other words, the sociotechnical toolbox is currently at what Argent (2004) would term a Level III integration while what is in principle desirable is Level IV. To progress towards this, the current thinking of the toolbox's developers is to focus on the implementation of standards to the socio-technical toolbox modules including, but not limited to OpenMI (http://www.openmi.org) (as a way to couple models in runtime, Gregersen *et al.* 2007) and WaterML (http://www.opengeospatial.org/standards/waterml) (as a way to exchange time series). Additional common 'infrastructure' such as the underlying database concept (based on the Enhydris system (Christofides *et al.* 2011) further supports these ongoing integration efforts.

Discussion

The integrated vision that the socio-technical toolkit offers can be seen in Figure 19. As suggested in Figure 19, the CA model simulates urban growth

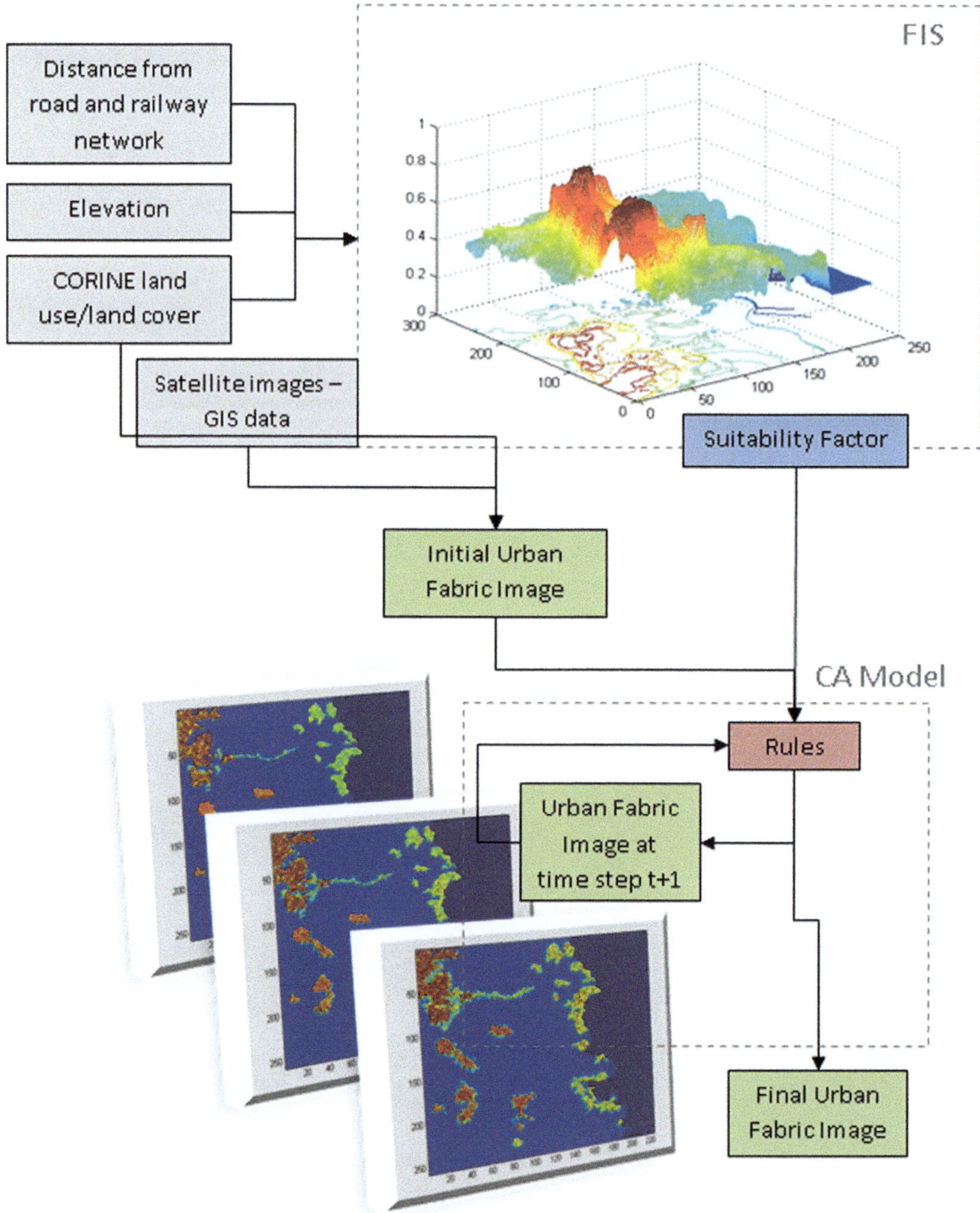

Fig. 16. The flow chart of the fuzzy constrained cellular automata (CA) model (Rozos *et al.* 2011).

and accounts for the spatial variability of demand. The ABM simulates choices and actions by the households of these areas, which decide to conserve water (or not), adopt devices and systems and whose demand for water, given those choices is calculated by UWOT which also simulates the provision of the water demanded all the way to the source. All the tools operate within a specific socio-economic and natural environment, some of whose higher order variables (such as, for example, gross

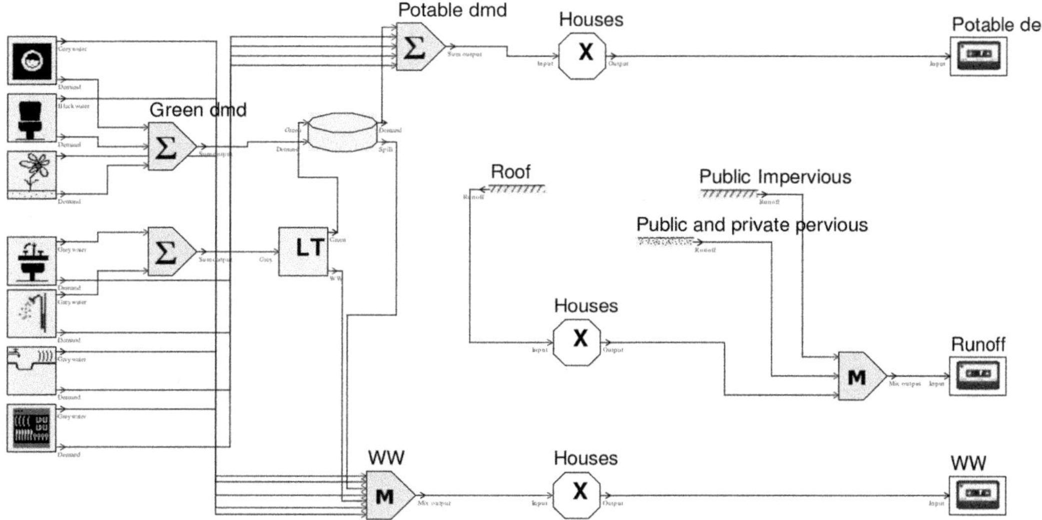

Fig. 17. Household network topology of greywater reuse.

domestic product) are simulated within the SD platform. As suggested in Figure 19, the toolkit can also:

- be driven by a multi-objective optimization algorithm (such as Genetic Algorithms, see Makropoulos & Butler 2005) to identify, for example, 'optimal' technology and policy mixes, under different scenarios (for example of climatic or economic uncertainty);
- accept data from distributed (smart) sensors, monitoring the performance of distributed assets;
- be used as a 'roadmapping' tool (e.g. Hinkebein & Price 2005), to enable a phased implementation of alternative interventions.

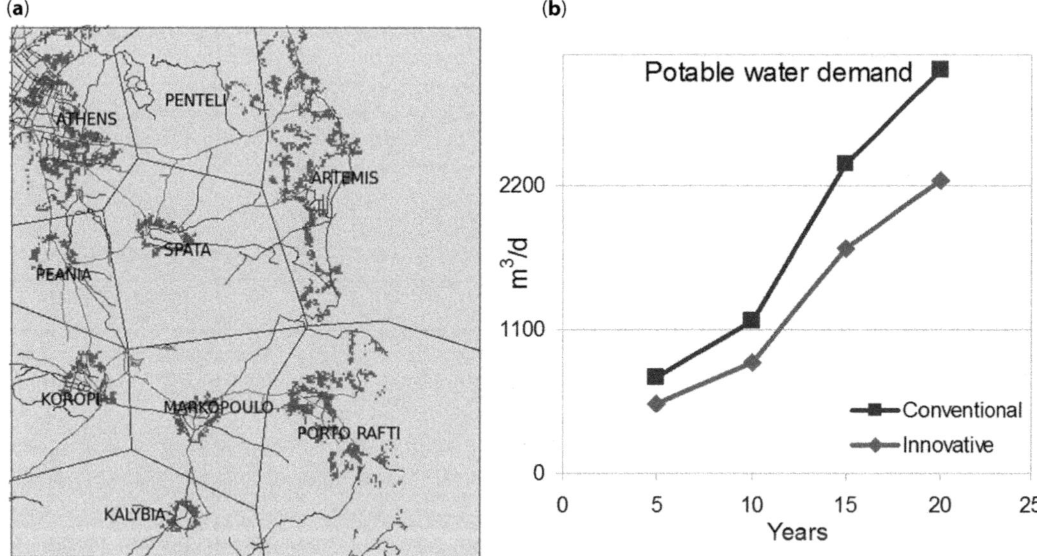

Fig. 18. Comparison of the evolution of potable water demand over the study area for different water demand management measures. (**a**) Snapshot of the cellular automata simulation of growth in the eastern suburbs of Athens at given point within the 20-year simulation. (**b**) Calculation by UWOT of the total water demand with conventional and innovative appliances.

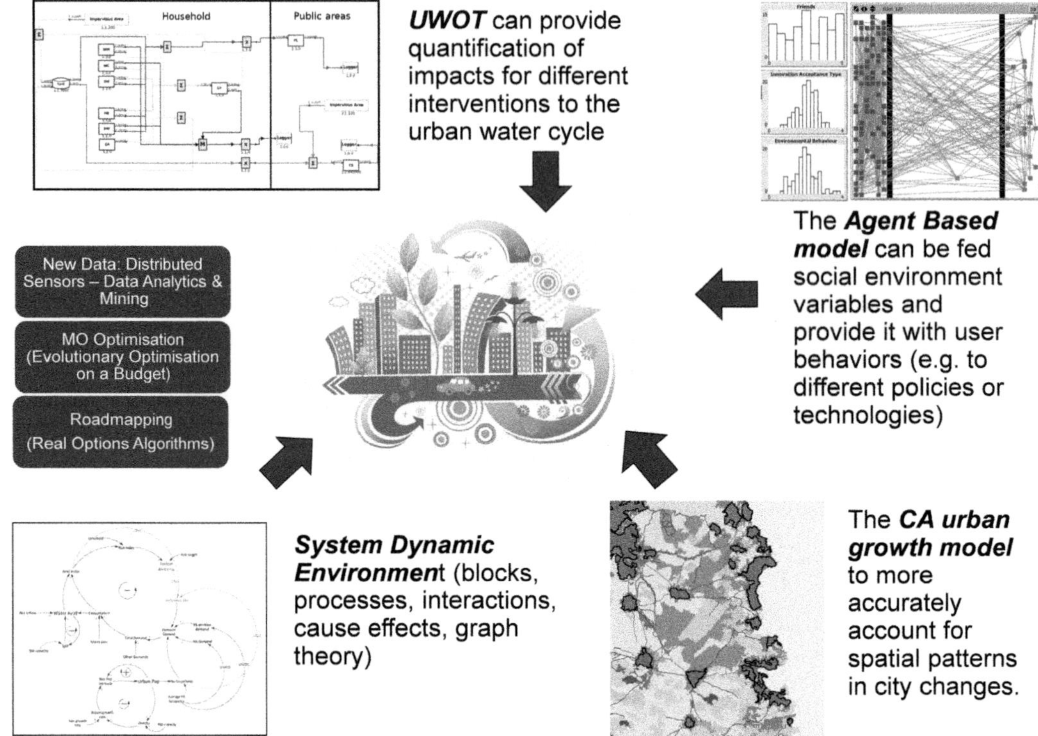

Fig. 19. Interaction between toolkit components. UWOT, Urban Water Optioneering Tool.

Examples of such uses of the toolkit (with smart sensors as a more operational tool and within a road-mapping context as a planning tool) are currently under way.

Using the toolkit as a 'thinking platform' (Makropoulos *et al.* 2008*b*), decision maker(s) can investigate technical and policy options and assess synergies between different parts of the socio-technical system in order to (a) negotiate and (b) co-develop blueprints for (hopefully more sustainable) urban water services. Some of the questions that need to be posed (and can be partly answered using the toolkit described in this chapter) within the process of planning for more 'sustainable'

Table 1. *Developing a blueprint for an urban water system: (some) questions that (could) be answered*

Technical measures	Policy	Synergies
• Centralized or decentralized wastewater recycling (GWRC)? • What is the impact of water demand management (WDM) on reliability of supply (under climatic and socio-economic uncertainty)? • What is the impact of combined rainwater harvesting (RWH) and GWRC on demand? (are there regional answers to this?)	• What will be the impacts of alternative awareness raising measures to demand management? • When to trigger a campaign and what is the link to other policies? • What would be the impact of different subsidies to technology adoption rates? • How to design a strategy that allows flexibility (keeping my options open for when I know more?	• What are the energy/cost/risk implications of decentralized systems? • What is an 'optimal' mix of hard and soft engineering interventions for sustainable urban water management (UWM)? • How resilient can a given mix make my city under a range of scenarios (climatic and socio-economic)?

urban water services can be seen in Table 1. They are by no means the only ones: Additional research on urban hydrology processes (Fletcher *et al.* 2013) is certainly needed to calibrate, for example, UWOT for specific areas, while more detailed (and sophisticated) field surveys of public opinion and attitudes (Willis *et al.* 2011) are certainly required to allow for better calibration of ABM models (currently a source of significant uncertainty) – to name but a few.

Since the toolbox is primarily intended for feasibility/scoping/roadmapping work, in other words for the early stages of project conceptualization and negotiation, existing detailed models (of the various physical processes involved) can then be used 'downstream' of the toolkit to actually design the solutions assessed and prioritized. Although, as is evident from the description so far, the toolkit is not a formal decision support system (DSS), it does address a number of the recommendations by van Delden *et al.* (2011) on policy-relevant DSS, including: a synthesis of relevant drivers and processes; an integrated approach to modelling different domains (economic, environmental/technical and social); and an ability (it is hoped) to provide added value to the current decision making practice. This is due to the creation of a common environment for discussion between:

- engineers (experts on the systems that UWOT simulates);
- social scientists (focusing on the social system aspects that the ABM simulates);
- urban planners (with an understanding of the processes addressed in the CA model);
- economists and policy makers (with an interest on the link between high level parameters (such as water pricing), which the SD platform captures and the evolution of options on the ground.

At its current stage of development, the toolkit needs to be set up and operated by expert facilitators, able to translate the questions posed by the stakeholders (see above) from linguist statements into model parameters. (Note that a limited functionality ABM model for investigating the impact of awareness raising campaigns on urban water demand is currently under development as a standalone web-based service.) This is certainly a limitation, albeit one common to most of the decision making environments to date. This is why the development of a 'decision theatre' front end to the toolbox (see, for example, Bok & Ruve 2007), ranks high on our (near) future development agenda. These limitations notwithstanding, it is suggested that the 'thinking platform' presented in this chapter could be particularly important as a means to investigate uncertain impacts and potential side effects of new interventions (such as distributed technologies) in urban water management under different 'futures' (Makropoulos *et al.* 2008*b*). This is because, as illustrated by Moglia *et al.* (2010), a sustainable (in theory) intervention aiming to improve the state of a particular system may have unintended and undesirable effects on another. When diverse interventions such as pricing, new centralized infrastructure or rebates for low water using appliances need to be examined in combination for planning horizons of often more than a couple of decades, the examination of possible side effects (feedback loops) across systems becomes crucial and, it is argued, gives the toolbox its main reason for uptake by research and practice.

Conclusions

This chapter presented a toolkit, currently under development, for thinking about and designing urban water systems. The main tools of the kit are briefly presented and some examples of their applications are illustrated. It is argued that this type of model fusion, which attempts to explicitly model elements of the overall socio-technical system that are beyond what has traditionally been thought of as 'integrated modelling' in the engineering domain, is in fact required by policy makers faced with the challenge of designing and deploying new 'options' for urban water management and the challenge of capturing (some of) the possible side effects of their decisions. Although the toolkit is not meant as a forecasting tool or even a DSS, it is suggested that this (loose) thinking platform approach is well suited to the unstructured nature of much of current decision making under uncertainty where, often, 'unknown-unknowns' dominate the transition pathway and additional (unforeseen), elements of the problem are brought to the table with every additional stakeholder.

References

Abbott, M. 2001. The democratisation of decision-making processes in the water sector I. *Journal of Hydroinformatics*, **3**, 23–34.

Albert, R., Jeong, H. & Barabási, A. L. 1999. Diameter of the world wide web. *Nature*, **401**, 130–131.

Argent, R. 2004. An overview of model integration for environmental applications—components, frameworks and semantics. *Environmental Modelling & Software*, **19**, 219–234.

ASCE 2012. *The Failure to Act: The Economic Impact of Current Investment Trends in Water and Waste Treatment Infrastructure*. American Society of Civil Engineers, Reston, VA.

Bahr, D. & Passerini, E. 1998. Statistical mechanics of opinion formation and collective behaviour:

micro-sociology. *The Journal of Mathematical Sociology*, **23**, 1–27.

BAKI, S., KOUTIVA, I. & MAKROPOULOS, C. 2012. A hybrid artificial intelligence modelling framework for the simulation of the complete, socio-technical, urban water system. 6th International Congress on Environmental Modelling and Software – iEMSs 2012, 1–5 July 2012, Leipzig, Germany.

BOK, B. & RUVE, S. 2007. Experiential foresight: Participative simulation enables social reflexivity in a complex world. *Journal of Futures Studies*, **12**, 111–120.

BOS, J. & BROWN, R. 2012. Governance experimentation and factors of success in socio-technical transitions in the urban water sector. *Technological Forecasting & Social Change*, **79**, 1340–1353.

BRUNNER, P. H. 2007. Reshaping urban metabolism. *Journal of Industrial Ecology*, **11**, 11–3.

CHRISTOFIDES, A., KOZANIS, S., KARAVOKIROS, G., MARKONIS, Y. & EFSTRATIADIS, A. 2011. Enhydris: A free database system for the storage and management of hydrological and meteorological data. *In*: *European Geosciences Union General Assembly Geophysical Research Abstracts, Vol. 13, Vienna*. European Geosciences Union. https://www.itia.ntua.gr/en/docinfo/1120/

DOMÈNECH, L. & SAURÍ, D. 2010. Socio-technical transitions in water scarcity contexts: Public acceptance of greywater reuse technologies in the Metropolitan Area of Barcelona, Resources. *Conservation and Recycling*, **55**, 53–62.

EEA 1999. *Environmental Indicators: Typology and Overview*. Technical Report, No. 25. European Environment Agency, Copenhagen.

ERDÖS, P. & RÉNYI, A. 1960. On the evolution of random graphs. *Magyar Tudományos Akadémia Matematikai Kutató Intézete Közl*, **5**, 17–61.

EUROPEAN COMMISSION 2012. *A Blueprint to Safeguard Europe's Water Resources*. COM(2012)673). European Commission, Brussels.

FLETCHER, T., ANDRIEU, H. & & HAMEL, P. 2013. Understanding, management and modelling of urban hydrology and its consequences for receiving waters: a state of the art. *Advances in Water Resources*, **51**, 261–279.

GLAUBER, R. 1963. Time-dependent statistics of the Ising model. *Journal of Mathematical Physics*, **4**, 294–307.

GONZÁLEZ, A., DONNELLY, A., JONES, M., CHRYSOULAKIS, N. & LOPES, M. 2013. A decision-support system for sustainable urban metabolism in Europe. *Environmental Impact Assessment Review*, **38**, 109–119.

GREGERSEN, J., GIJSBERS, P. & WESTEN, S. 2007. OpenMI: open modelling interface. *Journal of Hydroinformatics*, **9**, 175–191.

HINKEBEIN, T. & PRICE, M. K. 2005. Progress with the desalination and water purification technologies US roadmap. *Desalination*, **182**, 19–28.

JEFFREY, P. & MCINTOSH, B. S. 2006. Description, diagnosis, prescription: a critique of the application of co-evolutionary models to natural resource management. *Environmental Conservation*, **33**, 281–293.

KENNEDY, C., CUDDIHY, J. & ENGEL-YAN, J. 2007. The changing metabolism of cities. *Journal of Industrial Ecology*, **11**, 43–59.

KOSSIDA, M., KAKAVA, A., TEKIDOU, A. & MIMIKOU, M. 2012. *Vulnerability to Water Scarcity and Drought in Europe*. Thematic Assessment for EEA 2012 Report. ETC/ICM Technical Report **3/2012**, European Environment Agency, Copenhagen.

KOUTIVA, I. & MAKROPOULOS, C. 2012. Linking tools for social simulation and urban water network modelling for supporting an adaptive approach of urban water resources management. 6th International Congress on Environmental Modelling and Software – iEMSs 2012, 1–5 July 2012, Leipzig, Germany.

KOUTIVA, I., MAKROPOULOS,, C. & VOULVOULIS, N. 2012. Modelling the combined socio-technical system to support an adaptive approach for integrated water resources management. 12th International Conference on Hydroinformatics 2012, Hamburg.

KOUTSOYIANNIS, D. & ECONOMOU, A. 2003. Evaluation of the parameterization-simulation-optimization approach for the control of reservoir systems. *Water Resources Research*, **39**, 1170.

LATANE, B. 1981. The psychology of social impact. *American Psychologist*, **36**, 343–365.

LIU, Y. 2009. *Modelling Urban Development with Geographical Information Systems and Cellular Automata*. CRS Press, New York.

MAKROPOULOS, C. & BUTLER, D. 2005. A multi-objective evolutionary programming approach to the 'object location' spatial analysis and optimisation problem within the urban water management domain. *Civil and Environmental Systems*, **22**, 85–107.

MAKROPOULOS, C. & BUTLER, D. 2006. Spatial ordered weighted averaging: incorporating spatially variable attitude towards risk in spatial multicriteria decision-making. *Environmental Modelling & Software*, **21**, 69–84.

MAKROPOULOS, C. K. & BUTLER, D. 2010. Distributed water infrastructure for sustainable communities. *Water Resources Management*, **24**, 2795–2816.

MAKROPOULOS, C., BUTLER, D. & MAKSIMOVIC, C. 2003. Fuzzy Logic Spatial Decision Support System for Urban Water Management. *Journal of Water Resources, Planning and Management*, **129**, 69–78.

MAKROPOULOS, C. K., NATSIS, K., LIU, S., MITTAS, K. & BUTLER, D. 2008*a*. Decision support for sustainable option selection in integrated urban water management. *Environmental Modelling and Software*, **23**, 1448–1460.

MAKROPOULOS, C., MEMON, F. A., SHIRLEY-SMITH, C. & BUTLER, D. 2008*b*. Futures: an exploration of scenarios for sustainable urban water management. *Water Policy*, **10**, 345–373.

MANTELAS, L., PRASTACOS, P., HATZICHRISTOS, T. & KOUTSOPOULOS, K. 2012. A linguistic approach to model urban growth. *International Journal of Agricultural and Environmental Information Systems*, **3**, 35–53.

MOGLIA, M., PEREZ, P. & BURN, S. 2010. Modelling an urban water system on the edge of chaos. *Environmental Modelling & Software*, **25**, 1528–1538.

NALBANTIS, I. & KOUTSOYIANNIS, D. 1997. A parametric rule for planning and management of multiple reservoir systems. *Water Resources Research*, **33**, 2165–2177.

PAHL-WOSTL, C. 2006. Transitions towards adaptive management of water facing climate and global change. *Water Resources Management*, **21**, 49–62.

RAILSBACK, S. F. & GRIMM, V. 2009. *A Course in Individual-based and Agent-based Modelling*. Princeton University Press, Princeton, New Jersey.

ROZOS, E. & MAKROPOULOS, C. 2012. Assessing the combined benefits of water recycling technologies by modelling the total urban water cycle. *Urban Water Journal*, **9**, 1–10.

ROZOS, E. & MAKROPOULOS, C. 2013. Source to tap urban water cycle modelling. *Environmental Modelling & Software*, **41**, 139–150.

ROZOS, E., MAKROPOULOS, C. & BUTLER, D. 2010. Design robustness of local water-recycling schemes. *Water Resources Planning and Management*, **136**, 531–538.

ROZOS, E., BAKI, S., BOUZIOTAS, D. & MAKROPOULOS, C. 2011. Exploring the link between urban Development and water demand: the impact of water-aware technologies and options. Proceedings of the 11th International Conference on Computing and Control for the Water Industry, CCWI 2011, September 2011, Exeter, UK.

SOBKOWICZ, P. 2003. *Opinion formation in networked societies with strong leaders*. Cornell University, Ithaca, New York, http://arxiv.org/pdf/cond-mat/0311521

SU, S. & DEININGER, R. 1972. Generalization of White's method of successive approximations to periodic Markovian decision processes. *Operations Research*, **20**, 318–326.

UN 2012. *World Urbanization Prospects: The 2011 Revision*. United Nations Department of Economic and Social Affairs/Population Division, NY. www.unpopulation.org

VAN DELDEN, H., SEPPELT, R., WHITE, R. & JAKEMAN, A. J. 2011. A methodology for the design and development of integrated models for policy support. *Environmental Modelling & Software*, **26**, 266–279.

VAN DEN BULTE, C. & LILIEN, G. 1999. *A Two-Stage Model of Innovation Adoption with Partial Observability: Model Development and Application*, ISBM Report **20–1999**, Institute for the Study of Business Markets, Pennsylvania State University.

WATTS, D. & STROGATZ, S. 1998. The small world problem. *Collective Dynamics of Small-World Networks*, **393**, 440–442.

WHITE, R. & ENGELEN, G. 1997. Cellular automata as the basis of integrated dynamic regional modeling. *Environment and Planning B: Planning and Design*, **24**, 235–246.

WILLIS, R., STEWART, R., PANUWATWANICH, K., WILLIAMS, P. & HOLLINGSWORTH, A. 2011. Quantifying the influence of environmental and water conservation attitudes on household end use water consumption. *Journal of Environmental Management*, **92**, 1996–2009.

WOLFRAM, S. 1984. Universality and complexity in cellular automata. *Physica D: Nonlinear Phenomena*, **10**, 1–35.

WONG, T. 2006. Water sensitive urban design – the journey thus far. *Australian Journal of Water Resources*, **1**, 213–222.

WOOLDRIDGE, M. 1999. Intelligent agents. *In*: WEISS, G. (ed.) *Multiagent Systems*. MIT Press, Cambridge.

WRAGG, T. 2006. *Modelling the Effects of Information Campaigns Using Agent Based Simulation*, Public report DSTO-TR-1853. Department of Defence, Australian Government, Defence Science and Technology Organisation, Edinburgh, South Australia.

YOUNG, H. P. 2009. Innovation diffusion in heterogeneous populations: Contagion, social influence, and social learning. *American Economic Review*, **99**, 1899–1924.

The use of data-mining techniques for developing effective decision support systems: a case study of simulating the effects of climate change on coastal salinity intrusion

PAUL A. CONRADS[1]* & EDWIN A. ROEHL, JR.[2]

[1]*US Geological Survey – South Atlantic Water Science Center, 720 Gracern Road, Columbia, SC 29210, USA*

[2]*Advanced Data Mining International, LLC, 3620 Pelham Road, PMB 351, Greenville, SC 29615, USA*

**Correspondence: pconrads@usgs.gov*

Abstract: Natural-resource managers and stakeholders face difficult challenges when managing interactions between natural and societal systems. Potential changes in climate could alter interactions between environmental and societal systems and adversely affect the availability of water resources in many coastal communities. The availability of freshwater in coastal streams can be threatened by saltwater intrusion. Even though the collective interests and computer skills of the community of managers, scientists and other stakeholders are quite varied, there is an overarching need for equal access by all to the scientific knowledge needed to make the best possible decisions. This paper describes a decision support system, PRISM-2, developed to evaluate salinity intrusion due to potential climate change along the South Carolina coast in southeastern USA. The decision support system is disseminated as a spreadsheet application and integrates the output of global circulation models, watershed models and salinity intrusion models with real-time databases for simulation, graphical user interfaces, and streaming displays of results. The results from PRISM-2 showed that a 31-cm and 62-cm increase in sea level reduced the daily availability of freshwater supply to a coastal municipal intake by 4% and 12% of the time, respectively. Future climate change projections by a global circulation model showed a seasonal change in salinity intrusion events from the summer to the fall for the majority of events.

Natural-resource managers face the difficult problem of controlling the interactions between hydrological and societal systems in ways that preserve resources while optimally meeting the needs of disparate stakeholders (Fig. 1). The sustainable development of water resources requires the understanding of the interaction between natural processes of ecological systems and the social and economic drivers of societal systems. These two systems interact by societal consumption of goods and services from the natural ecosystem and human alterations to the natural system.

This general framework becomes especially complex for coastal communities. Coastal rivers are constantly integrating naturally complex and changing forces such as droughts, floods, tide and coastal storms. Coastal communities that depend on the rivers for municipal and industrial water supply and wastewater assimilative capacity face increased demographic pressures related to economic development, land use and population growth. These ecological and societal systems may become especially stressed with potential climate change.

To effectively regulate environmental systems, water-resource managers need direct access to scientific information, including complex computer models, to make informed decisions. Too often there is a large disparity between the technical abilities of the scientific and resource-management communities, which limits the dissemination of critical scientific information. Effective water-resource decision making depends on obtaining and utilizing detailed scientific information concerning the cause–effect relations that govern the dynamics of these hydrological systems. This information, often from complex computer models, is most credible when derived from large field-based datasets that encompass the wide range of variability in the parameters of interest. The role of models is to simulate the consequences of alternative management practices and guide resource managers towards the best path forward.

Hydrological systems are typically modelled using computer programs that implement traditional, generalized, physics equations, which are calibrated to match the field data as closely as possible. This type of mechanistic model commonly is

From: Riddick, A. T. Kessler, H. & Giles, J. R. A. (eds) 2017. *Integrated Environmental Modelling to Solve Real World Problems: Methods, Vision and Challenges*. Geological Society, London, Special Publications, **408**, 221–234.
First published online February 16, 2015, https://doi.org/10.1144/SP408.8

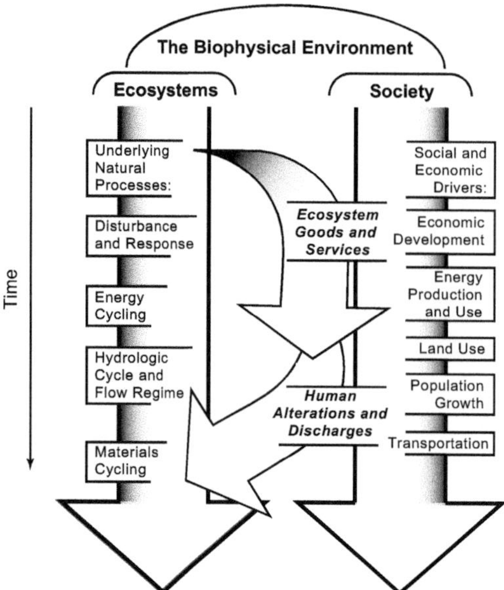

Fig. 1. General conceptual framework for the driving and underlying processes for sustainable water resources. From Sustainable Water Resources Roundtable (2006).

limited in terms of demonstrable predictive accuracy, development time and cost.

The science of data mining presents a powerful complement to physics-based models. Data mining is a relatively new science that assists in converting large databases into knowledge and is uniquely able to leverage the real-time, multivariate data now being collected for hydrological systems. Weiss & Indurkhya (1998, p. 1) describe data mining as 'the search for valuable information in large volumes of data', with its defining characteristic being large datasets, or 'big data'. In every field, big data's ubiquitous, massive databases comprise millions and even billions of records, and are the consequence of the powerful, low-cost monitoring and computing technology that has evolved over the last three decades. While large databases for environmental systems are common, they are usually under-interpreted and under-utilized.

In this paper, an application is presented that demonstrates how data-mining techniques can be applied to existing environmental databases to address regional concerns of long-term consequences. The case study describes the development of empirical artificial neural network (ANN) models to simulate salinity dynamics in the Waccamaw River and Atlantic Intracoastal Waterway (AIW) in South Carolina, USA (Conrads *et al.* 2013). The models are then used to evaluate the effect of climate change and sea-level rise at a specific conductance gauge located near two municipal water intakes. Specific conductance is a field measurement used to compute salinity.

Case study

Potential climate change could present water-resource management challenges along the East Coast of the United States. A large number of people live along the Atlantic seaboard, and an increasing population will further strain resources, particularly in the Southeast. Increased air and water temperatures, changes in regional precipitation regimes, and increased sea level may have a great impact on hydrological systems in the region.

The balance between streamflow conditions within a coastal drainage basin and sea level governs the characteristics and frequency of salinity intrusion into coastal rivers. Saltwater intrusion into freshwater coastal rivers and aquifers has been, and continues to be, one of the most important global challenges for coastal water-resource managers, industries and agriculture (Bear *et al.* 1999). Major economic and environmental consequences could include the degradation of natural ecosystems and the contamination of municipal, industrial and agricultural water supplies (Rice *et al.* 2011, 2012). Coastal communities must find approaches for sustaining their freshwater supplies while minimizing impacts on natural systems.

There are many municipal intakes along the southeastern coast of the United States with a potential for being affected by climate change (Fig. 2, Furlow *et al.* 2002). Two of the intakes are in the Waccamaw River and AIW along the South Carolina coast. Increases in the frequency and magnitude of salinity intrusion into the Waccamaw River and AIW could threaten the potability of intakes as well as the biodiversity of freshwater tidal marshes.

Description of the study area

The Waccamaw–AIW study area is located along the coast of South Carolina, an area of rapidly growing coastal communities from Little River Inlet to the north to Winyah Bay to the south (Fig. 3). The Pee Dee River Basin, with *c.* 32 900 km^2 of drainage area in eastern North and South Carolina, supplies freshwater along this portion of the South Carolina coast (US Geological Survey 1986). Six reservoirs in North Carolina discharge into the Pee Dee River, which flows 260 km through South Carolina to the coastal communities near Myrtle Beach. During the drought between 1998 and 2002, salinity intrusion forced a municipal intake near the US Geological Survey (USGS) Pawleys Island specific conductance gauge (station 021108125, Fig. 3)

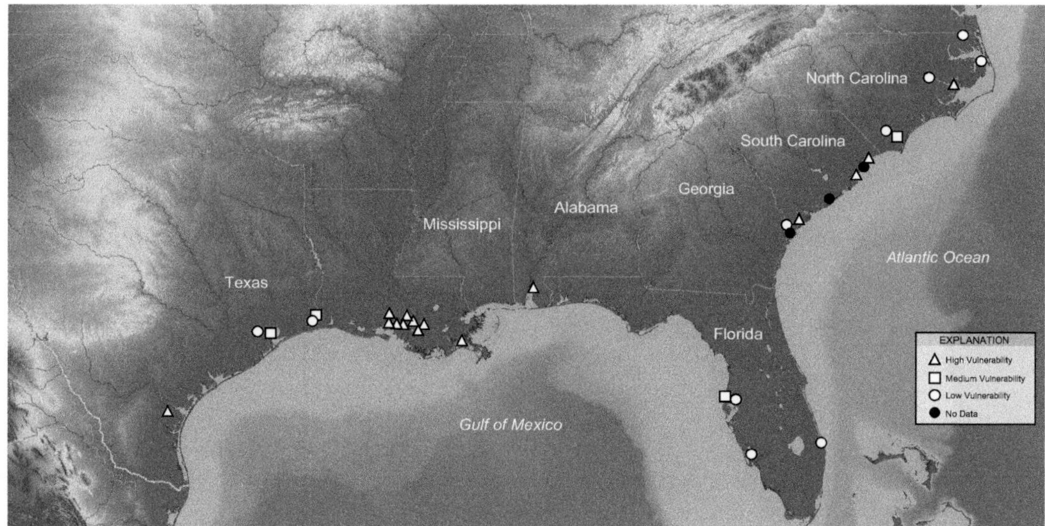

Fig. 2. Vulnerability ratings for potential salinity intrusion for selected major municipal water-supply intakes. Modified from Furlow *et al.* (2002).

to close until increased streamflow moved the freshwater–saltwater interface downstream from the intake.

Approach

Estuary salinity variability is largely driven by streamflow and tidal water levels, and climate change is expected to affect both. The location of the saltwater–freshwater interface is a balance between upstream river streamflow and downstream tidal forcing (Fig. 4). During periods of high streamflow, it is difficult for salinity to intrude upstream, and thus, the saltwater–freshwater interface moves downstream towards the ocean. During periods of low streamflow, salinity is able to intrude upstream; subsequently, the saltwater–freshwater interface moves upstream.

Empirical salinity intrusion models were developed for seven specific conductance gauging station locations in the Waccamaw–AIW study area (Fig. 3). These models use streamflow inputs from the upland rivers and water levels along the coast to simulate salinity response to changes in streamflow and/or sea-level conditions. To simulate changing streamflow in the rivers flowing to the coast, a watershed model of the Pee Dee River Basin was developed that uses precipitation and air temperature to simulate streamflow for inputs to the salinity models.

To forecast future streamflow and the effects of climate change on salinity intrusion, global circulation, regional circulation, watershed runoff and salinity intrusion models were implicitly integrated (Fig. 5). Global circulation models (GCM) make large-scale (>250-km^2 grid) estimations of precipitation and air temperature conditions for various carbon emission scenarios. These scenarios are typically 100-year projections into the future. To generate precipitation and air temperature predictions for a watershed (approximate 12-km^2 grid), the GCM output is downscaled to local scales using statistical or dynamic downscaling techniques (Vrac *et al.* 2007). The downscaled precipitation and air temperature data are input to a watershed model. The watershed model then simulates the streamflow inputs to an estuary salinity intrusion model. Below is a description of the salinity intrusion, watershed and GCMs used in the study.

Salinity intrusion models

Results from previously developed ANN-based models of estuaries in Georgia and South Carolina (Roehl *et al.* 2000; Conrads *et al.* 2003, 2006; Conrads & Roehl 2007, 2010; Conrads *et al.* 2011) have shown that ANN models, combined with data-mining techniques, are an effective approach for simulating complex estuarine systems. An ANN model is a flexible mathematical structure capable of describing complex nonlinear relations between input and output datasets. The architecture of ANN models is loosely based on the biological nervous system (Hinton 1992). Although there are numerous types of ANNs, the most commonly used type, and the one used in this study, is the multilayer perceptron (MLP)

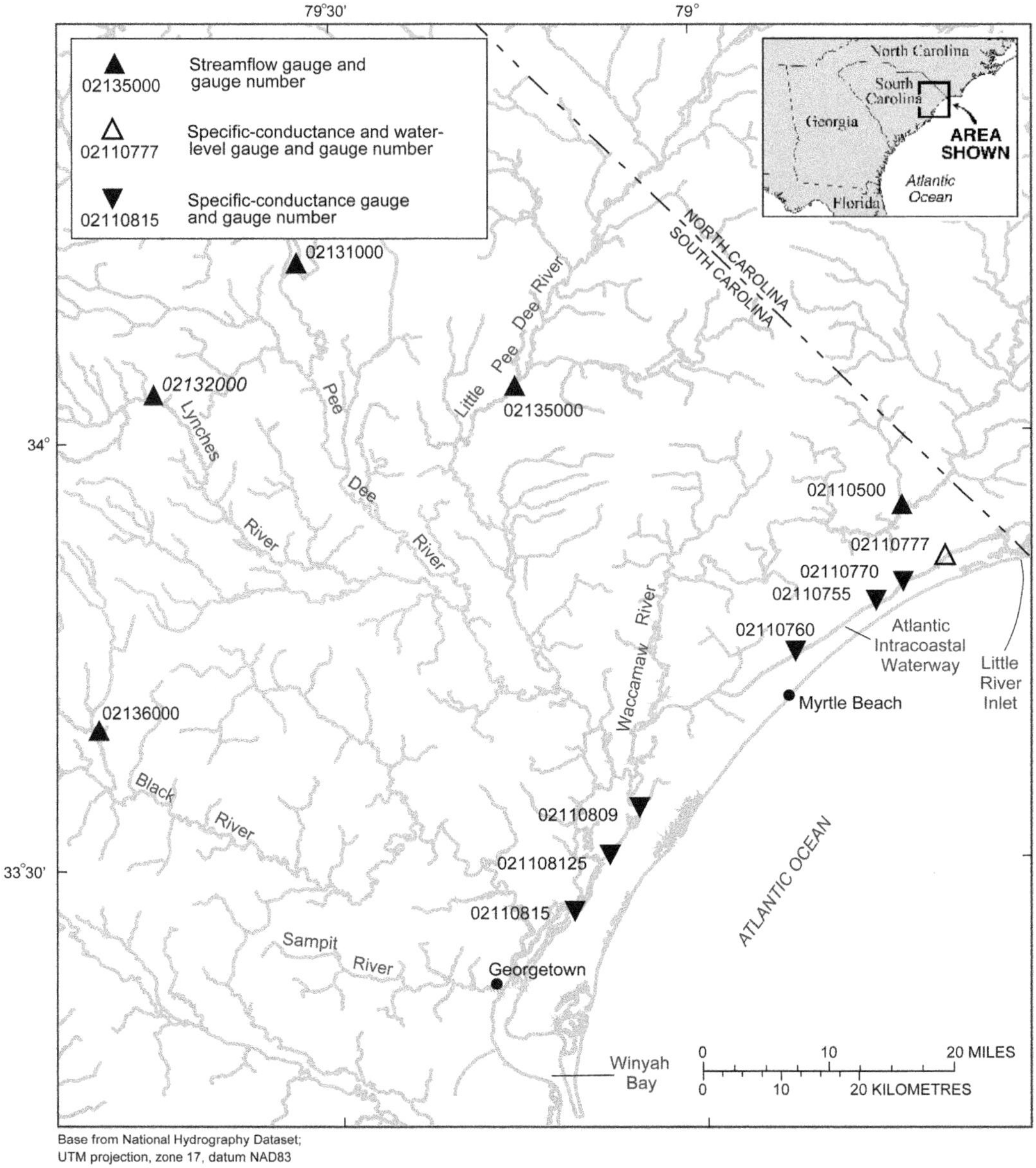

Base from National Hydrography Dataset;
UTM projection, zone 17, datum NAD83

Fig. 3. Location of the Waccamaw River and Atlantic Intracoastal Waterway study area and selected gauges in the Pee Dee River Basin, South Carolina. Modified from Conrads & Roehl (2007).

(Rosenblatt 1958; Jensen 1994), which is a multivariate, non-linear regression method based on machine learning.

The authors had previously developed ANN specific conductance models of the Waccamaw River and AIW (Conrads & Roehl 2007), which

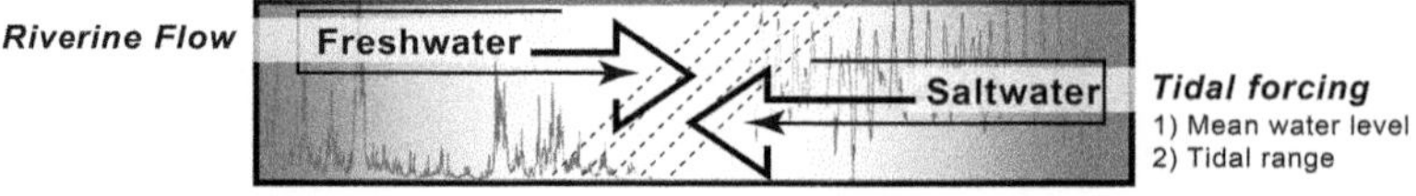

Fig. 4. Conceptual model of the major factors controlling the freshwater–saltwater interface.

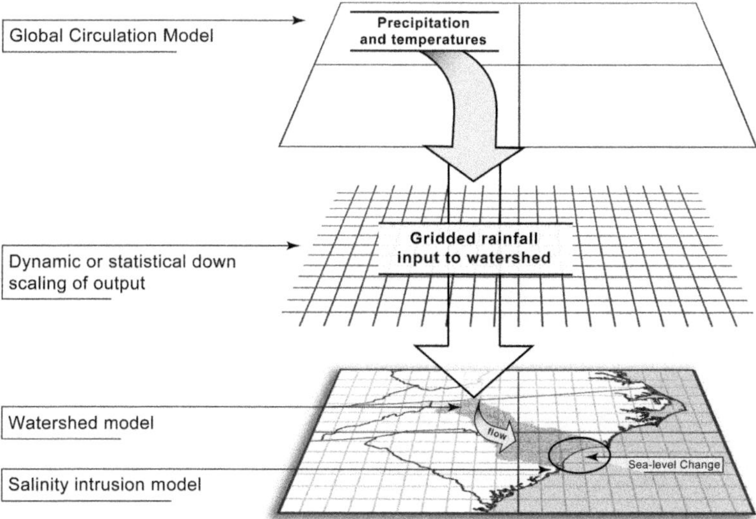

Fig. 5. Schematic diagram showing the conceptual modelling approach for evaluation of the effects of climate change on salinity intrusion, North and South Carolina.

were integrated into a decision support system (DSS). The DSS integrates the models, the databases for running simulations, a graphical user interface and supporting graphics into a spreadsheet application. The DSS as a spreadsheet application makes it immediately useable by people with varying degrees of computer skills, and provides equal access to the scientific knowledge needed to make the best possible decisions (Fig. 6, Roehl *et al.* 2006).

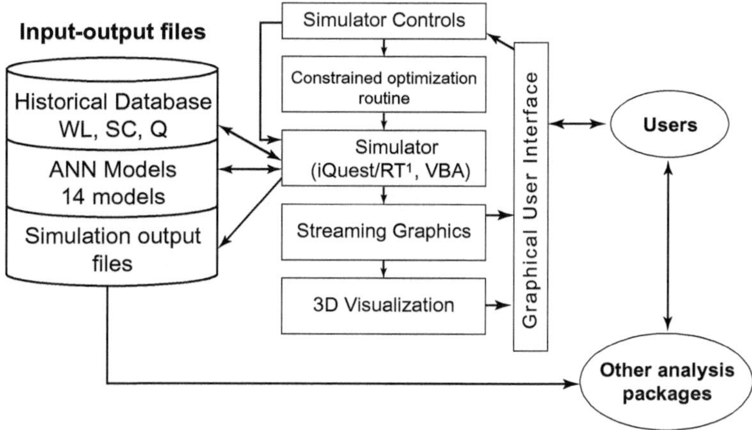

Fig. 6. Architecture of the PRISM-2 decision support system (PRISM-2 DSS). Modified from Conrads & Roehl (2007). WL, water level; SC, specific conductance; Q, flow; ANN, artificial neural network; RT, run-time; VBA, Visual Basic for Applications.

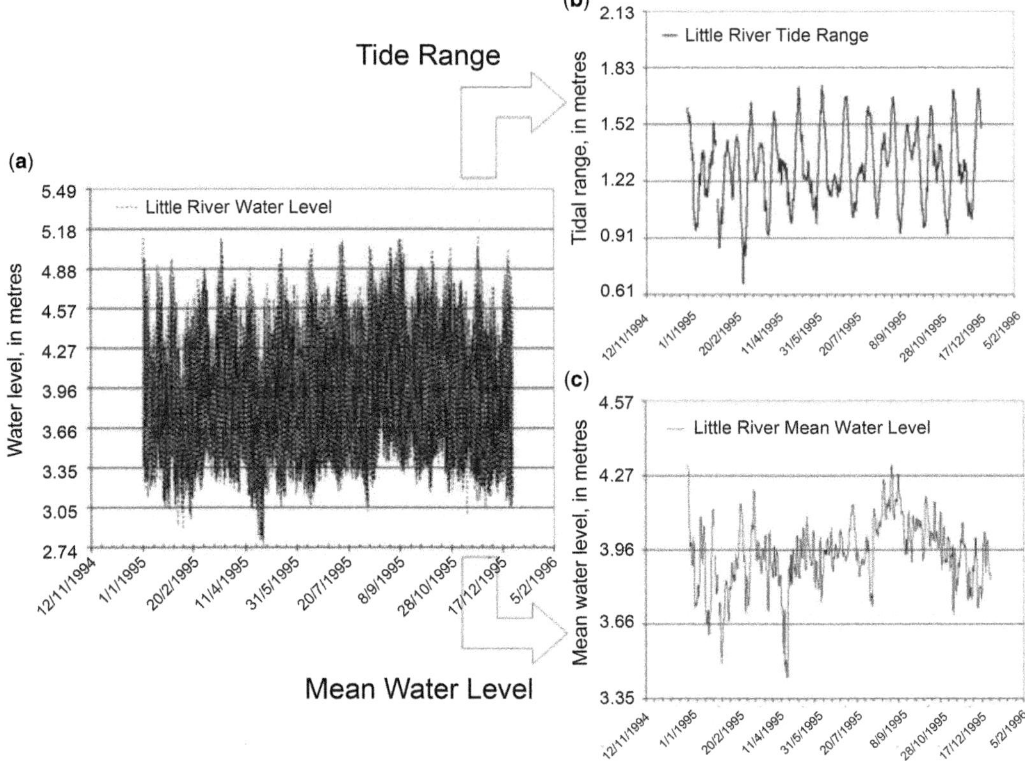

Fig. 7. Little River Inlet (**a**) tidal water level signal, (**b**) decomposition into a periodic signal of tide range and (**c**) a chaotic signal of mean water level.

The ANN models convert inputs that primarily represent freshwater streamflow and sea level into simulations of specific conductance at several locations. The USGS monitoring sites on the Waccamaw River and AIW that provided much of the data used to develop the specific conductance models in the DSS are shown in Figure 3. The DSS is named the **P**ee Dee **R**iver and Atlantic **I**ntracoastal Waterway **S**alinity Intrusion **M**odel DSS (PRISM DSS) and uses streamflows from five streamflow gauges and water levels from one coastal gauge for inputs to the salinity intrusion models. The DSS was developed to support the process of relicensing of the hydroelectric facilities, which occurs at 50-year intervals.

To evaluate the effect of potential climate change on salinity intrusion along the coast, the original PRISM DSS was redeveloped using additional field data, and modified to allow users to adjust the sea-level input and unregulated streamflow. The modified DSS was renamed PRISM-2 (Fig. 6). The PRISM-2 data span 14 years from 1995 to 2009, and include a broad range of climate and sea-level behaviours, including record droughts, high rainfall periods of El Niño climate events and hurricanes, which can greatly increase sea levels. Developing the empirical ANN models from this broad range of data behaviour makes the models

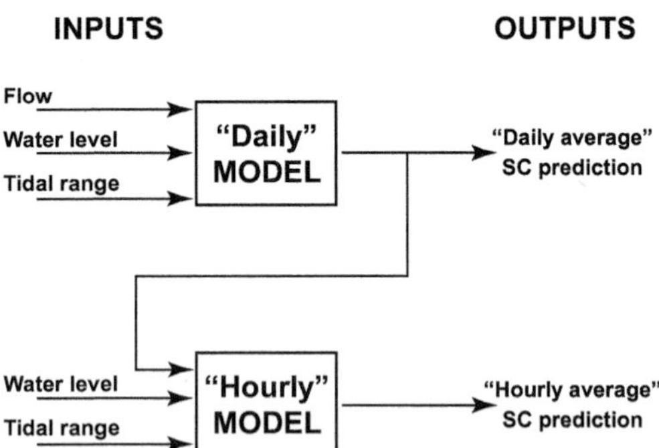

Fig. 8. Schematic diagram showing the two-stage model architecture to simulate specific conductance (SC) at a gauging station. This is an example of a 'cascading' model where the output from the first model (daily average SC) is used as an input to the second model.

more applicable to studies involving climate change and sea-level rise scenarios.

Data sets and data preparation. Tidal systems are dynamic and exhibit complex behaviours that evolve over multiple time scales. The hydrodynamic and water-quality behaviours observed in estuaries are superpositions of behaviours forced by periodic planetary motions and chaotic meteorological disturbances. The primary periodic input to the system is the tide. The primary chaotic inputs to this system, strongly influenced by meteorological disturbances, are the oceanic disturbances and are represented in the model by the water level in Little River Inlet.

Signals were decomposed into chaotic and periodic components using numerical filtering techniques. To filter the semi-diurnal tidal signal to extract the chaotic signal, nested 13- and 25-hour moving window averages were applied to the water level and specific conductance time series. The resulting time series represents the daily change (computed on a 60-minute time increment) in the tidal signal for water level and specific conductance.

Tidal dynamics are a dominant force for estuarine systems, and tidal range is an important variable for determining the lunar phase of the tide. Tidal range is calculated from water level and is defined as the water level at high tide minus the water level at low tide for each semi-diurnal tidal cycle. As shown in Figure 7, the measured water level at Little River Inlet (station 02110777, Fig. 7a) was decomposed into its periodic signal of tidal range time series (Fig. 7b) and its chaotic signal of mean water level time series (Fig. 7c).

Simulation of salinity intrusion. Subdividing a complex modelling problem into subproblems and then addressing each one separately is a means to developing an accurate simulation model. For the Waccamaw River–AIW study, individual ANN models for simulating daily specific conductance were developed for each coastal specific conductance gauge. The models were developed in two stages (Fig. 8). The first stage modelled the chaotic, lower-frequency part of the signal, as represented by the filtered specific conductance signals. The second stage modelled the periodic, higher-frequency, hourly specific conductance, using the daily simulated specific conductance as a carrier signal. Each model uses three general types of signals, or time series: streamflow signal(s), water level signal(s), and tide range signal(s). The signals may be of the measured series values, filtered values, and/or a time derivative of the signals. Most of the datasets that were used to develop the models were randomly bifurcated into training and testing datasets. Some datasets were too small to provide data for testing but were fitted conservatively. All ANN models were carefully evaluated to ensure the models did not 'overfit' the data. Generally, the daily models had coefficients of determination (R^2) values ranging from 0.62 to 0.96. An example of the measured and simulated daily and hourly specific conductance is shown in Figure 9. Note, the model is able to simulate the sharp specific conductance spikes.

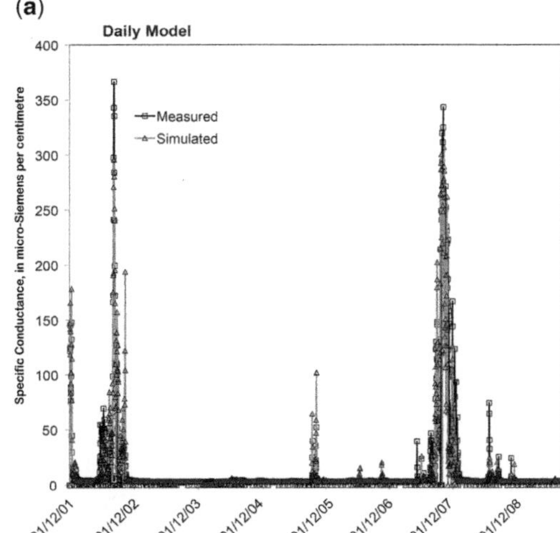

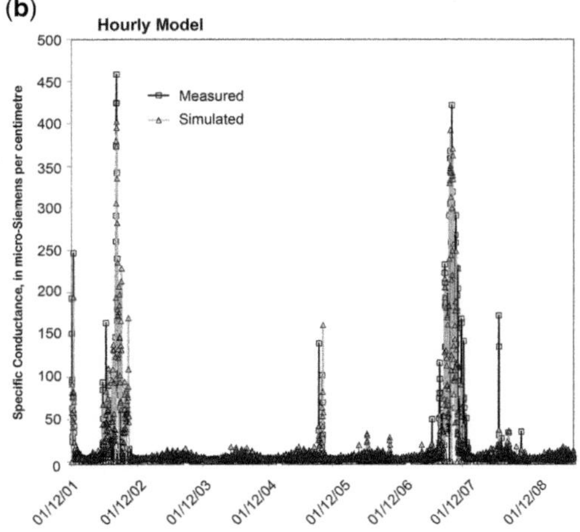

Fig. 9. Graphs showing measured and simulated specific conductance for the Pawleys Island gauge location (station 021108125) for (**a**) the daily model and (**b**) the hourly model.

Pee Dee River Basin watershed model

The University of South Carolina applied and calibrated the Hydrologic Simulation Program-Fortran (HSPF) watershed model from the US Environmental Protection Agency for the 32 900-km^2 Pee

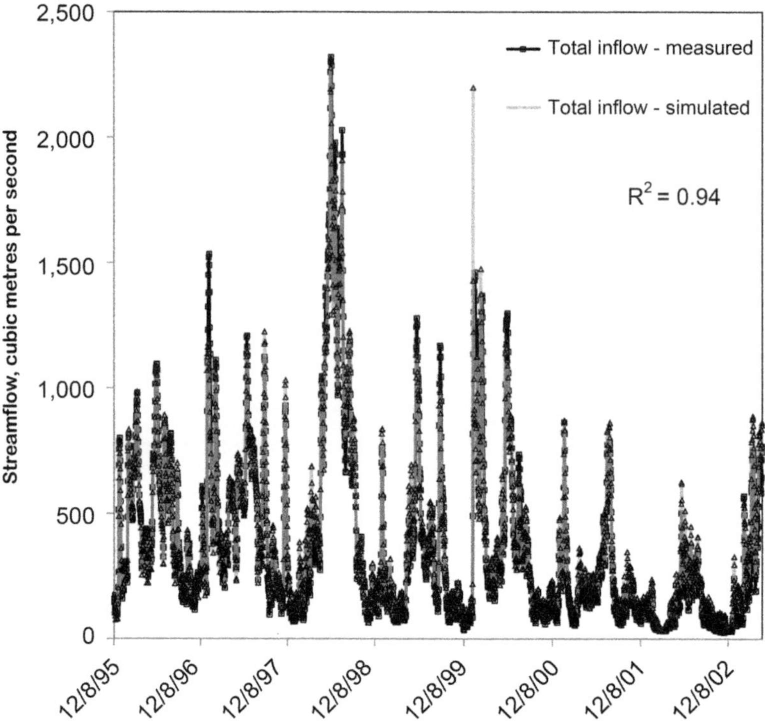

Fig. 10. Measured and simulated total daily inflow to the Waccamaw River and Atlantic Intracoastal Waterway study area. Streamflow simulated using the Hydrologic Simulation Program-Fortran (HSPF) watershed model.

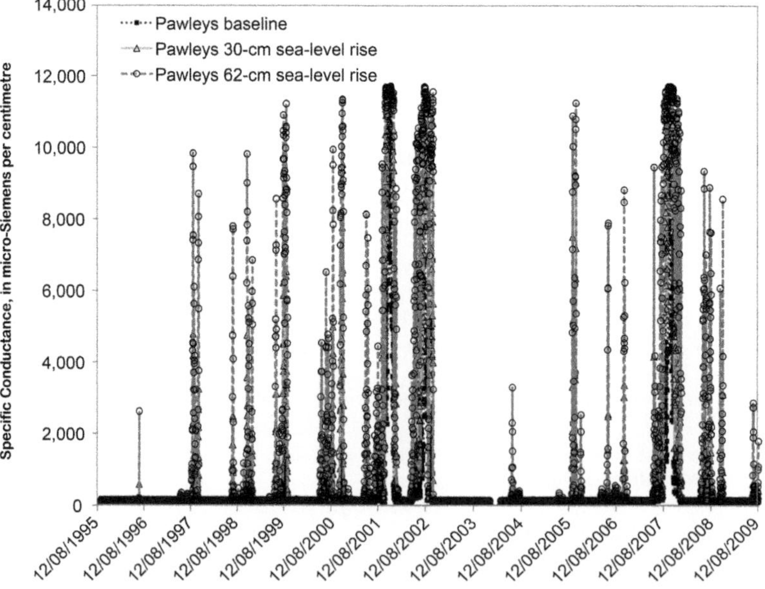

Fig. 11. Simulated specific conductance at the Pawleys Island gauge location (station 021108125) for baseline (historical) conditions and for a 31-cm and 62-cm sea-level rise for the period 12 August 1995 to 21 August 2009.

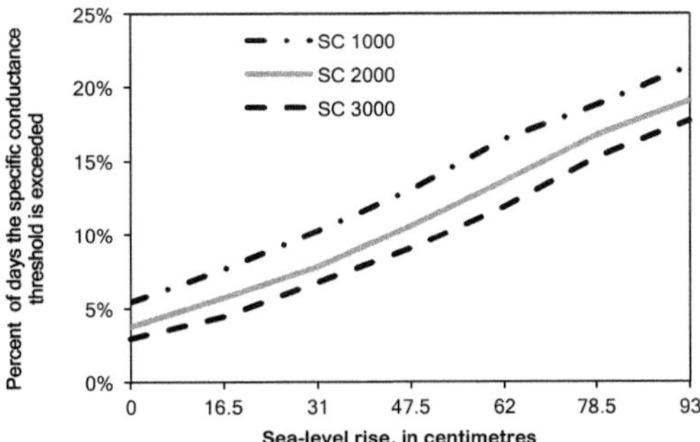

Fig. 12. The percent of days specific conductance (SC) thresholds of 1000, 2000, and 3000 microsiemens per centimetre are exceeded for incremental sea-level rise at the Pawleys Island gauge location (station 021108125) for the simulation period of 1995–2009.

Dee watershed. The HSPF model was calibrated using *c.* 30 years of historical climate (precipitation and air temperature) and streamflow data. The HSPF model simulates streamflow at the five streamflow gauging stations used for input to PRISM-2. Measured and simulated daily total inflow (summation of the five streamflow gauging stations shown on Fig. 3) is shown in Figure 10.

Global circulation model

There are many existent GCMs, four of which were evaluated for this study. The models included the Community Climate System Model 3 (CCSM3), the Parallel Climate Model (PCM), the Geophysical Fluid Dynamics Laboratory model (GFDL) and a hybrid of the European Center atmospheric GCM (ECHAM4) and Hamburg Primitive Equation ocean GCM (HOPE-G) models, the ECHO-G model. Statistical downscaling techniques were used to transform the GCM output to the scale of local interest (Dalton & Jones 2010). The ECHO-G model (Legutke & Voss 1999) was chosen because it simulates historical low-streamflow conditions more accurately than the other evaluated GCMs when coupled with the HSPF application.

Alternative emission scenarios cause GCMs to forecast different climate conditions. To simulate future streamflow, ECHO-G was run with the A2 future carbon emissions scenario, which assumes that nations will continue to pursue their interests individually rather than cooperate in dealing with climate change (IPCC 2000).

The ECHO-G A2 scenario projection was used as input to the Pee Dee River Basin HSPF model to generate streamflow inputs for the PRISM-2 DSS. The climate projection evaluated was for the

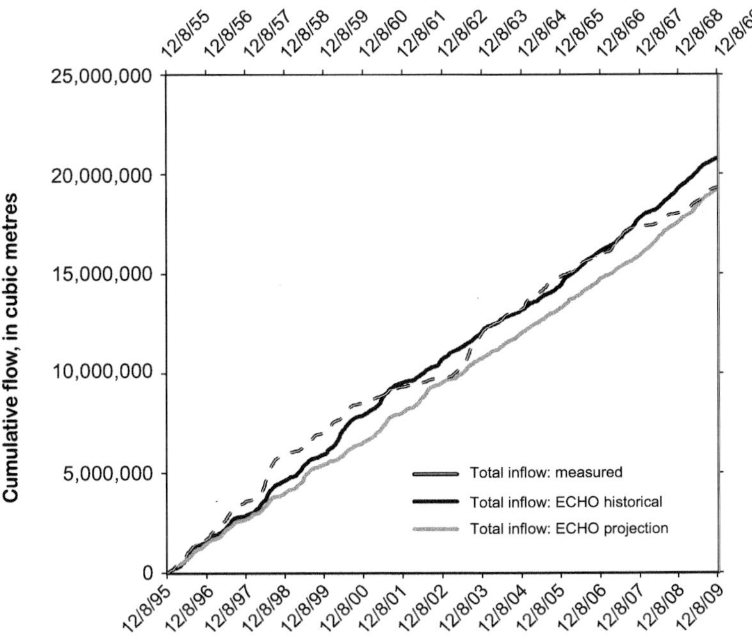

Fig. 13. Plots showing the cumulative flow for measured and simulated total inflow to the Waccamaw River and Atlantic Intracoastal Waterway study area for the period 1995–2009 and projected simulated total inflow for the period 2055–2069. Total inflow simulated using the ECHO-G global circulation model and the Hydrologic Simulation Program-Fortran (HSPF) watershed model for the historical and projected periods.

years 2055–2069, 60 years from the start of the PRISM-2 study period.

PRISM-2 DSS results

The PRISM-2 DSS provides a tool for water-resource managers from state and local agencies to evaluate changes in the timing, magnitude, frequency and duration of salinity at streamflow gauging station locations as the result of potential climate change. Prior to the final determination of potential effects of climate change on salinity intrusion near municipal intakes, several issues concerning protection of coastal water intakes must be addressed. What is the maximum specific conductance value that is acceptable? What level of protection from salinity intrusion is required? Is an intake required to stay online 100% of the time, or are limited intrusion events acceptable? Can the increase of salinity at the intake be minimized by increases in regulated flow? The DSS allows users to simulate changes in sea level and streamflow, separately and in combination, and analyse the specific conductance response at seven streamflow gauging stations along the AIW; the DSS is therefore an aid to understanding the complex interaction of sea level, streamflow and salinity dynamics in these estuarine systems. The following sections describe the application of the DSS to evaluate salinity intrusions for various climate change scenarios including sea-level rise and changes in streamflow.

Sea-level rise. To simulate the effects of sea-level rise, the mean coastal water levels were increased to 31 cm and 62 cm to simulate sea-level rises of up to 93 cm in PRISM-2. The specific conductance response to a 31-cm and a 62-cm sea-level rise at the Pawleys Island specific conductance gauge (station 021108125, Fig. 3), just downstream from a municipal freshwater intake, is shown for the period 12 August 1995 to 21 August 2009 in Figure 11. The baseline is the model simulation of the measured data for the period 1995–2009. In this simulation, the magnitude and duration of the large intrusion

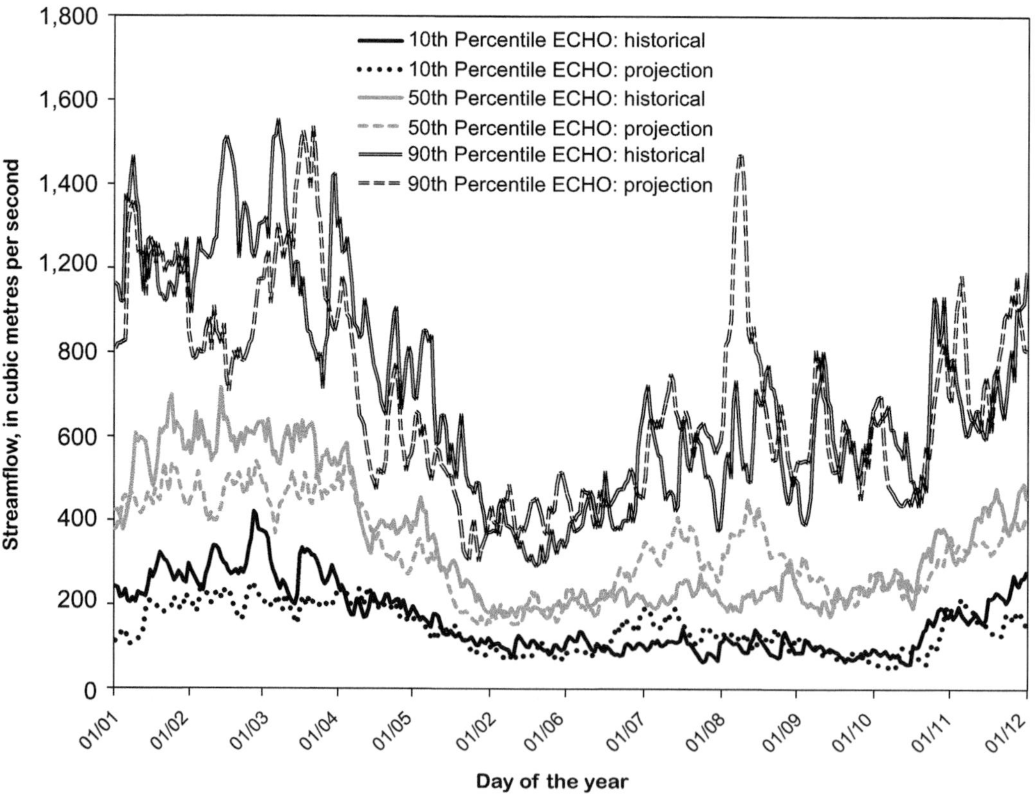

Fig. 14. Daily duration hydrographs for simulated total inflow to the Waccamaw River and the Atlantic Intracoastal Waterway study area for historical (1980–2010) conditions, and future (2040–2070) projections of climatic conditions. Historical and projected total inflow simulated using the ECHO-G global circulation model and the Hydrologic Simulation Program-Fortran (HSPF) watershed model for the historical and projected periods.

events of 2002 and 2007 are increased with a 31- and 62-cm sea-level rise. The frequency and magnitude of the smaller intrusion events between 2002 and 2007 increase from one event greater than 2000 microsiemens per centimetre ($\mu S\ cm^{-1}$) to four events over 2000 $\mu S\ cm^{-1}$ with a 31-cm sea-level rise, and five events over 6000 $\mu S\ cm^{-1}$ with a 62-cm sea-level rise.

For the operators of municipal water-treatment plants, often the concern with the source water is not the magnitude of the salinity intrusion but whether the source water at the intake exceeds a predetermined threshold level. When the specific conductance values of source water are greater than 1000–2000 $\mu S\ cm^{-1}$, additional treatment (and/or) blending with lower-conductance water is required to eliminate taste problems and potential health concerns. Graphs were generated from six sea-level rise scenario simulations with PRISM-2. The mean coastal water levels were increased in increments of 15.5 cm to simulate sea-level rises of up to 93 cm and the number of days the simulated specific conductance values exceeded thresholds of 1000, 2000, and 3000 $\mu S\ cm^{-1}$ for each 14-year simulation period was computed. Figure 12 shows the percent of days that a specific conductance threshold was exceeded for sea-level rises from 0 to 93 cm for the Pawleys Island gauge. For example, under historical conditions, the simulated historical daily specific conductance at the Pawleys Island gauge exceeded 2000 $\mu S\ cm^{-1}$ for 3.7% of the days. Simulations of the same conditions incorporating a 31-cm sea-level rise indicate that the number of days the municipal water at the intake is unsuitable for supply would double to 7.8% of days, and a 62-cm rise would increase the number of days where the specific conductance of water at the intake exceeded 2000 $\mu S\ cm^{-1}$ by 16.6%.

Future climate projection using a global circulation model. The downscaled air temperature and precipitation data for the ECHO-G GCM for the A2 carbon emission climate scenario were used as input to the HSPF watershed model. Simulations of streamflows for the Waccamaw, Little Pee Dee, Pee Dee, Lynches and Black Rivers for the periods 1980–2010 and 2040–2070 were obtained from the HSPF watershed model and used as input to PRISM-2 (total inflow, QTOTAL). For the simulation period of PRISM-2 (1995–2009), cumulative streamflow curves (Fig. 13) were generated for (1)

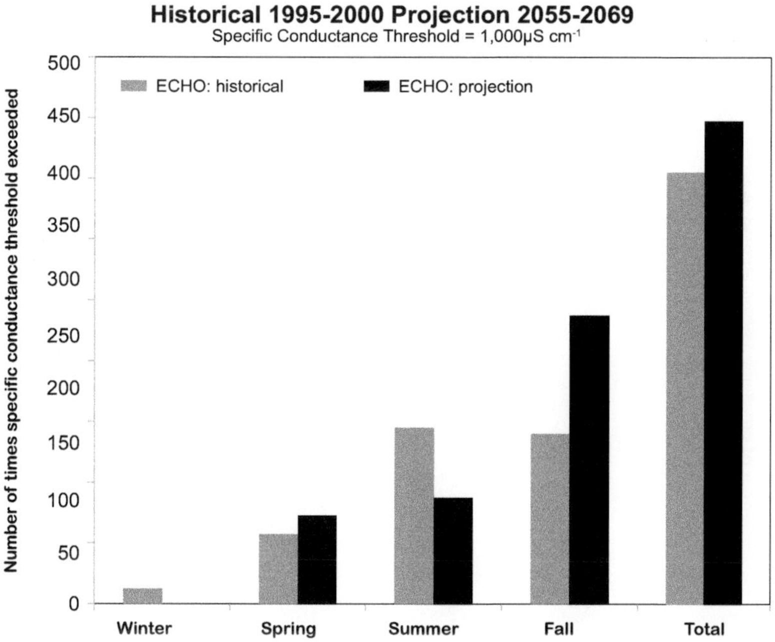

Historical 1995-2000 Projection 2055-2069
Specific Conductance Threshold = 1,000μS cm-1
ECHO: historical
ECHO: projection
Number of times specific conductance threshold exceeded
Winter
Spring
Summer
Fall
Total

Fig. 15. Number of days per season in which the daily specific conductance threshold of 1000 microsiemens per centimetre is exceeded at the Pawleys Island gauge location (station 021108125) for the simulated historical (1995–2009) and projected (2055–2069) daily specific conductance values. Simulated historical and projected conditions based on the total of the Waccamaw River and Atlantic Intracoastal Waterway study area using the ECHO-G global circulation model and the Hydrologic Simulation Program-Fortran (HSPF) watershed model.

the total streamflow for the measured data, (2) the ECHO-HSPF simulated streamflow for the period 1995–2009 and (3) the streamflow projected 60 years in the future for the period 2055–2069. The ECHO-HSPF streamflow volume estimates (Total Inflow: ECHO-G historical, Fig. 13) are fairly good compared to the measured data considering the number of model simulation and data processing steps to simulate streamflow (large-scale ECHO-G simulation output of precipitation and air temperature, statistically downscaled precipitation and air temperature inputs to HSPF, HSPF simulated streamflow output). The cumulative streamflow curves show that the resulting ECHO-HSPF streamflow for the period 1995–2009 simulate approximately the same order of magnitude of cumulative flow based on the measured data. The ECHO-HSPF streamflow projection for 2055–2069 shows a decrease in the cumulative total streamflow to the coast.

To evaluate seasonal shifts in the projected streamflow, daily duration hydrographs were plotted using the simulated ECHO-HSPF streamflow for the periods 1980–2010 and 2040–2070 for the total streamflow to the coast (Fig. 14). Although the simulated historical and simulated projected streamflows appear similar there are important differences, especially with respect to potential changes in salinity intrusion. Salinity intrusion along the coast is a low-streamflow phenomenon; therefore, changes in the distribution of low streamflow are the most critical for assessing the potential for salinity intrusion. During the winter months (January–March), the simulated projected low streamflow (10th percentile) are less than the corresponding simulated historical low streamflow, and during the spring months (April–June), the projected low streamflow are only slightly less than simulated historical low streamflow. During the summer months (July–September), simulated projected low streamflow are lower in the beginning of July but are higher than simulated historical low streamflow in August and September. In the fall months (October–December), the simulated projected low streamflow are less than the corresponding historical simulated streamflow.

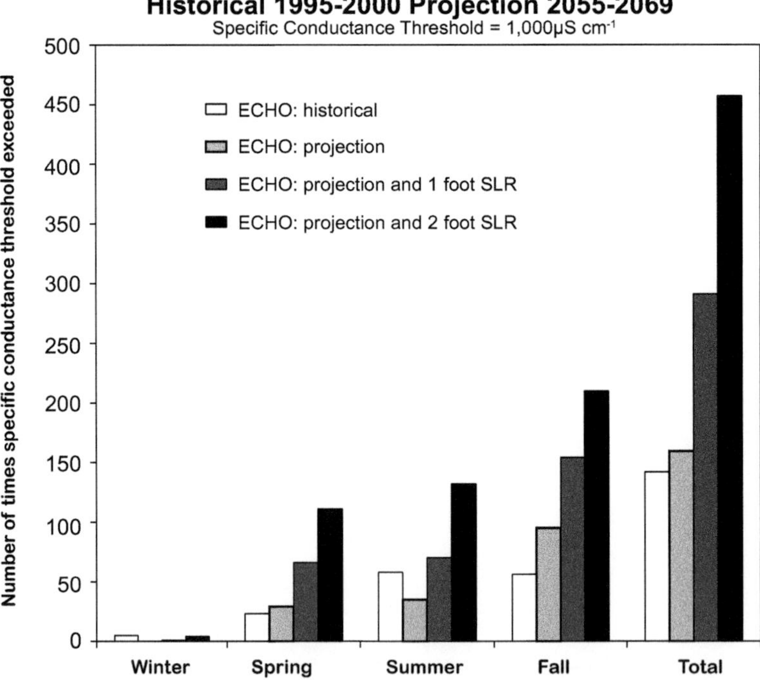

Fig. 16. Number of days each season the specific conductance threshold of 1000 microsiemens per centimetre is exceeded at the Pawleys Island gauge location (station 021108125) for the scenarios of simulated historical (1995–2009), projected (2055–2069) climate changes and projected climate changes in combination with a 31-cm sea-level rise (SLR) and a 62-cm SLR. Simulated historical and projected conditions based on the simulated total inflow to the Waccamaw River and Atlantic Intracoastal Waterway study area using the ECHO-G global circulation model and the Hydrologic Simulation Program-Fortran (HSPF) watershed model.

One approach to comparing the simulations of historic and projected specific conductance for the Pawleys Island gauge is to determine if there are seasonal shifts in the occurrence of salinity intrusion events that correspond to the seasonal changes in the low streamflow seen in the ECHO-HSPF simulated projected streamflow. To do this, the number of days in each season for which the specific conductance exceeded a threshold of 1000 μS cm^{-1} were counted (Fig. 15). The simulation of the historical specific conductance shows the largest number of days over 1000 μS cm^{-1} occurs in the summer (July–September) with slightly fewer days in the fall (October–December). The specific conductance simulation for the projected streamflow condition shows an interesting change in the occurrence of intrusion events greater than 1000 μS cm^{-1}. Relative to the historical data, the number of intrusion events increases in the spring, decreases in the summer, and substantially increases in the fall. Under potential future climate change, there will still be the convergence of conditions of low streamflow and high coastal water level to initiate an intrusion event. The decrease in the number of summer intrusion events seen in Figure 15 corresponds to an increase in the 10th percentile projected streamflow seen in Figure 14. Likewise, the increase in fall intrusion events corresponds to the slight decrease in the low streamflow in November and December. The decrease in the winter 10th percentile streamflow, although substantial, is not sufficient to cause an intrusion event.

The ECHO-HSPF streamflows also were simulated in PRISM-2 in combination with a 31-cm and 62-cm sea-level rise (Fig. 16). The seasonal shift of the majority of days with specific conductance greater than the 1000 μS cm^{-1} threshold from the summer to the fall with the projected streamflow conditions also occurs and increases in conjunction with 31-cm and 62-cm increases in sea level.

Discussion

The general problem for water-resource managers who need to plan for future circumstances, such as climate change and sea-level rise, is acquiring knowledge about how anticipated changes might specifically affect 'their' resources, so that they can make informed decisions. The problem can seem intractable because the details about what, when and how much is not known with any certainty. The case study presented here demonstrates a pathway to producing a knowledge-acquisition tool in the form of a DSS. The case study also demonstrates how a DSS can be developed to address a particular issue, such as the effect of reservoir flow releases on coastal salinity intrusion, and can be adapted with new data and information to address related issues, such as potential climate change effects on salinity intrusion.

A DSS's credibility is based on its integration of the best available science, including:

(1) *Long-term time-series data*, which capture a broad range of complex system behaviours. For the case study, the available data inherently spanned much of the range of change that experts predict will occur with climate change and sea-level rise.

(2) *Predictive models*, which accurately represent the physical processes captured by the long-term time-series data.

The fact that the DSS can be readily distributed and used directly by all stakeholders, regardless of computer skills, makes it a vehicle for sharing information, and supporting negotiations among competing stakeholders. For example, the original PRISM DSS revealed to reservoir operators, downstream water users, coastal utilities, and state and federal regulators that, under some rainfall and sea-level conditions, controlled streamflow could mitigate intake inundation, and under other conditions it cannot. A shared knowledge of how the hydrological system actually works, and reasonable estimates about the consequences of alternative management scenarios, increases the likelihood for the emergence of a well-informed management plan.

The submitted paper has been made possible through cooperation and funding from Beaufort-Jasper Water and Sewer Authority (BJWSA) and the Water Research Foundation (WaterRF). The authors thank Mr C. Sexton, Director of Engineering with BJWSA, and Dr K. Ozekin, Senior Account Manager with WaterRF for their technical assistance and coordination.

References

Bear, J., Cheng, A. H. D., Sorek, S., Ouazar, D. & Herrera, I. (eds). 1999. *Seawater Intrusion in Coastal Aquifers – Concepts, Methods and Practices*. Kluwer Academic Publishers, Dordrecht, The Netherlands.

Conrads, P. A. & Roehl, E. A., Jr. 2007. *Analysis of Salinity Intrusion in the Waccamaw River and the Atlantic Intracoastal Waterway near Myrtle Beach, South Carolina, 1995–2002*. US Geological Survey, Scientific Investigations Report 2007-5110, 2 apps.

Conrads, P. A. & Roehl, E. A., Jr. 2010. *Analysis and Simulation of Water-Level, Specific Conductance, and Total Phosphorus Dynamics of the Loxahatchee National Wildlife Refuge, Florida, 1995–2006*. US Geological Survey, Scientific Investigations Report 2010-5244.

Conrads, P. A., Roehl, E. A., Jr. & Martello, W. P. 2003. Development of an empirical model of a complex, tidally affected river using artificial neural networks. Paper presented at the Water Environment Federation TMDL Specialty Conference, Chicago, Illinois, 16–18 November 2003.

Conrads, P. A., Roehl, E. A., Jr., Daamen, R. C. & Kitchens, W. M. 2006. *Simulation of Water Levels and Salinity in the Rivers and Tidal Marshes in the Vicinity of the Savannah National Wildlife Refuge, Coastal South Carolina and Georgia.* US Geological Survey, Scientific Investigations Report 2006-5187.

Conrads, P. A., Roehl, E. A., Jr. & Davie, S. R. 2011. *Simulation of Specific Conductance and Chloride Concentration in Abercorn Creek, Georgia, 2000–2009.* US Geological Survey, Scientific Investigations Report 2011-5074.

Conrads, P. A., Roehl, E. A., Jr, Daamen, R. C. & Cook, J. B. 2013. *Simulation of Salinity Intrusion Along the Georgia and South Carolina Coasts using Climate-change Scenarios.* US Geological Survey, Scientific Investigations Report 2013-5036.

Dalton, M. S. & Jones, S. A. (eds) 2010. *Southeast Regional Assessment Project for the National Climate Change and Wildlife Science Center.* US Geological Survey Open-File Report 2010–1213.

Furlow, J., Scheraga, J. D., Freed, R., Rock, K. & Joel, D. 2002. The vulnerability of public water systems to sea level rise. *In*: Lesnik, J. R. (ed.) *Proceedings of the Coastal Water Resource Conference.* American Water Resources Association, Middleburg, Virginia, 13–15 May 2002, TPS-02-1, 31–36.

Hinton, G. E. 1992. How neural networks learn from experience. *Scientific American*, September issue, 145–151.

IPCC (Intergovernmental Panel on Climate Change) 2000. *Special Report on Emissions Scenarios.* Cambridge University Press, UK.

Jensen, B. A. 1994. *Expert Systems – Neural Networks, Instrument Engineers' Handbook.* 3rd edn. Chilton, Radnor PA.

Legutke, S. & Voss, R. 1999. *The Hamburg Atmosphere-Ocean Coupled Circulation Model ECHO-G.* Technical Report No. 18, German Climate Computer Centre (DKRZ), Hamburg.

Rice, K. C., Bennett, M. R. & Shen, J. 2011. *Simulated Changes in Salinity in the York and Chickahominy Rivers from Projected Sea-level Rise in Chesapeake Bay.* US Geological Survey Open-File Report 2011-1191.

Rice, K. C., Hong, B. & Shen, J. 2012. Assessment of salinity intrusion in the James and Chickahominy Rivers as a result of simulated sea-level rise in Chesapeake Bay, East Coast, USA. *Journal of Environmental Management*, **111**, 61–69, https://doi.org/10.1016/j.jenvman.2012.06.036

Roehl, E. A., Jr., Conrads, P. A. & Roehl, T. A. 2000. Real-time control of the salt front in a complex, tidally affected river basin. *In*: *Proceedings of the Artificial Neural Networks in Engineering Conference*, St Louis, Mo, 5–8 November 2000, 947–954.

Roehl, E. A., Jr., Conrads, P. A. & Daamen, R. C. 2006. Features of advanced decision support systems for environmental studies, management, and regulation. *In*: *Proceedings of the International Environmental Modeling and Software Society Summit.* Burlington, VT, 9–13 July 2006, http://www.iemss.org/iemss2006/papers/s10/132_Roehl_1.pdf.

Rosenblatt, F. 1958. The perceptron: a probabilistic model for information storage and organization in the brain. *Psychological Review*, **65**, 386–408.

Sustainable Water Resources Roundtable 2006. Factsheet, http://acwi.gov/swrr/swrr-fs.pdf

US Geological Survey 1986. National water summary 1985; hydrologic events and surface water resources. US Geological Survey Water-Supply Paper 2300.

Vrac, M., Stein, M. & Hayhoe, K. 2007. Statistical downscaling of precipitation through a nonhomogeneous stochastic weather typing approach. *Climate Research*, **34**, 169–184.

Weiss, S. M. & Indurkhya, N. 1998. *Predictive Data Mining – A Practical Guide.* Morgan Kaufmann Publishers, Inc., San Francisco.

Fusing and disaggregating models, data and analysis tools for a dynamic science–society interface

E. C. ROWE[1]*, D. G. WRIGHT[2], N. BERTRAND[2] & S. REIS[3]

[1]*Centre for Ecology & Hydrology, Environment Centre Wales, Deiniol Road, Bangor, LL57 2UW, UK*

[2]*Centre for Ecology & Hydrology, Lancaster Environment Centre, Library Avenue, Bailrigg, Lancaster, LA1 4AP, UK*

[3]*Centre for Ecology & Hydrology, Bush Estate, Penicuik, Midlothian, EH26 0QB, UK*

**Correspondence: ecro@ceh.ac.uk*

Abstract: Society requires rapid, most-probable predictions for specific and/or multifaceted questions related to environmental and geological science. In principle, models that encapsulate disciplinary knowledge are useful tools for making predictions and testing theory, but academic rewards favour disciplinary specialism and a proliferation of often insufficiently tested models. Decision makers have to assess the quality and robustness of predictions for complex environmental issues, and may prefer a model that performs accurately in a case study to a more parsimonious and generalizable model. Predictive ecosystem models tend to grow, as more processes are considered, even when a simpler model may be more appropriate and give results that are easier to interpret within a policy-relevant timeframe. Model fusion provides a practical way to combine knowledge from different disciplines, but can accelerate model growth. How then can we facilitate the evolution of useful predictive models? Coherent design is essential. When combining models it is often necessary to resolve overlapping scope, so tools need to allow for the disaggregation of model implementations as well as their fusion. Modelling software and integration frameworks can help resolve technical constraints, but to make models useful and used it is essential to involve stakeholders in their design and interpretation.

Aspirational statements abound as to the importance of integrated, multidisciplinary solutions for environmental problems. The website for the UK Centre for Ecology & Hydrology, for example, opens with a call to arms: 'Joined-up solutions are demanded for the complex environmental challenges facing humankind' (CEH 2012), a statement that it would be hard to disagree with. The importance of integrating scientific understanding with policy initiatives is also recognized (McIntosh *et al.* 2005). However, it is useful to expose the nature of the solutions, and the methods used for joining them up, to some critical thinking. Often the solutions required are predicted outcomes, whether these are presented as rules-of-thumb or detailed visualizations. Although some authors have presented typologies that distinguish projections, estimates, explorations, extrapolations, speculations, etc., from predictions (e.g. van Ittersum *et al.* 1998), our view is that all of these are predictions, with varying degrees of uncertainty. Scientists aim to reduce or at least quantify uncertainties associated with predictions, but policy decisions have to be made in the light of uncertain information, with the aim of anticipating likely change and managing risks. In the current study we examine the function and state of predictive modelling in environmental science, and review some of the methods proposed for improving the integration of models, databases, and tools for analysis and visualization.

Although measurements and observations are considered empirical, predictive modelling is also of central importance to science. Hypotheses are essentially predictions about as-yet-unobserved properties and are tested by matching these predictions against observations. Predictions may be highly uncertain or conjectural, for example future weather patterns, or highly certain, for example the mass of an as-yet-unobserved proton. Scientific understanding is a scaffold of predictions, in which new struts are continually being tried out and old struts are sometimes found to need replacement. However, there have been relatively few initiatives to compare predictive models and still fewer to reject those that are inaccurate or over-parameterized. Publishing a new model is comparatively easy and brings academic rewards more readily than working with existing models. Although the steep rise in the proportion of international journal articles terms related to 'model' (Fig. 1) suggests an increasing

From: Riddick, A. T. Kessler, H. & Giles, J. R. A. (eds) 2017. *Integrated Environmental Modelling to Solve Real World Problems: Methods, Vision and Challenges*. Geological Society, London, Special Publications, **408**, 235–244.
First published online June 23, 2014, https://doi.org/10.1144/SP408.1

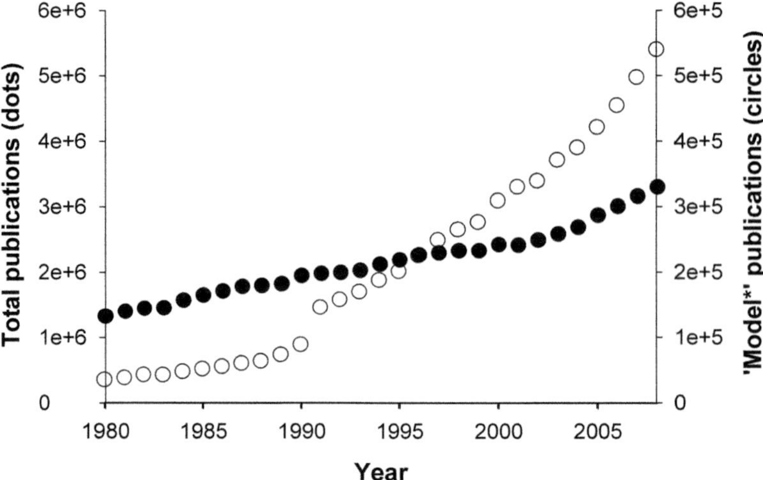

Fig. 1. Annual total number of articles on Web of Science (dots; left-hand scale) and number of articles on Web of Science with 'model*' in the topic (circles; right-hand scale), 1980–2008. Searches were carried out on 19 November 2012.

recognition of the importance of prediction, it may also reflect this bias. The problems in ecological science identified by Peters (1991) of weakly operationalized concepts and untestable or truistic hypotheses remain widespread. The potential of integrated models to address multifaceted issues must be set against the advantages of using simpler models that are more easily testable. The principle of parsimony, in which a model that is accurate despite having a simple form or few parameters is preferable to a less efficient model, was set out in the fourteenth century by William of Occam and has since been emphasized by leading scientists such as Newton and Einstein.

Another important scientific principle is that results should be reproducible, for example, that experiments should be described with sufficient detail to allow their outcomes to be verified independently. Despite the deterministic nature of many ecological models, it is debatable how far modelling activities fulfil this requirement. Models often keep evolving after publication, and there may be discrepancies between the model documentation and its current implementation. In a study of 150 modelling papers in a journal with strict rules requiring code and data to be archived, results could be independently verified for only 10% (McCullough *et al.* 2006; quoted in Merali 2010). This suggests that more care needs to be taken with the documentation and archiving of models.

Ecological systems are inherently complex. Although physical and chemical processes that operate in abiotic systems are not simple, in most cases the underlying theory is well-established and remaining uncertainties are due to difficulties with measurement or to stochastic and chaotic behaviour. The difficulties of predicting the behaviour of biotic systems, in which organisms use and store energy to drive endothermic reactions, and where interactions among multiple trophic levels affect the populations and activities of organisms, are orders of magnitude more complex. Most ecological systems are also influenced by human actions and there is debate over the extent to which these can be predicted. Nevertheless, we need to understand and predict the behaviour of environmental systems to foresee the outcomes of policy and management decisions.

Many successful predictive models, such as regression models, are based on static relationships. Perhaps the most useful predictive models are those that can be summarized as rules-of-thumb; there are many examples in the 'common sense' lore of land managers and some researchers have aimed to produce similar simple rules (e.g. Palm *et al.* 2000). Common sense can be misleading, however, and much of the success of the scientific method is due to its mechanisms for rejecting models that fail to predict well. Improving the accuracy of model predictions is often achieved by introducing new detail. When model designers realise the shortcomings of their model, there is a natural inclination to add processes or case-specific parameters, or attempt to generate additional indicators, and so models have a tendency to grow. This is useful, since ecological systems typically have many interactions and feedbacks, and models may need to be mechanistically rich to represent these (de Angelis

& Mooij 2003). The development of integrated models allows understanding from different areas of science to be combined. However, the behaviour of large and complex models can be difficult to interpret and they can have data requirements that are impractical to meet. To be useful, the outputs from such models may need to be summarized in meta-models, that is, simple relationships or rules-of thumb. Discussions of where to focus model detail and where to use simple models are often extremely useful in developing understanding (Crout *et al.* 2009).

In traditional or normal science, scientists generate and test hypotheses, and then 'speak truth to power' by providing quantified information for use by policy decision makers (Hoppe 1999). Increasing recognition that environmental problems urgently need to be addressed despite incomplete knowledge has led to the characterization of environmental science as post-normal (Funtowicz & Ravetz 1990). Societal actors other than scientists can bring useful knowledge to bear on policy development.

Conclusions and the credibility of model results often depend strongly on the framing questions and concepts (Aumann 2011). It is therefore useful to consider predictive models as part of an interactive and participatory process in which stakeholders can influence model design, scenario development and the forms in which model outputs are presented (Fig. 2). The outer box in this diagram represents a viewpoint in which scientists are also seen as stakeholders, and questions are best resolved through communication and humility (Nicolson *et al.* 2002). Humility, in this sense, means accepting that scientists are not the only arbiters of truth and that other stakeholders have important roles, for example, in setting questions and in constructing meaning.

The challenges of improving the use of scientific models within societal decision-making processes are sociological as much as technical (McIntosh *et al.* 2005). Nicolson *et al.* (2002) set out a practical guide to interdisciplinary modelling, emphasizing early problem definition and prototyping, sensitivity analyses, and above all communication among modellers, other scientists and stakeholders. To this end, they preferred transparent and accessible models such as those developed within spreadsheets.

In many cases stakeholders require information about different aspects of the socio-ecological system. Integrated assessment models attempt to summarize effects of environmental change on aspects such as human health, carbon sequestration, biodiversity conservation and water quality (Parker *et al.* 2002). Here, the challenges are often to enable an efficient identification of optimal or pareto-optimal solutions for policy decision making by combining and integrating a wide range of data and multidisciplinary scientific knowledge (Reis *et al.* 2005). However, it is important not to over-summarize disparate responses, because societal actors may attach widely different importances to different components of ecosystem response (McIntosh *et al.* 2005), and because summary terms such

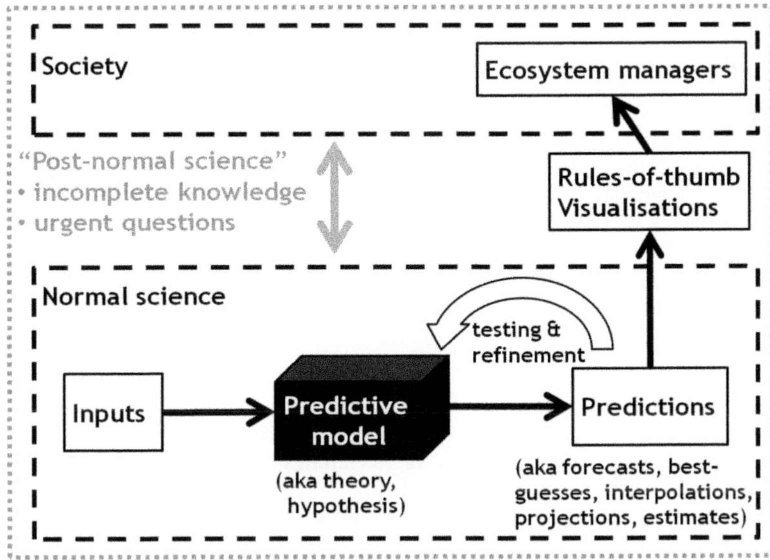

Fig. 2. Conceptual model of interactions between science and society. Policy decision makers, land managers and decision makers at other scales are collectively denoted 'Ecosystem managers'.

as 'ecosystem health' often cannot be predicted and are largely devoid of meaning (Suter 1993; Wolman 2006).

The distinction between a theoretical predictive model and its implementation is important (Nordstrom 2012). While any scientist worth their salt is in the business of making predictions, those who work on implementing, testing and combining models can be seen as having technical skills somewhat removed from cutting edge science. Keeping abreast of developments in information technology is not easy, and programmers have a natural tendency to stick to the technologies and language with which they are familiar. Important questions for theory development such as 'does this model predict with more accuracy than that?' or 'which model is the most parsimonious?' are consequently often difficult to answer, since different models are implemented using different technologies. This technical difficulty is on top of more substantive differences in temporal or spatial scale (Oxley & ApSimon 2007), parameter requirements, objectives and scope. Initiatives to compare models against common datasets encounter many difficulties, but are rewarding in terms of improved insight into the functioning of and commonalities among models (e.g. Smith *et al.* 1997; Simunek *et al.* 2003; Adams *et al.* 2004; Smith *et al.* 2004; Palosuo *et al.* 2012).

The scope of a model is in fact a major and poorly considered obstacle to comparison and integration. Models are constructed by imposing a boundary upon reality, in terms of spatiotemporal extent, but also by including and excluding particular processes and subsystems. For example, one model may consider soil hydrological conditions as boundary or forcing variables, whereas another includes routines to calculate hydrological fluxes. If the soil carbon cycling sub-models within these two models are to be compared, or used within another ecosystem model, the code for the soil carbon processes must be separated from the hydrology algorithms. Integrating models is therefore rarely a case of simply fusing the code for two models, but involves the careful disaggregation and extraction of sub-models.

Methods for improving the design and testing of ecosystem models

Methods for improving modelling practice have focused on creating standards for describing and documenting models and their parameters and outputs (Grimm *et al.* 2006; Jakeman *et al.* 2006; Crout *et al.* 2009; Schmolke *et al.* 2010). Standardizing definitions is the purpose of semantic technologies, including vocabularies/ontologies and tools for their creation and use, which have become the state of the art for web applications aiming to facilitate integration and re-use of resources (Williams *et al.* 2006; Latre *et al.* 2012). However, an alternative school of thought related to 'agile programming' (Williams & Cockburn 2003), emphasizes the importance of innovation and the use of models to advance theory and test its implications (Carpenter 2003). Maintaining the compatibility of documentation with code requires extra resources, and it may be more efficient to treat the implementation of a model as its definitive documentation and endeavour to make the code accessible and transparent. Nevertheless, when integrating models with each other and with tools for databasing, analysis and visualization, some standardization is usually necessary.

Defining quantities

Attempts to reliably combine knowledge encounter the issue of quantity definition. A soil scientist accustomed to using the term 'runoff' to describe the flow of water over the soil surface when infiltration rate is exceeded may be baffled, or make basic errors, when using data provided by a hydrologist for whom 'runoff' represents drainage flux. Moot unit definitions provide constant work for model and database developers. These issues are perhaps most dangerous when using different units results in only small differences in quantity which are not easy to spot, as for example when comparing C/N element ratios using mass ratio ($g\,C\,g^{-1}\,N$) or atomic ratio ($Mol\,C\,Mol^{-1}\,N$). However, even large differences in quantity can escape attention. The Mars Climate Orbiter, which broke up in the Martian atmosphere after programmers confused metric and imperial units (Stevenson *et al.* 1999), is a famous example. Even when units are reliably known, differences in measurement method may mean that the quantities are not comparable. These concerns have led to the development of semantic technologies for quantity definition.

The development of ontologies has followed a course similar to the original evolution of dictionaries. The first dictionaries of the English language had major benefits in terms of standardizing what had become very disparate written forms, but inevitably included errors and idiosyncratic spellings. Efforts to resolve these resulted in different dictionaries and the consolidation of rival and distinct spellings in Britain and the USA. The first computer ontologies (also called thesauri or vocabularies) were likewise milestones in the development of reliable standards, but if the same quantity is defined differently in different ontologies, the question of precedence arises. Computer technologies have, however, allowed an elegant solution to this

Table 1. *Analytical results, with detailed metadata providing links to a SKOS formatted thesaurus*

Sample number	Result name	Result URI	Formatted result	Method short name	Method URI	Instrument	Instrument URI
5383	Li corrected	http://onto.nerc.ac.uk/CAST/54	0.285	Waters ICPMS	http://onto.nerc.ac.uk/CAST/179	ICP_MS	http://onto.nerc.ac.uk/CAST/135
5383	Be corrected	http://onto.nerc.ac.uk/CAST/30	0.007	Waters ICPMS	http://onto.nerc.ac.uk/CAST/179	ICP_MS	http://onto.nerc.ac.uk/CAST/135

problem in the form of triple storage. As well as storing a definition for a quantity or concept, links to other ontologies can be stored to define broader and narrower concepts. If a concept is linked to a broader definition, it can be assumed to inherit all aspects of that definition. This allows the development of a body of knowledge in which ontologies can be combined, and draw on each other's strengths. Differing definitions may still exist and must be resolved using time-honoured criteria such as reputation or popularity; a definition that is linked to by numerous users, or that appears in a moderated ontology, is likely to be more definitive.

Currently one of the leading semantic technologies is the Simple Knowledge Organization System (SKOS) W3C standard (Semantic Web Deployment Working Group 2009). SKOS is built on the Resource Description Framework (RDF) (RDF Core Working Group 2004) and uses triples to describe the relationships (also called predicates) between subjects and objects, but with a simple, defined vocabulary of predicates describing all the possible relationships permitted within SKOS. There are several available applications capable of creating and editing of SKOS-formatted thesauri, such as PoolParty, which was used by the Centre for Ecology & Hydrology (CEH) to create and host the CEH Analytical Services Thesaurus (CAST) for the description of chemical determinands, instruments, units of measurement and methods. Within the SKOS standard, each 'concept' has a preferred label or name for an idea or quantity, alternative labels and a definition, and may also include relationships to broader/narrower concepts and related concepts. The key resource is the Uniform Resource Locator (URL), which both identifies an individual concept and provides a link to the location of the information associated with the concept. This enables succinct storage of metadata describing, for example, the analytical method and instrument used (Table 1).

One key area where standardization has developed considerably is for the definition of species names. Although a first impression is that species names are fixed, in fact taxonomic revisions regularly result in changes to names. Misspellings compound this problem. The precision with which taxa are recorded also varies and it is useful, for example, to link names of microspecies or subspecies to aggregate names. Initiatives to standardize species names have been led by institutions with taxonomic expertise and include the Catalogue of Life (2012) (Fig. 3).

Tools for designing integrated models

Tools for integrating models have focused on two main areas (Argent 2004): the combination of spatial datasets and spatially distributed models using technologies based around geographical information systems (GIS); and the integration of models describing different processes at a single location, often a one-dimensional (1D) vertical line through an ecosystem. There is overlap between these approaches, but tools for spatially explicit modelling are often geared towards representation rather

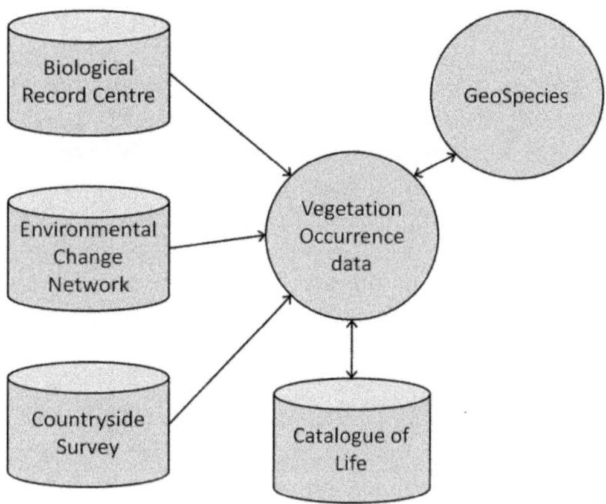

Fig. 3. Example of ontology-based storage of species records. Existing databases are combined into a virtual database of all occurrence records, which obtains definitions from the Catalogue of Life.

than facilitating clear design of integrated dynamic models, and tools for integrating 1D models are not easily scaled to two-dimensional (2D) and three-dimensional (3D) maps. Frameworks for spatial modelling are becoming more routinely used in a geological context (Jackson 2001; Wang *et al.* 2009) and here the discussion will focus rather on tools for integrating dynamic models. At its simplest, this involves running one model for the study period, generating a time-series of outputs which are then used as inputs for another model. This sequential or static approach cannot, however, represent dynamic feedbacks, where quantities predicted by the second model also affect processes within the first model.

Modelling frameworks have mainly emphasized the definition of components and of interactions among these components (Argent 2004; van Evert *et al.* 2005). Early attempts to standardize dynamic modelling components led to the definition of the Continuous System Modelling Program (CSMP) standard (IBM Corporation 1975), which requires components to track and if necessary manage memory allocation for their own internal variables, and to be callable in either 'Initial', 'Dynamic' or 'Terminal' mode. This principle also applies to more recent component-based frameworks, but the focus has been on defining quantity passes in a way that is unambiguous and preferably allows quantity passing among components written in different languages. This allows component developers to continue to work in a familiar language.

Options for passing quantities among component models, databases and analysis and visualization tools have diversified, and there is an increasing emphasis on associating quantities with formalized definitions. A first pass at integrating different models might use existing output formats such as ASCII files, but there are time costs associated with opening, closing, writing and reading such files. Input from and output to binary files can be considerably faster and binary standards such as NetCDF are now being used beyond their original climate modelling domain. Passing arguments to subroutines implemented in a common language, or as DLL 'handles', is also fast. In larger and more complex integrated models, however, it can be difficult to keep track of argument definitions, whether in terms of the scientific meaning of the quantity or in terms of variable type. Metadata describing units and type help considerably and can be incorporated into files, including the NetCDF binary format. However, the state of the art is now to use semantic definitions, similar to those outlined in the previous section, to ensure unambiguous quantity passes.

In a comparative study of three component modelling frameworks (Tarsier, SME and ICMS), the test model was not implemented in any of the frameworks within the time available, but each demonstrated different strengths and facilitated different aspects of integrated model design (Argent *et al.* 2006). Other frameworks are available, and the OpenMI framework (Gregersen *et al.* 2005) in particular has received much attention and support under EU Framework Programmes. OpenMI schedules quantity exchanges among dynamic models, databases and other tools, and incorporates semantic definitions of quantities passed. The framework was initially developed for the hydrological domain, but is finding applications beyond this. Graphical user interfaces have been developed to facilitate linking of OpenMI-compliant components, although the development of these components remains a task requiring some programming expertise in both the source language for the component and the OpenMI implementation language, that is, C# or Java.

While model integration can be seen in terms of the connection of 'black box' models, considering only their inputs and outputs approach, the importance of being able to access and modify internal model structure quickly becomes apparent when a real example is considered (Fig. 4a; adapted from Rowe *et al.* 2014). The models to be integrated overlapped in two respects: carbon and nitrogen dynamics were simulated by both the VSD and N14C models; and hydrological processes were represented by both the VSD and DyDOC_lite models. Although these process representations could conceivably have been left intact in the original models, this would have led to discrepancies and divergence, for example, of the hydrological quantities simulated by VSD and DyDOC_lite. It was therefore essential before integrating these models to identify overlaps, decide which of the process-representations would be used and which discarded, and disaggregate the model containing the latter process-representation (Fig. 4b). If the original model had been developed using a component-based framework and its individual process-representations (e.g. hydrology) implemented as components, this disaggregation would have been relatively straightforward. However, it is difficult to predict future requirements for disaggregation, and so models in which the structure is accessible and relatively transparent are generally more suitable for integration.

Constructing models is fundamentally a design process, in which scientific considerations of optimal detail must be weighed together with objectives such as fast processing speed, transparency and reproducibility (despite changing software standards) for the useful lifetime of the model. Scientists are best placed to judge where simpler models or metamodels can be used and where more process

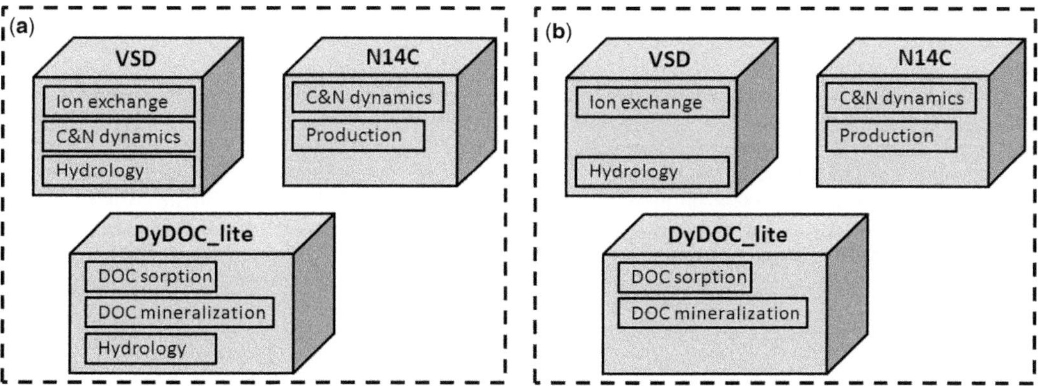

Fig. 4. Fusion of three dynamic models: **(a)** original structure of models in terms of process sub-models; and **(b)** models after disaggregation, ready to be integrated. C&N, carbon and nitrogen; DOC, dissolved organic carbon.

detail or more parameters are needed. Graphical tools such as Vensim™, Stella™, Simile™ and Simulink™ can reduce the requirement for scientists to learn programming languages, although some understanding of mathematical concepts is still required. These tools are perhaps best suited for initial or high-level design, since they lead users gently towards a system-analytical view of the world. Although models built using graphical software are arguably less flexible than models developed in procedural languages, they have distinct advantages in terms of transparency to non-programmers and successful examples exist of scientific models developed using graphical packages (e.g. van Noordwijk & Lusiana 1999). Requirements for integration with other code or with analysis tools that are not included within the graphical package may be met by automated conversion to code. Several of these packages include the facility to produce code in low-level languages such as C++ and tools for converting graphical models to code that can be analysed using high-level languages such as R have recently been developed (Naimi & Voinov 2012; Simulistics Ltd. 2012). Conversion in the opposite direction, from code to a coherent diagram, is more difficult. Muetzelfeldt (2004) advocates model description within a semantic framework that would allow implementation in, and conversion between, a variety of representations in a 'Declarative Modelling' approach. Domain-specific markup schemas such as MathML and CellML represent progress towards such an ontological definition of model structure.

Tools for statistical analysis and calibration

Both statistical analysis and calibration of models require quantity passes between the model implementation and an evaluation tool. The aim of statistical analysis is to characterize the accuracy, and preferably also the efficiency of models, as an aid to model evaluation and choice. Calibration tools also evaluate model accuracy, but seek to increase it by adjusting one or more model parameters. In the same way, optimization packages can be used to adjust parameters to find pareto-optimal solutions for multiple objectives. Numerous methods have been developed for model evaluation and parameter-optimization. Bennett *et al.* (2012) provide a useful review and suggest using workflows tailored to the model purpose and data availability. The choice of method also depends, however, on its ease of application. Analysis and optimization tools are available within low-level languages such as Fortran and higher level packages such as R, Simile™ and MATLAB™. Hitherto less attention has been paid to including formal analytical tools within model integration frameworks than links with databases and visualization tools. As such tools become more accessible within component-based modelling frameworks, it will become easier to select parsimonious models from a set of possible representations of the modelled process (cf. Cox *et al.* 2006).

Tools for visualization

Good visualizations of model outputs are invaluable for plausibility testing by scientists and stakeholders. Modern GIS tools have made it easier to generate colourful maps and projections of 3D spatial data for use in scientific publications, and 3D visualization and animation tools are beginning to provide a more immersive encounter with spatio-temporal data. Many people find maps accessible, since it is possible to relate the patterns presented to known places and spatial patterns. These factors can also lead to overconfidence in mapped data, perhaps because the spatial precision with which landscape features can be mapped suggests that overlaid predictions are also precisely known. Although the merit of visualizations for model testing is variable, they undoubtedly help with engaging people and can contribute to the success of proposals, articles and reports. Visualization tools driven by outputs from models provide powerful methods for illustrating scenarios (e.g. Jackson 2001; Li & Duffy 2012).

Communication between stakeholders and modellers

Environmental issues and societal concerns are continually changing, and so requirements for quantitative and qualitative model outputs also change. Models have been shown to have better uptake and a longer use-life if they were conceived and designed in collaboration with stakeholders (Sterk *et al.* 2007). Models represent aspects of reality in a manner that is separate from both the theoretical and real worlds, and have a role in the mediation process whereby stakeholders can be 'confronted with evidence and research' (Aumann 2011). This facilitates communication between stakeholders and researchers as well as among stakeholders (e.g. policy makers, regulators, non-governmental organizations and other interest/lobby groups, and the general public).

One of the key challenges for the communication between stakeholders and modellers, as well as among researchers working in different disciplines, is the lack of common vocabularies. Time should be allowed for informal discussion and explanation of different perspectives in the early stages of model development and it may help to develop an agreed glossary or formal ontology. Often, an inherently different understanding of environmental regimes in natural, social and political sciences proves to be the main barrier to the use of results from ecological models in policy decision making. De Vos *et al.* (2012) postulate that for meaningful integrated assessment modelling it is necessary – in addition to integrating information from various sources and scientific domains – to explicitly connect these to the relevant policy context, including interest groups and end-users. However, they conclude that 'the integration of knowledge on environmental regimes in integrated assessment is still in its early days, and requires further attention in the future' (de Vos *et al.* 2012).

Future directions

Software designers have made an enormous contribution to the analysis of environmental problems. Without tools such as spreadsheets, compilers, statistical packages and GIS, scientists would be working far more slowly and would be less able to share insights with policymakers and the public. More recently, mobile apps have opened up many potential areas for citizen science, allowing the public to make observations available to scientists and to access scientific outputs, for example, MySoil (BGS & CEH 2011). It is difficult to foresee what new possibilities will be opened up by next-generation technologies. Nevertheless, some suggestions can be made as to ways in which new techniques and technologies could improve the design of tools for integrating data, models, visualizations and analyses:

(1) improved graphical tools for model design;
(2) auto-translation among model implementations;
(3) use of the semantic web for quantity definition and verification;
(4) accessible tools for model analysis, for example, to locate uncertainty;
(5) reusable workflows;
(6) accessible tools for 2D, 3D and 4D visualization.

Conclusions

This study provides a brief overview of the issues associated with model integration. We have hitherto avoided drawing definitive conclusions on the merits of standardization, documentation, modularization, etc. This is in part because the diversity of software technologies used by model designers and the rate of change of these technologies make it difficult to make specific recommendations. In general, however, the integration of models and other tools is made easier if these are designed with modularization in mind, for example separating the 'engine' of a model from its user interface. Coding and documentation styles are personal, but we strongly approve of the statement by Abelson *et al.* (1996): 'programs must be written for people to read, and only incidentally for machines to execute'. Including comments and using meaningful variable names can greatly improve the readability of code.

Advances in information technology seem likely to continue to break down barriers and reduce the entry level of knowledge required to assemble, analyse and visualize environmental models. Provided that these abilities are accompanied by normal scientific scepticism and the guiding principle of parsimony, there is every reason to believe that new generations of environmental models with sufficient accuracy and generality will continue to provide useful supports to societal decision making. Models are 'tools to think with' (McIntosh *et al.* 2007), and if these tools are made accessible to a wider range of scientists and other actors, the benefit of more thinking can go into decisions concerning the environment.

This study was funded by the UK Natural Environment Research Council (NERC) under the SIMDAT (Strategies for Improving Model Design, development and Testing) project of the Centre for Ecology & Hydrology (CEH). The authors would like to acknowledge the contributions of K. Brock to the development of the manuscript and figures, and of the two anonymous reviewers whose suggestions we have attempted to incorporate.

References

ABELSON, H., SUSSMAN, G. J. & SUSSMAN, J. 1996. *Structure and Interpretation of Computer Programs,* 2nd edn. MIT Press.

ADAMS, B., WHITE, A. & LENTON, T. M. 2004. An analysis of some diverse approaches to modelling terrestrial net primary productivity. *Ecological Modelling*, **177**, 353–391.

ARGENT, R. M. 2004. An overview of model integration for environmental application–components, frameworks and semantics. *Environmental Modelling & Software*, **19**, 219–234.

ARGENT, R. M., VOINOV, A. ET AL. 2006. Comparing modelling frameworks – A workshop approach. *Environmental Modelling & Software*, **21**, 895–910.

AUMANN, C. A. 2011. Constructing model credibility in the context of policy appraisal. *Environmental Modelling & Software*, **26**, 258–265.

BENNETT, N. D., CROKE, B. F. W. ET AL. 2012. Characterising performance of environmental models. *Environmental Modelling & Software*, https://doi.org/10.1016/j.envsoft.2012.09.011

BGS & CEH 2011. MySoil iPhone app. http://www.bgs.ac.uk/mysoil/ [accessed 11 June 2014].

CARPENTER, S. R. 2003. The need for fast-and-frugal models. *In*: CANHAM, C. D., COLE, J. J. & LAUENROTH, W. K. (eds) *Models in Ecosystem Science*. Princeton University Press, Princeton, New Jersey, 455–460.

CATALOGUE OF LIFE 2012. http://www.catalogueoflife.org/ [accessed 11 June 2014].

CEH 2012. Centre for Ecology & Hydrology website. http://www.ceh.ac.uk/ [accessed 11 June 2014].

COX, G. M., GIBBONS, J. M., WOOD, A. T. A., CRAIGON, J., RAMSDEN, S. J. & CROUT, N. M. J. 2006. Towards the systematic simplification of mechanistic models. *Ecological Modelling*, **198**, 240–246.

CROUT, N. M. J., TARSITANO, D. & WOOD, A. T. 2009. Is my model too complex? Evaluating model formulation using model reduction. *Environmental Modelling & Software*, **24**, 1–7.

DE ANGELIS, D. L. & MOOIJ, W. M. 2003. In praise of mechanistically rich models. *In*: CANHAM, C. D., COLE, J. J. & LAUENROTH, W. K. (eds) *Models in Ecosystem Science*. Princeton University Press, Princeton, New Jersey, 63–82.

DE VOS, M. G., JANSSEN, P. H. M. ET AL. 2012. Formalizing knowledge on international environmental regimes: a first step towards integrating political science in integrated assessments of global environmental change. *Environmental Modelling & Software*. https://doi.org/10.1016/j.envsoft.2012.08.004

FUNTOWICZ, S. O. & RAVETZ, J. R. 1990. *Uncertainty and Quality in Science for Policy*. Kluwer Academic Publishers, Dordrecht.

GREGERSEN, J. B., GIJSBERS, P. J. A., WESTEN, S. J. P. & BLIND, M. 2005. OpenMI: the essential concepts and their implications for legacy software. *Advances in Geosciences*, **4**, 37–44.

GRIMM, V., BERGER, U. ET AL. 2006. A standard protocol for describing individual-based and agent-based models. *Ecological Modelling*, **198**, 115–126.

HOPPE, R. 1999. Policy analysis, science and politics: from 'speaking truth to power' to 'making sense together'. *Science and Public Policy*, **26**, 201–210.

IBM CORPORATION 1975. Continuous Simulation Modelling Programme III (CSMP III). Programme Reference Manual.

JACKSON, C. R. 2001. The development and validation of the object-oriented quasi three-dimensional regional groundwater model ZOOMQ3D. Internal Report IR/01/144, British Geological Survey, Keyworth, Nottingham. http://nora.nerc.ac.uk/9298/ [accessed 11 June 2014].

JAKEMAN, A. J., LETCHER, R. A. & NORTON, J. P. 2006. Ten iterative steps in development and evaluation of environmental models. *Environmental Modelling & Software*, **21**, 602–614.

LATRE, M. A., HOFER, B., LACASTA, J. & NOGUERAS-ISO, J. 2012. The development and interlinkage of a drought vocabulary in the EuroGEOSS interoperable catalogue infrastructure. *International Journal of Spatial Data Infrastructures Research*, **7**, 225–248.

LI, S. & DUFFY, C. J. 2012. Fully-coupled modeling of shallow water flow and pollutant transport on unstructured grids. *Procedia Environmental Sciences*, **13**, 2098–2121.

MCCULLOUGH, B. D., MCGEARY, K. A. & HARRISON, T. D. 2006. Lessons from the JMCB Archive. *Journal of Money Credit and Banking*, **38**, 1093–1107.

MCINTOSH, B. S., JEFFREY, P., LEMON, M. & WINDER, N. 2005. On the design of computer-based models for integrated environmental science. *Environmental Management*, **35**, 741–752.

MCINTOSH, B. S., SEATON, R. A. F. & JEFFREY, P. 2007. Tools to think with? Towards understanding the use of computer-based support tools in policy relevant research. *Environmental Modelling & Software*, **22**, 640–648.

MERALI, Z. 2010. Computational science: ... error. *Nature*, **467**, 775–777.

MUETZELFELDT, R. 2004. *Declarative modelling in ecological and environmental research*. European Commission Directorate-General for Research. EUR20918.

NAIMI, B. & VOINOV, A. 2012. StellaR: a software to translate Stella models into R open-source environment. *Environmental Modelling & Software*, **38**, 117–118.

NICOLSON, C. R., STARFIELD, A. M., KOFINAS, G. P. & KRUSE, J. A. 2002. Ten heuristics for interdisciplinary modeling projects. *Ecosystems*, **5**, 376–384.

Nordstrom, D. K. 2012. Models, validation, and applied geochemistry: issues in science, communication, and philosophy. *Applied Geochemistry*, **27**, 1899–1919.

Oxley, T. & ApSimon, H. M. 2007. Space, time and nesting integrated assessment models. *Environmental Modelling & Software*, **22**, 1732–1749.

Palm, C. A., Gachengo, C. N., Delve, R. J., Cadisch, G. & Giller, K. E. 2000. Organic inputs for soil fertility management in tropical agroecosystems: application of an organic resource database. *Agriculture, Ecosystems and Environment*, **1692**, 1–16.

Palosuo, T., Foereid, B. *et al.* 2012. A multi-model comparison of soil carbon assessment of a coniferous forest stand. *Environmental Modelling & Software*, **35**, 38–49.

Parker, P., Letcher, R. *et al.* 2002. Progress in integrated assessment and modelling. *Environmental Modelling & Software*, **17**, 209–217.

Peters, R. H. 1991. *A Critique for Ecology*. Cambridge University Press, Cambridge

RDF Core Working Group 2004. RDF Primer. *In*: Manola, F., Miller, E. (eds) *W3C Recommendation, World Wide Web Consortium*. Latest version: http://www.w3.org/TR/rdf-primer/ [accessed 11 June 2014].

Reis, S., Nitter, S. & Friedrich, R. 2005. Innovative approaches in integrated assessment modelling of European air pollution control strategies – implications of dealing with multi-pollutant multi-effect problems. *Environmental Modelling & Software*, **20**, 1524–1531.

Rowe, E. C., Tipping, E. *et al.* 2014. Predicting nitrogen and acidity effects on long-term dynamics of dissolved organic matter. *Environmental Pollution*, **184**, 271–282.

Schmolke, A., Thorbek, P., DeAngelis, D. L. & Grimm, V. 2010. Ecological models supporting environmental decision making: a strategy for the future. *Trends in Ecology & Evolution*, **25**, 479–486.

Semantic Web Deployment Working Group 2009. SKOS Simple Knowledge Organization System Reference. *In*: Miles, A., Bechhofer, S. (eds) *W3C Recommendation, World Wide Web Consortium*. Latest version: http://www.w3.org/TR/skos-reference/ [accessed 11 June 2014].

Simulistics LTD 2012. *Package 'Simile'* [online]. http://cran.r-project.org/web/packages/Simile/.

Simunek, J., Jarvis, N. J., van Genuchten, M. T. & Gardenas, A. 2003. Review and comparison of models for describing non-equilibrium and preferential flow and transport in the vadose zone. *Journal of Hydrology*, **272**, 14–35.

Smith, M. B., Seo, D. J. *et al.* 2004. The distributed model intercomparison project (DMIP): motivation and experiment design. *Journal of Hydrology*, **298**, 4–26.

Smith, P., Smith, J. U. *et al.* 1997. A comparison of the performance of nine soil organic matter models using datasets from seven long-term experiments. *Geoderma*, **81**, 153–225.

Sterk, B., van Ittersum, M. K., Leeuwis, C. & Wijnands, F. G. 2007. Prototyping and farm system modelling – Partners on the road towards more sustainable farm systems? *European Journal of Agronomy*, **26**, 401–409.

Stevenson, A. G., Mulville, D. R. *et al.* 1999. *Mars Climate Orbiter Mishap Investigation Board Phase I Report, November 10*, 1999. Unnumbered report to NASA, Washington DC, USA. ftp://ftp.hq.nasa.gov/pub/pao/reports/1999/MCO_report.pdf [accessed 11 June 2014].

Suter, G. W. 1993. A critique of ecosystem health concepts and indexes. *Environmental Toxicology and Chemistry*, **12**, 1533–1539.

van Evert, F. K. v., Holzworth, D., Muetzelfeldt, R. M., Rizzoli, A. E. & Villa, F. 2005. *Convergence in Integrated Modeling Frameworks*. Modelling and Simulation Society of Australia & New Zealand. http://edepot.wur.nl/40236

van Ittersum, M. K., Rabbinge, R. & van Latesteijn, H. C. 1998. Exploratory land use studies and their role in strategic policy making. *Agricultural Systems*, **58**, 309–330.

van Noordwijk, M. & Lusiana, B. 1999. WaNuLCAS, a model of water, nutrient and light capture in agroforestry systems. *Agroforestry Systems*, **43**, 217–242.

Wang, B. J., Shi, B. & Song, Z. 2009. A simple approach to 3D geological modelling and visualization. *Bulletin of Engineering Geology and the Environment*, **68**, 559–565.

Williams, L. & Cockburn, A. 2003. Agile software development: it's about feedback and change. *Computer*, **36**, 39–43.

Williams, R. J., Martinez, N. D. & Golbeck, J. 2006. Ontologies for ecoinformatics. *Journal of Web Semantics*, **4**, 237–242.

Wolman, A. G. 2006. Measurement and meaningfulness in conservation science. *Conservation Biology*, **20**, 1626–1634.

Future of technology in NERC data models and informatics: outputs from InformaTEC

ANDREW KINGDON[1]*, JEREMY R. A. GILES[1] & JONATHAN P. LOWNDES[2]

[1]*British Geological Survey, Keyworth, Nottingham, NG 12 5GG, UK*

[2]*British Geological Survey, Murchison House, West Mains Road, Edinburgh EH9 3LA, UK*

**Correspondence: aki@bgs.ac.uk*

Abstract: The 'Big Data' paradigm will revolutionize understanding of the natural environment. New technologies are revolutionizing our ability to measure, model, understand and make robust, evidence-based predictions at increasingly spatial and temporal resolutions. Realising this potential will require reengineering of environmental sciences in the observation infrastructure, in data management and processing, and in the culture of environmental sciences. Collectively these will deliver vibrant, integrated research communities. Manipulating such enormous data streams requires a new data infrastructure underpinned by four technologies. Pervasive environmental sensor networks will continuously measure suites of environmental parameters and transmit these wirelessly to scientists, regulators and modellers in real time. Integrated environmental modelling will process data, streamed from sensor networks, using components synthesizing natural systems developed by domain experts, each of which will be linked at runtime to other expert developed components. Semantic interoperability will facilitate cross-disciplinary working, as has already happened within the biosciences so that data items can be exchanged with unambiguous, shared meaning. Cloud computing will revolutionize data processing allowing scalable computing close to observations on an as-needed basis. Leveraging the full potential of these technologies requires a major culture change in the environmental sciences where national and continental scale observatories of sensors networks become basic scientific tools.

The nature of information about the natural environment is changing radically. Both data volumes and the processing power required to manipulate the data are increasing rapidly. Genomics is leading with the way with the imminent need for much higher performance computing. The challenges can only increase to meet the needs of multinode integrated environmental models, running in parallel across wide area networks, processing real-time data from national or continental scale observatories composed of tens of thousands of individual sensor nodes. For example, IBM has recently stated that 'Every day we create 2.5 quintillion bytes of data – so much that 90% of the data in the world today has been created in the last two years alone' (IBM 2013).The InformaTEC project was established to 'future-scope' the requirements of data, models and informatics with the aim of informing technology policy, and by extension funding, within the Natural Environment Research Council (NERC). The information resources held to facilitate environmental science vary enormously from large warehouses of physical samples (held at the UK National Geological Repository) through to the petabyte (10^9 MB of storage space) scale data holding of the British Atmospheric Data Centre. But regardless of the environmental science discipline, a set of common issues are being faced by all environmental informaticians. Given the scale of capital investment required to develop the infrastructure, it is essential that NERC informaticians participate in the design and development enabling them to share the common infrastructure created.

The InformaTEC process

The InformaTEC project engaged with community leaders to inform the project of the major issues in their sectors. These community leaders in turn selected key players for a large pan-discipline meeting held in Lancaster in May 2012 (Kingdon *et al.* 2012). This was informed by a series of presentations of the major issues. The principal mechanism for discovering the views of the participants was a large, facilitated discussion on the future of environmental informatics to establish the key challenges both within each sector and also for the wider NERC remit.

These challenges span NERC's scientific requirements and also require long-term structural changes to how environmental informatics operates within NERC. They will also require different timescales for delivery. Four technologies were seen to be ripe for delivering new functionality to environmental researchers and placed significant requirements on the informatics community. These have

From: Riddick, A. T., Kessler, H. & Giles, J. R. A. (eds) 2017. *Integrated Environmental Modelling to Solve Real World Problems: Methods, Vision and Challenges*. Geological Society, London, Special Publications, **408**, 245–253.
First published online September 17, 2014, https://doi.org/10.1144/SP408.5

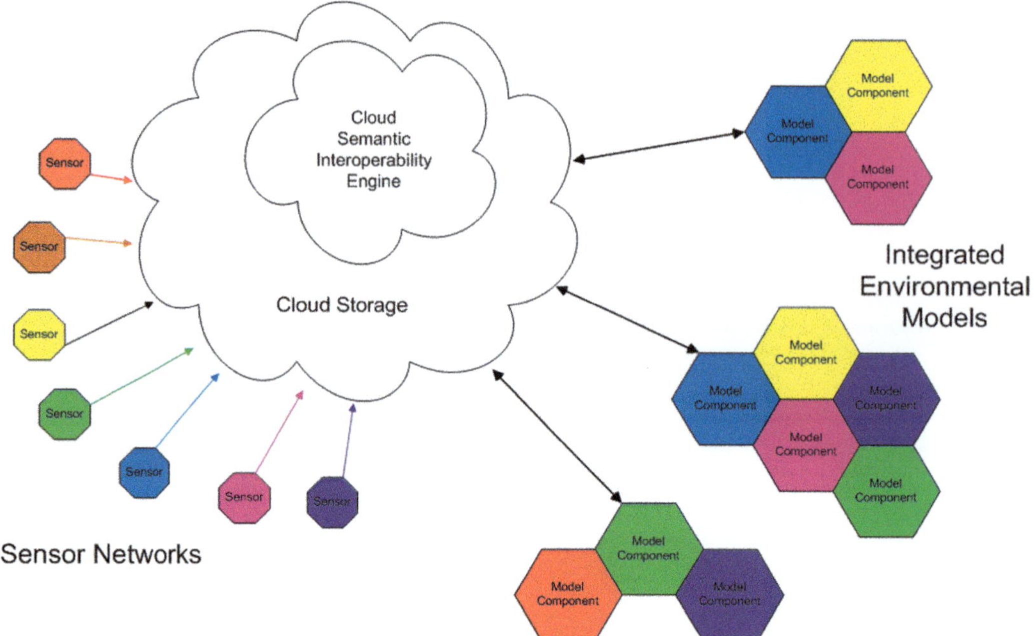

Fig. 1. A vision of models to aid environmental understanding in a 'Big Data' world.

been used to cast the following vision for environmental informatics.

InformaTEC vision

The British Isles have a high population density. This means that their land area as well as the seas that surround them are subjected to multiple uses and consequently affected by multiple pressures. If the British Isles are to respond effectively to these pressures, there is a need for mechanisms to model these issues and to scenario test possible responses.

The InformaTEC Vision is that in the near future an integrated environmental informatics infrastructure will be built that will provide the capability to model the environment and test scenarios using four main technological components:

- pervasive environmental sensor networks that are monitoring multiple parameters in real time and move these data to cloud storage;
- cloud storage makes these data readily available to an extended community;
- domain concepts are managed at interdisciplinary boundaries by a mature semantic interoperability architecture;
- integrated environmental models will exploit these data to deliver short-, medium- and long-term environmental forecasts of high societal value.

This vision (Fig. 1) represents a step change in the capabilities of environmental scientists to understand the effects of changes and to predict their societal impact. Therefore there is a huge incentive to deliver the technical improvements needed to make this happen. A key driver for making this happen is NERC's requirement for the delivery of understanding of the impacts of environmental change at a range of temporal and physical scales.

Components of the vision

The four key technologies identified by the Lancaster meeting in May 2012 are discussed briefly discussed. But while these technologies are maturing rapidly as independent technologies there is a need for a coordinated approach to achieve the interoperability required.

Pervasive environmental sensors

Environmental sensors are not new. Progressively, sites for environmental data acquisition have been automated to allow repeated data collection without manual intervention. For example, stream gauges have been automated with data served in real time for the Environment Agency in England and Wales for many years. However, such installations are major facilities requiring connection to mains electricity and wired data transfer, and are therefore typically only at locations where such services are available. Sensors can be placed in remote locations, as the ARGO floats (Feder 2000) have amply demonstrated by providing salinity and temperature data about ocean currents since 1991, though they suffer from a lifespan limited to around four years of autonomous operation (Kobayashi *et al.* 2012).

With changes in solid state electronics it is possible to create sensors that can now be built that fit on a single chip. Such sensors can therefore be manufactured comparatively cheaply, allowing a much wider emplacement within in the environment.

The US National Ecological Observatory Network (NEON) program (Kampe *et al.* 2010; NEON 2011) is installing multiple sensors at 60 sites across the United States to gather and synthesize data on the impacts of climate change, land use change and invasive species on natural resources and biodiversity. This provides an exemplar of the possibilities such sensor networks could provide in understanding and quantifying continental scale environmental change.

Two controlling factors exist in the emplacement of sensors through the wider environment: access to communication and power. Fixed line communications are of course restricted to urban areas and along transport routes. In the modern environment, however, wireless communications are pervasive with almost unlimited access to (in order to increase the area covered) WIFI networks (over a local area), GSM/3G mobile phone signals across much of the populated areas of the developed world, or (provided the project budget allows) satellite communication which are essentially global in coverage. The low cost and power wireless network ZigBee may facilitate longer battery life. However, sensing and communications require power which has proven a more serious restriction. Once installations are made off-grid and the requirements for power are too long-lived to be easily undertaken using batteries, there is an immediate increase in the complexity of the installations required. Maintenance of power is a significant issue in rural areas remote from dwellings and thus mains power. In low power use conditions, even with solar or wind installations as principal power supply, battery back-up will be required. If higher power installations are required, generator capacity or access to mains electricity will be necessary (Marsic *et al.* 2012).

There will certainly be issues with mass emplacement of sensors in terms of ensuring the safety and security of sensor installations. These will include issues such as permissions to implement sensors, emplacement of sensors in areas where they do not interfere with local ecosystems, access to adequate power, concerns over data security, and so on.

Once these issues have been dealt with then sensors will be sited in the natural environment and data will start being transmitted into data centres through data networks. Given that obtaining permission to get sensors mounted in the wider environment and the constraints placed by power requirements, it is very likely that multiple sensors will be mounted at common locations, thereby ensuring that a swathe of environmental data will be supplied from common single locations.

Cloud computing

'Cloud computing' is a popular term used to describe computing resources (processing power, storage, software) that are available in a remote location and accessible over a network. The term describes the abstraction involved in hiding the complexities involved in managing the computing resources from the end-user. It is a major step on the road to providing computing resources as a utility, similar to an electrical power supply. It will also change the way computing is purchased. Currently it is a large capital item with associated management and infrastructure costs. In the cloud model, computing resources change from a large capital item to a recurrent expenditure. This makes it easier to pay for what is needed rather than having underutilized resources available waiting to be used only at local peak demand.

Environmental information ranges from satellite observation, air- and ship-borne sensors, meteorological data, geospatial information and genomic data, through to 3D and 4D seismic beneath the ground. The growth in data holdings has consistently been at sixty per cent year on year over the last decade. This rate of growth represents an estimated 12 000-fold growth in NERC data holding between 2000 and 2020 (Giles, J. R. A., pers. comm., 2013). This represents a range of challenges in the areas of resource discovery, data storage, network bandwidth and computational resources. Collating, manipulating and managing the information will present serious challenge to the environmental community in their attempts to imbibe this information in such a way as to increase understanding of environmental change. Cloud computing offers a flexible solution to the ever increasing demands for storage.

Within this vision the cloud services likely to be used are storage, sharing and preprocessing the data before they are used for environmental modelling.

The sensor networks will transmit the data they have collected to cloud storage so that they are both secure and sharable. The timing and rate of data transfer to the cloud will depend on the power conservation model that individual sensor networks are using. Once in the cloud, various automatic quality and completeness checks can be performed and reports supplied to data managers. Data can then be released for use by the research community.

Prior to using the data collected by sensor networks in integrated environmental modelling it will be necessary to ensure that the data are semantically interoperable. This could be done by a cloud-based semantic interoperability engine with access to the necessary domain vocabularies to ensure metadata are understandable across discipline boundaries without reference to the data's originator.

Using cloud computing also has the potential to provide scalability, allowing for flexibility in how this is enabled without requiring large-scale capital expenditure and limitless expansion as sensor networks develop through time. Thus scalability is really important. Whilst it is clear that the availability of environmental sensors will provide a huge potential opportunity for the establishment of new data streams, it will require a new architecture to deliver this capability to scientists.

In the cloud computing paradigm there will be a different way of working for environmental scientists. The starting point for many scientific projects is the capabilities of the facilities and infrastructures already available to those scientists and how these are then applied to the next scientific challenge. In a cloud computing paradigm such limitations will not apply. Instead the limiting factor is the imagination of the scientists, their ability to develop code/sensor technology to deal with the problem under investigation and the resources that can be afforded to provide the computing power to achieve this.

Semantic interoperability

Whenever multiple scientific disciplines work together there will inevitably be misunderstanding arising as a result of the different terminologies used within each discipline. Semantic interoperability can help address these challenges. This is always a major constraint on manipulating data and information from other areas of scientific expertise. Without clear mapping between the meanings of scientific concepts across discipline boundaries it is difficult to identify linkable variables and to develop real understanding between these concepts.

The advent of the technologies required to implement the vision provides increased incentive to deliver semantic interoperability. In the science of integrated environmental modelling, discipline specific concepts embedded in component models may map in a number of ways. For example:

- Related concepts in different disciplines are identified by different names.
- Related concepts in two disciplines have a common identifier but the definition in one discipline has broader or narrower meaning than that in the other discipline.
- Unrelated concepts in different disciplines are identified by the same name.
- A concept in two or more disciplines has both a common identifier and a matching definition.

The effects of all these issues are felt whenever scientific concepts cross discipline boundaries. For example, whenever water passes between the ground surface and the subsurface, the terminology which describes the water and the processes that occur to it are described completely separately. The issues of this change in terminology already cause confusion. These issues are significantly magnified in the science of integrated environmental modelling where, almost as soon as model components are linked across discipline boundaries, these semantic issues create confusion and/or can lead to incorrect concept linkages. Therefore mechanisms need to be in place to identify the specific meaning intended by each community and translations between them to ensure that concepts are only linked when it is meaningful. Once these translations are codified then processes to achieve automated, or at least semi-automated, transition across these discipline boundaries is possible but will need investment to deliver this complex functionality. Once environmental sensors are implemented pervasively, there will be a strong requirement for translation of sensors output data so that they are ready to act as inputs into model components.

The InformaTEC process highlighted an enormous amount of progress in research and the workflows that underpin this from the bioinformatics community. The disaggregated structure of research groups within the biosciences community allows maximal flexibility, but mediates against the creation of major national data centres of the type already implemented within the environmental sciences. Data are recorded in the peer review literature and associated data table deposits. Therefore flexible tools are required to extract data from these systems to enable them to be linked and used for analysis. Semantic tools are used to make connections between such datasets allowing inferences to be drawn.

A key component in the delivery of the science of integrated environmental modelling will be implementation of semantic interoperability. The various levels of semantic interoperability described above can only be delivered by having a well-developed understanding of the concepts modelled in each of the component models in integrated environmental modelling. These allow dynamic relationships to be defined between concepts which can be used to establish complex relationships between these variables and thus allow us more complex relationships to be defined. These will inevitably be defined at the boundaries of different disciplines where the concepts used by those disciplines have to interface.

Clearly considerable work needs to be done to develop the semantic interoperability that will be required to underpin integrated environmental modelling. Standards are emerging. For example, the World Wide Web Consortium has published standards for Simple Knowledge Organisation System (SKOS) whilst the International Standards Organization (ISO) recently published ISO 25964, the international standard for thesauri and interoperability with other vocabularies. The major task facing integrated environmental modelling science is the creation of widely accepted interoperable vocabularies which are published in SKOS or ISO 25964 format. NERC centres are ideally placed to make the investment in such unglamorous but vital work building on large-scale European projects including the INSPIRE legislation (European Union 2007). This is not a trivial task but is an essential component of transforming large data volumes into knowledge and insight.

Integrated environmental modelling

All attempts to move scientific understanding from observations of what is to the understanding of what might be, or even simply that which is not directly observable, require models to develop a conceptual understanding, (Laniak *et al.* 2013). Modelling of the natural environment precedes the advent of digital computing but, as the availability of computing resources has increased, there has been an inexorable rise in the power and complexity of modelling outputs.

Most modelling procedures are developed to understand a single element of a natural system. Attempts to develop an understanding of wider natural systems requires, in its simplest incarnation, chaining of natural process models in serial to understand the capabilities of the entire system. In recent years frameworks have been developed which allow sets of model outputs to be linked at run time, that is, the models are operating in parallel. This allows the relevant property information to be exchanged throughout the modelling process wherever the processes interact. This concept has developed into standards such as OpenMI and academic endeavours such as those listed below.

The history of integrated environmental modelling is reported elsewhere (e.g. Laniak *et al.* 2013), but a number of key initiatives are now in place to seek to provide integration between the various integrated environmental modelling initiatives internationally. Different modelling frameworks are increasingly standardizing their outputs to allow the integration of multiple strands of modelling from different disciplines.

The main current successful integrated environmental modelling initiatives are:

- OpenMI: derived as a response to the requirements of the European Union's Water Framework Directive and originally focused on the objectives of whole catchment management;
- Community Surface Dynamics Modeling System (CSDMS): a wholly academic endeavour derived from National Science Foundation funding in the United States and its requirement to maximize the value of modelling undertaken from its science funds. CSDMS maintains model codes, makes them linkable, maintains an infrastructure and coordinates community to support these endeavours;
- EPA Frames: a framework for linking models and a mechanism for coordination of the modelling outputs of the US Environmental Protection Agency.

International discussions are ongoing to ensure that the various integrated environmental modelling initiatives work in a coordinated way to prevent needless duplication of effort and to eliminate opportunities for further separation of the integrated environmental model. Therefore an informal coordination structure between the key global integrated environmental modelling initiatives has been established with an agreed set of common priorities.

It is not intended that these initiatives be subsumed into some higher whole but that they should work collectively to achieve a 'standard of standard' which will ensure that there is no further divergence of the existing standards and that common approaches are taken to deal with common problems such as the management of uncertainty and wider interoperability. It has been recognized that making more model components available to more potential uses is essential for achieving the critical mass that this technology requires. Without this it is unlikely that these scientific opportunities of this technology will be maximized.

Integrated environmental modelling is not the solution to all scientific modelling endeavours. The linkage action adds an overhead on the processing time which means that these models cannot run as efficiently on computing architectures to deliver modelling outputs. Where a modelling action requires cross-disciplinary interaction, however, there are very significant advantages to utilizing integrated environmental modelling architecture. Integrated environmental modelling approaches allow multiple aspects of the system to be modelled to understand the full range of effects that could be felt from environmental changes.

Bringing the technologies together

The four technologies listed in the vision and described briefly above are distinct standalone technologies that can operate entirely independently from each other. The key to the successful implementation of the vision is managing the interfaces between the four technologies. Planning these interfaces between the four technologies is the most important step in bring them together. The nature of these interfaces is discussed briefly below.

The sensor networks will be streaming data into cloud storage constantly. Some of the data will be immediately incorporated into models. Other models will draw long runs of time series data from the cloud store months or years after collection. This will depend on the aims and objectives of the different research groups mining the data resource. Many grants come with a standard embargo period to give the researchers sole use of the data they collect for a limited period. In the case of NERC, the data policy permits a maximum of two years exclusive use of data by a grant holder.

The sensor networks are designed to collect specific parameters and to transmit them to a specified storage location. It is absolutely essential that the raw sensor data are associated with rich metadata at this stage so that future users will know where the sensor was located, what methods were used to acquire the data, what were the tolerances of the data, what was the calibration history, and so on. The aim is to prevent ambiguity emerging when the data are used by modellers. Whilst a research team may know much of these metadata whilst active, this knowledge is soon lost when the research team members move to other projects. Much of this information is common to data collected by a single sensor and can be associated immediately after the data have been added to the cloud store.

Where the data collected by a senor network are to be used in a cross-disciplinary integrated environmental model, the role of semantic interoperability becomes critical. The importance of this interface cannot be overestimated. The work involved is considerable so it should only be done once and shared as widely as possible. The embryonic foundations of this work are already in place. Emerging standards are being published and initial implementations are in place. NERC runs an open vocabulary server for environmental sciences (http://vocab.nerc.ac.uk/). But there is still considerable work to be done to provide the semantic mapping required by the postulated cloud semantic interoperability engine.

A wide array of environmental processes have been implemented as models. A much smaller number have been converted into components for integration with other environmental models. The emerging standards, such as OpenMI, make this much easier than it was in the past. Using such model components it will be possible to scenario plan how environmental systems affect the major challenges to human society. This presents an opportunity to look at the stresses and risks our society faces at an appropriate level of detail and including understanding the effects of specific environmental variables upon these risks. The outputs of integrated model runs will be returned to the cloud store for sharing with other research groups. The outputs themselves may also be processed by the postulated cloud semantic interoperability engine to make the information available to multiple disciplines.

Delivering the vision

In addition to having the technologies available there are a number of other actions required to deliver the vision.

NERC formally defines environmental informatics within its grant application process as the 'research and system development focusing on the environmental sciences relating to the creation, collection, storage, processing, modelling, interpretation, display and dissemination of data and information'. Although NERC clearly recognizes environmental informatics as a research topic within its purview, it has failed to develop as a separate environmental informatics discipline with sufficient critical mass. Contact between the various environmental communities with the UK is largely conducted from host disciplines (e.g. Geoscience) with only limited direct contacts between environmental informaticians within the main environmental disciplines within NERC. InformaTEC was established as a response to these problems and a set of challenges defined below.

Despite the significant long-term investment by NERC in developing an informatics science infrastructure there is little real evidence of a discipline of 'environmental informatics' within NERC and few NERC staff would describe themselves as environmental informaticians. Rather the various disciplines within the NERC science remit each support independent, and often isolated, informatics subdisciplines. Whilst the processes, tools and techniques for managing information sciences across the NERC scientific remit are in each case similar, the linkages between information scientists take place between scientific disciplines rather than between informaticians. One of the aims of the NERC

InformaTEC project was to seek to find common issues and common solutions across a range of terrestrial environmental informatics problems.

The natural sciences have until recently been principally a mechanism for describing what the status of natural phenomena are and by extension how natural phenomena have arisen. Only very recently has the availability of large validated datasets facilitated the change from simply describing what is there and how it got there, through to understanding how these phenomena have changed and will continue to change into the future. Predicting the responses of natural systems to the various pressures (climate change, population growth, depletion of finite natural resources and restricted land-use) means that there is a growing requirement for integrated approaches that cut across traditional discipline boundaries within the environmental science.

There is a widespread acceptance that the natural environment is being stressed by anthropogenic environmental change and that the effects need to be reversed to ensure the well-being of the human population and natural ecosystems. Understanding the impact of the multiple stresses on natural systems means that integrated environmental modelling tools are needed, to both comprehend and quantify possible scenarios and to make predictions of the rates of change. For example, global climate models have thus far concentrated on establishing global or continental scale trends of future climate. Less attention has been applied to understanding the impact of these changes on human spatial and temporal scales; for example, what are the likely effects on my city in the next five years.

Recognizing this issue NERC has previously set itself the challenge of understanding climate prediction on regional and local as well as on daily and decadal scales. The effects of both drought and flooding in 2012 and 2014 demonstrate that environmental informaticians have a long way to go to meet these expectations. Government and their regulators lack predictions in which they can place sufficient confidence to justify the financial costs and social change required to mitigate against the impacts of climate change at the local scales, or adapt the infrastructure so such events have less deleterious impact over the life of one or two parliaments. Additional research is required into these effects with associated data discovery required to adequately inform this debate.

Data centres responsible for the earth sciences, particularly but not limited to the National Geosciences Data Centre (NGDC), hold large databases of hardcopy data and/or scanned but not yet digitally captured data. These provide a huge archive of scientific material and provide historical archive information which can act as ground-truthing for time series data analyses. Whilst metadata have been collected, the full richness of the substantive data are often expensive to capture and thus this is unlikely to be undertaken in the near future. Capture of the full datasets will substantially increase the available data for both geological and other subsurface investigations. Resourcing the capture of this material will facilitate the best possible understanding of time-variant data problems by providing the longest single record that can be provided.

Reuse of environmental models, built for one specific purpose, as a component of an integrated environmental model is an important aspect of the vision. But a mechanism is required to ensure that environmental models developed by research groups are maintained beyond the lifespan of the owning projects and can be developed over a longer timescale. In the United States, the National Science Foundation has ensured that such models have a long-term future by funding the Community Surface Dynamics Modeling System (Syvitski *et al.* 2011) which facilitates (among other functions) the preservation of all modelling items which are then componentized and allows them to be preserved within this architecture. This allows such components to be reused within this componentized space.

Delivering the vision will be a cultural challenge to parts of the environmental sciences community. Part of the community gathers their data from shared instruments. These may be ship-mounted instruments, ARGO floats, satellites, seismometer networks, and so on. In these communities it is routine for data to be shared soon after collection. In other areas of the environmental science there is a more 'cottage industry' approach. Individual scientists collect data and make observations for themselves, and tend to be reluctant to share these data. In many cases the sensor networks will be replacing the more 'cottage industry' side of the environmental sciences. The immediacy of data sharing that the vision implies will be a major culture change for some communities in the environmental sciences.

The commoditisation of computing resources, as implied by cloud computing, will be a further culture change. The conventional large capital investments required to purchase servers and data storage will be replace by a monthly or quarterly 'utility' bill. Culturally it has been easier to obtain funding for new infrastructure that to 'rent' resources to use from third parties. This will mean changes to the way that resources are allocated to scientists for pilot projects and also when these activities are converted to full proposals. In addition there will need to be a reappraisal of the mechanisms by which data are stored and archived as there is a need to ensure that data are held securely and managed for the long-term digital continuity. Issues relating to ownership and intellectual property must be clear to all parties concerned.

The development of a centralized NERC informatics function is an essential output from this process. A whole series of new data streams and data types are to be collected from new sensors and new process models. These will be ultimately integrated with model outputs to produce new insights into environmental variability. However, this will only be accomplished if datasets are interoperable, data are supplied in immediately ingestible formats and the semantic issues are resolved. NERC's scientific outputs from these developments are thus wholly dependent upon delivering an integration of these outputs. This can only be achieved if NERC actively undertakes to deliver this integration as a core component of developing this new area of science. This will require a new centralized NERC informatics function to be established. This will need to incorporate the expertise of knowledgeable domain experts from the sensors and models developers. It will also need input from the NERC data infrastructure to achieve:

- standard data formats and data transmission standards;
- controlled vocabularies and ontologies linked by semantic interoperability;
- long-term archiving and historic data access tools;
- creation of a workbench of standardized processing algorithms.

Delivering this coordination will be aided by the creation of a centralized NERC informatics function. Scientists from many of the existing NERC data centres will need to collaborate to deliver properly cross-discipline data integration.

Conclusion and discussions

'Big Data' is placing new demands on both NERC's information infrastructure and those scientists seeking to deliver understanding from these new technologies. However, the delivery of the vision of pervasive sensors driving predictive environmental models is now within grasp. This will have many positive outcomes for understanding the human and natural environment. It will allow real-time understanding of the natural conditions, collect long time series of *in situ* conditions which can be used for calibrating models under specific conditions and provide long-term records of cyclic change and long-term trends. When combined with predictive models, these effects can be repeated to identify long-term consequences and to scenario test the effects of changes in climate on the natural environment or in human responses to these effects. This will advance testing of remediation or management strategies, and give protection from the law of unintended consequences by examining the far field effects of such strategies in time and space.

Where InformaTEC has been less successful has been in creating a community of interest in developing these objectives collectively. Though goodwill has been identified among these participants, practical action has been difficult to deliver. However, it seems likely that significant investment in the activities identified above would facilitate these approaches being adopted. NERC needs to ensure that national capability funding is prioritized to resolve those issues within NERC's direct purview. Most notably it is essential that the challenge of semantic interoperability is addressed with substantial funding to convert dictionaries to controlled vocabularies and to seek cross-disciplinary linkages.

The second area where targeted NERC funding will deliver is in the promotion of standards to ensure that, for example, sensors funded from one discipline area can be utilized immediately by scientists from another area without any issues apart from the issues of semantic interoperability. This will require standard formats for data transmission from sensor outputs and a common processing architecture, whether that is centralized or distributed. Preventing the fragmentation of environmental information and facilitating its open access is already a key NERC objective. Eliminating cross-disciplinary barriers to this vision would be the lasting legacy of the InformaTEC process.

Published with permission of the Executive Director of the British Geological Survey. The authors acknowledge assistance from G. Rees, G. Blair and N. Morrison. The authors also thank the referees for constructive suggestions for improvements. This work was supported by Natural Environment Research Council grant number NE/H003339/1 'InformaTEC: Informatics Technologies for the Environment Cluster'.

References

EUROPEAN UNION 2007. Directive 2007/2/EC of the European Parliament and of the Council of 14 March 2007 establishing an Infrastructure for Spatial Information in the European Community (INSPIRE). *Official Journal of the European Union*, **L108**, 1–14.

FEDER, T. 2000. Argo begins systematic global probing of the upper oceans. *Physics Today*, **53**, 50–51, https://doi.org/10.1063/1.1292477

IBM 2013. Every day we create 2.5 quintillion bytes of data – so much that 90% of the data in the world today has been created in the last two years alone.

http://public.dhe.ibm.com/software/data/sw-library/big-data/ibm-big-data-success.pdf [last accessed 1 March 2014]

Kampe, T. U., Johnson, B. R., Kuester, M. & Keller, M. 2010. NEON: the first continental-scale ecological observatory with airborne remote sensing of vegetation canopy biochemistry and structure. *Journal of Applied Remote Sensing*, **4**, 043510, https://doi.org/10.1117/1.3361375

Kingdon, A., Giles, J. R. A., Rees, G., Blair, G., Morrison, N. & Lowndes, J. 2012. *InformaTEC Conference 'Word-for-Word' Report*. NERC Technology Programme Open Report, **OR/13/020**.

Kobayashi, T., King, B. A. & Shikama, N. 2012. An estimation of the average lifetime of the latest model of APEX floats. *Journal of Oceanography*, **65**, 81–89, https://doi.org/10.1007/s10872-009-0008-x pp. 81-89al

Laniak, G. F., Olchin, G. *et al.* 2013. Integrated environmental modeling: a vision and roadmap for the future. *Environmental Modelling & Software*, **39**, 3–23, https://doi.org/10.1016/j.envsoft.2012.09.006

Marsic, V., Zhu, M. & Williams, S. 2012. Wireless sensor communication system with low power consumption for integration with energy harvesting technology. *Telfor Journal*, **4**, 89–94.

NEON 2011. The National Ecological observatory Network 2011 Science Strategy 2011. http://www.neoninc.org

Syvitski, J., Hutton, E., Peckham, S. D. & Slingerland, R. 2011. CSDMS – a modeling system to aid sedimentary research. *The Sedimentary Record*, **9**, 4–9, http://www.sepm.org/CM_Files/SedimentaryRecord/SedRecord_9-1.pdf [last accessed 28 August 2014]

The uncertainty cascade in model fusion

KEITH BEVEN[1,2,3]* & ROB LAMB[4]

[1]*Lancaster Environment Centre, Lancaster University, Lancaster LA1 4YQ, UK*

[2]*Department of Earth Sciences, Uppsala University, Uppsala, Sweden*

[3]*Centre for Analysis of Time Series, London School of Economics, London, UK*

[4]*JBA Trust, Broughton Hall, Skipton BD23 3AE, UK*

**Correspondence: k.beven@lancaster.ac.uk*

Abstract: There are increasing demands in assessing the impacts of change on environmental systems to couple different model components together in a cascade, the outputs from one component providing the inputs to another with or without feedbacks in the coupling. Each model component will necessarily involve some uncertainty in its specification and simulations that can be conditioned using some observational data. Taking account of this uncertainty should result in more robust decision making and may change the nature of the decision made. The difficulty in environmental decision making is in making proper estimates of uncertainties when so many of the sources of uncertainty result from lack of knowledge (epistemic uncertainties) rather than uncertainty that can be treated as random variability (aleatory uncertainty). This is particularly the case for problems that involve cascades of model components. Examples are the use of UKCP09 climate scenarios in impact studies, flood risk assessment involving models of runoff generation and their impact on hydraulic models of flood plains, and integrated catchment management involving upstream to downstream surface and subsurface routing of water quality variables. The uncertainties are such that, even for relatively simple problems, they can result in wide ranges of potential outputs. This poses the questions that will be considered in this paper: how to take account of knowledge uncertainties in cascades of model components; and how to constrain the potential uncertainties for use in making decisions. In particular we highlight the difficulties of defining statistical likelihood functions that properly reflect the non-stationary uncertainty characteristics expected of epistemic sources of uncertainty.

Integrated environmental management is currently fashionable for many different types of problems for very good reasons. Meeting the demands of legislation such as the European Union's Water Framework Directive, Habitats Directive or Floods Directive requires an integrated approach that includes the interactions between policy, society (in the form of different types of stakeholders) and various levels of (more or less scientific) process representations. Assessing the costs and benefits of different types of management decisions within this framework then requires the simulation of the impacts of those decisions into the future. Such simulation will often be based on process-based models, coupled together within a model cascade, the outputs from one component providing the inputs to another with or without feedbacks in the coupling. This is one type of model fusion.

Within such a model cascade there are many different sources of uncertainty that will affect the predicted outcomes (see, for example, Beven 2009; 2012*a*). These include:

- uncertainty in the input and boundary condition data;
- uncertainty in the model structure, including feedback mechanisms between different model components in the cascade;
- uncertainty in estimates of parameter values for all the different components in the cascade;
- commensurability of modelled and observed variables and parameters;
- uncertainty in the observations used to calibrate or evaluate models.

In all these different types of uncertainty there is potential for both random variability and uncertainty and error arising from a lack of knowledge. The terms 'aleatory' (from the Latin for a die, or game of dice) and 'epistemic' (from the Greek ἐπιστήμη, for knowledge or science) are often used to distinguish these types of uncertainty. One feature of this classification of uncertainties is often expressed in the form that aleatory uncertainties arise from irreducible natural variability, whereas epistemic uncertainties could be reduced by further study, observation or research to improve our knowledge. Similar concepts have been expressed elsewhere in different ways. An interesting example is that of Frank Knight, who in his 1921

From: Riddick, A. T. Kessler, H. & Giles, J. R. A. (eds) 2017. *Integrated Environmental Modelling to Solve Real World Problems: Methods, Vision and Challenges*. Geological Society, London, Special Publications, **408**, 255–266.
First published online July 17, 2014, https://doi.org/10.1144/SP408.3

book on *Risk, Uncertainty and Profit*, looked at uncertainties from the perspective of the insurance industry (Knight 1921). He distinguished risks that could be expressed in terms of odds, and which consequently an insurance broker would be prepared to issue a contract on at a suitable premium, and the *real* uncertainties which could not be expressed in such a way. The real uncertainties are still sometimes referred to as 'Knightian' uncertainties. They are the result of not having enough knowledge on which to define the odds of a future occurrence. New or better knowledge could turn real uncertainties into odds, but in many cases might also be irreducible. It is also possible that uncertainties that currently appear to be aleatory in nature could be reduced further in the future by improved measurement techniques or model representations.

Beven & Young (2013) outline the types of epistemic uncertainties encountered in hydrological modelling. These include: the neglect of, or incorrect corrections for, gauge errors and radar estimates in rainfall observations; errors associated with lack of knowledge of spatial heterogeneity of rainfalls and the choice of interpolation method; biases in meteorological variables in estimating effective evapotranspiration in a heterogeneous catchment; model structural errors arising from the choice of assumptions in process representations; errors in calibration data of both inputs and estimated values of discharges (especially when extrapolating to peak flows); inappropriate conversion algorithms in obtaining hydrologically relevant variables from remote sensing data; and the commensurability errors associated with state variables measured locally relative to values output by a model (see also Beven 2012*b*).

There may also be epistemic errors that arise from what might be called 'sins of omission'. Because of the complexity of the system there may well be things that have been left out of the model fusion (see, for example, Beven & Germann 2013). These might be 'unknown unknowns', that is factors that affect the response of the system but which have not been yet recognized as important. In fact, the modeller does not actually have to worry about the unknown unknowns until some evidence comes to light that suggests that the model representation is incomplete for some reason. More serious, perhaps, is the omission of the 'known unknowns' that is those factors that we think might well have an impact on the system but which we are not yet sure how to represent and so it is easier to leave them out. These appear in most environmental modelling studies.

Aleatory and epistemic uncertainties remain easier to distinguish in concept than in practice. In practice we see only differences between some observable (with its particularities of time and space scales and measurement errors), or some quantity derived from an observable through a calibration or rating curve, and some model simulation, that is, some modelling residual. It has been pointed out many times that it is impossible to disaggregate that residual into contributions from different sources of uncertainty (e.g. Beven 2005, 2006). The problem is ill-posed; at least without strong additional information on the nature of those uncertainties, information which will not usually be available (for good epistemic reasons). Thus, it is difficult to learn about different sources of uncertainty from a simple model residual series.

Consequently, inference is difficult in these types of system and there is an argument that modelling uncertainties are so dominated by such epistemic uncertainties that we do not know how to represent them properly, that they are magnified by propagation through a model cascade, and that any uncertainty estimation will only be confusing to decision makers. But ignoring them is not, we believe, at all a sensible strategy. Decision makers make decisions under uncertainty all the time. Not only is it intellectually honest to present the decision maker with estimates of uncertainty associated with model simulations, but it might also change the way in which a decision is made (Pappenberger & Beven 2006; Beven 2011; Juston *et al.* 2012).

In what follows we take an operational view of the difference between aleatory and epistemic uncertainties. Aleatory uncertainties will be defined as those that can be represented by a formal error model that can be assumed to have stationary parameters (albeit that such a model might be formulated to represent heteroscedasticity or seasonality or other identifiable structure in the residual error series). Aleatory uncertainties can then, with an appropriate choice of error model, provide probabilistic estimates of uncertainty in simulating or forecasting a future observation. Epistemic uncertainties will be defined as those arising from effects that do not have stationary characteristics, but are rather arbitrary in their occurrence and may contain an element of surprise in forecasting and simulation. They are thus much more difficult to treat using probabilistic models. The essence of this difference, as will become evident below, is that treating all uncertainties as if they can be represented by a simple probabilistic error model will lead to over-confidence in inference.

Statistical estimates of uncertainty in model fusion

The question that then arises is whether it might be possible to constrain the growth of uncertainty in a model cascade. In general the uncertainty associated with any model simulations will increase as it is propagated through the cascade. Statisticians have a way around this problem where there is some evidence with which to evaluate model simulations. Within a formal hierarchical Bayesian statistical framework, uncertainties can be estimated, constrained and propagated through a cascade of components using Bayes equation. In its simplest form the Bayes equation calculates the posterior probability distribution of a variable based on some estimate of prior probability distribution for that variable (and covariance with other variables) and an estimate of the odds for that variable expressed as a likelihood function (e.g. Rougier 2013).

In general, the likelihood function is based on a structural model of the residuals associated with the model simulations of that variable. This requires making formal probabilistic assumptions about the nature of the residuals (e.g. that they have a mean bias, some spatial or temporal autocorrelation or variogram structure, and a Gaussian distribution with constant variance). Where the assumptions hold, then the posterior distribution will have a formal probabilistic interpretation. And, of course, given some observed evidence that can be compared with the model simulations, the assumptions can be checked to see if they hold. If they do not, then the assumptions can be changed (e.g. to allow for heteroscedasticity of the variance, or to transform a skewed distribution into a more manageable Gaussian distribution, e.g. Montanari & Brath 2004).

The problem is that this requires treating all uncertainties as if they were fundamentally aleatory (after identifying parameters for bias, autocorrelation, or more complex model discrepancy functions, e.g. Kennedy & O'Hagan 2001). This means treating all epistemic uncertainties and errors as if they are random in choosing a definition of a likelihood function this way. The residual series still can be checked to see if the assumptions of a particular likelihood function model hold to justify this, but it is not at all clear whether this is a satisfactory way of dealing with epistemic errors as recent discussions in hydrological modelling have illustrated (see Clark *et al.* 2011, 2012; Beven *et al.* 2012*a*).

The nature of epistemic errors in modelling environmental systems

Formal statistical approaches to the uncertainty problem in model fusion are attractive in providing a formalism that can claim some objectivity in the sense that, if the assumptions made in carrying out the analysis hold, then the resulting estimates of uncertainty have a probabilistic interpretation. That is something that can be tested. The inverse also holds true, however, that if the assumptions made in carrying out an analysis do not hold, then the resulting estimates of uncertainty do not have a formal probabilistic interpretation. Yet we can be sure that the epistemic errors associated with each one of the sources of uncertainty listed above are unlikely to be fundamentally random in nature. This does not preclude the possibility of a probabilistic representation of all uncertainties and errors; it does, however, have implications for estimating an appropriate likelihood when some of those uncertainties arise from epistemic sources.

Consider some examples that arise in a model fusion aimed at assessing the impact of future climate change and land management change on meeting the requirements of the Water Framework Directive into the future. This would require a variety of different model components aimed at predicting the system response at the scale of a designated water body which might be a river reach, a lake, a groundwater body or an estuary. Thus this would require: a component to estimate the effects of future climate on rainstorms, evapotranspiration and temperatures; a component to predict the future changes in land management in response to changing climate and economic factors; a component to predict surface and subsurface runoff generation on the catchment and the way in which it might respond to changing land management (including urban development); and a component to predict the impacts of runoff and land management on catchment geochemistry. And, because the Water Framework Directive is concerned with good ecological status, some means of predicting the response of the ecological system to changes in hydrological regime and catchment geochemistry would also be required.

This is typical of the types of problems being posed for both short-term and longer term impact studies in integrated catchment management. Such simulations are required to make sensible decisions about both local and national investment strategies to improve water quality to meet the Water Framework Directive when some of the mitigation measures that might be used may have an effect only over longer time scales. The model fusion here will involve: multiple (uncertain) inputs, including some from other climate and economic model components; model structures that may or may not have (uncertain) representations of all the relevant processes; multiple (uncertain) model parameters in the different components; and (uncertain) commensurability effects such as the difference in meaning between the rainfall provided by a climate model projection and that required at a different scale by a hydrological model. It is clear that many of these uncertainties are fundamentally epistemic and that poor estimates or bias in the representations of uncertainty at any level in the cascade will produce poor estimates or biased uncertainty associated with the predicted outputs. A particular issue, where model components are only loosely coupled in such a cascade, is that of consistency of model components. The hydrology of the land surface parameterization that has been used in predicting the climate futures of UKCP09 (UK Climate Projections), for example, might be quite different to the representation of hydrological processes of a model used in assessing the impacts of those climate changes. Clearly any errors involved would then be unlikely to be aleatory in nature.

The question then is how to constrain the propagation of the uncertainty through the model fusion. In general, this can only be done in a similar way to the aleatory uncertainties, by evaluating the model simulations against any available observations. This cannot be done for simulations of the future (where there are no observations), except in the special case of using expert judgments about the reasonableness of future simulations (which might, of course, introduce its own epistemic uncertainties associated with the judgments of the selected experts). It can be done, however, in evaluating the model cascade against observations in the past, even if this does not necessarily test all the model components in the cascade.

In doing so the difficulty of distinguishing aleatory from epistemic uncertainties in practice becomes clear. Keeping with the catchment example, we might have a runoff generation and water quality model that is being forced by meteorological and other inputs, and predicting river discharges and water quality variables as loads being exported from the catchment. We have some past data on both input variables and catchment outputs with which to calibrate the model and constrain the associated uncertainties. The rainfall inputs might be generated from point raingauge measurements or from a weather radar. Both need an interpretative model to provide catchment inputs; in the first case to interpolate between the raingauges and in the second to interpret the radar reflectivity in terms of rainfall intensities. In both cases we expect the uncertainties to be both random and epistemic in nature and non-stationary in time, varying from event to event with factors such as whether a rainfall cell passes over a raingauge or not, whether there is snow or only liquid precipitation, whether a radar is properly corrected for anomalies, and whether there is a strong difference between what the radar sees at elevation and what reaches the ground as precipitation. Similarly in terms of the outputs, river discharge is not actually measured but requires a way of converting from an observed water level or limited Doppler velocity measurements (Beven *et al.* 2012*b*), while water quality concentrations may be sampled only infrequently leading to additional uncertainties in the estimated of loads for comparison with model simulations. In principle, of course, some of the associated epistemic errors could be reduced by additional monitoring or improved observation techniques. In practice, this may not be the case for past data, which are often presented as if they were both precise and accurate without any assessment of uncertainty associated with the data themselves.

In extreme cases, there may be a mismatch between the observed inputs and outputs to the extent that certain periods of data might be *disinformative* in model calibration and evaluation (Beven & Westerberg 2011). Beven *et al.* (2011) demonstrate this in showing how storm periods where apparent runoff coefficients are greater than one (more output that input) might be misleading in model calibration. Hydrological logic also suggests that, if the inputs for one event are poorly estimated and used to force a runoff model, then this will also have an effect on how accurate that model can predict in the subsequent event (or events) because of the impact of such an epistemic error on the antecedent conditions prior to that next event.

Similar considerations might hold in water quality data or ecological data. This complex nonstationarity of response can also hold in nature. Newson (1980) gives the example of two extreme events on the Plynlimon experimental catchments in mid-Wales in the 1970s. Both produced high flood discharges but were the result of quite different rainfall patterns. The first also generated a number of hillslope failures but little channel change. The second produced few hillslope failures (some sites that might have been susceptible had already failed in first event) but significant channel change, in part because of the materials transported to the channels by the landslides of the first event. With hindsight we have understanding of what was involved, but such uncertain responses would be difficult to incorporate into a model for predicting the future; what will be the geomorphological impact of the next event of such a magnitude (see also Beven 2012*a*)? Epistemic uncertainties imply limitations in predictability (e.g. Smith 2001). There is a large literature on limits of predictability in coupled models in the field of modelling weather and climate (e.g. Thielen *et al.* 2009) but also examples from ecology (e.g. Melbourne & Hastings 2009; Banas 2011) and hydrology (e.g. Zehe & Blöschl 2004; Blöschl & Zehe 2005; Zehe *et al.* 2007; Collier 2007; Koutsoyiannis 2010, 2011). We should, it seems, expect surprises.

Why surprise should not be a surprise in uncertainty estimation

But this should not really be a surprise in itself. A little careful thought about the epistemic uncertainties inherent in environmental problems would suggest that it is impossible to assess properly the probabilities associated with future outcomes. Thus, for example, climate change simulations are now being produced as ensembles of different model runs, but these are not presented as simulations but as projections, even where those outcomes are based on some probabilistic interpretation of a large ensemble of model runs as in UKCP09 (see http://ukclimateprojections.metoffice.gov.uk). UKCP09 describes the probabilities associated with the scenarios as subjective not objective probabilities (see link to 'Probability in UKCP09') that do not indicate the probability of a scenario occurring in the future. This then leaves the onus on the user as to how to interpret the probabilities given.

One of the most important implications of epistemic uncertainties in the modelling process is that past uncertainties may not be a good guide to the future. For aleatory uncertainties, we would expect that a model of the error structure that is fitted during model calibration will hold, with more or less constant parameters into the future. This is the basis for formal statistical methods of uncertainty estimation. That stationarity may not hold for epistemic errors; errors in predicting the next event might be rather different from those that been seen in the past, either because of the spatial or temporal patterns of the error, or in terms of magnitude, or because small sample sizes of extraordinary events have not sampled the full distribution. Taleb (2007, 2010) makes the point that the Gaussian assumptions that underlie many statistical analyses will tend to underpredict the occurrences of high magnitude, high impact events if the underlying distributions are heavy-tailed. This can apply to the magnitudes of errors as well as the magnitudes of the events themselves. Thus, there is always the possibility of surprise in predicting the future, sometimes with high impact such as in the 2011 Tōhoku tsunami in Japan or the 2009 L'Aquila earthquake in Italy. In the latter case the surprise resulted in several scientists being prosecuted and found guilty of misleading the public and causing additional deaths as a result (see, for example, Beven 2012*a*).

And yet . . . the only guide we have to future uncertainties is what we can assess during model calibration and evaluation. Some form of stationarity assumption is generally thought to be necessary, even if it is not the stationary error model assumption of formal statistics. The issue is whether treating epistemic uncertainties as if they are aleatory can, to a first approximation provide useful assessments of uncertainty for real simulations. Our experience would suggest that, at least in hydrological modelling and perhaps much more widely, this is not generally the case.

Communicating the meaning of uncertainty estimates

If the user is to be left to interpret the meaning of the possibilities associated with some model projects, then it is necessary that the user should be informed in order to enable an intelligent interpretation. This then implies that the assumptions of the analysis should be open to scrutiny. This is not always the case. It is not easy, for example, to track down the assumptions that underlie the UKCP09 probability estimates, nor the fact that nearly half of the high resolution dynamic Regional Climate Model runs which were made for the analysis were not included in the final analysis because they were not considered satisfactory. This might not have been a surprise to the modellers, but might be a great surprise to many users of the UKCP09 products.

While there may be increasingly visual ways to convey the outcomes of an uncertainty analysis to users (e.g. Fig. 1), it is more difficult to approach the problem of communicating the *meaning* of uncertainty estimates to the user, especially when different sources of uncertainty cascade through a sequence of coupled model components (e.g. Faulkner *et al.* 2007). Figure 1 shows the estimated probabilities of an area being flooded during the event with an annual exceedance probability (AEP) of 0.01 (the visualization software can also show the distribution of predicted depths for such an event at any point in the space as well). Any estimate of risk constructed from such a map could also take account of the uncertain damages at any point, depending on the distribution of depths (and velocities if necessary).

Many different sources of uncertainty have gone into the construction of such a map. There is the uncertainty in the input discharge for the AEP0.01 event, the uncertainty in the hydrograph associated with that event, the uncertainty in how the infrastructure on the floodplain will affect the pattern of flooding, the uncertainty in the conveyance characteristics of the flood plain and channel are specified, and the effect of any historical uncertainties in constraining the ensemble of models that needs to be run (in this case, observed maximum inundation depths around the flooded area in the flood of 2007 were used in the model evaluation). All of these different sources of uncertainty interact in producing the final outcome. The outcome is therefore a pattern of *conditional* probabilities,

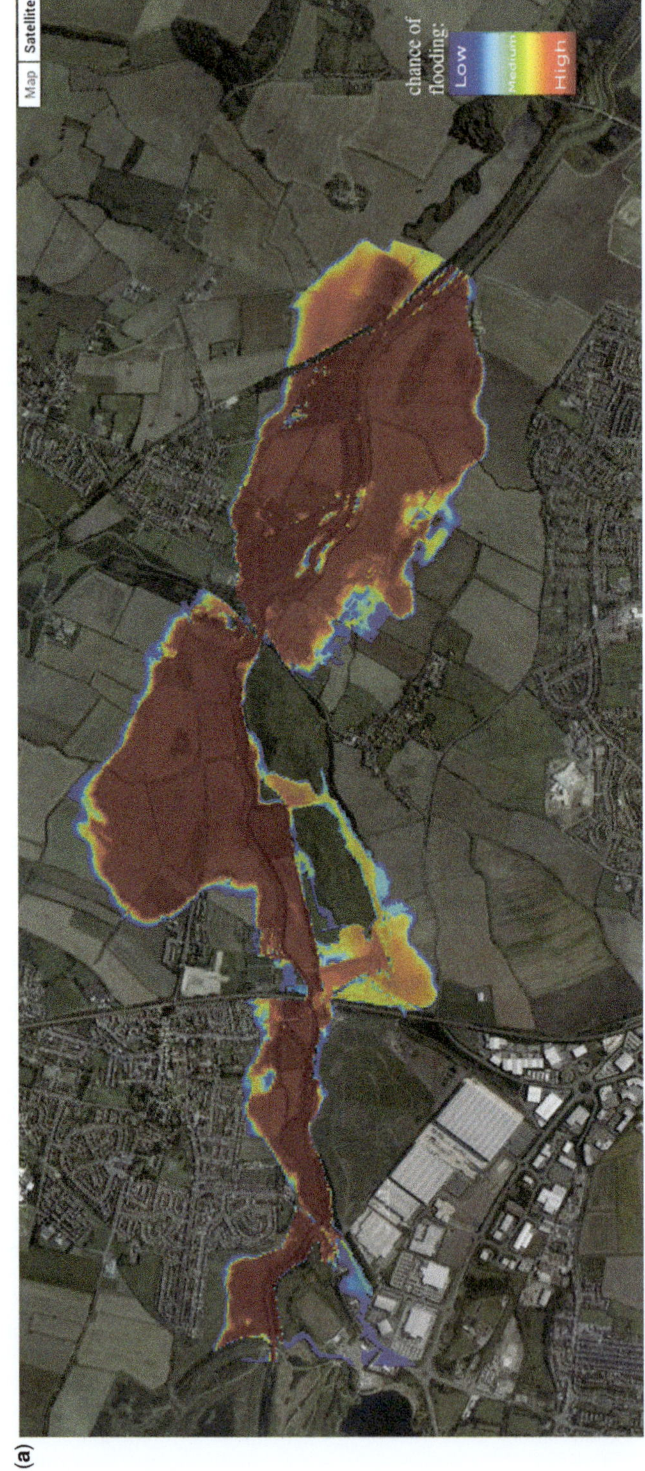

Fig. 1. Different ways of presenting uncertain flood inundation simulations: **(a)** as a colour coded probability of inundation for a given annual exceedance probability (AEP).

(b)

Fig. 1. (*Continued*) **(b)** as the area inundated at a given exceedance probability for the same flood, zoomed in to show more detail. In this case is the uncertain estimate of the AEP0.01 event for the town of Mexborough in Yorkshire (from Beven *et al.* 2014).

conditional on all the assumptions that have been made in producing those outcomes. The question is how to convey the assumptions that are being made to the users, in a way that allows the user to both understand and question those assumptions.

One framework for doing so has been suggested in the context of flood inundation modelling by Beven *et al.* (2014). This involves the specification of assumptions for each element of a condition tree defining the problem and sources of uncertainty. An overview of such a condition tree for the flood inundation problem is shown in Figure 2. The decisions involve: how to treat uncertainty in the input discharges and flood conveyance estimates; how to deal with the effects of infrastructure, etc.; and how to use any available historical observations to constrain the effects of uncertainty. In this case the different decisions are organized into a Source–Pathway–Receptor structure commonly used in flood analysis, with additional decisions associated with conditioning on any observations, visualization and managing uncertainty.

This condition tree framework has two important features in communicating the meaning of uncertainty to users. The first is that the different decisions about assumptions should ideally involve a dialogue between the modeller and user before a final decision is made. Some decisions might have default options or be institutionalized to some extent (such as in the use of the AEP0.01 flood event as the design discharge in the design of fluvial flood defences and flood plain planning). Others may require more thought, such as how best to estimate the uncertainty associated with any design discharge when the records are short or lacking. Even the standard UK *Floods Estimation Handbook* (Institute of Hydrology 1999) is somewhat ambiguous on what methods to use and how in this respect. Part of the discussion should be about how to represent the epistemic nature of certain uncertainties as well as any aleatory part. The discussions themselves act as a mechanism for communication with the end-user of any analysis. One important feature of this framework follows from the fact that the decisions must be made explicit, and if they are explicit then there is an audit trail that can be evaluated later. Audit trails concentrate the mind and are also a useful mechanism for making the discussions more meaningful.

Figure 1 is, of course, only one of many different potential ways of presenting uncertainty information when estimated as conditional probabilities. Other more linguistic interpretations have been preferred by the Intergovernmental Panel on Climate Change (IPCC), which suggests using classes from the highly likely to the highly unlikely. UKCP09

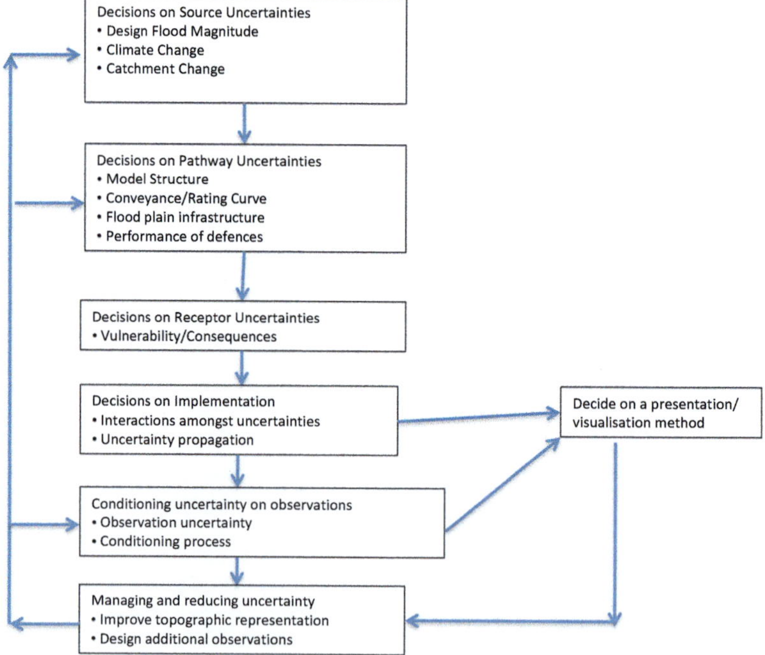

Fig. 2. Flow diagram for the condition tree framework. Feedback arrows refer to use of visualization in guiding additional data collection to constrain uncertainties (after Beven *et al.* 2014).

use both percentage probabilities derived from their ensemble projections integrated using a statistical interpolation, and the IPCC linguistic classes. It is also possible to institutionalize the uncertainty into forms considered easier to use. The concept of the AEP0.01 could be considered an example of this. It is an institutional way of dealing with all the uncertainties in potential floods by trying to define some 'rare' event. The AEP0.001 event, used in the UK for defining a planning zone, is a way of defining a 'very rare' event. The AEP0.0001 event, used for dam safety and levels of protection for some nuclear plants, is an institutionalized 'very rare' event. It is quite impossible to evaluate the accuracy or even uncertainty in such an event with the data available but that is not necessary. Different institutions will define their standards in different ways depending on perceived vulnerability (design standards in the Netherlands for both river and coastal flood defences are, for example, much higher than in the UK).

Such decisions can also depend on political expediency: the AEP0.01 standard defences designed for Carlisle in northern England were redesigned to a higher AEP0.005 standard following the flood of 2005, which would have overtopped the original design. Interestingly, this prevented flooding in Carlisle in 2009 and 2012, so this could also be interpreted as an exercise in adaptive management (although this then has implications for flood risk locations in the rest of the country which have standards of defences to AEP0.01 or lower).

So the question then is whether the type of evaluation of uncertainty in Figure 1 is adding useful information or not. This might still be the case. In the centre of Figure 1b, there is a dip in the upper flood boundary. The dip follows the line of the defences that surround the Mexborough Waste Water Treatment Works. Under the deterministic simulation this is not flooded by the AEP0.01 event, but taking uncertainty into account there is a probability that the Works will be flooded. This might suggest that for this site or some other area of high vulnerability, or indeed for the whole area at risk of flooding, it might be worth being more risk averse or precautionary than the deterministic simulation. An important advantage of this type of visualization is that local stakeholders can focus on their local outputs–and importantly feedback information when the simulations appear to be quite wrong (see Beven 2007; Beven & Alcock 2011).

Some critical questions

The condition tree concept is a useful framework for making explicit the assumptions about different sources of uncertainty in the modelling process in a way that allows communication of those assumptions and buy-in to those assumptions from relevant stakeholders (albeit that we do not consider here how that agreement might be reached). There remain some important critical questions in creating realistic representations of epistemic sources of uncertainty, including both data and model structural uncertainties. This will require a new way of thinking.

Traditionally statistical representations of uncertainty are derived under an assumption that a particular model (together with any structure representing model discrepancy, see above) is a correct representation of the system of interest. Within a Bayesian framework, there are ways of combining different model outputs through an averaging process, but each model is then treated as if it was a correct representation of the system. In fact, using formal Bayesian likelihoods no model will ever be rejected, only given a low likelihood. Comparison of different model structures then generally considers the likelihood ratios associated with each model as a test of whether one model is more likely than another (albeit that the likelihoods might also be dependent on treating the epistemic errors in the forcing and evaluation data as if they were aleatory). Nowhere in this process is there any objective test as to whether a model structure is fit for purpose, although there are some heuristic guidelines as to whether a likelihood ratio might be great enough to accept one model over another (e.g. Kass & Rafferty 1995). It can be a good thing to reject all the model runs as adequate predictors. This means that either the model or the data that are being used to run it need to be improved.

But we do not expect a model to be perfect in its simulations. All models are wrong, after all, but we hope that some models might be useful in simulation. It is then necessary to be rather careful about model rejection: we do not want to make a Type II (false negative) error of rejecting a model that might be useful in simulation just because of errors in the calibration data. We also do not want to make a Type I (false positive) error of accepting a model that would be poor in simulation just because of errors in the calibration data. However, the latter is somewhat less serious since, as more data are collected and used to evaluate the simulations in inference, we would hope to refine the choice of models. There are alternative, rejectionist approaches to inference about models available (e.g. such as the GLUE methodology, Beven *et al.* 2011, 2014; Beven 2012*b*), but in the past these have arguably not taken sufficient recognition of the possibility of making Type II errors.

The possibility of epistemic error in model evaluation therefore places significant emphasis on assessing the real information content of calibration and evaluation data. As noted above, in hydrology, it is quite possible to have inconsistencies in the observations of catchment inputs and outputs, such that some periods of data might not be informative. Thus there is a real need for a methodology to evaluate the information content of data used in the modelling process, but independently of a model run (see Beven 2006, 2010; Beven & Smith 2014). This should be part of any uncertainty assessment.

Another critical question is how to evaluate a likelihood measure for a model run when the residual series may be structured in complex non-stationary ways and where there may be commensurability issues between observed and predicted variables. This also requires a different way of thinking. The GLUE methodology allows the use of informal likelihood measures that can sometimes be given a probabilistic interpretation (e.g. Smith *et al.* 2008) but has been criticized as being too subjective in the choice of likelihood measures used.

The most critical question is, however, how to deal with future surprise in simulation. Identification of inconsistent data or other unusual features during model calibration implies that similar occurrences might happen in the future. There is a difficulty in distinguishing between the unusual that is misleading in calibration and the unusual that might be the most informative in calibration. But the characteristics of future surprises might be quite different to those seen in the past, particularly where the system is expected to be changing in response over time. Elimination of any non-informative periods of data from the calibration process might help identify a model or set of models properly representative of the system under study (e.g. Beven & Smith 2014), but will mean that any estimation of the associated uncertainties should not be expected to cover new surprises in simulation. We cannot know what those surprises might be, of course, and the unusual parts of the calibration data will necessarily remain the only guide to future surprise. This might then be an argument for defining different types of simulation limits; those associated with reproducing the dominant response of the catchment and those (wider) limits that might indicate the possibility of future surprise (Beven & Smith 2014). These would only be indicative of the possibilities of potential future scenarios; it would necessarily be difficult to give such estimates a probabilistic interpretation.

Epistemic uncertainties and decision making

The element of surprise might be important in decision making, particularly for systems that might be vulnerable to catastrophic consequences of the occurrence of extreme events (the consequences of the Tōhoku tsunami at the Fukushima Daiichi nuclear plants is a recent example, Beven 2012*a*). Makridakis & Taleb (2009*a*) give a discussion of why it is necessary to embrace the nature of uncertainty and unpredictability in such circumstances. It will generally result in rather different strategies and decision making than relying on model simulations alone (see also Makridakis & Taleb 2009*b*; Wright & Goodwin 2009; Wilby & Dessai 2010; Beven 2011; Brown *et al.* 2012).

But if the characteristics of future surprises might be different to those seen in the past, how can we plan for them in decision making? The only answer seems to be to make some assessments of scenarios that might scope out the full range of future possibilities, even if it is not possible to infer much about the probabilities with which those scenarios might be expected since they will necessarily involve subjectively chosen assumptions. The requirement of communicating the meaning of those assumptions to any decision makers is then even more important. The potential for epistemic uncertainties can be included in those scenarios and can be used as a basis for evaluating various decision options. Rougier & Beven (2013) show how this can be expressed in terms of an exceedance probability (or, given the subjectivity inherent in the process, perhaps more correctly possibility) of risk, where the assessment of risk includes the consequences of different potential outcomes.

Conclusions

Uncertainty is a real issue in model fusion and cascades of model components. The different sources of uncertainty have been outlined and the importance of epistemic uncertainties arising from lack of knowledge of different types has been stressed. Epistemic uncertainties raise some interesting issues about model inference, the value of data in model calibration, the definition of likelihood measures, communicating the assumptions and meaning of uncertainty estimates, and dealing with the potential for surprise in future simulation and forecasting. Taking account of such uncertainties is likely to change the nature of decision making processes, and where the consequences or risks are high, suggests that adaptive and precautionary decision approaches would be appropriate. These are all difficult problems and require the development of new methodologies, particularly in assessing the information content of data independent of a model, in constraining the propagation in uncertainty cascades, in estimating the uncertain potential for future surprises, and in extending the scenario approach and its use in decision making.

The visualization tool used to produce Figure 1 was written by D. Leedal at Lancaster University. The JFlow inundation model used to produce the maps was run by N. Hunter at JBA Consulting. The thoughts expressed owe much to discussions with J. Rougier of Bristol University. The comments of two anonymous referees are acknowledged in improving the presentation. The work was funded by the UK Flood Risk Management Research Consortium (www.floodrisk.org.uk) funded primarily by the UK Engineering and Physical Sciences Research Council, and the SAPPUR, Catchment Change Network (www.catchmentchange.net) and CREDIBLE projects funded by the UK Natural Environment Research Council.

References

BANAS, N. S. 2011. Adding complex trophic interactions to a size-spectral plankton model: emergent diversity patterns and limits on predictability. *Ecological Modelling*, **222**, 2663–2675.

BEVEN, K. J. 2005. On the concept of model structural error. *Water Science and Technology*, **52**, 165–175.

BEVEN, K. J. 2006. A manifesto for the equifinality thesis. *Journal of Hydrology*, **320**, 18–36.

BEVEN, K. J. 2007. Working towards integrated environmental models of everywhere: uncertainty, data, and modelling as a learning process. *Hydrology and Earth System Science*, **11**, 460–467.

BEVEN, K. J. 2009. *Environmental Modelling: An Uncertain Future?* Routledge, London.

BEVEN, K. J. 2010. Preferential flows and travel time distributions: defining adequate hypothesis tests for hydrological process models. *Hydrolgical Processes*, **24**, 1537–1547.

BEVEN, K. J. 2011. I believe in climate change but how precautionary do we need to be in planning for the future? *Hydrological Processes*, **25**, 1517–1520, https://doi.org/10.1002/hyp.7939

BEVEN, K. J. 2012*a*. So how much of your error is epistemic? Lessons from Japan and Italy. *Hydrological Processes*, **27**, 1677–1680, https://doi.org/10.1002/hyp.9648

BEVEN, K. J. 2012*b*. Causal models as multiple working hypotheses about environmental processes. *Comptes Rendus Geoscience, Académie de Sciences*, **344**, 77–88, https://doi.org/10.1016/j.crte.2012.01.005

BEVEN, K. J. & ALCOCK, R. 2011. Modelling everything everywhere: a new approach to decision making for water management under uncertainty. *Freshwater Biology*, **56**, 124–132, https://doi.org/10.1111/j.1365-2427.2011.02592.x

BEVEN, K. J. & GERMANN, P. F. 2013. Macropores and water flow in soils revisited. *Water Resources Research*, **49**, https://doi.org/10.1002/wrcr.20156

BEVEN, K. J. & SMITH, P. J. 2014. Concepts of information content and likelihood in parameter calibration for hydrological simulation models. *ASCE Journal of Hydrologic Engineering*, in press.

BEVEN, K. J. & WESTERBERG, I. 2011. On red herrings and real herrings: disinformation and information in hydrological inference. *Hydrological Processes*, **25**, 1676–1680, https://doi.org/10.1002/hyp.7963

BEVEN, K. J. & YOUNG, P. C. 2013. A guide to good practice in modelling semantics for authors and referees. *Water Resources Research*, **49**, 5092–5098, https://doi.org/10.1002/wrcr.20393

BEVEN, K., SMITH, P. J. & WOOD, A. 2011. On the colour and spin of epistemic error (and what we might do about it). *Hydrology and Earth System Science*, **15**, 3123–3133, https://doi.org/10.5194/hess-15-3123-2011

BEVEN, K. J., BUYTAERT, W. & SMITH, L. A. 2012*a*. On virtual observatories and modeled realities (or why discharge must be treated as a virtual variable). *Hydrological Processes*, **26**, 1905–1908.

BEVEN, K. J., SMITH, P. J., WESTERBERG, I. & FREER, J. E. 2012*b*. Comment on "Pursuing the method of multiple working hypotheses for hydrological modeling" by P. Clark et al. *Water Resources Research*, **48**, W11801, https://doi.org/10.1029/2012WR012282

BEVEN, K., LEEDAL, D. & MCCARTHY, S. 2014. *Framework for assessing uncertainty in fluvial flood risk mapping*. CIRIA Report No. C721.

BLÖSCHL, G. & ZEHE, E. 2005. On hydrological predictability. *Hydrological Processes*, **19**, 3923–3929.

BROWN, C., GHILE, Y., LAVERTY, M. & LI, K. 2012. Decision scaling: linking bottom-up vulnerability analysis with climate projections in the water sector. *Water Resources Research*, **48**, W09537, https://doi.org/10.1029/2011WR011212

CLARK, M., KAVETSKI, D. & FENICIA, F. 2011. Pursuing the method of multiple working hypotheses for hydrological modeling. *Water Resources Research*, **47**, W09301, https://doi.org/10.1029/2010WR009827

CLARK, M., KAVETSKI, D. & FENICIA, F. 2012. Reply to comments by Beven et al. on Pursuing the method of multiple working hypotheses for hydrological modeling. *Water Resources Research*, **48**, https://doi.org/10.1029/2012WR012547

COLLIER, C. G. 2007. Flash flood forecasting: what are the limits of predictability? *Quarterly Journal of the Royal Meteorological Society*, **133.622,** 3–23.

FAULKNER, H., PARKER, D., GREEN, C. & BEVEN, K. J. 2007. Developing a translational discourse to communicate uncertainty in flood risk between science and the practitioner. *Ambio*, **16**, 692–703.

INSTITUTE OF HYDROLOGY 1999. *Flood Estimation Handbook*. Institute of Hydrology, Wallingford, UK.

JUSTON, J. M., KAUFFELDT, A., MONTANO, B. Q., SEIBERT, J., BEVEN, K. J. & WESTERBERG, I. K. 2012. Smiling in the rain: seven reasons to be positive about uncertainty in hydrological modeling. *Hydrological Processes*, **27**, 1117–1122, https://doi.org/10.1002/hyp.9625

KASS, R. E. & RAFFERTY, A. 1995. Bayes ratios. *Journal of American Statistical Association*, **90**, 773–795.

KENNEDY, M. C. & O'HAGAN, A. 2001. Bayesian calibration of mathematical models. *Journal of the Royal Statistical Society*, **D63**, 425–450.

KNIGHT, F. H. 1921. *Risk, Uncertainty and Profit*, Houghton-Mifflin Co. (reprinted University of Chicago Press, 1971).

KOUTSOYIANNIS, D. 2010. HESS Opinions "A random walk on water". *Hydrology and Earth System Science*, **14**, 585–601.

Koutsoyiannis, D. 2011. Hurst-Kolmogorov dynamics and uncertainty. *Journal of American Water Resources Association*, **47**, 481–495.

Makridakis, S. & Taleb, N. 2009*a*. Living in a world of low levels of predictability. *International Journal of Forecasting*, **25**, 840–844.

Makridakis, S. & Taleb, N. 2009*b*. Decision making and planning under low levels of predictability. *International Journal of Forecasting*, **25**, 716–733.

Melbourne, B. A. & Hastings, A. 2009. Highly variable spread rates in replicated biological invasions: fundamental limits to predictability. *Science*, **325**, 1536–1539, https://doi.org/10.1126/science.1176138

Montanari, A. & Brath, A. 2004. A stochastic approach for assessing the uncertainty of rainfall–runoff simulations. *Water Resources Research*, **40**, W01106, https://doi.org/10.1029/2003WR002540

Newson, M. D. 1980. The geomorphological effectiveness of floods: a contribution stimulated by two recent events in mid-Wales. *Earth Surface Processes and Landforms*, **5**, 1–16.

Pappenberger, F. & Beven, K. J. 2006. Ignorance is bliss: 7 reasons not to use uncertainty analysis. *Water Resources Research*, **42**, W05302, https://doi.org/10.1029/2005WR004820

Rougier, J. 2013. Quantifying natural hazard risk. *In*: Rougier, J., Sparks, S. & Hill, L. (eds) *Risk and Uncertainty Assessment for Natural Hazards*. Cambridge University Press, Cambridge, UK, 19–39.

Rougier, J. & Beven, K. J. 2013. Model limitations: the sources and implications of epistemic uncertainty. *In*: Rougier, J., Sparks, S. & Hill, L. (eds) *Risk and Uncertainty Assessment for Natural Hazards*. Cambridge University Press, Cambridge, UK, 40–63.

Smith, L. A. 2001. Disentangling uncertainty and error: on the predictability of nonlinear systems. *In*: Mees, A. I. (ed.) *Nonlinear Dynamics and Statistics*. Birkhäuser, Boston, USA, 31–64.

Smith, P. J., Tawn, J. & Beven, K. J. 2008. Informal likelihood measures in model assessment: theoretic development and investigation. *Advances in Water Resources*, **31**, 1087–1100.

Taleb, N. N. 2007. *Fooled by Randomness*. Penguin, London.

Taleb, N. N. 2010. *The Black Swan*, 2nd edn. Penguin, London.

Thielen, J., Bogner, K., Pappenberger, F., Kalas, M., Del Medico, M. & de Roo, A. 2009. Monthly-, medium-, and short-range flood warning: testing the limits of predictability. *Meteorological Applications*, **16**, 77–90.

Wilby, R. L. & Dessai, S. 2010. Robust adaptation to climate change. *Weather*, **65**, 180–185.

Wright, G. & Goodwin, P. 2009. Decision making and planning under low levels of predictability: enhancing the scenario method. *International Journal of Forecasting*, **25**, 813–825.

Zehe, E. & Blöschl, G. 2004. Predictability of hydrologic response at the plot and catchment scales: role of initial conditions. *Water Resources Research*, **40**, W10202.

Zehe, E., Elsenbeer, H., Lindenmaier, F., Schulz, K. & Blöschl, G. 2007. Patterns of predictability in hydrological threshold systems. *Water Resources Research*, **43**, W07434.

Index

Page numbers in *italic* denote figures. Page numbers in **bold** denote tables.